ANNUAL REVIEW OF ASTRONOMY AND ASTROPHYSICS

EDITORIAL COMMITTEE (1979)

ANNUAL REVIEW OF ASTRONOMY AND ASTROPHYSICS

GEOFFREY BURBIDGE, *Editor*
Kitt Peak National Observatory

DAVID LAYZER, *Associate Editor*
Harvard College Observatory

JOHN G. PHILLIPS, *Associate Editor*
University of California, Berkeley

VOLUME 17

1979

ANNUAL REVIEWS INC.   4139 EL CAMINO WAY   PALO ALTO, CALIFORNIA 94306

ANNUAL REVIEWS INC.
Palo Alto, California, USA

REPRINTS The conspicuous number aligned in the margin with the title of each article in this volume is a key for use in ordering reprints. Available reprints are priced at the uniform rate of $1.00 each postpaid. The minimum acceptable reprint order is 5 reprints and/or $5.00 prepaid. A quantity discount is available.

International Standard Serial Number: 0066-4146
International Standard Book Number: 0-8243-0917-0
Library of Congress Catalog Card Number: 63-8846

FILMSET BY TYPESETTING SERVICES LTD, GLASGOW, SCOTLAND
PRINTED AND BOUND IN THE UNITED STATES OF AMERICA

PREFACE

This volume of the *Annual Review of Astronomy and Astrophysics* was planned at the meeting of the Editorial Committee held April 30, 1977 in La Jolla. Committee members present were: Editor Geoffrey Burbidge; Associate Editors David Layzer and John G. Phillips; Committee Members W. Ian Axford, Wallace L. W. Sargent, and Leo Goldberg; Guests Judith Perry and B. T. Soifer; and Production Editor Rosalie West. Marshall Cohen, Peter Conti, and James Pollack also attended in place of absent Committee Members Tobias C. Owen, Arno A. Penzias, and George Wallerstein.

I would like to thank the Associate Editors for carrying out the scientific work very effectively. I particularly would like to thank the past Production Editor, Rosalie West, and the current Production Editor, Elizabeth Brower, for their excellent work.

THE EDITOR

SOME RELATED ARTICLES APPEARING
IN OTHER ANNUAL REVIEWS

Annual Review of Astronomy and Astrophysics
Volume 17, 1979

CONTENTS

P. Swings

Ann. Rev. Astron. Astrophys. 1979. 17: 1–7

A FEW NOTES ON MY CAREER AS AN ASTROPHYSICIST

*2143

P. Swings

23, Avenue Léon Souguenet, 4050 Esneux, Belgium

Beginnings

Nothing in my childhood had prepared me to become an astrophysicist. But in 1922, as a high school prize, I received a wonderful book, *Astronomie Populaire* by Camille Flammarion. I read this admirable book from the first page to the last, and I decided to do my utmost to become an astrophysicist, forgetting about the plans of my parents (and myself) who were thinking in terms of agriculture.

I was born on September 24, 1906, in Ransart, Belgium, a small industrial town, near Charleroi. My parents were of very modest conditions. I was the best pupil every year in grammar and high school. My teachers were wonderful and I am most grateful to them. At the time I began going to the Charleroi high school in September 1917 Belgium was occupied by the German army. The winter 1917–1918 was very cold and the food rather scarce. Happily I received regularly some bread from the Hoover Foundation (C.R.B., Commission for Relief in Belgium). I used to walk about three miles to high school in the early morning and back home in the late afternoon. My clothing was very mediocre. Despite the material difficulties I was first in my class at Charleroi high school and received the highest government reward every year.

My parents had hardly dared hope that I could go to the University, but my grades in high school were so good that my father decided to continue his sacrifices and to work extra hours. In addition, I also did a great deal of tutoring every Sunday and holiday so that I could start at the University in Liège in 1923.

After obtaining a PhD in four years—the title of my thesis was "Corrections to the Newton Law and the orbits with moving perihelion"—I was afraid I would be unable to find a job on account of the depression; indeed several of my friends in the University who had obtained PhDs

0066-4146/79/0915-0001$01.00

could not find jobs, even low-paying ones. In order to increase my opportunities, I took a number of practical university courses in survey topography, in mathematics applied to finances and insurance problems, and in insurance legislation. Happily, I got a modest scholarship beginning in 1927.

It was not an easy matter to specialize in astrophysics as there was neither an astrophysicist nor a spectroscopist in the University of Liège. Indeed there was no astrophysicist in any Belgian university! Hence I decided to work by myself in a field, celestial mechanics, which was fairly close to astrophysics. I studied an important part of the four volumes of Tisserand's *Celestial Mechanics* which I had again received as a prize.

As soon as I had obtained my PhD in Liège (1927) I went to Paris, with a Belgian scholarship, for a one year stay and I took a number of courses in Paris: calculus with Goursat and Picard, celestial mechanics with Andoyer and Chazy, astrophysics with Bruhat, geophysics with Brillouin, optics with Fabry, and spectography with Croze. I also attended occasional lectures by Mme. Curie and P. Langevin. I spent half of my time, however, at the Astrophysical Observatory in Meudon in the solar, cometary, planetary, and stellar services.

The academic year, 1927–1928, was fruitful, but I began to realize that I needed a much more thorough knowledge of spectroscopy if I wanted to do reasonably good work in astrophysics. Thus, I accepted the invitation of Professor Stefan Pieńkowski to go to Warsaw, Poland, for a couple of years to learn spectroscopy. There, in 1931, I obtained another DSc in physics, writing a thesis on the fluorescence of diatomic sulfur. In doing so, I was one of the first Belgian physicists to apply spectroscopy to practical problems, especially metallurgy.

Other practical problems were also being solved. The University of Liège asked me to give a course on applied spectroscopy; I did this in 1933 and published a book in 1935. Professor Charles Fabry, in particular, encouraged me by writing the Preface to this book, the first on applied spectroscopy in the French language.

In 1931 I thus began to devote the major part of my investigations to astronomical spectroscopy. I met the genial Dr. Otto Struve in 1931 at the Yerkes Observatory and I started with him a collaboration that lasted more than 20 years. In addition to my collaboration with Struve I also worked with S. Chandrasekhar, B. Edlén, I. S. Bowen, P. W. Merrill, A. H. Joy, A. S. King, Ch. Fehrenbach, L. Rosenfeld, and J. Kaplan. My astrophysical activity was slowed down from 1943 to 1946 because of a war research assignment in optics in the U.S. Nevertheless, I was able to devote to astrophysics the rare moments of leisure found during this period.

At first sight, my publications seem to cover a great disparity of topics, since they range from mathematical astronomy to geometrical optics and spectrochemistry. Yet it is clear that my principal interest and, I believe, my main contributions concerned astrophysics. Otto Struve insisted, on several occasions, that my work in physics had prepared me well for the interpretations of astronomical observations.

Importance of the Identifications in Stellar Spectra

Throughout my career as an astrophysicist, I have been quite unabashedly interested in identifying atoms and molecules in stellar and cometary spectra. For certain astrophysicists, including myself, pure identification work is lots of fun. This point of view, of course, has been criticized for being too descriptive and for entering into too many observational details. However, the contributions of Struve, as well as those of outstanding stellar spectroscopists such as W. S. Adams, P. W. Merrill, A. H. Joy, and R. F. Sanford, have grown out of such studies. Our progress in the theoretical understanding of stellar and other atmospheres could not have gone very far if we had not had identification and detailed description as a sound and solid basis. To take two extreme cases: without the identification of the nebular lines by I. S. Bowen, and his successors, and of the coronal lines by B. Edlén we would not know about forbidden transitions in cosmic bodies. In a more modest case, Bengt Edlén and I devoted two full years of hard steady work to the laboratory analysis of the Fe III spectrum. Forty years later we both still feel that our endeavors paid excellent dividends later on, when it turned out that the forbidden as well as the permitted lines of Fe III explained many features in novae, peculiar bright line stars, and even normal hot stars, including dilution effects in shell stars and other mechanisms affecting the stellar atmospheres.

Unidentified Emission Lines in Of-Stars

A simple example of how pure identification must be accompanied by studies of intensities and profiles may be found in the work on unidentified emission lines in Of-stars. Two unassigned emission lines were and still are unidentified. Their wavelengths are $\lambda 4485.7 \pm 0.2$ and $\lambda 4503.7 \pm 0.2$ Å. These lines were confirmed by R. J. Wolff (1963). No convincing assignment has been discovered thus far. Dr. Roy H. Garstang and Dr. Michael J. Seaton have computed transition probabilities (even of very complicated spectra, such as the forbidden transitions of Fe II, Fe III, Ni II, etc. . . .) or collisional cross sections. In P Cygni stars strange behaviors are observed in different multiplicities, for example the sextets and quartets in the Fe II multiplets, and the triplets and quintets in O I.

Attempts to identify the unexplained lines of the Of-stars have been made without success by several spectroscopists, including B. Edlén, L. Goldberg, R. J. Wolff, A. Underhill, and myself. If these two lines belong to the same element—which is likely—and have one common level, the wavenumber separation between the other two levels should be between 87 and 91 cm^{-1}. We have found no convincing assignment in searching for levels of such separation in C I to IV, N I to V, O I to VI, Si I to IV, but we should not become discouraged!

Individual spectroscopic studies still carry considerable importance. We could not have gotten very far in our understanding of our own milky way and other galaxies if we had not made great progress in individual detailed spectroscopic studies at the observatories of Mt. Wilson, Palomar, McDonald, Lick, Victoria, Kitt Peak, La Silla, Cerro Tololo, and Haute Provence.

Commission 29 (Stellar Spectra) of the International Astronomical Union

Commission 29 of the International Astronomical Union has, throughout the years, played an important role in stellar spectroscopy. Walter S. Adams was the first chairman of the Commission, organized to discuss stellar spectra. In 1935, when I first attended the I.A.U. meeting at the Paris General Assembly, Commission 29 had only 19 members. Henry Norris Russell, one of the greatest astrophysicists of this century, was then chairman of this small body.

I remember the 1935 meeting, not only because it was the first I attended—and I have attended all the I.A.U. meetings of the Commission since then—but because it was one of the most interesting. There we discussed, and adopted, Beals' classification of Wolf-Rayet stars. Finally, I remember the 1935 meeting because Russell appointed me secretary, and entrusted me with the responsibility of writing the report on the sessions.

Later, in 1948, Otto Struve accepted chairmanship of Commission 29 on condition that I be secretary during his term, and then succeed him, in 1952, as chairman, at which time he would become secretary. Such arrangements were possible then. Members of the Commission who know how close were our collaboration and friendship may have forgotten about this "deal", but they certainly would not be surprised.

The Commission membership has greatly increased, and its responsibilities have grown since those early days. The six eminent stellar spectroscopists who have chaired the Commission since my term, Drs. J. L. Greenstein, L. H. Aller, J. Sahade, Y. Fujita, M. W. Feast, and Mrs. Hack, have done an excellent job adopting standards of comparison for the tremendously increased body of data that has accumulated from stellar spectroscopic studies.

Major Research

In 1975, the 20th Liège International Symposium was devoted to "Astrophysics and Spectroscopy", the fields in which I have primarily worked. There first was a ceremony of hommage in the presence of His Majesty King Baudouin of Belgium who is greatly interested in astronomy and owns several small telescopes. Then, during their introductory speeches, the following personalities stressed the importance of my contributions in various fields of astrophysics: C. Townes (Nobel prizewinner), on intersellar molecules; G. Herzberg (Nobel prizewinner), on cometary spectra and related topics; A. Unsöld, on the history of interrelations between spectroscopic laboratory work and astrophysical problems; P. Conti, on the relation between Of- and WR-stars; L. Aller, on central stars of planetary nebulae; J. Sahade, on symbiotic stars; B. Donn, on problems of cosmic chemistry; G. Münch, on planetary spectroscopy; F. Whipple, on space missions to comets; J. Greenstein, on subluminous stars, post-novae, white dwarfs, and hot subdwarfs. It is very gratifying to see that one's work has had application to such a diversity of problems.

And it is, of course, a compliment to have various phenomena carrying one's name. In my case, this has happened an inordinate number of times. Nevertheless, using the Swings name as a device, I shall try to trace what I consider my major contributions to astrophysics, or at least to outline those areas which have occupied my thought and time throughout my career.

COMETARY PHYSICS   The peculiar intensity distribution within the bands had been a puzzle for over half a century. This distribution was interpreted as being due to the presence of the Fraunhofer lines in the exciting solar radiation. The fluorescence mechanism in the comets, called the "Swings effect," is affected by the radial velocity of the comet relative to the Sun. The "grazing comets" show also a very striking "Swings effect." Actually the Swings effect is applicable to all cases of fluorescence excited in astronomical bodies by radiation presenting discrete absorptions or emissions.

PHENOMENA IN THE TWILIGHT AND IN SUNLIT AURORAE   The peculiar intensity distribution in the bands of the twilight radiation (especially the N_2^+ spectrum) is interpreted in the same way as in the comets. It is now also called the "Swings effect," for example by Dalgarno and by Vallance Jones; the "Swings effect" is also observed in "sunlit aurorae."

INTENSE COMETARY EMISSION IN THE NEIGHBOURHOOD OF $\lambda 4050$ This phenomenon had been a puzzle for many decades. On account of my research on this emission Otto Struve suggested that the 4050 group be named the "Swings bands." Many investigators have adopted this recommendation.

The bands near $\lambda 4050$ have been found by G. Herzberg and his associates to be due to the C_3 radical. These bands are also observed in the coolest carbon stars and in the phenomena of pyrolysis, solid particles (soot), and hydrocarbons.

THE "SWINGS MECHANISM" IN THE O-STARS WITH EMISSION LINES   The excitation mechanism which I discovered in 1948 is explained by the "pumping" by continuum radiation within the multiplets of N III. This mechanism, named the "Swings mechanism" by Mihalas, gives the interpretation of the selectivities among the N III and other emission lines in the Of-stars. It may be presumed that this mechanism will explain other selectivities in hot stars.

THE FIRST FORBIDDEN LINE IN COMETS   This phenomenon has been found by J. L. Greenstein and myself to be due to the red doublet of [O I]. This [O I] cometary emission has played an outstanding role, for example in Biermann's cometary investigations.

Conclusion, of Sorts

I have been fortunate, since obtaining the modest scholarship in 1927, to have found employment as an astrophysicist, and have not had to turn to a career in survey topography or in mathematics as applied to finances and insurance problems. Throughout the years I have taught at the University of Liège, the University of Chicago, and the University of California at Berkeley, and have taken temporary positions in many American universities. I have been a research associate at various institutions, Lick Observatory and Harvard University among them, and have had visiting fellowships in many countries (France, Germany, Poland, USSR, Sweden, Denmark, Norway, U.K., Canada, Eire, Holland, Italy, Switzerland, Turkey, the Vatican). I was Darwin Lecturer at the Royal Astronomical Society, U.K., in 1964.

The graduate courses which I gave to many students in a number of astronomy departments were a source of real satisfaction. My activities with the I.A.U., organizing various symposia, chairing various commissions, and as vice-president (1952–1958) and president (1964–1967) have been rewarding. Perhaps most importantly, my research and collaboration with my colleagues over the years have been most gratifying.

I and my collaborators continue investigations in a number of areas: cometary physics (including investigations by space vehicles); twilight phenomena and sunlit aurorae, in relation to the profile of the exciting solar spectrum; fluorescence phenomena in hot stars, symbiotic objects, novae, and long period variables; pumping effects in Of-stars; comparative behavior of different multiplicities; far ultraviolet region studies with space vehicles; extension of the observed spectral region in the near infrared; and continuation of the investigations which I started with Otto Struve half a century ago.

Ann. Rev. Astron. Astrophys. 1979. 17:9–41

INFRARED SPECTROSCOPY ✻2144
OF STARS

K. M. Merrill

Astronomy Department, University of Minnesota, Minneapolis, Minnesota 55455

S. T. Ridgway

Kitt Peak National Observatory, P.O. Box 26732, Tucson, Arizona 85726

1 INTRODUCTION

Cool matter emits thermal radiation primarily in the infrared (IR). The past decade has seen substantial advances in IR spectroscopy of astronomical sources, leading to expanded observational possibilities for study of cool stars, dust, and gas. IR investigations have identified new classes of astronomical entities, revealed new astrophysical processes, and probed hitherto unobservable regions of space.

IR astronomy is currently experiencing a period of rapid technological expansion (Soifer & Pipher 1978). Special purpose ground, airborne, and space telescopes are in various phases of planning, construction, or operation. Significant improvements in IR detector sensitivities have been recently achieved, and spectral and spatial multiplexing techniques are now available. Important IR sky surveys are complete or in progress. The exploitation of the IR for the benefit of many areas of astronomy is under way.

Spinrad & Wing (1969) thoroughly reviewed early developments in IR spectroscopy of stars. To update their discussion with comparable detail would require a book-length endeavor. We have made our current assignment tractable by excluding the photomultiplier IR $\lambda \lesssim 1.1$ μm, greatly restraining discussion of circumstellar shells, and emphasizing the current status of topics at the expense of not documenting historical development. Many valuable contributions could not be cited due to space limitations. Our review covers the application of IR techniques in stellar classification, studies of stellar photospheres, elemental and isotopic abundances, and the nature of remnant and ejected material in the near circumstellar region.

9

0066-4146/79/0915-0009$01.00

2   SPECTRAL CLASSIFICATION

Photographic spectral sequences for distinguishing luminosity, temperature, and relative elemental abundances have been determined for the most common types of visible stars. Similar IR spectral sequences must be established if we are to fully utilize the wealth of information available. IR classifications will often parallel their optical counterparts, but for cool stars numerous molecular vibration-rotation bands in the $\lambda\lambda 1$–8-μm region provide additional information. At longer wavelengths, detection of thermal emission of dust and gas from circumstellar envelopes adds a measure of mass loss.

Once the IR characteristics of standard stars are known, highly reddened stars can be properly classified from the IR spectrum. Hence stars deep within the galaxy, obscured by dense molecular clouds, or buried within massive circumstellar envelopes, can still be studied. For cool stars, approximate spectral types can now be assigned on the basis of low resolution observations. For hot stars, where detection of hydrogen lines is needed for classification, moderate to high resolution spectra are required.

The qualitative nature and general characteristics of infrared stellar spectra in the range $\lambda\lambda 1$–3.3 μm are well illustrated in an important atlas prepared by Johnson & Méndez (1970). The usefulness of this atlas may be traced to several features: the resolution is adequate to resolve molecular bands; the wavelength span is large; a good selection of spectral types is represented; and telluric absorption has been removed. Correction of IR spectra for atmospheric extinction is particularly important if spectra are to be useful to other than the specialist.

Atlases of low to moderate resolution spectra already exist for a wide range of objects with reasonable wavelength coverage through the $\lambda\lambda 1$–13-μm region. The information content of published IR spectra is higher than generally realized, owing at least in part to the qualitative rather than quantitative emphasis given by the observers. The quality of available spectra merits more detailed analyses. The real work of establishing a reliable IR classification scheme has just started.

2.1   Cool Stars

Qualitative IR spectral classification of cool stars has been established at low spectral resolution ($\lambda/\Delta\lambda \lesssim 200$). The presence of numerous molecular absorption bands—some of which remain unidentified due to the difficulty of obtaining high resolution spectra for adequate confirmation—permits unique assignments of IR spectral types consistent with photographic spectral types. Spectral coverage from the near IR out to 13 μm has demonstrated that for cool stars IR spectra clearly distinguish among the abundance classes M, S, and C, indicate rudimentary temperature classes, and, in the case of M stars, differentiate between luminosity classes in an elementary way. In addition to such classical parameters, mass loss can be determined (as measured by the observed excess at $\lambda \gtrsim 5$ μm above that due to a cool stellar photosphere). The progressive extension of IR spectral sequences from "normal" stars without substantial IR excess to visually faint, IR bright stars from the IRC *Two Micron Sky Survey* (herein denoted IRC; Neugebauer & Leighton 1969) and to the optically invisible stars from the *AFCRL/AFGL Infrared Sky Survey* (herein denoted AFGL; Walker & Price 1975, Price & Walker 1976) has been demonstrated (Merrill & Stein 1976a,b,c). Stars in dense shells can be identified and assigned at least approximate spectral types.

Since the review of IR spectroscopy by Spinrad & Wing (1969), which of necessity dealt primarily with data at wavelengths less than 3 μm, considerable effort has been devoted to obtaining longer wavelength stellar spectra: for cool stars in general (Hyland et al. 1972, Merrill & Stein 1976a,b, Noguchi et al. 1977); for carbon stars (Frogel & Hyland 1972, Goebel et al. 1978a,b); and for S stars (J. D. Bregman and F. C. Witteborn, in preparation). Airborne observations at wavelengths between $\lambda\lambda 1$–8 μm inaccessible from the ground are now available (Puetter et al. 1977, Strecker et al. 1978, Goebel et al. 1978a,b).

Many of the criteria used in assigning IR spectral types have been reviewed by Merrill (1977). Figure 1 of that review and the figures in Johnson & Méndez (1970) summarize the salient absorption features for the late spectral types K, M, S, and C. CO is present in all these stars, with the 2.3-μm vibration-rotation first overtone becoming progressively more prominent towards later (cooler) M stars and showing a strong positive luminosity effect. For S stars the CO bands are very deep. Absorption bands at 1.8 and 2.7 μm due to H_2O steam become prominent near M6 (giants) or early M (dwarfs) and show a negative luminosity effect due to the strong pressure dependence of polyatomic association. CO is also present in carbon stars, but CN blanketing produces marked changes in the spectrum. For carbon stars the C_2 molecule is often seen near 1.8 μm, but the 3.1-μm absorption band is the most reliable IR indicator for an N type carbon star. The 3.1-μm band is composite, with varying contribution from HCN and/or C_2H_2 (Section 4.1).

During the pioneering stage of IR spectroscopy, 3-μm absorption bands were reported in a few M stars, but subsequent spectra (Merrill & Stein 1976a, Noguchi et al. 1977) have not confirmed this result. The similarity between the 3-μm C star band and the "ice" band associated with objects in molecular clouds has led to interesting speculation about the carrier of the "ice" band (Mukai et al. 1978). (High resolution spectra of the "ice" band are clearly necessary.) However the two bands can be clearly distinguished at low resolution given a well-defined reference level (Merrill & Stein 1976a). Further, to date, the "ice" band is seen only in conjunction with the 10-μm absorption band (e.g. Merrill et al. 1976), so that minimal confusion should arise.

At longer wavelengths the broad 9.7-μm emission band due to silicates and the narrower 11-μm emission band attributed to SiC provide an additional discriminant between stars with $C/O < 1$ (called oxygen-rich but possibly with near solar abundances), and stars with $C/O > 1$ (called carbon-rich) (Treffers & Cohen 1974, Forrest et al. 1975, Merrill & Stein 1976a). For the most dense circumstellar shells the dust signatures may be the only diagnostics, and in the absence of additional information (e.g. the type of radio emission present) confusion with nonstellar sources such as HII regions is possible.

Until recently use of the IR for spectral classification has been limited to very rough discriminations. It should be possible to define a detailed IR classification scheme. A comprehensive set of standards and quantitative analysis would provide such a system. One approach is to determine equivalent widths for spectral features. Alternatively, maximum depths at band center can be measured. In either case, it is important that the spectrum cover a sufficient range in wavelength to show a meaningful flux reference level. The reference level will not generally represent a continuum, but perhaps a useful apparent continuum. For example, Goebel et al. (1978b) have found that the high points in carbon star spectra typically follow a blackbody curve. By reference to an apparent continuum interpolated in this fashion they have defined a system of narrow band indices. Such indices are convenient for empirical studies, but may present formidable problems of interpretation. Contamination by circumstellar emission or absorption should also be considered.

Another approach, especially useful for observing large numbers of faint objects, is to establish a narrow bandpass filter system calibrated against stellar standards [hopefully studied at high resolution (see Section 4.1)]. This has been done successfully for the 1.9-μm H_2O and 2.3-μm CO bands (Baldwin et al. 1973, Persson et al. 1977, Aaronson et al. 1978). Measures of SiO near 4 μm are under development (Rinsland et al. 1977, Wing et al. 1977). Correlations of these indices with each other and various established optical parameters (e.g. the data of Gow 1977) are improving as the data base increases. Extensive studies (e.g. Bergeat et al. 1976a) based on early IR spectra have reconfirmed that the CN, C_2, C_3 band strengths increase with IR excess in carbon stars, and that for CO there is an inverse correlation with IR excess. This is a result of blanketing by other molecules (as suggested by Mertz 1972a), especially CN, and dilution with thermal reemission from a dust shell (Frogel & Hyland 1972).

The reconnaissance stage for cool star classification is over. Low resolution IR spectroscopy provides an important tool, but the necessary stage of quantification has just begun. Instrumentation now under development will permit the extension of IR spectral studies throughout the galaxy and in the Magellanic clouds.

2.2 Hot Stars

Relatively few IR spectra of early-type (hot) stars have been published. In general both photometric reconnaissance of O, B, A stars (Barlow & Cohen 1977) and theoretical analysis of hot stellar atmospheres have suggested that (with a few notable exceptions for the case of emission-line stars) classification cannot be done with low resolution IR spectra. The pioneering search by Cohen (1975) for molecular features in the $\lambda\lambda 2$-4-μm spectra of hot young stars yielded generally featureless continua for most of the stars observed. Normal early-type stars (i.e. those without large IR excesses) are faint in the IR and hence difficult to study with sufficient resolution to detect expected atomic absorption and emission lines. Wolf-Rayet stars, however, are readily identified by strong atomic emission from the circumstellar region (Section 5.2). Much exploratory high resolution work remains to be done.

2.3 Composite Spectra

The optical spectra of many stars contain features indicative of differing, often disparate, physical conditions for what is otherwise considered to be one star. Such composite spectra suggest that a circumstellar shell or disc, binary companion, or possibly more exotic arrangement is contained in the stellar system. Recent IR spectra of several classes of such composite systems have revealed unexpected spectral features near 2 μm.

Optical spectra of symbiotic stars present the puzzling combination of high excitation nebular emission lines with absorption line systems typical of a late-type star. An evolved binary system (Tutukov & Yungelson 1976) containing a hot dwarf star (dominating the visual spectrum) and a cool giant (producing the near IR spectrum) might have these characteristics. Detection of soft X-ray emission in addition to IR excess from the short-period symbiotic variable AM Her (Stockman et al. 1977) has focused renewed attention on symbiotic phenomena in general. Photometric studies of representative symbiotic stars have shown large IR excesses in many VV Cephei stars (Swings & Allen 1972), high excitation stars (Webster & Allen 1975), supergiant stars (Humphreys & Ney 1974a,

Humphreys 1976), dwarf novae (Szkody 1977), and slow novae (Glass & Webster 1973) as well as the class as a whole. Thermal emission from gas and dust is present, but often near-IR spectral features are also clearly seen. Optical and IR spectra of the Be star XX Oph (Lockwood et al. 1975, Humphreys & Gallagher 1977) show TiO and the 2.3-μm CO band, suggesting the presence of a cool M companion. Similarly, detection of CO and H_2O absorption at 2 μm in the spectrum of the slow nova RR Tel (Allen et al. 1978) indicates a cool M companion.

IR spectra of FU Orionis class eruptive variable stars provide a clear example of the importance of IR observations in establishing a comprehensive physical description of what is usually considered to be an optical phenomenon. These stars are generally believed to represent an early stage in stellar evolution (Welin 1976, Herbig 1977) with a rapid five magnitude visual brightening due to the abrupt dissipation of a circumstellar shell or changes in stellar structure. In contrast with photographic spectral classifications, typically F or G supergiant, low resolution $\lambda\lambda 2$-4-μm spectra by Cohen (1975) suggested the presence of absorption band systems typical of late M giant stars. Detailed study at high spectral resolution (Mould et al. 1978) of strong H_2O steam and weak CO absorption in FU Ori and V1057 Cyg led to the consideration of models involving a flattened circumstellar envelope or extreme rotational darkening of a very rapidly rotating star.

The emission-line eruptive variable star HM Sge provides another example (Davidson et al. 1978). HM Sge and V1016 Cyg are considered possible precursors to planetary nebulae (e.g. Kwok et al. 1978, Ciatti et al. 1977). Well studied optically (Ciatti et al. 1978, Wallerstein 1978), HM Sge appears to be a hot star. Moderate resolution IR spectra of both HM Sge and V1016 Cyg, however, show strong optically thin 10-μm silicate emission and 2.3-μm CO absorption of the type found in cool stars (Puetter et al. 1978) superposed on a blackbody-like 900 K continuum. V1016 Cyg undergoes periodic IR flux variations like a Mira variable (Harvey 1974). Although the CO identifications await confirmation at higher resolution, clearly models of HM Sge (cf. Kwok & Purton 1979) must account for the presence of such cool star characteristics. It may be significant that, while the optical spectra of HM Sge and V1016 Cyg show many similarities with planetary nebulae, the IR spectra do not (cf NGC7027 in Russell et al. 1978).

Note the intriguing emergence of a shared IR spectral characteristic in what are generally considered distinct stellar phenomena. The evolutionary states of these objects remain an open question. In contrast to the FU Orionis stars which are thought to be young objects, symbiotic stars are considered to be highly evolved binaries and HM Sge objects possibly preplanetary nebulae. The significance of the indicated cool companion remains a puzzle. Further IR studies of candidate systems may well add other objects to this category.

2.4 Obscured Stars

IR spectra provide a unique means of establishing the identities of stars inaccessible for classical optical study due to the presence of large amounts of either interstellar or circumstellar dust along the line of sight. Cool stars and emission-line stars are clearly identifiable in the IR even in the presence of substantial reddening. A number of bright near-IR sources at the galactic center have been identified as luminous M stars on the basis of their 2-μm spectra (Soifer et al. 1976, Neugebauer et al. 1976, Treffers et al. 1976). Late-type evolved stars undergoing extensive mass loss may be hidden from optical analysis by their circumstellar envelopes, but can be clearly identified on the basis of IR spectra (Hyland et al. 1972, Frogel et al. 1975, Merrill & Stein 1976b,c). In particular, many so-called OH/IR stars, discovered during radio surveys of microwave OH maser emission, can be unambiguously identified as M stars. OH26.5 + 0.6 (AFGL2205) is a case in point (Forrest et al. 1978). IR spectra of the invisible objects AFGL3068 (Jones et al. 1978) and AFGL3099 (Gehrz et al. 1978) have shown these sources to be carbon-rich stars.

Elias (1978a,b,c) has utilized the 2-μm spectra of obscured objects in the IC5146, ρ Oph, and Taurus dark cloud complexes to distinguish cool stars from more exotic objects. The ability to discriminate members of the cloud from foreground and background field stars is crucial to establishing the active sites for star formation and the characteristics of young stellar populations. By providing a method for sorting the contributions of circumstellar, intracloud, and general interstellar extinction, IR spectra can then be utilized to compare the properties of grains in these varied environments.

Recent sky surveys at wavelengths greater than 3 μm have revealed many IR-bright sources with no visible counterpart bright enough for optical spectroscopic study. Particularly large numbers of such sources appear in the AFGL Catalog (Price & Walker 1976) produced by the rocket-borne AFCRL Infrared Sky Survey (see Allen et al. 1977, Lebofsky et al. 1978, Merrill 1977 for discussion). Many of the sources appear to be cool, heavily obscured stars for which IR spectroscopy offers the only means of spectral classification. An astonishing fraction of the cool, obscured stars are carbon-rich (probably Mira) variables buried in clouds of their own ejecta. Apart from unforeseen selection effects the initial implication is that these stars are a more significant component of the stellar population than would have been expected from photographic surveys. An interesting hypothesis (Lo & Bechis 1976, Zuckerman et al. et al. 1976) persuasively summarized by Zuckerman et al. (1978) suggests that many of these objects (which tend also to be associated with thermal CO emission at millimeter wavelengths) are progenitors of planetary nebulae. However, at least for the prototype AFGL2688 some caution is still necessary in making such an interpretation (Humphreys et al. 1976). We note with interest that if both these objects and the HM Sge objects (Section 2.3) are indeed "proto-planetary nebulae" we may be facing a surplus.

At present the detection of obscured hot stars entails either establishing a lack of cool star features or detection of Brackett series hydrogen emission lines (which require a source of ionizing radiation) and, from the resulting measure of ionizing photons (Lyman luminosity), establishing a plausible spectral type. For example, detection of Brα in η Car, where the high electron density prevented radio detection, has allowed Aitken et al. (1977) to suggest an early O star as the central object. Working from Brγ detection, Thompson and his collaborators have assigned early spectral types to the obscured emission-line stars MWC349 and LkHα101 (Thompson & Reed 1976) and to the embedded object found by Allen (1972) in NGC 2264 (Thompson & Tokunaga 1978).

Since the IR spectra of hot stars are nearly featureless (Section 2.2), the spectra of highly obscured early-type stars are invaluable as probes of the interstellar medium. VI Cyg No. 12 (Cyg OB 2 No. 12), which suffers 10 magnitudes of visual extinction, has been particularly useful in this regard. Low resolution $\lambda\lambda$2–4-μm spectra have set severe upper limits on the "ice" to silicate absorption ratio (Gillett et al. 1975b) of $\lesssim 0.02$, and high-resolution 2-μm spectra have set upper limits to the column densities of H_2 and CO gas (Kleinmann et al. 1978).

2.5 *Stellar Populations*

Although the acquisition of detailed IR spectra of globular clusters and galaxies is a difficult task representing the limit of current technology, such studies provide valuable criteria for the synthesis of stellar populations and for the determination of the source of the observed IR flux.

The 2-μm spectra of the nuclear bulge in our own galaxy (Neugebauer et al. 1978) as well as the active infrared nucleus in the galaxy M82 (Willner et al. 1977) clearly show the presence of a cool giant population. On the other hand, IR spectra of the nuclear region of the Seyfert galaxy NGC1068 (Thompson et al. 1978) and NGC5128 (Grasdalen & Joyce 1976a) show no such indicators of cool stars. The complicated nature of the M82 spectrum, which includes strong band emission at 3.3, 6.3, 7.9, and 11.2 μm, indicates the necessity of obtaining representative IR spectra before relying too heavily on filter photometry.

Photometric systems established to measure CO and H_2O at 2 μm in cool stars (Section 2.1) have been used to study globular clusters (Aaronson et al. 1978, Cohen et al. 1978, Pilachowski 1978a). Direct comparison of stellar populations has been made including the effects of luminosity and metallicity (see also Section 4.1). There is general agreement between the properties of individual cluster members and the integrated spectra. Detailed observation of E and S0 type galaxies using the same indices (Frogel et al. 1978, Aaronson et al. 1978) have imposed constraints on the coolest stellar components of these systems.

3   THE PHYSICS OF STELLAR ATMOSPHERES

Progress toward a detailed understanding of cool star spectra is inextricably linked to theoretical modeling of the stellar atmospheres (for a review see Carbon 1979). Infrared spectroscopy provides important calibrations and diagnostics for atmospheric temperature, pressure, mass motions, and line formation mechanisms.

3.1   *Stellar Temperatures and Temperature Profiles*

In some cool star studies selection of an atmospheric temperature profile (T-τ relation) has been based on broad-band color(s). A more stringent criterion is to require that the model reproduce in detail the spectrophotometric flux distribution over the largest possible wavelength range at resolution $\gtrsim 100$. The value of the IR for constraining model temperatures may be attributed in part to the strong wavelength dependence of H^- continuous opacity.

This constraint has been employed in recent studies of α Ori (M2Iab) by Faÿ & Johnson (1973) and Tsuji (1976a,b). Both studies yielded relatively high temperatures, $T_{eff} \sim 3500$–3800 K and ~ 4000 K, respectively. These temperatures indicate an angular diameter substantially smaller than measured in numerous interferometric observations. The discrepancy and its possible resolution by modeling of scattering on dust grains has been discussed by T. Tsuji (preprint). In a recent study of the flux distribution of α Boo (K2III) Johnson et al. (1977) obtained a good match of computed and observed fluxes for $T_{eff} \sim 4250$ K. This temperature was subsequently used by Ayres & Johnson (1977) in a reexamination of the surface gravity and mass of α Boo. Allen (1978) has used high spectral resolution IR studies of the sun to determine true continuum intensities. Through the wavelength dependence of model opacity and observed limb darkening he has succeeded in accurately specifying the solar temperature structure.

Two studies have extended the flux distribution modeling technique to a representative sample of bright stars in an attempt to better define cool star effective temperatures. Tsuji (1978) estimated T_{eff} for eight M giant and Mira stars. J. D. Scargle & D. W. Strecker (preprint) present a similar analysis for 24 giant, Mira, and carbon stars. The results of both studies show favorable agreement with a recent determination of empirical effective temperatures based on angular diameters measured by the occultation technique (S. T. Ridgway and co-workers, in preparation). While the recent modeling results are encouraging, they must still be considered preliminary. More spectrophotometry with wide spectral coverage is needed (Honeycutt et al. 1977). High altitude airborne observations in the IR are particularly valuable owing to substantially higher transparency of the atmosphere in the telluric H_2O bands.

The presence of a stellar chromosphere may be indicated by either broad-band fluxes or spectral details. Lambert & Snell (1975) have proposed that the IR excesses of luminous stars such as α Ori and W Hya (M7–9) may include a chromospheric component (Section 5.3). In their model, the α Ori chromosphere would be optically thick at 14 μm and ~ 5000 K. Schmitz & Ulmschneider (1977) predict that in cool stars a temperature minimum in the stellar temperature profile will produce a brightness temperature minimum in the IR flux distribution.

High resolution spectra of the 5-μm CO fundamental vibration-rotation bands in α Boo have led Heasley et al. (1977) to suggest an inhomogeneous chromosphere for this star. Earlier work on the CaII K line by Ayres & Linsky (1975) demonstrated the probable existence of a strong temperature rise in the low chromosphere. However, such a temperature structure leads to the prediction of emission cores in the strongest CO lines. This effect was not observed, so Heasley suggested that localized regions of strong chromospheric rise may dominate the K line profile, while a weaker temperature rise characterizes the disk average and determines the CO line shapes.

A technique for extracting a detailed model temperature profile from high-resolution spectra of molecular vibration-rotation bands or atomic multiplets has been proposed, and demonstrated for the atomic case, by Ramsey & Johnson (1975) and Ramsey (1977). They propose the eventual application of their technique to CO when observational material becomes available.

The IR spectrum offers several observational tools for study of stellar temperatures and temperature profiles. Studies of the continuous flux distributions and molecular bands have already extracted temperature information for continuum optical depths in the range $\tau \sim 1$ to $\tau \sim 10^{-6}$. Such applications will be extended as the techniques are perfected.

3.2 Dynamics in Stellar Photospheres

Mass motions in Mira-type stars have been inferred from variations in apparent radial velocity noted in photographic spectra. Gross spectral changes are also evident at low resolution (Strecker et al. 1978). The high-resolution, high wavelength-precision spectra of molecular bands obtainable with an IR Fourier transform spectrometer (FTS) greatly extend the possibilities for study of mass motions. Since IR spectra can be recorded in the daytime and at visual minimum, the time series need not be interrupted as is necessary in photographic work. Maillard (1974) discovered line doubling in the CO first overtone bands of the Mira R Leo (M7-M9e). Subsequent time series studies (Hinkle 1978, D. N. B. Hall unpublished) of the Miras R Leo and χ Cyg (S7e) show that multiple absorption components in CO, OH, H_2O, and atomic lines exhibit cyclic behavior with the same period as the visual brightness. Changing relative doppler shifts reveal absorption from several layers of varying excitation temperature in the range 150 K to 3500 K. While the Mira phenomenon is impressively complex, the wealth of new observational material tightly constrains the atmospheric dynamics (Wing 1978). Hinkle has shown that the observed mass motions are qualitatively consistent with the general idea of a shockwave-pulsational model.

Low amplitude, irregular mass motions were observed in the supergiant α Ori by Brooke et al. (1974). Various atomic and molecular species exhibit differential velocities which are presumably associated with bulk motions of the atmosphere. The variations might be produced by large-scale convective phenomena of the type predicted by Schwarzschild (1975) for cool, luminous stars. In a possibly related development, Peery & Wojslaw (1977) collected a time series of IR spectra for the small amplitude, irregular variable TX Psc (C6,2). They interpret variations of the well-resolved CO line profiles in terms of differential motions of atmospheric layers.

The usefulness of the IR for these studies of cool star atmospheric dynamics may be traced to the presence of molecular bands with convenient distribution of excitation characteristics, the clean spectrum with a well-defined continuum, the possibility of very high spectral resolution, and the relatively weak (linear) dependence of the Planck function on temperature (so that the hottest region does not necessarily dominate the spectrum). For the brighter stars observations are readily acquired; the interpretive problems are likely to present a continuing challenge.

3.3   Departures from Local Thermodynamic Equilibrium (LTE)

Several studies have treated the conditions for ionization, dissociation, and excitation LTE relevant to line formation in cool stars. Thompson (1973) and Hinkle & Lambert (1975) have considered CO and other important diatomic molecules. They expect non-LTE(NLTE) effects especially for electronic transitions and in the outer atmospheric layers of the coolest, highest luminosity stars. Carbon et al. (1976) formulated the CO vibrational NLTE problem for solution in a model atmosphere computation. By spectrum synthesis they predicted observable NLTE effects in the fundamental CO bands of α Ori (M2Iab). The onset of CO NLTE effects depends critically upon collision excitation rates which are not adequately determined, and as a result it is necessary to consider possible NLTE effects even in α Boo (K2III) (Heasley et al. 1977).

A serious question has been raised by Auman & Woodrow (1975). They predict strong departures from ionization LTE in models with $T_{\rm eff} \lesssim 3000$ K. Over-ionization has been observed in spectral types as early as K2III (Ramsey 1977). For studies of cool stars this problem clearly requires careful attention.

Most of the work reviewed in this section is of a preliminary, first-look nature. There is no question that the IR spectrum provides an abundance of information about a cool star atmosphere. It is also clear that continued development and exploitation of both observational and modeling techniques is required to extract that information with confidence. As an example of how far we have yet to go, no single cool star has yet been treated by a full model-atmosphere analysis incorporating substantial information from the IR spectrum. This is not for lack of interest, but reflects the magnitude of effort required to mobilize simultaneously all the computational tools, the atomic and molecular data, and the observational material. A proper absolute calibration of IR fluxes is also required for confident interpretation of photospheric, chromospheric, and grain emission sources.

4   THE COMPOSITION OF STELLAR ATMOSPHERES

Since the near IR offers spectral transitions of abundant diatomic molecules, including CO, OH, and CN, quantitative abundance analysis of cool stars has long been thought to be one of the important potentials of the IR. Furthermore, the study of bright, cool stars is one of the outstanding capabilities of IR Fourier transform spectroscopy. At present, however, high-resolution IR spectroscopy of stars is still in a reconnaissance phase. Some spectral regions and some stellar types have only recently (or not yet) been studied. Sufficient work has been carried out, however, to specify the prominent atomic and molecular features observable in much of the IR for the most common types of cool stars.

4.1 Atomic and Molecular Spectral Features

Most atomic lines identified in the IR spectrum are assigned to high excitation transitions of neutral species. Many line assignments have been provided by Hall (1970) in his atlas of the solar photospheric and sunspot spectra. In general, atomic lines detected in the solar IR spectrum become much more prominent in cool giants as a result of reduced continuum opacity and reduced ionization. Chauville et al. (1970) tabulate numerous atomic identifications for transitions of 14 neutral atoms observed in M stars. However, many atomic spectral lines in solar and stellar IR spectra remain unidentified. Line lists computed by Kurucz (1975), though intended primarily for statistical treatment of stellar opacities, have some use in line identification.

Molecular equilibrium abundances can be computed for specified conditions of temperature, pressure, and elemental abundances if the required thermochemical parameters are available (e.g. Tsuji 1964, 1973). The availability and accuracy of the required parameters is a continuing problem, but at least for temperatures $\gtrsim 2000$ K the qualitative predictions of molecular equilibria appear to be reliable. A useful parameter for a stellar atmosphere is the column abundance of a species. This is defined as the number density per unit area in a column above optical depth "unity." Johnson et al. (1975) have prepared an extensive table of molecular column abundances for a substantial grid of models. This table is useful for determining the probable detectability of a given band.

To fully justify attribution of a molecular feature to a particular species it is necessary to resolve unambiguous structure, most commonly individual lines or band heads. At low resolution this criterion can not usually be attained, so caution is required. It is important to obtain excellent correction for any telluric extinction, and wide spectral coverage is necessary to clearly define a useful apparent continuum on both sides of the band. Reference should be made to appropriate auxiliary bands that may be present, and expected abundances of a species relative to other candidates should be considered.

The 3.1-μm C star band provides an interesting case study. Attempts to definitively specify its carrier from low resolution spectra were (in retrospect) futile. High resolution shows that the feature is composite, with variable contributions from many bands of HCN and C_2H_2

Table 1 Molecular bands in the infrared spectra of cool stars

Molecule	Bands	Wavenumber range (cm^{-1})	Spectral types	Selected references	Comments
C_2	$b^1\Pi_u - x^1\Sigma_g^+$ (Phillips)	$\sigma \gtrsim 4000$	C	Ballik & Ramsey 1963b Querci & Querci 1975	Branches throughout the near IR
	$A'^3\Sigma_g^- - X'^3\Pi_u$ (Ballik–Ramsey)			Ballik & Ramsey 1963a Thompson et al. 1973	
C_3	ν_3	1850	C	Treffers & Gilra 1975	Spectrum is apparently a pseudocontinuum
	$\nu_1 + \nu_3$	3200		Goebel et al. 1978a	
C_2H_2	ν_1 (C–H stretch)	3200–3400	C	Ridgway et al. 1978	Numerous combination bands; strong in circumstellar shells
CH	$\Delta v = 1$	2500–3000	C	Ridgway et al. 1977	No measured or computed f values
CN	$A^2\Pi - X^2\Sigma$ (red)	$\sigma \gtrsim 2500$	M,C	Fay et al. 1971 Thompson & Schnopper 1970	Dominates IR spectrum of C stars
CO	$\Delta v = 1$	2000	M,S,C	Geballe et al. 1977	Well-studied molecule
	$\Delta v = 2$	4000		Johnson & Méndez 1970 Ridgway 1974a	
	$\Delta v = 3$	6000			
CS	$\Delta v = 2$	2500–2600	C	Bregman et al. 1978 Ridgway et al. 1977	Fundamental not yet observed

Molecule	Transition	Range	Type	References	Comments
HCl	$\Delta v = 1$ $\Delta v = 2$	2500–4500	S	Ridgway et al. 1977 Hall & Ridgway 1979	Detected in S type Mira R And only
HCN	v_3 (C–H stretch)	3200–3400	C	Ridgway et al. 1978	Numerous combination bands; strong in circumstellar shells
HF	$\Delta v = 1$	4000–4400	M	Spinrad et al. 1971	—
H$_2$	$\Delta v = 1$ (quadrupole vib-rot)	4000–6000	Me,Se	Hall & Ridgway 1979	Positive detection only in Mira type stars
H$_2$O	$v_3, 3v_2$ $v_2 + v_3 - v_2$ $v_2 + v_3$ $v_1 + v_2$	2800–6200	> M6III > M0V	Flaud et al. 1976, 1977 Hinkle & Barnes 1979	Plus other bands throughout the IR H$_2$O is an asymmetric molecule with a very complex spectrum
NH	$\Delta v = 1$	2500–3400	MI	Lambert & Beer 1972	α Ori only. No measured or computed f values
OH	$\Delta v = 1$ $\Delta v = 2$	2500–3200 5000–6400	K,M	Beer et al. 1972 Mertz 1972b	—
SiO	$\Delta v = 1$ $\Delta v = 2$	1200–1250 2400–2500	K,M	Knacke et al. 1969 Beer et al. 1974	High resolution of the fundamental $\Delta v = 1$ obtained by Geballe, Lacy & Beck (private communication)

(Ridgway et al. 1978). Furthermore, the absorption may arise in the stellar photosphere at temperatures ~ 2000 K or in circumstellar shells ~ 600 K, or in both. The low-resolution spectra could not distinguish among the variety of phenomena.

Additional examples exist. A case has been made for CS at 3.9 μm based on low-resolution spectra (Bregman et al. 1978), and in fact high-resolution spectra display individual rotational lines (Ridgway et al. 1977). An argument can be made for C_3 at 5.2 μm (Goebel et al. 1978a). The suggested identifications of SiO in C stars (Fertel 1970, Bregman et al. 1978), however, are surprising considering the expected abundance of SiO and the strength of the overtone bands. Indeed, in the former case, alternate and more plausible identifications for the spectral features were proposed (Wing & Price 1970) and in the latter case, available high-resolution spectra show no evidence for SiO. It also is true, however, that some spectral features are more readily detected at moderate resolution than at high resolution.

The IR molecular bands for which detections have been reported are collected in Table 1. (The references cited were chosen to indicate illustrative and/or recent work—not necessarily first detections.) Purely circumstellar species are omitted. From a comparison of the work of Johnson et al. (1975) and Table 1 it is apparent that the principal expected molecular detections have been realized. Additions to the list should be forthcoming as high resolution spectroscopy reaches new wavelength regimes (e.g. the MgH vibration-rotation fundamental at 1500 cm^{-1}) and as detailed studies of the cooler stars permit discrimination of relatively weak bands against the background of H_2O (in oxygen stars) or of CN and C_2 (in carbon stars).

An urgent need exists for progress in laboratory and theoretical studies of many atomic and molecular species. In the coolest stars, identifications are far from complete. Parameters required include f values, collisional excitation rates, and dissociation energies. An FTS can routinely measure line positions to ± 0.001 cm^{-1}, and full exploitation of a spectrum requires supporting laboratory measurements of comparable quality. Completely adequate information is not available for even one molecule in Table 1. For abundance studies, the most important molecules are CO, CN, OH, CH, NH, C_2, and H_2O. Thus it is gratifying to note important recent laboratory results for OH (Maillard et al. 1976), C_2 (Chauville et al. 1977), and H_2O (Flaud et al. 1976, 1977). A regularly updated compilation of edited molecular data for cool stars would require a large effort, but would greatly facilitate exploitation of IR stellar spectra.

At present only rudimentary empirical evidence is available regarding the variation of molecular band strengths with abundance. One result that appears firm is a correlation of CO band strength with other band strength and abundance measures in oxygen-rich stars. For example, stars classified super-metal-rich and CN-strong also have enhanced CO (Ridgway 1974a,b). Photometric studies have shown that stars with a weak optical G band have weak CO (Hartoog et al. 1977). Studies of CO in metal-deficient stars (Pilachowski 1978b) suggest a range in CNO/Fe for metal-poor field giants. More extensive and detailed studies of these and other molecular bands are of substantial interest, since such measurements can be extended to faint stars (Section 2.1).

It is difficult to overemphasize the importance of using model simulations and high-resolution spectra to fully validate the use of low-resolution spectral or photometric indices. Bell et al. (1978) combined these tools to demonstrate a carbon deficiency in the bright giants of the globular clusters M92 and NGC 6397. In this study, high-resolution spectra of the moderately metal-deficient giant α Boo were examined to verify the model and spectrum synthesis techniques. These in turn were used to interpret the behavior of narrow band photometry for faint giants in the metal-poor clusters.

4.2   Elemental Abundance Analysis from IR Spectra

An isothermal absorbing layer model is generally satisfactory for estimating order-of-magnitude column abundances. To determine believable abundance ratios (relative to hydrogen) requires more detailed modeling of the atmospheric structure. While this requirement is generally recognized, the actual exploitation of modeling techniques is somewhat erratic. In short, the model atmosphere analysis of IR molecular spectra is not sufficiently systematized to define a standard technique.

Reduced to its essentials, a model atmosphere analysis requires specification of effective temperature, pressure, temperature profile, and line broadening. This information plus the related physics leads to the determination of abundances. In analysis of photographic spectra, the model parameters are deduced primarily from various characteristics of the line spectrum. In studies of the IR spectra, temperature is traditionally either guessed or estimated from flux distributions, pressure is always guessed, and temperature profiles are taken from models with the chosen effective temperature or scaled from a different (even solar) model. Spectral synthesis (almost universally employed) is used to compute the emergent spectrum and compared with observation. The comparison is evaluated by eye. Several parameters (including the abundances) are then varied to seek a best fit. Abundance determinations obtained in this way are not unique. Error estimates in these studies are usually highly subjective. Since improvement in the model specification techniques is foreseeable, existing results should be considered preliminary.

4.2.1 OXYGEN-RICH GIANTS   In spite of the limited sophistication of the work published to date some progress has been made. Studies of the molecules in Table 1 indicate that in oxygen-rich stars the assumption of solar abundances is usually consistent with the observational data. The most complete discussion available is for α Ori. Lambert (1974) reviews work by the Texas-JPL group. This study is particularly significant since the model used (Faÿ & Johnson 1973) was chosen to reproduce the spectral flux distribution. Lambert concludes that the C, N, O, and Si abundances are probably solar within a factor of two. The most serious problems in studies of α Ori (M2Iab) are associated with its high luminosity: the usual model atmosphere assumptions (plane parallel, etc) may not be valid. For example, the large line widths led Lambert to adopt the high microturbulent velocity of 7 km/sec, indicative of important, poorly understood phenomena in the stellar atmosphere.

In an analysis of CO bands only, Thompson & Johnson (1974) found the spectrum of α Her (M5II) consistent with a solar C abundance. They obtained the best fit for a microturbulent velocity of 2 km/sec. In a study of K giants based on the CO $\Delta v = 2$ bands, Ridgway (1974a,b) found that a solar C abundance satisfactorily reproduced the observed spectra of four field K giants. The CO spectrum of the high velocity giant α Boo was consistent with abundances deduced in previous model atmosphere analyses of photographic spectra. Lambert & Ries (1977) examined the ^{13}CO lines in α Boo to confirm C abundance results deduced from the [CI] line at 8727 Å.

The University of California Fabry-Perot group has recorded and studied spectra of the CO fundamental near 5 μm. Geballe et al. (1977) reported the results of that program for α Ori, α Sco, and α Her. They obtained a best fit to the observed spectra for reduced carbon (~ 0.1 solar), but with substantial uncertainty. The fundamental bands are certainly difficult to work with. The oscillator strengths are typically 10^4 times stronger than those for the second overtone studied by Lambert (1974) and Thompson & Johnson (1974). Line saturation, chromospheric, and NLTE effects may complicate the analysis of the fundamental (Carbon et al. 1976).

4.2.2 M DWARFS   Mould (1978) studied near-IR spectra (4000–6600 cm^{-1}) of six M dwarfs. He derived a correlation between luminosity (excess or deficiency relative to the main sequence) and metallicity (determined from numerous IR atomic lines of Na, Mg, Al, and Ca), and was able to set limits on the abundance dispersion in disk population M dwarfs. The CO and OH bands indicate that carbon correlates well with general metallicity, but that the metal-poor dwarfs show relatively less depletion of O.

4.2.3   CARBON STARS   It is quite possible that the existence of C stars would not have been predicted on purely theoretical grounds, and even with empirical knowledge the explanation of their genesis has been by no means obvious. Considering the depth of our initial ignorance, even order-of-magnitude abundances suffice to eliminate some theories. Unfortunately, the interpretive problem is unusually severe as well. The number densities of molecules contributing strongly to line blanketing opacities, especially CN, are dependent on C abundance. Consequently, the atmospheric temperature structure is very sensitive to C abundance (Carbon 1974). In addition, Scalo (1973a) has shown that the atmospheric gas pressure is a sensitive function of the C excess, $(C - O)$. These technical problems are quite aside from the fact that C star modeling seriously challenges available theoretical techniques.

In an effort to find a basis to prefer one of the principal classes of nucleosynthesis models for C stars, Thompson and his collaborators have reported a series of observations. The most recent work (Thompson 1974, 1977) indicates that the C stars studied have a relatively high C abundance (solar or greater). This is consistent with mixing theories which transport additional C to the stellar surface. Thompson is unable to match the observed spectra with low C models (e.g. 0.1 solar). This weighs against mixing theories which would cycle the surface material through the CNO process, depleting both C and O. In spite of uncertainties due to model selection, Thompson's result appears persuasive because the effect is so large.

Previously, other evidence for C enhancement was reported by Scalo (1973b) who reanalyzed early IR observations. The photographic spectral region has also produced evidence for carbon enhancement (Kilston 1975). These most recent results concur on carbon enhancement for the stars studied. It should be noted, however, that the C stars exhibit a substantial variety of observational characteristics, and even in the IR the spectra are often frightfully complex.

Continued effort in cool star abundance studies is very important, but progress will likely be slow as the modeling techniques must develop hand in hand with our understanding of the observations. At present the computational tools are available primarily to the individuals engaged in developing the techniques. Some recourse is possible to limited grids of models (Johnson 1974, Querci et al. 1974, Gustaffson et al. 1975). The long-term need, however, is for a cool star analogue of the computer program ATLAS (Kurucz 1970). ATLAS is a well-documented model atmosphere program in which established techniques have been systematically implemented and physical parameters (especially opacities) exhaustively pursued. Both the scope and limitations of the input physics are reasonably comprehensible to the user. The eventual availability of a well-documented code for cool stars, and of a molecular data bank suggested earlier, would open the field of elemental abundance studies by IR stellar spectroscopy to general use.

4.3   *Isotopic Ratios*

By comparison with absolute abundances, determination of isotope ratios should be relatively easy. While a number of early results were shown to be in error because of saturation (Thompson 1973), and several results are still disputed, some trends are now apparent.

The ^{13}C isotope is readily detected in the CO bands of cool giants. In carbon stars the ^{13}C isotope is also observed in infrared CN and C_2 bands. In addition, photoelectric scans of the CN red system have been heavily used, and in some cases CH (in the blue) also. The observed $^{12}C/^{13}C$ ratios are typically ~ 4–30 compared to the solar-terrestrial value ~ 89. Values $\gtrsim 20$ are in accordance with expected mixing in evolution off the main sequence. The low values ~ 4 are not understood, and consequently accurate specification of the lowest $^{12}C/^{13}C$ ratio observed in normal giants is of some importance. We cannot discuss all of the work in this area, but an excellent example is the study by Hinkle et al. (1976) of five cool stars: $^{12}C/^{13}C$ ratios in the range 7–25 were obtained from the CO overtone bands. The principal discrepancies in ^{13}C isotope studies are for the high luminosity stars. Thompson & Johnson (1974) suggested a limit $^{12}C/^{13}C > 20$ for α Her (M5II), Gautier et al. (1976b) estimated $^{12}C/^{13}C > 20$ for α Ori (M2Iab). Hinkle et al. (1976) found a ratio of 17 for α Her but 7 for α Ori. A more thorough understanding of the supergiant atmosphere would augment our confidence in these results.

Studies of ^{13}C in carbon stars from the photographic spectral region have been notoriously inconsistent, although some stars are definitely ^{13}C strong and some ^{13}C weak. Dominy et al. (1978), in a simultaneous analysis of ^{13}CO (near 1.7 μm) and ^{13}CN (near 8000 Å), obtained consistent results of $^{12}C/^{13}C \sim 30$ for V360 Cyg.

The oxygen isotopes ^{17}O and ^{18}O can be studied in the CO fundamental and first overtone. Several studies confirm a definite enhancement of ^{17}O. Maillard (1974) and Geballe et al. (1977) estimated the ratio $^{16}O/^{17}O$ in α Her at ~ 450 and 500, respectively, compared to the terrestrial value of 2700. Furthermore, both studies gave $^{17}O/^{18}O > 1$ (also reported by Lambert 1974), while the terrestrial ratio is $^{17}O/^{18}O \sim 0.18$. In a survey of numerous cool giants (still in progress), Lambert (1977) finds variable ^{17}O enrichment. An enhancement of the $^{17}O/^{18}O$ ratio is consistent with computed evolution models (Dearborn et al. 1976), but a quantitative comparison is not yet available.

Other available isotopic ratios are consistent with solar-terrestrial abundances. The bandheads of ^{30}SiO are faintly visible in the SiO first overtone region. Beer et al. (1974) and Ridgway et al. (1977) find $^{28}Si/^{30}Si$ consistent with a terrestrial ratio of 30 in several K and M giants. From a detection of HCl in the S type Mira R And, $^{35}Cl/^{37}Cl$ is found to be very close to the terrestrial ratio of 3.

A decade ago elemental and isotope ratios were expected to pinpoint the nuclear processes and mixing phenomena responsible for surface abundance changes in the cool giants. The goal is not yet achieved, but the hope remains. Intervening years have seen substantial progress in both observation and theory. We expect rapid expansion of isotopic studies by means of IR spectroscopy. Elemental abundance work will develop more slowly as the required stellar atmosphere models and molecular data become available.

5   STELLAR RECYCLING: THE RAW AND THE COOKED

Stars do not exist in splendid isolation. Dynamic interactions between a star and the environment occur both early and late in its life. Material accretes from the interstellar medium during star formation. Some fraction of this mass is processed in the stellar interior by nucleosynthesis. Late stellar evolution is marked by continuing, sometimes violent, episodes of mass loss. IR spectroscopy holds a unique position in the study of this mass exchange: both the molecular and particulate phases of the material can be studied. The goal, then, is to determine the nature of the material, the exchange process itself, and the interaction between interstellar and circumstellar matter.

Recent evidence for the role of various grain types in interstellar extinction and in molecular clouds enhances interest in identifying the possible sites and circumstances of grain formation (Salpeter 1977). The proceedings of IAU Symposium 52 (see Woolf 1973) and IAU Colloquium 42 (see Merrill 1977) comprise excellent general reviews of these subjects. We must, of necessity, limit discussion to the specific contribution of IR spectroscopy in the ongoing effort.

5.1 Young Stars

Strom et al. (1975) have reviewed the environment of young stellar objects. One of the most challenging goals of IR spectroscopy will be to establish the detailed structure of the circumstellar envelopes surrounding pre–main-sequence stars. The source(s) of the observed IR emission is presently unknown, in part because the available spectra do not cover a wide enough range in wavelength with sufficient detail to unambiguously determine the origin of the IR continuum. Hence, at least for the present, a detailed analysis of the anticipated effects of differential extinction and the relative contributions of photosphere, viscous heating, plasma, and dust emission to observed IR excess cannot be fully realized. High-resolution IR spectra are necessary to determine the kinematics and physical properties of both the obscured central source and the surrounding material. Unfortunately the general faintness of young stellar objects has to date made such spectroscopy difficult at best. Hence, no detection of gaseous molecular absorption has yet been attributed to the circumstellar material directly associated with these stars. The presence of dust in the vicinity of star formation is certainly reasonable, but the detailed interaction of the young star with the local circumstellar medium is only recently opening to observational study. To observe a star in the actual process of collapse—potentially the most exciting prospect of IR spectroscopy—will require going fainter still, but may be within reach.

For hot young stars detection of hydrogen emission from an associated compact HII region provides important information on otherwise obscured central stars. The detection of highly localized hydrogen emission lines and cold CO absorption lines coupled with the nondetection of H_2 emission in the IR spectrum of the Becklin-Neugebauer (BN) source in Orion (Hall et al. 1978) has provided significant information on the early stellar evolutionary phases: the data support the hypothesis that BN is a young B star within a compact HII region and an expanding dust envelope. Following the detection (Grasdalen 1976) and mapping (Joyce et al. 1978) of Brackett α emission near 4 μm, Simon et al. (1978) surveyed a number of compact and presumably young IR sources with similar success. In addition to providing information on the radiative mechanisms, studies of hydrogenic spectra permit determination of the extinction. Studies of hot (~ 1500–2000 K excitation) H_2 emission in the 2-μm spectra of the region around BN in the Orion Nebula (Gautier et al. 1976a) will be a useful probe of conditions in the immediate vicinity of active or recent star formation.

The T Tauri stars, usually identified with a later stage of pre–main-sequence evolution, have been studied at moderate spectral resolution: $\lambda\lambda 2$–4 μm (Cohen 1975) and $\lambda\lambda 8$–13 μm (Cohen, private communication). The spectra are generally featureless. In particular, the 10-μm silicate signature is usually absent in either emission or absorption (contrary to broad bandpass evidence for silicates summarized by Rydgren et al. 1976). The notable exception to date is HL Tau, which shows both 3-μm "ice" and 10-μm silicate absorption, presumably due to an associated molecular cloud. Detection of Brackett γ and molecular hydrogen emission in the high-resolution 2-μm spectrum of T Tau (Beckwith et al. 1978) demonstrates that some mechanism in the envelope is capable of ionizing hydrogen and exciting the H_2. Further information and analysis is necessary to distinguish between competing mechanisms such as UV radiation and shock fronts. Rydgren et al. (1976) have summarized many of the salient analytic and observational features of the T Tauri phenomenon.

Studies of Ae stars in young stellar aggregates suggest that thermal emission by dust and/or plasma might contribute to the observed radiation but other mechanisms are possible (Cohen 1973, Allen & Penston 1973, Warner et al. 1978). The analytical situation has been summarized along with recent optical data by Garrison (1978). Strong evidence exists in many instances for a circumstellar disc. Spectroscopic studies of the obscured emission-line stars MWC 349 and LkHα 101 (Thompson & Reed 1976) have led Thompson et al. (1977) to the intriguing hypothesis of thermal emission from a viscous circumstellar disc. The expected properties of such solar discs have been reviewed by Huang (1973) and Cameron (1978).

5.2   Hot Stars

Mass loss in very early high-luminosity stars (see Conti 1978 for a review) has long been recognized from photographic studies. Optical studies of hydrogen line profiles indicate expanding atmospheres, with expansion velocities as large as several thousand km/sec. Although one might reasonably expect the outflowing gas to condense into dust, an IR survey by Barlow & Cohen (1977) found little evidence for substantial IR excess among luminous O, B, and A stars. The small excesses observed are consistent with thermal bremsstrahlung radiation. In particular, detailed model fits to the observed emission from many Be stars (cf Gehrz et al. 1974) do not require the presence of dust. Since, as will be seen below (Section 5.4), dust appears to condense in the hot gas of novae, such a null result may represent an important constraint on the conditions leading to grain formation. Spectrophotometry of the classical Be star γ Cas at $\lambda\lambda 1$–4 μm has permitted a study of the energy balance in the stellar envelope (Scargle et al. 1978) using a self-consistent plasma model for the IR excess including bound-free emission and attenuation by the shell.

The large long wavelength IR excesses in a number of Be stars, however, do strongly suggest dust. Coyne & Vrba (1976) present a good case for emission from a circumstellar dust ring around HD45677 over the suggested plasma emission process of Milkey & Dyck (1973). Detection of an UV absorption band at 0.22 μm in the same star (Savage et al. 1978) suggests that when dust is present in hot stars it is possibly the same material associated with cool carbon stars (Section 5.3).

The powerful stellar winds from Wolf-Rayet stars are returning material processed by nucleosynthesis in stellar interiors to the interstellar medium. The IR excesses from most of these stars (WN and early WC) can be attributed to thermal bremsstrahlung; in marked contrast, Wolf-Rayet stars at the extreme end of the carbon sequence (WC9) show blackbody-like excesses typical of dust (Gehrz & Hackwell 1974, Webster & Glass 1974, Hackwell et al. 1974, M. Cohen et al. 1975). Since the WC7 star HD193793 undergoes irregular episodes of apparent dust formation as revealed by the rapid rise and decay of the blackbody-like excess (Hackwell et al. 1978), study of this star at high resolution may yield information on conditions favorable to dust formation. The sheer complexity of the rich emission-line spectrum dominating visual and ultraviolet spectra of WR stars makes abundance analysis quite difficult and severely model dependent in those spectral regions. However, low-resolution spectra have confirmed that strong He^+ and He^{+2} emission can be seen out to 4 μm in representative WN stars and that in addition C^{+3} emission occurs in WC stars (Bernat et al. 1977, Cohen & Vogel 1978, Merrill & Black 1978). Hydrogen line emission is either absent or very weak. In the IR, measures of hydrogenic recombination lines relative to free-free continuum for various elements may yield direct He, C, and N abundances. Study at higher spectral resolution of the structure of the continuum and detailed line profiles should permit accurate modeling of the extended stellar atmospheres (Hartmann & Cassinelli 1977). Since WC and WN stars are recognizable at large distances even though highly reddened, lines in the UV and the IR arising from the same atomic levels offer a powerful measure of interstellar extinction.

Detection of neutral atomic emission lines in the 2-μm spectrum of IRC+10420 (Thompson & Boroson 1977) along with atomic hydrogen absorption lines provide a unique probe of the circumstellar envelope of this puzzling highly luminous (F8Ia) IR source (Humphreys et al. 1973). In addition to confirming the optical spectral type, these spectra indicate a large column density of material above the photosphere. Other objects of this type associated with polarized symmetric reflection nebulae (see Calvert & Cohen 1978 for a summary) should also be studied spectroscopically.

5.3   Cool Stars

The discovery of dust particles in the circumstellar envelopes of cool stars was one of the first and remains one of the most important results of IR astronomy. Numerous reviews (e.g. Woolf 1973, Khozov 1976, Merrill 1977) have treated this subject. Optical information on mass loss from evolved stars has been reviewed by Reimers (1977). The current stage of our knowledge (or ignorance) on the theoretical aspects of mass loss mechanisms are extensively reviewed by Weymann (1977). For cool stars in particular, available observational evidence suggests that the mechanisms of grain formation and mass loss can be explored in detail. Further, it is clear that the type of dust returned to the interstellar medium depends critically on the chemical composition of the star.

Forrest et al. (1975) have shown spectroscopic evidence for three distinct grain types in cool stars. As expected on the basis of simple thermodynamic equilibrium calculations (Gilman 1969), the presence of a given species appears to follow the C/O abundance ratio. In stars with C/O < 1 the strong 10- and 20-μm bands of silicates dominate the IR spectra; for stars with C/O > 1 a narrower emission band near 11 μm usually attributed to silicon carbide appears (Treffers & Cohen 1974). Some S stars as well as many carbon stars also show a featureless continuum indicative of a third grain species with unknown opacity (apart from lack of IR band structure). The relative strength of the SiC band does not correlate with the degree of emission in the featureless continuum, indicating that two different materials are involved. Photometric studies of S stars suggest another material may be present (Thomas et al. 1976) but spectra are necessary to confirm this result. Photometry of M stars at 33 μm (Hagen et al. 1975) suggests the presence of additional excess possibly due to a silicate structural reasonance. The optical properties of small grains and radiative transfer effects tend to minimize the observed contrast between band and continuum at longer wavelengths. Thus moderate resolution spectra of both the 20- and 30-μm features are needed in M or K stars with large excess due to optically thin emission. The K and M supergiants IRC + 60370 (Humphreys & Ney 1974b) and RW Cyg would be excellent candidates for further study (see Merrill 1977 for IR spectra).

The 10-μm band in oxygen-rich stars is probably due to a single type or class of material. The few available high resolution spectra (more are needed) show no band substructure (Gammon et al. 1972, Treffers &

Cohen 1974). Distributions of particle size, shape, and composition may contribute to this lack of band structure, but the observed invariance of band shape and structure place severe constraints on such distributions.

At low resolution, no compelling spectroscopic evidence exists for a variation in band opacity from star to star. As noted by Forrest et al. (1975) in their comparative studies of grain emission in stars, uncertainties in placement of the continuum can account for all the apparent differences. Whenever a large 10-μm excess is observed, the excess is quite similar to that found from the interstellar material heated by the Trapezium stars in Orion. Further, simple models can be made using the same band opacity to fit both the observed optically thin emission in cool stars and the observed absorptions seen in other sources (Gillett et al. 1975a) even when the 10-μm optical depth is quite large (Willner 1977, Capps et al. 1978). Attempts to quantify observed silicate band strengths using a simple model to define equivalent width (Dorshner et al. 1978) seem to be meeting with some success. Such analyses are essential to arriving at an understanding of the mass loss mechanism itself.

Cool stars exhibit a continuous range of silicate optical depths (Merrill 1977). The classic excess usually involves optically thin emission. Thick shells may show absorption at the core of a broad emission, or in extreme cases net absorption only. Although implications regarding the detailed properties of circumstellar material are beyond the scope of this text it is important to realize that such a wide range in shell optical depth allows important grain and shell characteristics to be isolated for further study. Independent attempts to fit the observed range in silicate spectra (Jones & Merrill 1976, Hagen 1978, Bedijn 1977) indicate the 1- to 3-μm opacity (where photospheric emission peaks) of the circumstellar material around M stars must be comparable to the opacity near the 10-μm band center. Further, much of this near-IR opacity must be due to the band carrier. Bergeat and his co-workers (Bergeat et al. 1976a,b,c) have attempted detailed analyses of the more intractable carbon star envelopes without as yet incorporating IR spectral information.

In addition to thermal emission of dust at longer wavelengths, the nature of the radiative processes producing the flux distribution of luminous cool stars remains uncertain. On the basis of the apparent filling in or "veiling" of the optical lines and the overall flux distribution of the very luminous M supergiants S Per and VX Sgr, Humphreys (1974) and Gilman (1974) were led to consider stellar models including the effects of thermal bremsstrahlung from a circumstellar plasma. Lambert & Snell (1975) have suggested a more specific model for M supergiants including contributions from a chromosphere and photosphere with significant photospheric SiO opacity further complicating determination of the excess due to dust. Fawley (1977) has countered this model by suggesting that the optical veiling is due to line weakening in the photosphere, that near-IR excesses are not present and that thermal dust emission provides the IR excess. However, the $\lambda\lambda 2$–4-μm spectra of S Per and VX Sgr are qualitatively different from other supergiants of equivalent spectral type (Merrill 1977). Although molecular absorption bands seen in M stars are present, they are much weaker in these peculiar stars. The $\lambda\lambda 1$–4-μm spectrum of NML Tau shows a similar weakening of molecular bands compared to otherwise comparable stars (Strecker et al. 1978). Further, the slope of the continua are indeed roughly free-free in character at moderate resolution. Along similar lines T. Tsuji (private communication) has suggested that the moderate excesses at $\lambda\lambda 4$–8 μm (Puetter et al. 1977) and 33 μm are due to thermal emission of H_2O steam in the circumstellar envelope.

Many luminous M and C stars are amply bright for very high resolution IR spectroscopy. Such observations can provide a wealth of information about the physics of mass ejection. Preliminary studies of the carbon-rich object IRC + 10216 (Geballe et al. 1973, Barnes et al. 1977) have revealed cool (250–600 K) material in an expanding shell. Resolved line profiles for the CO 2-0 band provide the basis for a model of the shell (Hall 1977). This object is an important meeting ground for IR and radio studies of cool star envelopes (e.g. Morris 1975, Kwan & Hill 1977). In the Me star R Leo, Hinkle & Barnes (1979) distinguished circumstellar from photospheric components of the molecular bands.

5.4 Novae

The general properties of classical novae have been recently reviewed by Wolf (1977) and by Gallagher & Starrfield (1978). IR spectroscopy has revealed a number of interesting phases in the evolution of nova ejecta. The time dependence of the IR excess indicates that dust formation can occur rather abruptly (Ney & Hatfield 1978, Sato et al. 1978) with only a matter of days elapsing between the onset of formation and completion. The resultant conversion of visual flux to thermal re-emission in the IR accounts for the otherwise inexplicable abrupt decline seen in the light curves of many classic novae. Not all novae form dust, however. The physical parameters that account for this behavior have been examined by Clayton & Wickramasinghe (1976) and Gallagher (1977). The novae offer a unique opportunity to study grain formation and growth; IR spectroscopy may well provide decisive clues about the mechanism itself.

At higher resolution Grasdalen & Joyce (1976b) have identified numerous highly ionized "coronal" lines in the IR spectrum of the fast nova, Nova Cygni 1975 (V1500 Cyg), in addition to a rich hydrogen line spectrum (e.g. Strittmatter et al. 1977). The identification of these lines must not rely solely on line coincidences, however, as has been pointed out by Black & Gallagher (1976) who propose an alternative list of helium line identifications based on physical arguments. The more typical nova, Nova Vulpeculae 1976 (NQ Vul), also shows a rich hydrogen emission-line spectrum with strong band emission at 5 μm and a weaker excess near 2.4 μm (see Figure 10 of Merrill 1977, reporting work by many observers). The strong 5-μm excess was also seen in Nova Serpentis 1970 (FH Ser) (Geisel et al. 1970). Ferland et al. (1979) have identified the 2.4-μm emission with the first overtone band of CO on the basis of high resolution spectra. This indicates that the 5-μm excess would be due at least in part to CO. The 5-μm band seemed to abruptly disappear with the onset of grain formation. After grains formed, a featureless IR continuum ensued (R. D. Gehrz, private communication). Coordinated UV, visual, IR, and radio observations of future novae should greatly enhance our understanding of accelerated grain growth.

5.5   Grain Composition

The quest to identify the specific constituents of circumstellar and interstellar material has produced an extensive literature. Until recently, the emphasis has been on the determination of the optical constants of candidate materials—terrestrial minerals, meteoritic or lunar rocks—to be used in Mie theory calculations. Recent reviews by Aannestad & Purcell (1973), Huffman (1977), and Andriesse (1977) have thoroughly summarized the state of the art both in the laboratory and for the theory of interaction of small particles with radiation. The best analogues to observation for silicates appear to have a characteristic lack of high order crystal structure.

As has been stressed by Huffman (1977), the environment of the extrasolar minerals is quite unlike that found on earth. Many effects are possible that would alter the actual properties *in situ* from the strict predictions of Mie theory applied to standard materials. An artificial silicate proposed by Day (1974) has been particularly promising. Day & Donn (1978) have produced a greatly under-oxidized (Si, Mg, O) mineral which, independently of the detailed stochiochemistry of the compounds, reproduces the observed 10-μm band both in structure and position. Most recently Rose (1977, 1978) and Russell (1978) have measured the actual emissivities of grain-sized candidate minerals. Such studies promise to provide fresh inputs into the controversies concerning the nature of circumstellar and interstellar material.

6 CONCLUSION

The tools of IR spectroscopy provide basic astronomical information: temperature of cool material; stellar spectral types; atomic, molecular, and isotopic abundances; velocities. Special rewards of IR spectroscopy include: study of molecular species; information about particulate matter; penetration through obscuring material. We have reviewed several major topics within the expanding horizon of IR stellar astronomy. Many of the current results are suggestive, not final—current indications from novel observations and developing interpretation.

While IR astronomy is still in its infancy, a revolution in the observational possibilities can already be foreseen. New IR detector arrays with cryogenic instrumentation will extend the reach of high-resolution IR spectroscopy to the fringes of our galaxy and beyond.

ACKNOWLEDGMENT

Astronomy at the University of Minnesota is supported by grants from the US Air Force, NASA, and NSF. Kitt Peak is operated by the Association of Universities for Research in Astronomy, Inc., under contract with the National Science Foundation.

Literature Cited

Aannestad, P. A., Purcell, E. M. 1973. *Ann. Rev. Astron. Astrophys.* 11:309–62

Aaronson, M., Cohen, J. G., Mould, J., Malkan, M. 1978. *Ap. J.* 222:824–34

Aitken, D. K., Jones, B., Bregman, J. D., Lester, D. F., Rank, D. M. 1977. *Ap. J.* 217:103–7

Allen, D. A. 1972. *Ap. J. Lett.* 172:L55–58

Allen, D. A., Beattie, D. H., Lee, T. J., Steward, J. M., Williams, P. M. 1978. *MNRAS* 182:57p–60p

Allen, D. A., Hyland, A. R., Longmore, A. J., Caswell, J. L., Goss, W. M., Haynes, R. F. 1977. *Ap. J.* 217:108–26

Allen, D. A., Penston, M. V. 1973. *MNRAS* 165:121–31

Allen, R. G. 1978. *An Infrared Investigation at the Temperature Structure of the Solar Atmosphere.* PhD thesis. Univ. Ariz. 204 pp.

Andriesse, C. D. 1977. *Vistas Astron.* 21:107–90

Auman, J. R., Woodrow, J. E. J. 1975. *Ap. J.* 197:163–73

Ayres, T. R., Johnson, H. R. 1977. *Ap. J.* 214:410–17

Ayres, T. R., Linsky, J. L. 1975. *Ap. J.* 200:660–74

Baldwin, J. R., Frogel, J. A., Persson, S. E. 1973. *Ap. J.* 184:427–34

Ballik, E. A., Ramsey, D. A. 1963a. *Ap. J.* 137:61–83

Ballik, E. A., Ramsey, D. A. 1963b. *Ap. J.* 137:84–101

Barlow, M. J., Cohen, M. 1977. *Ap. J.* 213:737–55

Barnes, T. G., Beer, R., Hinkle, K. H., Lambert, D. L. 1977. *Ap. J.* 213:71–8

Beckwith, S., Gatley, I., Matthews, K., Neugebauer, G. 1978. *Ap. J. Lett.* 223:L41–43

Bedijn, P. J. 1977. *Studies of dust shells around stars.* PhD thesis. Univ. Leiden

Beer, R., Hutchinson, R. B., Norton, R. H., Lambert, D. L. 1972. *Ap. J.* 172:89–115

Beer, R., Lambert, D. L., Sneden, C. 1974. *Publ. Astron. Soc. Pac.* 86:806–12

Bell, R. A., Dickens, R. J., Gustafsson, B. 1978. *Astronomical Papers Dedicated to Bengt Strömgren,* ed. A. Reiz, T. Anderson, pp. 249–60. Denmark: Copenhagen Univ. Obs. 428 pp.

Bergeat, J., LeFevre, J., Kandel, R., Lunel, M., Sibille, F. 1976a. *Astron. Astrophys.* 52:245–61

Bergeat, J., Lunel, M., Sibille, F., LeFevre, J. 1976b. *Astron. Astrophys.* 52:263–72

Bergeat, J., Sibille, F., Lenel, M., LeFevre, J. 1976c. *Astron. Astrophys.* 52:227–44

Bernat, A. P., Barnes, T. G., Schupler, B. R., Potter, A. E. 1977. *Publ. Astron. Soc. Pac.* 89:541–45

Black, J. H., Gallagher, J. S. 1976. *Nature* 261:296–98

Bregman, J. D., Goebel, J. H., Strecker, D. W. 1978. *Ap. J. Lett.* 223:L45–46

Brooke, A. L., Lambert, D. L., Barnes, T. G. 1974. *Publ. Astron. Soc. Pac.* 86:419–28

Calvet, N., Cohen, M. 1978. *MNRAS* 182:687–704

Cameron, A. G. W. 1978. In *The Origin of the Solar System*, ed. S. P. Dermott, pp. 49–74. New York: Wiley. 668 pp.

Capps, R. W., Gillett, F. C., Knacke, R. F. 1978. *Ap. J.* 226:863–68

Carbon, D. F. 1974. *Ap. J.* 187:135–45

Carbon, D. F. 1979. *Ann. Rev. Astron. Astrophys.* 17. In press

Carbon, D. F., Milkey, R. W., Heasley, J. N. 1976. *Ap. J.* 207:253–62

Chauville, J., Maillard, J. P., Mantz, A. W. 1977. *J. Mol. Spectrosc.* 68:399–411

Chauville, J., Querci, F., Connes, J., Connes, P. 1970. *Astron. Astrophys. Suppl. Ser.* 2:181–99

Ciatti, F., Mammano, A., Vittone, A. 1977. *Astron. Astrophys.* 61:459–67

Ciatti, F., Mammano, A., Vittone, A. 1978. *Astron. Astrophys.* 68:251–57

Clayton, D. D., Wickramasinghe, N. C. 1976. *Astrophys. Space Sci.* 42:463–75

Cohen, J. G., Frogel, J. A., Persson, S. E. 1978. *Ap. J.* 222:165–80

Cohen, M. 1973. *MNRAS* 164:395–421

Cohen, M. 1975. *MNRAS* 173:279–93

Cohen, M., Barlow, M. J., Kuhi, L. V. 1975. *Astron. Astrophys.* 40:291–302

Cohen, M., Vogel, S. N. 1978. *MNRAS* 185:47–55

Conti, P. 1978. *Ann. Rev. Astron. Astrophys.* 16:371–92

Coyne, G. V., Vrba, F. J. 1976. *Ap. J.* 207:790–98

Davidson, K., Humphreys, R. M., Merrill, K. M. 1978. *Ap. J.* 220:239–44

Day, K. L. 1974. *Ap. J.* 192:L15–17

Day, K. L., Donn, B. 1978. *Ap. J. Lett.* 222:L45–48

Dearborn, D. S. P., Eggleton, P. P. 1976. *Q. J. R. Astron. Soc.* 17:448–56

Dominy, J. F., Hinkle, K. H., Lambert, D. L., Hall, D. N. B., Ridgway, S. T. 1978. *Ap. J.* 223:949–58

Dorschner, J., Friedemann, C., Gürtler, J. 1978. *Astrophys. Space Sci.* 54:181–85

Elias, J. H. 1978a. *Ap. J.* 223:859–75

Elias, J. H. 1978b. *Ap. J.* 224:453–72

Elias, J. H. 1978c. *Ap. J.* 224:857–72

Fawley, W. M. 1977. *Ap. J.* 218:181–94

Faÿ, T. D., Johnson, H. R. 1973. *Ap. J.* 181:851–64

Faÿ, T., Marenin, I., van Citters, W. 1971. *J. Quant. Spectrosc. Radiat. Transfer* 11:1203–14

Ferland, G. J., Lambert, D. L., Netzer, H., Hall, D. N. B., Ridgway, S. T. 1979. *Ap. J.* In press

Fertel, J. 1970. *Ap. J. Lett.* 162:L75–76

Flaud, J.-M., Camy-Peyret, C., Maillard, J.-P. 1976. *Mol. Phys.* 32:499–521

Flaud, J.-M., Camy-Peyret, C., Maillard, J.-P., Guelachvili, G. 1977. *J. Mol. Spectrosc.* 65:219–28

Forrest, W. J., Gillett, F. C., Houck, J. R., McCarthy, J. F., Merrill, K. M., Pipher, J. L., Puetter, R. C., Russell, R. W., Soifer, B. T., Willner, S. P. 1978. *Ap. J.* 219:114–20

Forrest, W. J., Gillett, F. C., Stein, W. A. 1975. *Ap. J.* 195:423–40

Frogel, J. A., Dickinson, D. F., Hyland, A. R. 1975. *Ap. J.* 201:392–96

Frogel, J. A., Hyland, A. R. 1972. *Mem. Soc. R. Sci. Liège Ser 6* 3:111–20

Frogel, J. A., Persson, S. E., Aaronson, M., Matthews, K. 1978. *Ap. J.* 220:75–97

Gallagher, J. S. 1977. *Astron. J.* 82:209–15

Gallagher, J. S., Starrfield, S. 1978. *Ann. Rev. Astron. Astrophys.* 16:171–214

Gammon, R. H., Gaustad, J. E., Treffers, R. R. 1972. *Ap. J.* 175:687–91

Garrison, L. M. 1978. *Ap. J.* 224:535–45

Gautier, T. N., Fink, U., Treffers, R. R., Larson, H. P. 1976a. *Ap. J. Lett.* 207:L129–33

Gautier, T. N., Thompson, R. I., Fink, U., Larson, H. P. 1976b. *Ap. J.* 205:841–47

Geballe, T. R., Wollman, E. R., Lacy, J. H., Rank, D. M. 1977. *Publ. Astron. Soc. Pac.* 89:840–50

Geballe, T. R., Wollman, E. R., Rank, D. M. 1973. *Ap. J.* 183:499–504

Gehrz, R. D., Hackwell, J. A. 1974. *Ap. J.* 194:619–22

Gehrz, R. D., Hackwell, J. A., Briotta, D. 1978. *Ap. J. Lett.* 221:L23–27

Gehrz, R. D., Hackwell, J. A., Jones, T. W. 1974. *Ap. J.* 191:675–84

Geisel, S. L., Kleinmann, D. E., Low, F. J. 1970. *Ap. J. Lett.* 161:L101–4

Gillett, F. C., Forrest, W. J., Merrill, K. M., Capps, R. W., Soifer, B. T. 1975a. *Ap. J.* 200:609–20

Gillett, F. C., Jones, T. W., Merrill, K. M., Stein, W. A. 1975b. *Astron. Astrophys.* 45:77–81

Gilman, R. C. 1969. *Ap. J. Lett.* 155:L185–87

Gilman, R. C. 1974. *Ap. J.* 188:87–94
Glass, I. S., Webster, B. L. 1973. *MNRAS* 165:77–89
Goebel, J. H., Bregman, J. D., Strecker, D. W., Witteborn, F. C., Erickson, E. F. 1978a. *Ap. J. Lett.* 222:L129–32
Goebel, J. H., Strecker, D. W., Witteborn, F. C., Bregman, J. D., Erickson, E. F. 1978b. *Bull. Am. Astron. Soc.* 10:407 (Abstr.)
Gow, C. E. 1977. *Publ. Astron. Soc. Pac.* 89:510–18
Grasdalen, G. L. 1976. *Ap. J. Lett.* 205: L83–85
Grasdalen, G. L., Joyce, R. R. 1976a. *Ap. J.* 208:317–22
Grasdalen, G. L., Joyce, R. R. 1976b. *Nature* 259:187–89
Gustafsson, B., Bell, R. A., Eriksson, K., Nordlund, Å. 1975. *Astron. Astrophys.* 42:407–32
Hackwell, J. A., Gehrz, R. D., Grasdalen, G. L. 1978. *Bull. Am. Astron. Soc.* 10: 407–8 (Abstr.)
Hackwell, J. A., Gehrz, R. D., Smith, J. R. 1974. *Ap. J.* 192:383–90
Hagen, W. 1978. *Ap. J.* 222:L37–40
Hagen, W., Simon, T., Dyck, H. M. 1975. *Ap. J. Lett.* 201:L81–84
Hall, D. N. B. 1970. *An Atlas of Infrared Spectra of the Solar Photosphere and of Sunspot Umbrae.* Tucson: Kitt Peak Natl. Obs.
Hall, D. N. B. 1977. *Bull. Am. Astron. Soc.* 9:604 (Abstr.)
Hall, D. N. B., Kleinmann, S. G., Ridgway, S. T., Gillett, F. C. 1978. *Ap. J. Lett.* 223: L47–50
Hall, D. N. B., Ridgway, S. T. 1979. *Mem. Soc. R. Sci. Liège.* In press
Hartmann, L., Cassinelli, J. P. 1977. *Ap. J.* 215:155–58
Hartoog, M. R., Persson, S. E., Aaronson, M. 1977. *Publ. Astron. Soc. Pac.* 89:660–62
Harvey, P. M. 1974. *Ap. J.* 188:95–96
Heasley, J. N., Ridgway, S. T., Carbon, D. F., Milkey, R. W., Hall, D. N. B. 1977. *Ap. J.* 219:970–78
Herbig, G. H. 1977. *Ap. J.* 217:693–715
Hinkle, K. H. 1978. *Ap. J.* 220:210–28
Hinkle, K. H., Barnes, T. G. 1979. *Ap. J.* In press
Hinkle, K. H., Lambert, D. L. 1975. *MNRAS* 170:447–74
Hinkle, K. H., Lambert, D. L., Snell, R. L. 1976. *Ap. J.* 210:684–93
Honeycutt, R. K., Ramsey, L. W., Warren, W. H., Ridgway, S. T. 1977. *Ap. J.* 215:584–95
Huang, S. S. 1973. *Icarus* 18:339–76
Huffman, D. R. 1977. *Adv. Phys.* 26:129–230
Humphreys, R. M. 1974. *Ap. J.* 188:75–85

Humphreys, R. M. 1976. *Ap. J.* 206:122–27
Humphreys, R. M., Gallagher, J. S. 1977. *Publ. Astron. Soc. Pac.* 89:182–84
Humphreys, R. M., Ney, E. P. 1974a. *Ap. J.* 190:339–47
Humphreys, R. M., Ney, E. P. 1974b. *Publ. Astron. Soc. Pac.* 86:444–47
Humphreys, R. M., Strecker, D. W., Murdock, T. L., Low, F. J. 1973. *Ap. J. Lett.* 179:L49–52
Humphreys, R. M., Warner, J. W., Gallagher, J. S. 1976. *Publ. Astron. Soc. Pac.* 88: 380–87
Hyland, A. R., Becklin, E. E., Frogel, J. A., Neugebauer, G. 1972. *Astron. Astrophys.* 16:204–19
Johnson, H. L., Méndez, M. E. 1970. *Astron. J.* 75:785–817
Johnson, H. R. 1974. *NCAR Technical Note, NCAR-TN/Str-95.* Boulder: Natl. Ctr. Atmos. Res.
Johnson, H. R., Beebe, R. F., Sneden, C. 1975. *Ap. J. Suppl.* 29:123–36
Johnson, H. R., Collins, J. G., Krupp, B., Bell, R. A. 1977. *Ap. J.* 212:760–67
Jones, T. W., Merrill, K. M. 1976. *Ap. J.* 209:509–24
Jones, B., Merrill, K. M., Puetter, R. C., Willner, S. P. 1978. *Astron. J.* 83:1437–39
Joyce, R. R., Simon, M., Simon, T. 1978. *Ap. J.* 220:156–58
Khozov, G. V. 1976. *Astrophysics* 12:468–85
Kilston, S. 1975. *Publ. Astron. Soc. Pac.* 87: 189
Kleinmann, S. G., Hall, D. N. B., Ridgway, S. T., Wright, E. L. 1978. *Astron. J.* 83: 373–75
Knacke, R. F., Gaustad, J. E., Gillett, F. C., Stein, W. A. 1969. *Ap. J. Lett.* 155:L189–92
Kurucz, R. L. 1970. *SAO Special Report 309.* Cambridge, Mass.: Smithsonian Astrophys. Obs. 291 pp.
Kurucz, R. L. 1975. *SAO Special Report 362.* Cambridge, Mass.: Smithsonian Astrophys. Obs. 219 pp.
Kwan, J., Hill, F. 1977. *Ap. J.* 215:781–87
Kwok, S., Purton, C. R. 1979. *Ap. J.* In press
Kwok, S., Purton, C. R., FitzGerald, M. P. 1978. *Ap. J. Lett.* 219:L125–27
Lambert, D. L. 1974. *High Lights of Astronomy,* ed. G. Contopoulos. Dordrecht: Reidel 3:237–54
Lambert, D. L. 1977. *Bull. Am. Astron. Soc.* 9:573 (Abstr.)
Lambert, D. L., Beer, R. 1972. *Ap. J.* 177: 541–45
Lambert, D. L., Ries, L. M. 1977. *Ap. J.* 217: 508–20
Lambert, D. L., Snell, R. L. 1975. *MNRAS* 172:277–88
Lebofsky, M. J., Sargent, D. G., Kleinmann,

S. G., Rieke, G. H. 1978. *Ap. J.* 219:487–93
Lo, K. Y., Bechis, K. P. 1976. *Ap. J. Lett.* 205:L21–25
Lockwood, G. W., Dyck, H. M., Ridgway, S. T. 1975. *Ap. J.* 195:385–89
Maillard, J. P. 1974. See Lambert 1974, pp. 269–84
Maillard, J. P., Chauville, J., Mantz, A. W. 1976. *J. Mol. Spectrosc.* 63:120–41
Merrill, K. M. 1977. *The Interaction of Variable Stars with their Environment, IAU Colloq. No. 42*, ed. R. Kippenhahn, J. Rake, W. Strohmeier. *Veröff. Bamberg* 11, No. 121:446–94. 649 pp.
Merrill, K. M., Black, J. H. 1978. *Bull. Am. Astron. Soc.* 10:407 (Abstr.)
Merrill, K. M., Russell, R. W., Soifer, B. T. 1976. *Ap. J.* 207:763–69
Merrill, K. M., Stein, W. A. 1976a. *Publ. Astron. Soc. Pac.* 88:285–93
Merrill, K. M., Stein, W. A. 1976b. *Publ. Astron. Soc. Pac.* 88:294–307, 808
Merrill, K. M., Stein, W. A. 1976c. *Publ. Astron. Soc. Pac.* 88:874–87
Mertz, L. 1972a. *Mem. Soc. R. Sci. Liege* 3:121
Mertz, L. 1972b. *Mem. Soc. R. Sci. Liege* 3:101–6
Milkey, R. W., Dyck, H. M. 1973. *Ap. J.* 181:833–39
Morris, M. 1975. *Ap. J.* 197:603–10
Mould, J. R. 1978. *Ap. J.* 226:923–30
Mould, J. R., Hall, D. N. B., Ridgway, S. T., Hintzen, P., Aaronson, M. 1978. *Ap. J. Lett.* 222:L123–26
Mukai, T., Mukai, S., Noguchi, K. 1978. *Astrophys. Space Sci.* 53:77–84
Neugebauer, G., Becklin, E. E., Beckwith, S., Matthews, K., Wynn-Williams, C. G. 1976. *Ap. J. Lett.* 205:L139–41
Neugebauer, G., Becklin, E. E., Matthews, K., Wynn-Williams, C. G. 1978. *Ap. J.* 220:149–55
Neugebauer, G., Leighton, R. B. 1969. *Two Micron Sky Survey—a Preliminary Catalog (NASA SP-3047)*. 309 pp.
Ney, E. P., Hatfield, B. F. 1978. *Ap. J. Lett.* 219:L111–15
Noguchi, K., Maihara, T., Okuda, H., Sato, S., Mukai, T. 1977. *Publ. Astron. Soc. Jpn.* 29:511–25
Peery, B. F., Wojslaw, R. S. 1977. *Bull. Am. Astron. Soc.* 9:365 (Abstr.)
Persson, S. E., Aaronson, M., Frogel, J. A. 1977. *Astron. J.* 82:729–33
Pilachowski, C. A. 1978a. *Ap. J.* 224:412–16
Pilachowski, C. A. 1978b. *Publ. Astron. Soc. Pac.* In press
Price, S. D., Walker, R. D. 1976. *The AFGL Four Color Infrared Sky Survey. Air Force Geophys. Lab. Tech. Rep. No. AFGL TR-76-0208.* 154 pp.
Puetter, R. C., Russell, R. W., Sellgren, K.,

Soifer, B. T. 1977. *Publ. Astron. Soc. Pac.* 89:320–22
Puetter, R. C., Russell, R. W., Soifer, B. T., Willner, S. P. 1978. *Ap. J. Lett.* 223:L93–95
Querci, M., Querci, F. 1975. *Astron. Astrophys.* 42:329–40
Querci, F., Querci, M., Tsuji, T. 1974. *Astron. Astrophys.* 31:265–82
Ramsey, L. W. 1977. *Ap. J.* 215:603–8
Ramsey, L. W., Johnson, H. R. 1975. *Sol. Phys.* 45:3–15
Reimers, D. 1977. See Merrill 1977, pp. 559–76
Ridgway, S. T. 1974a. See Lambert 1974, pp. 327–39
Ridgway, S. T. 1974b. *Ap. J.* 190:591–96
Ridgway, S. T., Carbon, D. F., Hall, D. N. B. 1978. *Ap. J.* 225:138–47
Ridgway, S. T., Hall, D. N. B., Carbon, D. F. 1977. *Bull. Am. Astron. Soc.* 9:636 (Abstr.)
Rinsland, C. P., Wing, R. F., Joyce, R. R. 1977. In *Symposium on Recent Results in Infrared Astrophysics*, ed. P. Dyal, pp. 23–25. *NASA Tech. Mem., TM X-73*, p. 190. 84 pp.
Rose, L. A. 1977. *Laboratory Simulation of Infrared Astrophysical Features.* PhD thesis. Univ. Minn. 140 pp.
Rose, L. A. 1979. *Icarus*. In press
Russell, R. W. 1978. *An Analysis of Infrared Spectra of some Gaseous Nebulae.* PhD thesis. Univ. Calif., San Diego. 160 pp.
Russell, R. W., Soifer, B. T., Willner, S. P. 1978. *Ap. J. Lett.* 217:L149–53
Rydgren, A. E., Strom, S. E., Strom, K. M. 1976. *Ap. J. Suppl.* 30:307–36
Salpeter, E. E. 1977. *Ann. Rev. Astron. Astrophys.* 15:267–94
Sato, S., Kawara, K., Kobayashi, Y., Toshinori, M., Oda, N., Okuda, H., Iijima, T., Noguchi, K. 1978. *Publ. Astron. Soc. Jpn.* 30:419–32
Savage, B. D., Wesselius, P. R., Swings, J. P., Thé, P. S. 1978. *Ap. J.* 224:149–56
Scalo, J. M. 1973a. *Ap. J.* 186:967–78
Scalo, J. M. 1973b. *Ap. J.* 184:801–13
Scargle, J. D., Erickson, E. F., Witteborn, F. C., Strecker, D. W. 1978. *Ap. J.* 224:527–34
Schmitz, F., Ulmschneider, P. 1977. *Astron. Astrophys.* 59:177–79
Schwarzschild, M. 1975. *Ap. J.* 195:137–44
Simon, T., Simon, M., Joyce, R. R. 1979. *Ap. J.* In press
Soifer, B. T., Pipher, J. L. 1978. *Ann. Rev. Astron. Astrophys.* 16:335–69
Soifer, B. T., Russell, R. W., Merrill, K. M. 1976. *Ap. J. Lett.* 207:L83–85
Spinrad, H., Kaplan, L. D., Connes, P., Connes, J., Kunde, V. G., Maillard, J. P. 1971. *Contrib. Kitt Peak Natl. Obs.* 554:59–75

INFRARED SPECTROSCOPY OF STARS   41

Spinrad, H., Wing, R. F. 1969. *Ann. Rev. Astron. Astrophys.* 7:249–99
Stockman,H.S.,Schmidt,G.D.,Angel,J.R.P., Liebert, J., Tapia, S., Beaver, E. A. 1977. *Ap. J.* 217:815–31
Strecker, D. W., Erickson, E. F., Witteborn, F. C. 1978. *Astron. J.* 83:26–31
Strittmatter, P. A., Woolf, N. J., Thompson, R. I., Wilkerson, S., Angel, J. R. P., Stockman, H. S., Gilbert, G., Grandi, S. A., Larson, H., Fink, U. 1977. *Ap. J.* 216–23
Strom, S. E., Strom, K. M., Grasdalen, G. L. 1975. *Ann. Rev. Astron. Astrophys.* 13:187–216
Swings, J. P., Allen, D. A. 1972. *Publ. Astron. Soc. Pac.* 84:523–27
Szkody, P. 1977. *Ap. J.* 217:140–50
Thomas, J. A., Robinson, G., Hyland, A. R. 1976. *MNRAS* 174:711–23
Thompson, R. I. 1973. *Ap. J.* 181:1039–54
Thompson, R. I. 1974. See Lambert 1974, pp. 255–68
Thompson, R. I. 1977. *Ap. J.* 212:754–59
Thompson, R. I., Boroson, T. A. 1977. *Ap. J. Lett.* 216:L75–77
Thompson, R. I., Johnson, H. L. 1974. *Ap. J.* 193:147–50
Thompson, R. I., Johnson, H. L., Forbes, F. F., Steinmetz, D. L. 1973. *Publ. Astron. Soc. Pac.* 85:643–52
Thompson, R. I., Lebofsky, M. J., Rieke, G. H. 1978. *Ap. J. Lett.* 222:L49–53
Thompson, R. I., Reed, M. A. 1976. *Ap. J. Lett.* 205:L159–61
Thompson, R. I., Schnopper, H. W. 1970. *Ap. J. Lett.* 160:L97–100
Thompson,R.I.,Strittmatter,P.A.,Erickson, E. F., Witteborn, F. C., Strecker, D. W. 1977. *Ap. J.* 218:170–80
Thompson, R. I., Tokunaga, A. T. 1978. *Ap. J.* 226:119–23
Treffers, R. R., Cohen, M. 1974. *Ap. J.* 188:545–52
Treffers, R. R., Fink, U., Larson, H. P., Gautier,T. N. 1976. *Ap. J. Lett.*209:L115–18
Treffers, R. R., Gilra, D. P. 1975. *Ap. J.* 202:839–43

Tsuji, T. 1964. *Ann. Tokyo Astron. Obs. 2nd Ser.* 9:1–54
Tsuji, T. 1973. *Astron. Astrophys.* 23:411–31
Tsuji, T. 1976a. *Publ. Astron. Soc. Jpn.* 28:543–65
Tsuji, T. 1976b. *Publ. Astron. Soc. Jpn.* 28:567–86
Tsuji, T. 1978. *Astron. Astrophys.* 62:29–50
Tutukov, A. V., Yungelson, L. R. 1976. *Astrophysics* 12:342–48
Walker, R. G., Price, S. D. 1975. *AFCRL Infrared Sky Survey. Air Force Cambridge Res. Lab. Tech. Rep., AFCRL-TR-0373.* 158 pp.
Wallerstein, G. 1978. *Publ. Astron. Soc. Pac.* 90:36–38
Warner, J. W., Strom, S. E., Strom, K. M. 1978. *Ap. J.* 213:427–37
Webster, B. L., Allen, D. A. 1975. *MNRAS* 171:171–80
Webster, B. L., Glass, I. S. 1974. *MNRAS* 166:491–97
Welin,G.1976. *Astron. Astrophys.*49:145–48
Weymann, R. J. 1977. See Merrill 1977, pp. 577–90
Willner, S. P. 1977. *Ap. J.* 214:706–11
Willner, S. P., Soifer, B. T., Russell, R. W., Joyce, R. R., Gillett, F. C. 1977. *Ap. J. Lett.* 217:L121–24
Wing, R. F. 1978. *Conference on Current Problems in Stellar Pulsation Instabilities.* Goddard, June 12, 1978
Wing, R. F., Price, S. D. 1970. *Ap. J. Lett.* 162:L73–74
Wing, R. F., Rinsland, C. P., Joyce, R. R. 1977. See Rinsland et al 1977, pp. 26–28
Wolf, B. 1977. See Merrill 1977, pp. 151–81
Woolf, N. J. 1973. *Interstellar Dust and Related Topics, IAU Symp. No. 52,* ed. J. M. Greenberg, H. C. Van de Hulst, pp. 485–504. Dordrecht: Reidel. 584 pp.
Zuckerman, B., Gilra, D. P., Turner, B. E., Morris, M., Palmer, P. 1976. *Ap. J. Lett.* 205:L15–19
Zuckerman, B., Palmer, P., Gilra, D. P., Turner, B. E., Morris, M. 1978. *Ap. J. Lett.* 220:L53–56

Ann. Rev. Astron. Astrophys. 1979. 17: 43–71

ADVANCES IN ASTRONOMICAL PHOTOGRAPHY AT LOW LIGHT LEVELS

*2145

Alex G. Smith

Department of Physics and Astronomy, University of Florida, Gainesville, Florida 32611

A. A. Hoag

Lowell Observatory, Flagstaff, Arizona 86002

1 INTRODUCTION

Upon seeing his obituary in a newspaper, Mark Twain commented that the rumors of his death had been greatly exaggerated. So it is with astronomical photography. Periodically we hear that photography is being replaced by electronic detectors, and that analog recording is giving way to digital techniques. In the face of this adversity astrophotography perversely continues to flourish, with a new generation of large wide-field telescopes recently coming on line. What is the reason for this stubborn refusal to die?

Certainly, simplicity and low cost are among the virtues of direct photography. Many observatories cannot afford the initial costs of electronic imaging detectors, to say nothing of the expenses of maintenance. Related to the notion of simplicity is the fact that photography displays information in a direct visual format, the perceptual mode to which nearly all workers relate most easily; it is interesting that many types of numerical data are finally presented in pseudo-photographic form to aid comprehension.

Even more fundamental is the fact that the photographic emulsion is a compact solid-state information storage device of enormous capacity easily adapted to large formats. Whereas it has proven difficult to produce electronic detectors with linear dimensions exceeding about 50 mm, photographic plates with areas two orders of magnitude greater are in regular use. Since a typical fine-grain emulsion has a storage capacity

0066-4146/79/0915-0043$01.00

of $\sim 10^6$ bits cm^{-2} (Kowaliski 1972), such a plate can store several billion bits of information. And if the plate is properly processed and preserved, such storage is essentially permanent. The serendipitous possibilities of wide-field, high-density, long-term information storage have been repeatedly illustrated by the seemingly endless applications of the Palomar Sky Survey to problems undreamed of a quarter-century ago when the plates were taken.

Not the least of the reasons for photography's present good health is the introduction of new high-performance emulsions and the discovery of techniques whereby the user can hypersensitize these emulsions to practical speeds. Discussion of these techniques is the primary concern of the present article. Condensed reviews of this topic have been given recently by Barlow (1978), Sim (1978), and A. G. Smith (1977).

1.1   *The Latent Image and Reciprocity Failure*

A discussion of methods of increasing the sensitivities of photographic materials is aided by a minimal understanding of the photographic process itself. Curiously, we know of only one group of chemical compounds, the silver halides AgCl, AgBr, and AgI, that display enough sensitivity to be useful for most photographic purposes. These substances have been described as solid-state amplifiers with a gain of a billion, which is to say that when light releases an atom of silver it can result in the production of as many as 10^9 silver atoms in the developed image (Dainty & Shaw 1974). How is this tremendous gain achieved?

1.1.1   FORMATION OF A LATENT IMAGE   According to the mechanism proposed by Gurney & Mott (1938), absorption of a photon by a silver halide grain excites an electron into the conduction band, leaving behind a positively charged hole. The mobile photoelectron is trapped at a defect in the crystal lattice, where its negative charge attracts an Ag^+ ion. The electron and the ion combine to deposit an atom of neutral silver Ag^0. Because a single Ag atom is unstable, the process must be repeated several times at the same site to build up a stable *latent image* cluster of at least three or four silver atoms, which renders the grain developable. A smaller cluster, of perhaps two atoms, may be stable but too small to trigger development; such subdevelopable latent image is the target of efforts at latensification (see Section 3.7). The positive hole is important in a detrimental sense because it may compete with the foregoing process by destroying the photoelectron through recombination, or by oxidizing latent image silver back to Ag^+. Certain sensitizing and hypersensitizing procedures are successful precisely because they inhibit the destructive tendencies of the holes.

At least three methods are used to create the vital lattice defects or electron traps in the grains (cf Duffin 1966). In the oldest method a liquid emulsion containing *sulfur* is "ripened" by heating, the result being the formation of "sensitivity specks" of silver sulfide on the grains. In *reduction sensitization*, sensitivity centers composed of atoms of pure silver are created by heating liquid emulsion in the presence of a reducing agent such as sodium sulfite. Finally, *gold sensitization* is achieved by adding a gold compound during the latter stages of emulsion-making; it is not clear whether the active agent is gold or gold sulfide. It appears that the fastest modern emulsions are prepared by a synergistic addition of gold and sulfur sensitizers.

1.1.2 DEVELOPMENT   Development of an exposed emulsion is a process of selective chemical reduction, in which grains containing latent image silver are reduced entirely to metallic silver to form the final image; unreduced silver halide is later removed by a solvent during "fixation." How the latent image "tags" a grain to render it developable is more controversial than the Gurney-Mott theory itself. According to one hypothesis, development is an electrochemical process in which the silver latent image speck acts as an electrode to initiate the reaction (cf James 1977). Unfortunately, real emulsions contain "fog" grains that develop whether or not they have been exposed, introducing noise into the photographic system. Almost any treatment to which a finished emulsion is subjected, including merely letting it age, increases the number of fog grains.

1.1.3 RECIPROCITY FAILURE   The Gurney-Mott theory explains a phenomenon that is the bête noire of astronomical photography. In 1862 Bunsen & Roscoe propounded a "reciprocity law" for photochemical reactions which states that the total product of such a reaction depends only on the total energy absorbed, i.e., on the product It of the intensity of illumination I and the time of exposure t, and not on either factor alone. In everyday photography we invoke the reciprocity law when we make compensating adjustments of the shutter speed and lens aperture of a camera. When t becomes large, as in astronomical exposures, we encounter "failure of the reciprocity law," and a given It product yields less image silver than is the case for exposures in the usual snapshot range.

This loss in sensitivity at long exposures resides in the instability of latent images consisting of single Ag atoms. If photoelectrons are created in a grain at a low rate, thermal diffusion may cause the resulting Ag atoms to wander away from the sensitivity specks before they can be joined by additional silver atoms, so that formation of a stable latent image cluster is delayed or even prevented. That this effect is more than academic is illustrated by the curve for the II-O emulsion in Figure 1. For exposures of an hour, this typical emulsion requires about 20 times as much energy to produce a standard density as it does for snapshot exposures around 1/100 sec.

Fortunately, emulsion makers have discovered methods of reducing low-intensity reciprocity failure (LIRF). Kodak spectroscopic plates treated in this way are distinguished by an "a" in their designations: see the curve labelled IIa-O in Figure 1. While the sensitivities of II-O and IIa-O are similar for short exposures, IIa-O is nearly 10 times more sensitive than II-O at exposures of an hour. Much of the work described in this article is aimed at reducing still further the LIRF of special a-type emulsions.

1.2   Evolution of Photographic Sensitivity

Figure 2 shows the progress of photographic sensitivity since the invention of the daguerreotype in 1839. Major revolutions occurred around 1851 and 1875 with the introduction of the wet plate and dry plate processes. During the last 100 years progress seems to have occurred by slow evolution, rather than by revolution, and in recent decades the curve seems to be approaching an asymptote.

In astronomy, however, which is characterized by long exposures severely affected by reciprocity failure, a major breakthrough occurred

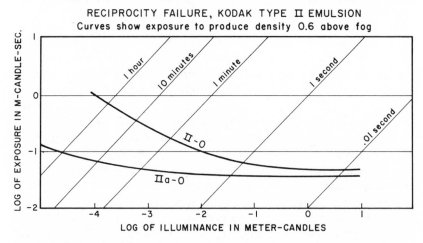

Figure 1   Reciprocity failure of a "normal" emulsion (II-O), compared with a second emulsion (IIa-O) that has been manufactured to reduce LIRF. Since the ordinate is the energy necessary to produce a standard density, the rise of each curve at the left indicates decreasing sensitivity at long exposures. Adapted from Eastman Kodak (1973).

during the 1970s. Stimulated by new fine-grain emulsions with superior resolving power and the ability to record fainter objects, new hyper-sensitization techniques applied by the user prior to exposure create gains that often reach an order of magnitude or more.

2   EVALUATING PHOTOGRAPHIC RESPONSE

With the attention now being given to optimum pre-exposure hyper-sensitization of high-quality photographic materials, there is an increasing need to compare results. These results have been most often expressed in terms of speed gains. This scheme does not work very well, as emulsion batches differ. The trend is toward repeatably standardized response evaluation.

2.1   The Characteristic Curve

Although a number of alternative schemes have been proposed for illustrating photographic response (cf de Vaucouleurs 1968 and references therein; Ables et al. 1971, Young 1977), the characteristic curve originated by Hurter & Driffield (1890) is familiar and useful. The H and D curve is constructed by plotting density as a function of the logarithm of the

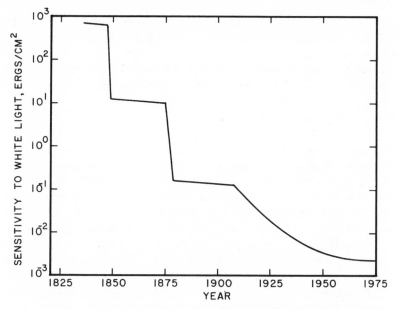

Figure 2   History of photographic sensitivity, adapted from Kirillov (1967). Since the ordinate is the energy required to create an image, sensitivity is inversely proportional to the height of the ordinate.

exposure. Unfortunately, density is not only a property of the exposed and processed emulsion, but its value is dependent on the properties of the density-measuring instrument as well. To further compound the difficulty of comparing performance data, the relative log exposure scale generally has an arbitrary zero-point.

In order to provide for uniform comparability of photographic performance, the American Astronomical Society's Working Group on Photographic Materials (WGPM) has recommended standards for H and D curve presentation (Schoening 1976, Hoag 1976).

2.1.1 DENSITY   The zero-point of the density scale is defined by "no plate" in the measuring beam—this means air as opposed to the plate or film with the emulsion removed. The scale of density recommended is a *diffuse density* defined by the American National Standards Institute (ANSI) standard PH 2.19-1959. As an aid in conforming to this scale, the Eastman Kodak Company (Sewell 1975) produced a number of sets of *Interobservatory Densitometer Calibration Plates* that have been widely distributed. These step wedges, having a range of emulsion types and granularity, are measured as a means of finding the relation between the experimenter's system and the standard scale (see, for example, Latham 1978a).

2.1.2 EXPOSURE   For a given photographic sample, the relation between exposure and density is dependent on a number of factors, such as time, environment, and spectral distribution of the light source. The WGPM has therefore recommended standards for sensitometric test exposures. Exposure is to be expressed in terms of photons per unit area for a specified time and for specified effective wavelengths. The unit area is to be 1000 μm^2, typical of scanning apertures used in microdensitometry of astronomical photographs. Twenty minutes has been recommended as representative of astronomical exposures. At the same time, this is an interval short enough to be practical for laboratory exposures. The time parameter may be diminished in importance if the effects of LIRF are overcome by pre-exposure treatment and the exposure environment (cf Miller 1970, Babcock 1976, Kaye 1976, Kaye & Meaburn 1977, A. G. Smith 1978b). Finally, because photographic response characteristics depend on the spectral distribution of the exposing light source (cf Kaler 1966, Miller 1969, Honeycutt & Chaldu 1970), it has been recommended that test light sources be limited by interference filters having full widths at half-maximum of about 10 nm. It has been proposed that emulsions having O, J, and G spectral sensitizations be tested at an effective wavelength of 460 nm. A 600-nm effective wavelength is suited to tests of D, E, and F sensitizations, while 800 nm is recommended for the N sensitization.

The zero-point of the log exposure scale can be established by calibrating the test sensitometer with a commercial radiometer (Hoag 1976). Direct comparisons of laboratory photographic sensitometers that have been checked with different radiometers have shown that calibration errors of less than 5% may be expected (A. G. Smith 1978a). Because most commercial radiometers have broad spectral sensitivity, it is especially important to avoid "red leaks" in filters.

With conventions of representing density and exposure as described, it is not only possible to directly compare photographic test results from standardized H and D curves, but one can also derive other comparable performance parameters such as *speed, output signal-to-noise ratio*, and *detective quantum efficiency*.

2.2   Speed

With the characteristic curve defined as in the preceding section, speed is simply the reciprocal of the exposure required to produce a specified photographic effect as, for example, a D_{ANSI} of 0.6 above fog. Otherwise, speed is related to a relative exposure scale of a particular test sensitometer. In the latter case, speed ratios or speed gains are meaningful as in, say, relating the effectiveness of various procedures applied to a specific material. However, comparison of photographic performance on an interobservatory basis depends on conformance with rigorous standards. The standards proposed are just now being implemented; thus pre-exposure treatment results are reported here primarily in terms of relative speed gains.

2.3   Signal-to-Noise

The precision with which a small increment in exposure can be measured will depend on the gradient of the H and D curve, γ, and the noise characteristics of the plate at the appropriate density and sample area. Output signal-to-noise, $[S/N]_{OUT}$, is a concept that has been rediscussed recently by Latham (1974, 1978b), Furenlid et al. (1977), Hoag (1978), Furenlid (1978), and Hoag et al. (1979). It has become increasingly clear that this parameter (originally termed Contrast Detectivity by R. Clark Jones 1959) is of great value in assessing the performance of photographic materials.

One tactical difficulty that has inhibited the use of the signal-to-noise concept in photography has been the complexity of the rms density noise, σ_D, determination. One must evaluate σ_D from detailed microdensitometer measures. However, Ables et al. (1971) showed that σ_D depends only on D for a given emulsion. Furenlid et al. (1977) have determined σ_D as a function of D for a variety of emulsions of astronomical interest. To get

$[S/N]_{OUT}$, one has only to extract the gradient, γ, and use published values of σ_D applicable to the emulsion being used for a range of values of D_{ANSI} in the standard H and D curve. A representative relation between log exposure and $[S/N]_{OUT}$ is illustrated in Figure 3. In this example, $[S/N]_{OUT}$ is normalized to a sample area of 1000 μm^2. Information of this kind tells you exactly what precision you can expect to achieve in quantitative photographic photometry. Greatest precision is obtained at a density near the beginning of the straight-line portion of the H and D curve. Here the gradient approaches a maximum, while the noise is small. $[S/N]_{OUT}$ falls off as the exposure increases because the noise increases while the gradient remains constant and then decreases at the shoulder of the characteristic curve. For decreasing exposure the signal-to-noise falls off as the gradient of the H and D curve decreases.

2.3.1 DYNAMIC RANGE   Referring again to Figure 3, we observe that the dynamic range is graphically shown by the range in exposure represented by any chosen value of output signal-to-noise. In our example the dynamic range at an output signal-to-noise ratio of 10 to 1 for a 1000-μm^2 sample area is about a factor of 200.

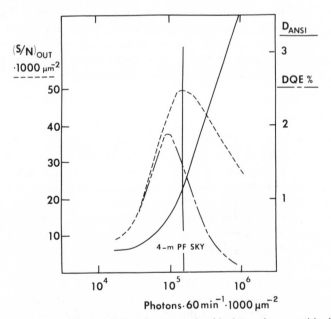

Figure 3 Performance characteristics of one sample of hydrogen-hypersensitized Kodak IIIa-J. The exposure level of the sky at the prime focus of the KPNO 4-m telescope is shown by the vertical bar. The H and D (*solid*), output signal-to-noise (*dotted*), and detective quantum efficiency (*broken*) relations are shown. From Hoag (1978).

2.4   Detective Quantum Efficiency

Detective quantum efficiency, DQE (Rose 1946, Jones 1958), is simply the ratio of the square of the output signal-to-noise to the square of the input signal-to-noise. $[S/N]_{IN}$ is $\sqrt{n}$, where n is the number of photons impinging on the sample area throughout the exposure. Consequently, the DQE as a function of exposure can be derived from $[S/N]_{OUT}$ and absolutely calibrated exposure data. In principle, the DQE relation indicates the exposure level at which the photographic material may be used with greatest efficiency. In practice, however, it may be more advantageous to expose to the level corresponding to maximum $[S/N]_{OUT}$ because of the difficulty of making and combining separate exposures.

The DQE relation is shown for one sample of hypersensitized Kodak IIIa-J in Figure 3. Peak DQE values of 2–4% have been reported for optimally treated IIIa-materials. This performance, rather than the 0.1% so often erroneously quoted for photography, when combined with the other advantages outlined in our introduction, is what keeps astronomical photography alive and well.

2.5   Application Classes

Marchant (1964) and Millikan (cf 1976), in discussing response of emulsions, have pushed the notion of two classes of faint signal detection. They refer to signal-limited cases as class I.

Sky-limited exposures, class II operation, on the other hand, require the sky exposure to be at an optimum density or $[S/N]_{OUT}$ level. For best detection at the sky level, one chooses material having the steepest gradient and lowest density noise that can be appropriately exposed to the maximum $[S/N]_{OUT}$ density. These, in fact, are the considerations that went into the creation of the Kodak IIIa-J and IIIa-F emulsions. In class II operation one can often improve $[S/N]_{OUT}$ at the sky exposure level by hypersensitization, as is shown in Figure 4. However, the price paid for performance at low exposure levels is a degradation of peak $[S/N]_{OUT}$.

2.6   Uniformity of Response

Information that relates to precision has been derived from sensitometric tests using relatively small areas of emulsion. Latham (1978b) and Millikan & Tinney (1977) have taken a look at uniformity on an extended spatial scale and find that response is typically uniform to within 2% on a scale of approximately 10 cm if care is used in processing.

Because photographic uniformity is best on a small scale (0.5% over millimeter spacings), Latham (1978b) has applied grid masking and flat-

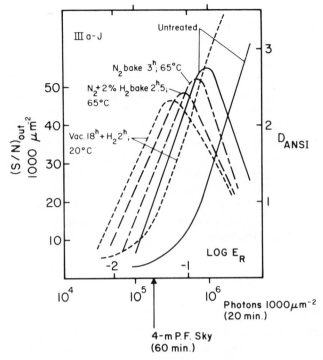

Figure 4 Evaluation of various hypersensitization treatments for a sample Kodak type IIIa-J emulsion. Note the change in maximum $[S/N]_{OUT}$ as the response to lower-level exposure is increased by hypersensitizing. From Hoag et al. (1979).

field calibration to problems in precision surface photometry. His technique is to expose the plate over half its area through a grid having bars and spaces about 0.5 mm in width. The grid is then shifted by the bar width, and a flat-field exposure is made. In this way one can correct for large-scale nonuniformities and measure surface brightnesses that are a very small fraction of the sky brightness.

3 CURRENT HYPERSENSITIZATION TECHNIQUES

3.1 *Chemical Pre-Exposure Treatment*

Chemical, or bathing, treatments have been reviewed by Allbright (1976) and by Jenkins & Farnell (1976). In practice, this type of hypersensitization procedure is almost entirely restricted to use with infrared-sensitive emulsions. Kodak IV-N plates have been preferred by most investigators who work in the $\lambda\lambda700$–870 nm range. The plates respond well to hyper-sensitizing and have better output signal-to-noise potential than I-N plates in many applications. In some signal-limited cases, however, I-N plates may be preferred.

Bathing treatments can be tricky and it is sometimes difficult to achieve the results that others report. Requirements in techniques include good agitation in all solutions, strict adherence to cleanliness in handling plates and solutions, and procedures for rapid and uniform drying.

3.1.1 WATER BATHING   Some observers have continued to make use of water hypersensitization for Kodak IV-N plates because speed gains (at D_{ANSI} 0.6 above fog) of the order of 10 can be achieved with good uniformity and low background. For class I operation, Kodak 098 plates have also been advantageously hypersensitized (Spinrad & Wilder 1972) in water.

3.1.2 AMMONIA   The most recent optimistic discussion of ammonia hypersensitization of IV-N plates (Young 1977) reports relative speed gains of the order of 50, as well as consistently uniform results and low background. Various practitioners using a variety of ammonia concentrations and a variety of techniques report relative speed gains for IV-N ranging from 15 to the order of 100. There are varying degrees of difficulty with nonuniform response, excessive fog, and keeping qualities. Ammoniation is generally recognized as a difficult technique but one that can give good results in practiced hands.

Ammoniation, or some equally drastic process, is mandatory for Kodak I-Z plates, where relative speed gains of several hundred are required for practical applications (Eastman Kodak 1973).

3.1.3 SILVER NITRATE BATHING   Since Jenkins & Farnell's (1976) report on chemical hypersensitization of infrared emulsions with dilute solutions of silver nitrate, there have been numerous reports of success in astronomical applications (J. F. Hoessel 1976 private communication, 1978; V. M. Blanco 1977 private communication; Schoening 1978a,b). $AgNO_3$ concentrations of 0.001 molarity have been used with good success. Plates are bathed in this solution for four minutes at room temperature, then are rinsed and dried rapidly. Reports agree in stating that larger speed gains and more uniform results are obtained than in the application of ammoniating techniques.

Bathing techniques increase the Ag^+ concentration in the emulsion but also seem to increase the effectiveness of dye sensitization, as speed gains are much greater for "red" exposures than for "blue" ones.

3.2   Evacuation

A second broad class of treatments immerses the emulsion in a gas or a vacuum, rather than in a liquid. That many photographic materials display increased sensitivity in a vacuum was discovered over fifty years ago by Masaki (1926). Masaki correctly assumed that the gain is due to removal of moisture and occluded gases, oxygen being the prime offender.

Lewis & James (1969) found that the combination of O_2 and moisture is particularly destructive to sensitivity, with the effect being most detrimental for low light levels and long exposures; they went so far as to comment that "Oxygen, either alone or in combination with water, appears to be the main source of low intensity reciprocity failure." Although the mechanism by which O_2 desensitizes an emulsion remains unknown, a good possibility is that the O_2 destroys photoelectrons by reacting with them to form O_2^- (James 1972). Since dry gelatin is relatively impermeable to gases, the synergetic role of moisture may be simply that of softening the gelatin so O_2 can reach the silver halide grains more easily.

The benefits of evacuation endure in the sense that if a thoroughly evacuated emulsion is returned to room air its sensitivity generally takes hours or even days to fall to its original value (Lewis & James 1969). If the evacuated emulsion is transferred to an inert atmosphere such as dry nitrogen the gain in sensitivity can be preserved for weeks or months, particularly under refrigeration. This lends itself to practical hypersensitization routines in which plates are prepared in the laboratory and stored for field use. Miller (1972) showed that a vacuum of 0.05 torr

Table 1   Representative results of vacuum treatment of Kodak spectroscopic plates[a]

Emulsion	Time of evacuation (hr)	Pressure (torr)	Time of exposure (min)	Speed gain factor	Reference
IIa-O	5	5×10^{-2}	120	1.6	Miller 1972
IIIa-J	16	1×10^{-4}	10	4	A. G. Smith et al. 1971
IIIa-J	16	5×10^{-7}	17	5	Babcock et al. 1974a
IIIa-J	15	5×10^{-2}	120	3	Miller 1972
098	60	5×10^{-2}	120	1.7	Miller 1972
IV-N	16	5×10^{-7}	17	4	Babcock et al. 1974a

[a] Speed gain is relative to an untreated plate of the same type.

produced by a fore pump yields good results, although some provision such as a cold trap or desiccant must be made for removal of water. There is in fact reason to believe that too rapid evacuation to very low pressures may be counterproductive because of the extreme hardening of the emulsion that occurs. Such hardening can seal in oxygen and water, or even result in fog because of pressure exerted on the silver halide grains (Millikan 1973).

Results for emulsions of astronomical interest are listed in Table 1. While simple evacuation has the merit of producing minimal increases in background fog, it does not appear to be widely used at present. More common is the use of evacuation in conjunction with other processes such as hydrogenation (see Section 3.4).

3.3   Treatment With Nitrogen

Once it is understood that the function of evacuation is to remove oxygen and water from the emulsion, other strategies suggest themselves. Extended flushing with an inert gas is one such strategy.

3.3.1   NITROGEN FLUSHING   Smith et al. (1971) showed that storage of IIIa-J plates in a flow of nitrogen for 16 hr at room temperature produced speed gains in excess of 2 for 10-min exposures. Argon was slightly

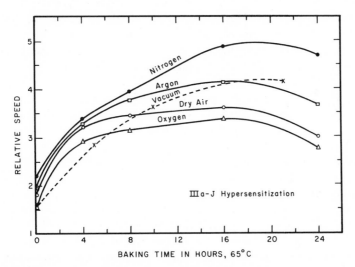

Figure 5   Relative speeds of Kodak type IIIa-J plates baked for varying times in five different dry environments. All exposures were 10 min to unfiltered 3100 K tungsten light. In each case the plate was stored 16 hr in the environment in which it had been baked, before being exposed. Note that the dry environments produced significant gains, even for 0 hr of baking. After A. G. Smith et al. (1971).

less effective, and even desiccated air or pure dry oxygen yielded some gain (see Figure 5).

Nitrogen soaking was pursued to its limit by Corben et al. (1974) and by Sim et al. (1976), who employed soaking times up to 100 days. The plates were stored at room temperature in gas-tight metal boxes that were flushed with fresh N_2 at regular intervals during the treatment period to eliminate waste products that outgassed from the emulsion. Because room-temperature N_2 soaking requires no ovens or vacuum components the method lends itself to handling quantities of large plates. Sim et al. (1976) describe a system that simultaneously treated as many as 30 boxes of 14 × 14-inch plates for a sky survey; they reported speed gains up to 2 for 5 to 80 days of soaking 103a-O and 098 emulsions, and gains from 2 to 12 when IIIa-J was soaked 30 to 100 days (M. E. Sim, personal communication 1978). The disadvantages of the nitrogen-soaking technique are the long times required to achieve useful gains, together with the possibility of unacceptable increases in fog from the prolonged storage at room temperature.

3.3.2 NITROGEN BAKING   In Section 1.1.1 we saw that modern photographic materials owe their speed to sensitivity centers created during the later stages of manufacture by a process of heating or ripening. For the product to have acceptable shelf life the manufacturer must terminate this process short of achieving maximum speed; increasing the size and number of sensitivity centers beyond this point accelerates the growth of unacceptable background fog. It is, however, feasible for the user to extend this sensitization with further gain in speed if he recognizes that the material must then be exposed rather promptly.

In a pioneering paper Bowen & Clark (1940) reported significant gains in speed when both dyed (panchromatic) and undyed plates were baked several days at 50°C. Of great interest to astronomers was the fact that the gains increased for long exposures, suggesting that LIRF was being reduced. Although baking plates in air became widespread, the process suffered from such liabilities as sensitivity to airborne contaminants, difficulty in duplicating results, and ineffectiveness for red-sensitized emulsions.

An obvious solution to some of these problems is to bake in a controlled atmosphere. Smith et al. (1971) showed that baking Kodak IIIa-J plates in a flow of nitrogen yielded higher speeds and less fog than identical baking in desiccated air or several other gases. The curves of Figure 5 are typical of baking experiments; at first the speed rises rapidly, but then it reaches a broad maximum and finally declines if baking is carried too far. As is common in astronomical and spectrographic work, speed is defined in these tests as the reciprocal of the exposure producing a density 0.6 above background fog; the speed of an untreated plate was taken as unity.

For a number of years nitrogen baking has been one of the most widely used methods of hypersensitizing astronomical plates (Smith & Leacock 1973, Scott & Smith 1974, Miller 1975, Schoening 1975, Sewell & Millikan 1975, Sim et al. 1976). A major attraction is that it yields higher speed gains with less fog than air baking; typical fog curves are shown in Figures 6 and 7. Reproducibility increases because problems with contaminated air are eliminated. Whereas red-sensitive emulsions generally respond poorly to air baking, several types have been successfully baked in nitrogen. Finally, nitrogen in commercial cylinders is cheap and widely available; it is completely safe if care is taken not to overpressurize any vessel that is being filled (safety valves are a wise precaution). Tests have shown that commercial dry nitrogen is so free of water that further drying is unnecessary.

The foregoing does not mean that nitrogen baking is without problems. While reproducibility is excellent within a single batch of emulsion, the optimum baking time may vary considerably from batch to batch; certain emulsions such as IIIa-J have been notorious in this regard. It is thus important to determine optimum parameters for each batch, under local

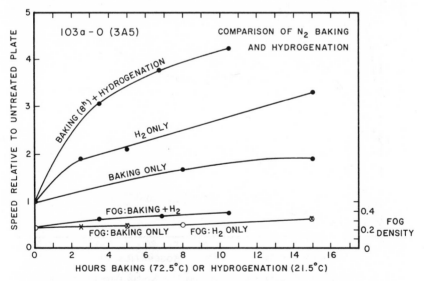

Figure 6 Comparison of speed and fog produced in Kodak type 103a-O emulsion by nitrogen baking, by hydrogenation, and by combining the two. All exposures were 10 min to unfiltered 3100 K tungsten light. After Scott & Smith (1976).

conditions. Since baked emulsion is hygroscopic, treated plates should be protected from moisture. Smith et al. (1971) found that nitrogen-baked IIIa-J lost most of its speed gain and rose in fog from overnight storage in air in a refrigerator. Most workers store treated plates in nitrogen in the same metal boxes in which they were baked; nearly all emulsions can be stored satisfactorily in this manner at freezer temperatures for weeks or even months. Observatories in humid climates may even find it advantageous to expose the plates in a nitrogen atmosphere (A. G. Smith 1977, Kaye & Meaburn 1977).

Schoening (1975) and Schrader et al. (1978) have given complete descriptions of systems for baking, storing, and exposing plates under nitrogen. Results for nitrogen baking of Kodak plates are listed in Table 2. Within limits baking time can be adjusted to a convenient value by varying the temperature. Miller (1970) found that 6 hr of baking at 75°C, 24 hr at 65°C, and 72 hr at 50°C all produced similar speeds and fog levels; above 75°C damage from softening of the emulsion was noted.

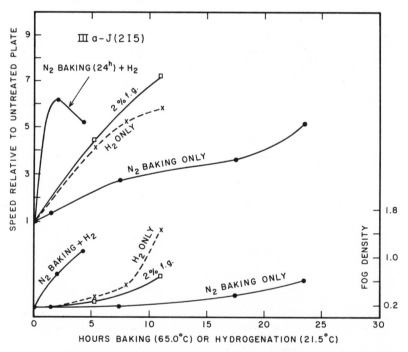

Figure 7 Comparison of speed and fog produced in Kodak type IIIa-J emulsion by nitrogen baking, by hydrogenation, by sequential combination of the two, and by 2% forming gas. All exposures were 10 min to unfiltered 3100 K tungsten light. Adapted from Scott et al. (1977).

3.4   Treatment With Hydrogen

Several years after the introduction of nitrogen baking, Babcock et al. (1974b) described another highly effective gas treatment: hypersensitization by room-temperature soaking in hydrogen. The mechanism appears to be that of initiating or extending the reduction sensitization described in Section 1.1.1, probably through reduction of Ag^+ to metallic silver (Babcock et al. 1975). A secondary benefit is derived from removal of water and oxygen from the emulsion, particularly since experience indicates that hydrogenation is generally most satisfactory if it is preceded by evacuation.

A typical hydrogen treatment involves evacuating the plates in a metal box at room temperature for at least an hour; if the emulsion was hardened during manufacture to resist abrasion, as much as 3 hr may be beneficial (Schoening 1977). However, excessive evacuation may itself harden the emulsion and inhibit later treatments. A good fore pump vacuum is adequate if provision is made to remove water vapor with a cold trap or a drying agent such as P_2O_5. Pure, dry hydrogen is then admitted to a pressure slightly above ambient, from a commercial cylinder equipped with a regulator. The box is sealed off and allowed to stand at room temperature for the prescribed soaking time. Finally, the box is thoroughly flushed with dry nitrogen, sealed off, and stored in a refrigerator. Since no flushing with hydrogen is required, this procedure involves the minimum volume of flammable gas. Needless to say, caution should be used in handling any quantity of this hazardous gas (Babcock 1976).

Table 2   Representative results for nitrogen baking of Kodak spectroscopic emulsions[a]

Emulsion	Temp (°C)	Time (hr)	Speed gain factor	Final fog level	Reference
103a-O	73	12	3	0.30	Scott & Smith 1974
IIa-O	65	8	3.2	0.13	Miller 1975
IIIa-J	65	16[b]	5	0.30	A. G. Smith et al. 1971
103a-D	65	26	2	0.30	Miller 1975
IIa-D	65	13	2	0.30	Miller 1975
098	55	5	1.7	0.30	A. G. Smith unpub.
IIa-F	65	2.5	1.8	0.30	Miller 1975
IIIa-F[c]	65	8	3.8	0.30	Smith & Leacock 1973

[a] There is considerable batch-to-batch variation in optimum baking times and in speed gains achieved.
[b] More recent batches require much less baking, perhaps 4 hr.
[c] Formerly designated Type 127. Tends to store poorly after baking.

A few workers have employed a preliminary nitrogen soak in place of evacuation (Sim et al. 1976). Limited tests with type 103a-O emulsion suggested that the speed was comparable to that achieved with evacuation, but the fog level was much higher (R. L. Scott 1975, unpublished). Babcock et al. (1975) showed that moisture in an emulsion at the time of hydrogenation retarded the action of the hydrogen; it is likely that evacuation is more effective in removing water than any reasonable period of nitrogen soaking.

Results for hydrogenation of various emulsions are given in Table 3. As is true of baking, hydrogenation can be accelerated by raising the temperature; Babcock et al. (1974b) report that an emulsion requiring hours to hydrogenate at 20°C can be treated in minutes at 85°C. Like

Table 3 Representative results for hydrogenation of Kodak spectroscopic plates by two observatory groups (SM and SG) and a laboratory group (B)

Emulsion	Filter	Exposure time (min)	Hours in H_2	Fog[a] Increase	Fog[a] Total	Speed gain factor[b]	Ref[c]
I-O	Wr.7[d]	60	2.5	0.04	—	2.9	B
103a-O	GG 387	60	4.5	0.03	—	2.2	SM
	Wr.7	60	3	0.04	—	2.0	B
IIa-O	GG 385	20	1.8	—	0.17	2.3	SG
	Wr.7	60	3	0.07	—	2.5	B
IIIa-J	GG 385	20	4	—	0.16	3.0	SG
	GG 385	60	8	—	0.12	3.2	SM
103a-D	Wr.7	60	3	0.06	—	2.2	B
IIa-D	GG 495	20	3	—	0.23	2.6	SG
	Wr.7	60	5	0.18	—	3.0	B
103a-E	RG 630	60	12	0.08	—	2	SM
IIa-F	Wr.22[e]	60	5	0.17	—	4.7	B
098-04	RG 610	20	4	—	0.21	1.7	SG
	RG 630	60	7	0.09	—	2.0	SM
IIIa-F[f]	RG 610	20	1.7	—	0.33	4.0	SG
	RG 630	60	16	—	0.30	3.5	SM
I-N	Wr.7	60	3	0.02	—	3.6	B
	Wr.22	60	3	0.02	—	2.8	B

[a] Reported in ANSI diffuse density. Some workers report final total fog, others the increase in fog due to treatment.

[b] Measured at density 0.6 above fog.

[c] B, Babcock et al. (1974b), Eastman Kodak; SM, Sim (1977), UK Schmidt Telescope Unit; SG, Schoening (1977), Kitt Peak. SM used a nitrogen purge rather than evacuation prior to hydrogenation; this may explain her consistently longer hydrogenation times.

[d] Wratten No. 7, passband 340–480 nm.

[e] Wratten No. 22, transmits above 510 nm.

[f] Formerly type 127.

nitrogen-baked plates, hydrogenated plates deteriorate rapidly upon contact with moist room air. Stored under refrigeration in a nitrogen atmosphere, nearly all hydrogenated emulsions can be kept satisfactorily for many weeks (an exception has been the IIIa-F emulsion, which deteriorates rapidly). Observers in humid climates should also give consideration to protecting treated plates during exposure.

For plates dye-sensitized to various regions of the visible spectrum, hydrogenation increases speed in the dye-sensitized region to about the same degree as in the region of inherent sensitivity. However, infrared emulsions show much less gain in the infrared than in the visible (Babcock et al. 1974b).

3.5   Combined Treatment with Nitrogen and Hydrogen

When hypersensitization treatments operate by different mechanisms it should be possible to achieve additive gains by cascading them. Unfortunately, the fog produced by each treatment is also additive, and this generally limits the possibilities.

3.5.1   NITROGEN BAKING PLUS HYDROGENATION   Since hydrogenation works through reduction, whereas baking extends chemical sensitization, the two might be expected to be complementary. Scott & Smith (1976) found this was true for Kodak emulsions 103a-O and IIIa-J. Figure 6 shows curves for 103a-O comparing speed and fog for nitrogen baking, for hydrogenation, and for the combination of the two. Similar results were obtained for IIIa-J, an advantage in this case being that the fog for a given speed gain was lower for the combined treatment than for hydrogenation alone. Hoessel (1978) later reported that the combined treatment was effective for the red-sensitive emulsions 127-04 and 098-04, yielding gains twice those of simple hydrogenation. Curiously, tests have shown that even nitrogen-baked plates benefit from evacuation prior to hydrogenation, and this has been standard in the above procedures.

3.5.2   BAKING IN FORMING GAS   There appears to be no reason why nitrogen baking and hydrogenation cannot be applied simultaneously as well as sequentially. Because of the rapidity of hydrogenation at elevated temperatures (Babcock et al. 1974b), use of a very dilute mixture of hydrogen in the nitrogen suggests itself (C. Denny, 1975 meeting of the AAS Working Group on Photographic Materials). Such mixtures are in fact used industrially under the name "forming gas."

Figure 7 shows results of tests conducted by Scott et al. (1977) on IIIa-J. Both 2% and 8% concentrations of H_2 were used, and the results were compared with nitrogen baking, with simple hydrogenation, and with the combination of nitrogen baking and hydrogenation. Following these tests the experimenters adopted baking in 2% forming gas as the preferred treatment for plates used on the sky. Similar tests on 103a-O plates yielded practical speeds higher than those produced by N_2 baking or hydrogenation, but the results were inferior to sequential nitrogen baking plus hydrogenation, which the experimenters adopted as their preferred treatment for 103a-O. To those who find bathing of I-N plates inconvenient, forming gas offers an attractive alternative. Scott et al. (1977) found that 8 hr at 65°C in 8% forming gas yielded speed gains of 3 or more for 25-min exposures with an RG 695 filter; this was more gain than they were able to achieve by ammonia bathing, and the baked plates were far more uniform, more consistent, and stored well.

As is generally true where hydrogen is used, baking in forming gas should be preceded by evacuation, and Scott et al. (1977) maintain a slow flow of gas through the box during baking. Forming gas has the obvious advantage of safety. Since at least 4% by volume of H_2 in air is needed to form a flammable mixture (Hodgman 1963), even 8% forming gas is unlikely to reach an explosive concentration. Nevertheless, it would be foolhardy deliberately to vent the gas into a laboratory. Where safety regulations or personal caution militate against the use of pure hydrogen, forming gas is an attractive alternative.

3.6   Pre-Flash

Pre-flashing is a useful procedure only when the astronomical exposure is signal-limited (photon starvation). One introduces background in order to get the exposure increment representing the desired signal on a more favorable gradient of the D-log E curve. As can be inferred from the discussion of $[S/N]_{OUT}$, the limit of useful background will be set by density-noise. Marchant (1964) outlined pre-flash theory in terms of output signal-to-noise, and these precepts have been frequently reiterated (cf Ables et al. 1971, Millikan 1976, Latham & Furenlid 1976, Allbright 1976).

If exposure levels are low because of flux limitations as in direct photography at long focal ratios or with narrow-band filters, or in some kinds of spectrography, pre-exposure can help. Theory and experience indicate that a short pre-exposure that produces a density increment of about 0.2 above fog may be near optimum for this kind of operation. Bird (1971) has defined a "short" exposure as being of the order of 10 seconds or less in the case of emulsions designed for long-exposure applications. As in other hypersensitization procedures, pre-exposure will degrade the maximum $[S/N]_{OUT}$ capability of the emulsion. However, detection and measurement of exposures made at low flux levels can be improved by this technique.

3.7   Latensification

The term "latensification" denotes treatments applied between exposure and development, the intent being to augment subdevelopable latent image centers to the point where they are developable. To date latensification has found little use in astronomy. Among the probable reasons are (a) the emulsions used in astronomy do not respond well to latensification, (b) latensification tends to be most effective after short, high-intensity exposures, and (c) the procedures are inconvenient in an observatory setting.

One frequently mentioned technique subjects the exposed emulsion to chemical fumes; mercury vapor, ammonia, hydrogen peroxide, bromine, sulfur dioxide, acetic acid, formic acid, oxalic acid, and propionic acid are among those mentioned (cf O'Hara & Osterburg 1951). Other treatments involve bathing in benzene, carbon tetrachloride, ammonia, or solutions of silver or gold salts (James et al. 1948).

A different approach is to give the emulsion a second exposure to a uniform light source. Since the objective is to build up existing image centers without creating new ones that would add to background fog, the second exposure should be in the region of extreme reciprocity failure; i.e. it should be "long" compared to the initial exposure. In the usual astronomical case this would involve very long times indeed! Suggestions for circumventing this problem include using a desensitizing dye between the first and second exposures to promote reciprocity failure (Kirillov 1967), making the second exposure in an atmosphere that favors reciprocity failure (e.g. moist oxygen), and using emulsions having naturally high reciprocity failure, with a supplementary method such as cooling being used during the first exposure to mitigate LIRF (Bird 1971, 1973).

After testing a variety of techniques James & Vanselow (1949) concluded that useful gains were achieved when the emulsions were given normal development. However, when they were fully developed in energetic developers latensification conferred no benefit. James & Vanselow concluded that latensification increased the rate of development of exposed grains, rather than increasing the number of developable grains. A similarly disappointing result is quoted by Kirillov (1967).

3.8   Push Development and Intensification

Photographic literature is rampant with schemes for obtaining more image silver in underexposed negatives. For scientific purposes one must distinguish carefully between methods that actually increase information and those that merely create a prettier picture or one that is easier to print.

3.8.1   PUSH DEVELOPMENT   "Push" development or forced development is generally achieved by extending development time or using a more vigorous developer. A well-known example of extending development was the discovery by astronomers that they obtained better results by processing spectroscopic plates 5 min in D-19, rather than the 4 min originally recommended by Kodak. Latham (1974) showed that developing IIa-O in D-76 for 30 min, rather than 5 min, triples the speed (measured as usual by the exposure required to produce a density 0.6 above fog); moreover, this gain is achieved without degrading the signal-to-noise ratio. However, similar experiments with IIIa-J (Latham 1976) indicated that extended development had little effect on speed and resulted in a lower signal-to-noise ratio. Clearly each emulsion—and probably each application—must be investigated individually to determine whether extended development is beneficial.

An extreme example of a vigorous developer is Kodak SD-19a, described by Miller et al. (1946). Under many circumstances this hydrazine-containing developer produces significant gains in effective speed by promoting "infectious development." Reaction by-products from the development of an exposed grain attack neighboring unexposed grains and catalyze their development, thus amplifying the production of image silver. Obviously such a process results in loss of resolution and an increase in graininess; it is also effective only when development is carried to fairly high fog levels.

3.8.2   INTENSIFICATION   Numerous chemical treatments have been described for increasing the density of an already developed and fixed negative (cf Wall & Jordan 1976). In general these methods add silver or another metal to the original image, often through a process of bleaching and intensive redevelopment. The result is usually similar to strongly forced development, with the same liabilities of increased graininess, loss of resolution, and enhancement of emulsion defects. While chemical intensification may make it easier to obtain an attractive print from a thin negative, it is questionable whether there is any real gain in information. Since chemical intensification is generally irreversible, caution is indicated in applying it to valuable originals.

Recently, autoradiography has received attention as a possible means of intensifying astronomical photographs. The silver of the developed and fixed image is made radioactive by placing it in the beam of a nuclear reactor (Perry 1976) or, more conveniently, by bathing it in a radioactive solution (Askins 1976). A secondary film is then exposed by contact-printing it against the now-radioactive original. The density and contrast of the secondary negative can be controlled over wide ranges through choice of emulsion, length of exposure to the radioactive image, and control of development. There is less hazard to the original image than in chemical intensification, since its appearance is unchanged by the temporary radioactivity of the silver. Early tests with astronomical plates have shown marked enhancement of liminal images, unfortunately accompanied by serious losses in resolution and marked increases in granularity. It is too soon to say whether these liabilities can be overcome.

3.9 Environment During Exposure

3.9.1 TEMPERATURE   King (1912) did some rudimentary experiments that concerned the relation between temperature and response of photographic plates. He found that cool plates responded better than warm ones for long exposures. Being a real experimenter, he tried to enhance the effect by pressing ice against the back of a plate while it was being exposed and thereby discovered the deleterious effect of moisture that condensed on the emulsion. The effects of temperature have been thoroughly investigated (cf Babcock et al. 1971, 1972, and references therein). Results of purposeful cooling of emulsions at the telescope have been to some degree confused by the practice of evacuating the film holder to avoid condensation effects. Cooling and evacuation each act to shift the minimum of the log I, log It reciprocity relation toward a beneficial long exposure value.

There are easier ways to correct reciprocity failure, so purposeful cooling may be justified only when special material with extreme LIRF, such as color film, is being used.

3.9.2 MOISTURE AND OXYGEN   Because of what we now know of the deleterious effects of oxygen and moisture (Section 3.2), there may be advantages to exposing plates in a controlled environment. Sealed cassettes have been in use at the Rosemary Hill Observatory since 1970 (Smith et al. 1971). They were designed to permit baking, storage, and exposure in a controlled environment. This closed system not only excludes moisture and oxygen but, in so doing, reduces or eliminates LIRF (Miller 1970, Babcock 1976, Kaye & Meaburn 1977, A. G. Smith 1978b). Filters act as gas-tight windows for these cassettes, thus permitting nitrogen gas filling in use. The benefit of exposing plates in an inert environment may be greater in a humid location such as in Florida than in drier locations. Plates that have been thoroughly dried and stored in nitrogen in pre-exposure treatment are slow to degrade in dry air. However, substantial advantages may accrue from exposing plates in a controlled environment, even in a dry area, as demonstrated by Kaye & Meaburn (1977). One not only improves speed and output signal-to-noise, but elimination of changes in sensitivity during exposure improves the comparability of photographic transfers (A. G. Smith 1978b).

4   PROSPECTS

Not only has photography in astronomy prospered by its own merits, but it has been greatly stimulated by new instruments and procedures used in measurement and analysis.

4.1   *Photographic Materials*

4.1.1   EMULSION TECHNOLOGY   Marchant & Millikan (1965) outlined the precepts that led to the production of the emulsions that have so improved limit detection with moderately fast focal ratio telescopes. These are the Kodak IIIa emulsions. The secrets are high DQE by appropriate sensitization, monodispersed (uniformly sized) fine grain for high gradient and low noise, and techniques of hypersensitization that tailor the material for appropriate applications.

Can further advances be made by extending these precepts? The answer is yes! However, there is a price to be paid. Emulsions with finer monodispersed grains than the IIIa emulsion have been produced and tested at the telescope (A. G. Millikan and W. C. Miller 1975, private communication). Maximum output signal-to-noise can be increased in this way. However, the gradient, or gamma, of the characteristic curve of such materials becomes steep, so the dynamic range diminishes because the maximum density or means of measuring it must remain about the same as for emulsions in current use. Moreover, as a practical matter, one must consider exposure conditions. Properly hypersensitized IIIa emulsions can be given optimum sky exposures for broad-band work at F/3 in convenient times. Faster focal ratio telescopes would be required to exploit finer-grained materials, and, to get improved performance in the "limit," the scale could not be diminished.

For existing telescopes, the advances that must come are in grain sensitization. The trend shown in Figure 2 indicates that significant advances in sensitivity will be hard won. Further, it is likely that hypersensitization will remain a challenge because of the need to produce photographic materials with adequate storage life.

4.1.2   APPLICATIONS   As is the case in this review, hypersensitization was a major topic in the recent conference on Modern Techniques in Astronomical Photography (West & Heudier 1978). A summary table of hypersensitization techniques and their effectiveness, adapted from one prepared by Millikan, A. G. Smith, and Sim at that conference, is presented as Table 4. Indications are that a repertoire of three simple techniques will be adequate for near-optimum application of available

Table 4 Summary of hypersensitization treatments for commonly used Kodak spectroscopic plates

Type of *Kodak* Spectroscopic Plate	Gas Treatments						Liquid Treatments				Others	
	Baking (usually 55 to 75°C)			Soaking (~20°C)			Bathing (usually 5 to 20°C)					
	Air	N₂	Forming gas	N₂	H₂	N₂ Bake + H₂ Soak	H₂O	NH₄OH	AgNO₃	Evacuate	Cool	H₂O Bathe + H₂ Soak
103a-O	se	E	E	E	[E]	[VE]	se			E		
103a-D		E				[VE]		se			se	
103a-E		E		se	E							
098	se	E		E	VE	[VE]	E			E		
IIa-O		[E]	[VE]	E	[E]		se	se				
IIa-D		E		E	VE					me		
IIa-F	se	E		E	VE		A			me	E	
IIIa-J	se	[VE]	[VE]	E	[VE]	VE	A			me	E	
IIIa-F		E	[VE]		[VE]	[VE]		me		me		
IV-N	se	se		se	E		me	[E]	[VE]	me	E	
I-N			me				E	E	[VE]			E
I-Z			[VE]				E	E	[VE]			E

Presumed Mechanism of Treatment

Mechanism of Treatment	Air	N₂	Forming gas	N₂	H₂	N₂ Bake + H₂ Soak	H₂O	NH₄OH	AgNO₃	Evacuate	Cool	H₂O Bathe + H₂ Soak
Further Chemical sensitization	✓	✓	✓	✓	✓	✓	✓					✓
Reduction sensitization	✓	✓	✓	✓	✓	✓						✓
Remove O₂		✓	✓	✓	✓	✓				✓		
Remove H₂O	✓	✓					✓	✓		✓		
Raise Ag⁺ conc.									✓			
Reduce recombination								✓			✓	✓
Increase Ag° stability											✓	✓

Key to top half of table

[ ] Preferred by at least one user   me Moderately effective   A To be avoided
VE Very effective   se Small effect (not very useful   (blank) No result known
E Effective   or better alternatives exist)

June, 1978

materials: pre-flash for signal-limited problems, $N_2 + 2\%$ H_2 bake for blue-sensitive and panchromatic "spectroscopic" emulsions, and $AgNO_3$ (0.001 molar) bathing for infrared-sensitive emulsions. None of these techniques requires elaborate equipment or poses a hazard in use.

4.2   Photographic Measures

Photography in astronomy has been revolutionized by the availability of machines that measure plates astrometrically and photometrically with unprecedented speed and precision. These developments have depended on the rise of computer technology both for control of the measuring machines and assimilation of the vast amounts of data generated in the digitization of high-resolution photographs. A number of special-purpose machines for astrometric and photometric measures are described in the proceedings of a conference on Image Processing Techniques in Astronomy (de Jager & Nieuwenhuijzen 1975). Fast general-purpose computer-controlled microdensitometers are also available (cf King 1976). For astrometric work, these machines are capable of fractional micron precision (van Altena & Auer 1975). For photometric work, precision exceeds that obtainable from iris photometers (cf McClure et al. 1978).

4.3   Photographic Projects

King & Hinrichs (1967) showed that stellar photometry could be done by calibrated microphotometry of images on photographic plates. At the same time, Walker & Kron (1967) and Lallemand et al. (1966) were doing microphotometric stellar photometry on electrograms, a procedure that was simplified because of the linear response of the electrographic systems. Now, with the advent of newer machine methods (de Jager & Nieuwenhuijzen 1975, F. G. Smith et al. 1978), microphotometric analysis of photographic material is coming into its own. Success with machine methods has to a large degree stemmed from pioneering work by Ables et al. (1969), Hewitt (1969), Walker (1970), Newell & O'Neal (1974), Wlérick et al. (1974), and Lynds (see Butcher 1977) on behalf of electrograph and video camera detectors.

Because photographic formats can be not only large but also relatively uniform (Latham 1978b, Dixon 1978), plates are unequaled as detectors for many astronomical projects. A few of many recent research examples that have emphasized microphotometric techniques applied to photographs are contributions by Kinman (1975), Butcher (1976, 1977), Herzog & Illingworth (1977), McClure et al. (1978), Lorre (1978), Lorre & Lynn (1978), and Strom & Strom (1978a,b). These examples include astrometry and photometry of direct star images by automatic means, analytical discrimination between star and faint galaxy images, and limit surface photometry of resolved galaxies and associated low surface brightness features. The nonlinear response of photographic materials is not an obstacle in getting results of reasonable precision, particularly in sky-limited direct work. The large format, resolving power, spatial stability, and information storage capacity available through photography are ideal for extensive projects requiring astrometric, morphological, and photometric measures.

ACKNOWLEDGMENT

The photographic research of one of the authors (AGS) has been supported by a series of grants from the National Science Foundation, and preparation of the present review was aided by NSF grant AST 77-24821.

Literature Cited

Ables, H. D., Hewitt, A. V., Janes, K. A. 1971. *AAS Photo-Bull.* 3:18–22

Ables, H. D., Hewitt, A. V., Kron, G. E. 1969. *Publ. Astron. Soc. Pac.* 81:530

Allbright, G. S. 1976. *J. Photogr. Sci.* 24:115–19

Askins, B. S. 1976. *Appl. Opt.* 15:2860–65

Babcock, T. A. 1976. *AAS Photo-Bull.* 13:3–8

Babcock, T. A., Ferguson, P. M., James, T. H. 1974a. *Astron. J.* 79:92–98

Babcock, T. A., Ferguson, P. M., Lewis, W. C., James, T. H. 1975. *Photogr. Sci. Eng.* 19:49–55, 211–14

Babcock, T. A., Lewis, W. C., James, T. H. 1971. *Photogr. Sci. Eng.* 15:297–303

Babcock, T. A., Lewis, W. C., McCue, P. A., James, T. H. 1972. *Photogr. Sci. Eng.* 16:104–9

Babcock, T. A., Sewell, M. H., Lewis, W. C., James, T. H. 1974b. *Astron. J.* 79:1479–87

Barlow, B. V. 1978. *Contemp. Phys.* 19:47–64

Bird, G. R. 1971. *Photogr. Sci. Eng.* 15:442

Bird, G. R. 1973. *AAS Photo-Bull.* 7:4–6

Bowen, I. S., Clark, L. T. 1940. *J. Opt. Soc. Am.* 30:508–10

Butcher, H. R. 1976. *Note to Members of IAU Comm. 25*

Butcher, H. R. 1977. *Ap. J.* 216:372–80

Corben, P. M., Reddish, V. C., Sim, M. E. 1974. *Nature* 249:22–23

Dainty, J. C., Shaw, R. 1974. *Image Science*, p. 35. New York: Academic. 402 pp.

de Jager, C., Nieuwenhuijzen, H. 1975. *Image Processing Techniques in Astronomy* (Astrophysics and Space Science Library Vol. 54). Dordrecht: Reidel. 418 pp.

de Vaucouleurs, G. 1968. *Appl. Opt.* 7:1513–18

Dixon, K. L. 1978. *Observatory* 98:103

Duffin, G. F. 1966. *Photographic Emulsion Chemistry*, pp. 83–98. New York: Focal. 239 pp.

Eastman Kodak. 1973. *Kodak Plates and Films for Scientific Photography, P-315*, pp. 14d–19d. Rochester. 45+39 pp. 1st ed.

Furenlid, I. 1978. In *Modern Techniques in Astronomical Photography*, ed. R. M. West, J. L. Heudier, pp. 153–64. Geneva: European Southern Observatory. 304 pp.

Furenlid, I., Schoening, W. E., Carder, B. E. Jr. 1977. *AAS Photo-Bull.* 16:14–16

Gurney, R. W., Mott, N. F. 1938. *Proc. R. Soc. London Ser. A* 164:151–66

Herzog, A. D., Illingworth, G. 1977. *Ap. J. Suppl.* 33:55–67

Hewitt, A. V. 1969. *Publ. Astron. Soc. Pac.* 81:541–42

Hoag, A. A. 1976. *AAS Photo-Bull.* 13:14–16

Hoag, A. A. 1978. See Furenlid 1978, pp. 121–39

Hoag, A. A., Furenlid, I., Schoening, W. E. 1979. *AAS Photo-Bull.* 19:3–6

Hodgman, C. D., ed. 1963. *Handbook of Chemistry and Physics*, p. 1941. Cleveland, Ohio: Chemical Rubber Publ. Co. 3604 pp.

Hoessel, J. F. 1978. *AAS Photo-Bull.* 17:10–11

Honeycutt, R. K., Chaldu, R. S. 1970. *AAS Photo-Bull.* 2:14–15

Hurter, F., Driffield, V. C. 1890. *J. Soc. Chem. Ind., London* 9:455–69

James, T. H. 1972. *J. Photogr. Sci.* 20:182–86

James, T. H., ed. 1977. *The Theory of the Photographic Process*, pp. 377–92. New York: Macmillan. 714 pp.

James, T. H., Vanselow, W. 1949. *J. Photogr. Soc. Am.* 15:688–93

James, T. H., Vanselow, W., Quirk, R. F. 1948. *J. Photogr. Soc. Am.* 14:349–53

Jenkins, R. L., Farnell, G. C. 1976. *J. Photogr. Sci.* 24:41–50

Jones, R. C. 1958. *Photogr. Sci. Eng.* 2:57–65

Jones, R. C. 1959. Presented at 102nd Meeeting, Am. Astron. Soc., Rochester, New York

Kaler, J. B. 1966. *Publ. Astron. Soc. Pac.* 78:537–41

Kaye, A. L. 1976. *Astrophys. Space Sci.* 45:477–82

Kaye, A. L., Meaburn, J. 1977. *AAS Photo-Bull.* 15:18–19

King, E. S. 1912. *Harvard Ann.* 59:81–88

King, I. R. 1976. In *Proc. IAU Working Group on Photographic Problems*, ed. J. L. Heudier, pp. 95–97. Nice

King, I. R., Hinrichs, E. L. 1967. *Publ. Astron. Soc. Pac.* 79:226–34

Kinman, T. D. 1975. *Observatory* 95:280

Kirillov, N. I. 1967. *Problems in Photographic Research and Technology*, pp. 11, 20. New York: Focal. 205 pp.

Kowaliski, P. 1972. *Applied Photographic Theory*, p. 308. New York: Wiley. 533 pp.

Lallemand, A., Canavaggia, R., Arniot, F. 1966. *C. R.* 262:838–40

Latham, D. W. 1974. In *Methods of Experimental Physics*, Vol. 12, Part A, ed. N. P. Carleton, pp. 221–35. New York: Academic. 587 pp.

Latham, D. W. 1976. *AAS Photo-Bull.* 13:9–13

Latham, D. W. 1978a. *AAS Photo-Bull.* 18:3–7

Latham, D. W. 1978b. See Furenlid 1978, pp. 141–52

Latham, D. W., Furenlid, I. 1976. *AAS Photo-Bull.* 11:11–14

Lewis, W. C., James, T. H. 1969. *Photogr. Sci. Eng.* 13:54–65

Lorre, J. J. 1978. *Ap. J. Lett.* 222:L99–103

Lorre, J. J., Lynn, D. J. 1978. *Jet Propulsion Lab. Publ. 78–17*. Pasadena. 55 pp.

Marchant, J. C. 1964. *J. Opt. Soc. Am.* 54:798–800

Marchant, J. C., Millikan, A. G. 1965. *J. Opt. Soc. Am.* 55:907–11

Masaki, O. 1926. *Mem. Kyoto Coll. Sci. Ser. A* 9:285–302

McClure, R. D., Newell, B., Barnes, J. V. 1978. *Publ. Astron. Soc. Pac.* 90:170–78

Miller, H. A., Henn, R. W., Crabtree, J. I. 1946. *J. Photogr. Soc. Am.* 12:586–609, 692

Miller, W. C. 1969. *Bull. Am. Astron. Soc.* 1:163

Miller, W. C. 1970. *AAS Photo-Bull.* 2:15–17

Miller, W. C. 1972. *AAS Photo-Bull.* 6:4–5, 19

Miller, W. C. 1975. *AAS Photo-Bull.* 9:3–7

Millikan, A. G. 1973. Comment in *AAS Photo-Bull.* 7:7

Millikan, A. G. 1976. In *Proc. IAU Working Group on Photographic Problems*, ed. J. L. Heudier, pp. 9–39. Nice

Millikan, A. G., Tinney, J. R. 1977. *AAS Photo-Bull.* 15:8–11

Newell, E. B., O'Neil, E. J. 1974. In *Electrography and Astronomical Applications*, ed. G. L. Chincarini, D. J. Griboval, H. J. Smith, pp. 153–75. Austin: Univ. of Texas. 364 pp.

O'Hara, C. E., Osterburg, J. W. 1951. *Practical Photographic Chemistry*, p. 71. Minneapolis, Minn.: American Photography. 189 pp.

Perry, M. P. 1976. *AAS Photo-Bull.* 12:9–11

Rose, A. 1946. *J. Soc. Motion Pict. TV Engrs.* 47:273–94

Schoening, W. E. 1975. *AAS Photo-Bull.* 10:18–20

Schoening, W. E. 1976. *AAS Photo-Bull.* 11:18–19

Schoening, W. E. 1977. *AAS Photo-Bull.* 14:3–7

Schoening, W. E. 1978a. *AAS Photo-Bull.* 17:12–14

Schoening, W. E. 1978b. See Furenlid 1978, pp. 63–82

Schrader, H. W., Graves, E. E., Smith, A. G., Leacock, R. J. 1978. *Fla. Scientist.* 41:207–14

Scott, R. L., Smith, A. G. 1974. *Astron. J.* 79:656–58

Scott, R. L., Smith, A. G. 1976. *AAS Photo-Bull.* 12:6–8

Scott, R. L., Smith, A. G., Leacock, R. J. 1977. *AAS Photo-Bull.* 15:12–15

Sewell, M. H. 1975. *AAS Photo-Bull.* 8:13–15

Sewell, M. H., Millikan, A. G. 1975. *AAS Photo-Bull.* 8:6–9

Sim, M. E. 1977. *AAS Photo-Bull.* 14:9–13

Sim, M. E. 1978. See Furenlid, pp. 23–41

Sim, M. E., Hawarden, T. G., Cannon, R. E. 1976. *AAS Photo-Bull.* 11:3–4

Smith, A. G. 1977. *J. Appl. Photogr. Eng.* 3:4:205–8

Smith, A. G. 1978a. Presented at Austin Meet. of AAS Working Group on Photographic Materials

Smith, A. G. 1978b. See Furenlid 1978, pp. 83–93

Smith, A. G., Leacock, R. J. 1973. *AAS Photo-Bull.* 7:18–19

Smith, A. G., Schrader, H. W., Richardson, W. W. 1971. *Appl. Opt.* 10:1597–99

Smith, F. G. 1978. *Observatory* 98:101–15

Smith, F. G. et al. 1978. *Observatory* 98: 101–15

Spinrad, H., Wilder, J. 1972. *AAS Photo-Bull.* 5: 14–15

Strom, K. M., Strom, S. E. 1978a. *Astron. J.* 83: 73–134

Strom, S. E., Strom, K. M. 1978b. *Astron. J.* 83: 732–63

van Altena, W. F., Auer, L. H. 1975. See de Jager & Nieuwenhuijzen 1975, pp. 411–18

Walker, M. F. 1970. *Ap. J.* 161: 835–44

Walker, M. F., Kron, G. E. 1967. *Publ. Astron. Soc. Pac.* 79: 551–68

Wall, E. J., Jordan, F. I. 1976. *Photographic Facts and Formulas* (revised by J. S. Carroll), pp. 168–76. Englewood Cliffs, New Jersey: Prentice-Hall. 480 pp.

West, R. M., Heudier, J. L., eds. 1978. *Modern Techniques in Astronomical Photography.* Geneva: European Southern Observatory. 304 pp.

Wlérick, G., Michet, D., Labeyrie, C. 1974. See Newell & O'Neal 1974, pp. 177–98

Young, A. T. 1977. *AAS Photo-Bull.* 15: 3–7

Ann. Rev. Astron. Astrophys. 1979. 17 : 73–111
Copyright © 1979 by Annual Reviews Inc. All rights reserved

OBSERVED PROPERTIES OF INTERSTELLAR DUST

×2146

Blair D. Savage and John S. Mathis

Washburn Observatory, University of Wisconsin, Madison, Wisconsin 53706

Introduction

Interstellar dust plays a significant role in influencing the physical and chemical state of the interstellar medium. It now appears well established that H_2, the most abundant molecule in the interstellar medium, forms on dust grains (Spitzer 1978). The dust contains a significant fraction of the interstellar heavy elements and thereby ties up important atomic species that can cool the interstellar gas through collisionally excited fine structure transitions. The dust, because of its large opacity, influences the diffuse interstellar radiation field and undoubtedly has a fundamental effect on the processes whereby interstellar clouds collapse to form stars. Unfortunately, the obscuring effects of dust can introduce very significant uncertainties in the interpretation of the energy distributions of astronomical sources situated behind dust clouds. Progress in understanding the nature of interstellar dust therefore not only provides important information about a significant constituent of the universe, but also should help observational astronomers to improve their correction procedures for the extinction produced by dust.

This review will concentrate on providing up-to-date information on the observed properties of interstellar dust. An attempt will be made to clarify some of the observational uncertainties associated with obtaining dust parameters. In some cases, our state of understanding is quite advanced, while, in other cases, observational and interpretative difficulties introduce large errors in the inferred parameters. About 300 research papers that are in some way related to cosmic dust are published each year. Obviously, it is impossible in a short review to adequately cover all such work, or to treat interesting side investigations. Some topics omitted include: the alignment of grains (see Spitzer 1978 and Aannestad & Purcell 1973); solar system grains (see McDonnell 1978 and Elsässer & Fechtig 1976); extragalactic dust; and circumstellar dust (see Ney

73

0066-4146/79/0915-0073$01.00

Table 1 Recent review papers or books discussing cosmic dust

Author(s)	Title	Comments
Aannestad & Purcell 1973	*Interstellar Grains*	Extensive review of interstellar dust
Andriesse 1977	*Radiating Cosmic Dust*	Theoretically oriented review
de Jong 1976	*Interstellar Dust: Invited Review*	General review of interstellar dust
Elsässer & Fechtig 1976	*Interplanetary Dust & Zodiacal Light*	Proceedings of IAU Colloq. No. 31, many papers on solar system dust
Field & Cameron 1975	*The Dusty Universe*	Review articles on interstellar and solar system dust
Greenberg & van de Hulst 1973	*Interstellar Dust & Related Topics*	Proceedings of IAU Symp. No. 52, many papers covering nearly all aspects of interstellar dust
Huffman 1977	*Interstellar Grains: The Interactions of Light with a Small Particle System*	Extensive review of interstellar dust, includes a review of the bulk optical properties of solids of astronomical importance
McDonnell (1978)	*Cosmic Dust*	Review articles on interstellar and solar system dust
Merrill 1977	*Infrared Observations of Late Type Stars*	Review of infrared observations of circumstellar dust
Ney 1977	*Star Dust*	Review of circumstellar dust
Salpeter 1977	*Formation and Destruction of Dust Grains*	Review of theories of dust formation and destruction
Spitzer 1978	*Physical Processes in the Interstellar Medium*	Chapters 7, 8, and 9 of this graduate level text concern interstellar grains
Stein 1975	*Recent Revelations of Infrared Astronomy*	Review of the infrared emission from circumstellar dust
Wesson 1974	*A Synthesis of Our Present Knowledge of Interstellar Dust*	General review of interstellar dust
Wickramasinghe & Morgan 1976	*Solid State Astrophysics*	Collection of short papers on interstellar and circumstellar particles
Wickramasinghe & Nandy 1972	*Recent Work on Interstellar Grains*	Extensive review of interstellar dust
Wynn-Williams & Becklin 1974	*Infrared Emission From HII Regions*	Review of the infrared emission from heated dust

1977 and Merrill 1977). However, the relationship between circumstellar and interstellar grains is briefly discussed. While we concentrate on the observational side of this subject, we do not totally ignore theory since real progress is only made when observations and theory are interrelated.

The recent literature reviewing dust is extensive. Table 1 lists some of this literature, along with a brief description of aspects of cosmic dust covered by each reference. The book by Spitzer (1978) is a particularly useful introduction to the subject of interstellar dust. Aannestad & Purcell (1973), Huffman (1977), and Greenberg (1978) include more extensive discussions.

1 Interstellar Extinction

Extinction refers to the sum of absorption and scattering and is generally determined by comparing observations of reddened and unreddened stars assumed to have identical intrinsic energy distributions. Observed differences are attributed to extinction but may also reflect a breakdown

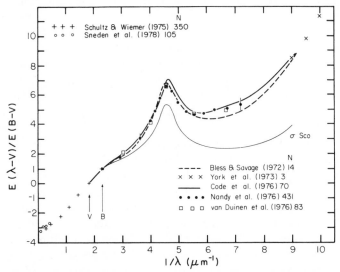

Figure 1   Average normalized interstellar extinction is plotted versus $1/\lambda$ in μm^{-1}. $E(\lambda-V)$ refers to the extinction in magnitudes between a wavelength λ and the photoelectric V band. The references for the various curves are provided along with an indication of how many stars were used to derive each average curve. One abnormal ultraviolet curve for σ Sco from Bless & Savage (1972) is also shown. The average curves plotted can be converted to total normalized extinction, $A_\lambda/E(B-V)$, by adding $R = 3.1$ to the quantity plotted. Note that the normalization to $E(B-V) = 1$ implies a corresponding hydrogen column density of $N(HI + H_2) = 5.8 \times 10^{21}$ atoms cm^{-2} (see Section 2). The error bars on two of the TD-1 points (Nandy et al. 1976) give an indication of the maximum observed variation in the *average* extinction curves derived for different galactic regions.

of the basic assumption above. Since the review of Aannestad & Purcell (1973) a considerable effort has gone into investigations of the wavelength dependence of extinction. While most observers appear to agree on the general characteristics of the extinction curve, the question of variations in the shape of the curve in localized galactic regions is still controversial, particularly in the IR. Average normalized extinction curves from a number of recent determinations are collected together in Figure 1. Here $E(\lambda-V)/E(B-V)$ is plotted against λ^{-1} in μm^{-1}. $E(\lambda-V)$ refers to the extinction in magnitudes between a wavelength, λ, and the V band of the photoelectric UBV system. $E(B-V)$ is the $(B-V)$ color excess in magnitudes. The relation between the quantity plotted and the total extinction, A_λ, at a wavelength λ is

$$A_\lambda/E(B-V) = E(\lambda-V)/E(B-V) + R,$$

where $R = A_V/E(B-V) \approx 3.1$ but is somewhat variable (see Sections 1.1 and 4.1).

In the following sections we discuss IR and visual extinction (Section 1.1) and UV extinction (Section 1.2). Within each section we consider the broadband aspects of extinction, fine structure in the extinction, and variation in the curve shape from region to region or from object to object. The extinction fine structure in the visual, commonly referred to as the diffuse interstellar features, is considered separately in Section 7. An average extinction curve representing a combination of IR, visual, and UV measurements is given in Section 1.3.

1.1 INFRARED AND VISUAL EXTINCTION   Extinction measurements in the IR are potentially subject to large systematic errors, since it is difficult to ensure that the reddened star and comparison star are intrinsically identical. The major problem is that some stars exhibit IR emission from heated circumstellar dust or free-free gaseous emission processes. A related but less severe problem can occur in the vicinity of 3600 Å if there are differences in the strength of the Balmer discontinuity between the reddened star and comparison star. A way around these problems is to exclude from the observing lists any object suspected of having a circumstellar envelope that might produce excess IR emission. Unfortunately, when such objects are excluded, the opportunities for studying the abnormal extinction characteristics of disturbed dust situated near hot or cool luminous stars are often eliminated. Also, it is not always easy to identify stars with excess IR emission and the criteria for identification often differ from one observer to another.

Several recent papers, which pay attention to these problems, present extinction measurements extending into the IR. Schultz & Wiemer (1975) analyze a large collection of published observations of O and B stars obtained in the standard BVRIJKL filter bands but reject peculiar objects such as Be stars, stars with abnormal spectra, and close visual binaries. From color-color plots for a group of 350 stars they derive the mean extinction values illustrated with the plus symbols in Figure 1. Approximately 90% of the objects investigated follow the illustrated extinction relation. The remaining 10% of the objects deviate significantly either because of the presence of a circumstellar IR excess or because of real differences in the character of their extinction curves. Schultz & Wiemer prefer the former explanation. In a similar investigation, Sneden et al. (1978) obtain IR measurements of 105 stars near 2.3, 3.6, 4.9, 8.7, 10, and 11.4 μm. By eliminating stars with known emission characteristics, these authors derive the mean extinction results shown with the open circles in Figure 1. The data for the three long wavelength bands were combined and plotted at $\lambda^{-1} = 0.1$ μm^{-1}. In the region of overlap, the Schultz & Wiemer (1975) and Sneden et al. (1978) results agree reasonably well and appear to define a reliable average extinction curve (see Section 1.3). In estimating an average value of $R = A_V/E(B-V)$ from the data illustrated in Figure 1, Schultz & Wiemer obtain $R = 3.14 \pm 0.10$ while Sneden et al. obtain $R = 3.08 \pm 0.15$. These values of R are in reasonable agreement with other recent determinations by the color difference method and by other methods (see Sneden et al. 1978 and Schalén 1975 for additional discussion and comparisons). The error estimates for the derived values of R are somewhat misleading since assumptions must be made about the extrapolation of extinction from the longest wavelength IR measurements to infinite wavelengths. Changes in the character of this extrapolation can appreciably influence the estimate of R. For instance, the difference between A_λ varying as λ^{-1} versus λ^{-4} for $\lambda > 2$ μm introduces a difference of ~ 0.2 in the derived value of R (Sneden et al. 1978). For the remainder of this paper we adopt the value $R = 3.1$ as being consistent with the highest quality modern determinations.

In addition to the broadband aspects of the visual and IR extinction discussed above, there exists fine structure. In the visual there are the diffuse interstellar features with half-intensity widths ranging from 1 to 20 Å (see Section 7) and the very broad structure sometimes referred to as the VBS. In the IR, absorption features are found at 3.07 μm and 9.7 μm.

Hayes et al. (1973) discuss the existence of very broad structure with widths of 500 to 1000 Å superimposed on the visible extinction curve. The structure is of low amplitude and therefore difficult to measure. However, a number of independent investigations reveal roughly similar results (Whiteoak 1966, Hayes et al. 1973, Schild 1977) and the reality of the structure appears established. In the published curves the most convincing feature is a broad depression in the extinction curve between $\lambda^{-1} \approx 1.5$ and $2 \ \mu m^{-1}$. For $E(B-V) = 1.0$, the amplitude of the depression is ~ 0.1 mag. Huffman (1977) notes the resemblance between the VBS and the absorption produced by terrestrial and meteoric magnetite (Fe_3O_4).

An absorption feature at 9.7 μm is found in many objects embedded within molecular clouds (Merrill, Russell & Soifer 1976). This feature also occurs in emission in stars and nebulae (Merrill & Stein 1976a,b) in which the dust is circumstellar or otherwise heated, and is almost certainly produced by silicates (see Section 8). Absorption at 9.7 μm has been detected in only two objects in which most of the intervening dust is in the ordinary interstellar medium (diffuse clouds as opposed to molecular clouds): the Galactic center and the star VI Cyg No. 12 (= Cyg OB2 No. 12). However, the dust obscuring these objects does not necessarily have a particularly large concentration of the carrier of the feature; one would not expect the feature to be observable in ordinary stellar spectra, because several magnitudes of extinction at V are required before the optical depth at 9.7 μm becomes appreciable. For the ordinary interstellar medium, this much extinction requires a large distance unless there is an atypically large concentration of dust along the line of sight. In general, the heavy extinction and great distance preclude detecting the features in even the brightest individual stars. However, the concentration of stars at the Galactic center (Oort 1977, Neugebauer et al. 1978) provides a suitably bright source, and the path length includes about 30 mag visual absorption (Becklin et al. 1978), almost all of which is in diffuse clouds not connected with the center itself. Similarly, the B8 Ia star VI Cyg No. 12 has $E(B-V) = 3.25$, or $A_V \approx 10$ mag, and no sign of an infrared excess (Johnson 1968). It is one of the intrinsically brightest stars in the Galaxy, with $M_V \approx -10$ (which is 3 mag brighter than a normal B8 Ia), and happens to lie behind a great concentration of apparently normal dust.

The 9.7 μm absorption feature has also been detected in several IR sources that lie behind very dense cold molecular clouds associated with normal or compact H II regions (Gillett et al. 1975a, Willner 1976, 1977, Persson, Frogel & Aaronson 1976). The feature is found in great strength in the spectrum of the Becklin-Neugebauer (BN) object in the Orion molecular cloud (Gillett & Forrest 1973). However, it is not possible to determine the optical depth of the feature very accurately by comparing the observed profile with that of the source, because the source's intrinsic emission profile is not known. One would expect the feature to be in emission when the dust is heated by luminous stars, as it is in the

Trapezium (Forrest, Gillett & Stein 1975). It is remarkable that the profiles of the 9.7 μm feature are very similar, in a wide variety of objects, over a considerable range of optical depths. The feature is broad ($\Delta\lambda/\lambda \sim$ 0.5) and has an asymmetric wing extending out to about 13 μm. It is broader than the feature in any naturally occurring terrestrial or lunar rock. However, its width agrees very well with laboratory measurements of disordered silicates produced by bombardment with energetic particles (Krätschmer & Huffman 1979) or produced in smoke (Day 1976).

Another IR absorption feature is found near 3.07 μm. It is commonly ascribed to solid H_2O and NH_3. In contrast to the 9.7 μm feature, it is (a) never found in emission, and (b) not found outside of molecular clouds, among which it has a highly variable strength relative to the 9.7 μm feature. Gillett et al. (1975b) have discussed the absence of absorption at 3.07 μm in the spectrum of VI Cyg No. 12. They find $\tau(3.07\ \mu m) \leq 0.02$, so $A_V/\tau(3.07\ \mu m) \geq 500$. Similarly, after a careful search, Soifer, Russell & Merrill (1976) failed to observe 3.07 μm absorption in the spectrum of the Galactic center, and find $A_V/\tau(3.07\ \mu m) \geq 110$. On the other hand, the feature is strong in BN (Becklin & Neugebauer 1967, Gillett et al. 1975b) and in several other sources in molecular clouds (Merrill, Russell & Soifer 1976). The ratio $\tau(3.07\ \mu m)/\tau(9.7\ \mu m)$ differs widely among these molecular clouds, from 0.2 to 2, too large a variation to be explained by errors in the estimates of the optical depths. Gillett et al. (1975b) find that $\tau(3.07\ \mu m)/\tau(9.7\ \mu m) < 0.04$ for VI Cyg No. 12, clearly quite different than for molecular clouds. NGC 2024 No. 1 and No. 2 (Merrill, Russell & Soifer 1976) convincingly show the association of the 3.07 μm feature with the inner regions of molecular clouds, and not with the outer regions or with the ordinary interstellar medium. No. 1, an early type star with $A_V \approx 8.3$ mag, lies in front of the core of the NGC 2024 molecular cloud, and has $\tau(3.07\ \mu m) \leq 0.05$, so $A_V/\tau(3.07\ \mu m) > 165$. No. 2, physically nearby No. 1 but behind the core of the cloud, has $A_V \sim 40 \pm 10$ mag (Hudson & Soifer 1976) and $A_V/\tau(3.07\ \mu m) < 45$. The interstellar component is presumably similar in the two stars.

The relative width ($\Delta\lambda/\lambda$) of the 3.07 μm feature is a factor of three narrower than that of the 9.7 μm feature and is asymmetric towards longer wavelengths. It is easily distinguished from the narrow, symmetrical feature seen in the spectra of cool carbon stars (Merrill & Stein 1976a; see their Figure 4). Hoyle & Wickramasinghe (1977) have suggested that the 3.07 and 9.7 μm features are both produced by interstellar polysaccharides. However, the very large variation in the relative strength of these two features speaks against this speculation (Egan & Hilgeman 1978).

The 18–22 μm flux distribution for several H II regions exhibits an emission or absorption feature with deviations from the nearby continuum of about 30% (Forrest & Soifer 1976, Forrest, Houck & Reed 1976, Frogel, Persson & Aaronson 1977). Silicates show such a feature and this is taken to confirm the identification of the 9.7 μm feature with silicates.

The question of differences in the shape of the IR and visual extinction curve from region to region or object to object is controversial because of the observational complications outlined in the first paragraph of this section. Often highly contradictory conclusions are obtained because at the outset one kind of study eliminates objects surrounded by disturbed dust while another includes such objects. Different conclusions are sometimes obtained because different stars and regions are sampled. For example, Johnson (1977) still contends that the extinction curve for the Orion Trapezium region is abnormal while Penston, Hunter & O'Neill (1975) claim a normal reddening law in the Orion nebula cluster. Penston, Hunter & O'Neill base their conclusions on stars more than ~ 1.5 min from the Trapezium. However, scattered light measurements near θ^1 Orionis imply a gas-to-dust ratio that varies with position, so the dust near the Trapezium must have been modified (Schiffer & Mathis 1974). The Orion Trapezium extinction curve derived by Johnson (1977) exhibits large deviations from the normal curve for $\lambda < 1$ μm and significant contamination from free-free emission and/or heated dust seems unlikely at these wavelengths. In addition, Johnson notes that the hydrogen Paschen lines for θ^1C Ori and θ^1D Ori are of normal strength, which would appear to preclude a substantial amount of filling in by continuous IR emission. While it is possible to dispute the far IR portion ($\lambda > 2$ μm) of the Orion Trapezium extinction curve because of potential nebular and circumstellar contamination (Ney, Strecker & Gehrz 1973), it appears established that the Orion Trapezium extinction is abnormal for $\lambda < 1$ μm.

ρ Oph and HD 147889 are two other objects for which the case for abnormal IR and visual extinction seems reasonably secure (Whittet, van Breda & Glass 1976, Whittet & van Breda 1975). Both stars are situated in the complex of clouds that includes the ρ Ophiuchus dark cloud. The data suggest values of R significantly larger than 3.1. McMillan (1978) shows for HD 147889 that the 3.07 μm ice feature is absent [$\tau(3.07) < 0.01$]. Therefore, the large value of R for this particular line of sight is not produced by ice mantles.

Small differences in the visual and IR extinction from region to region appear in the average curves Johnson (1977) presents for Scorpius, Cygnus, Ophiuchus, and Perseus. Unfortunately, it is difficult to evaluate from the published results the extent to which stars with potential IR emission problems were rejected. However, the reasonably good correlations between IR extinction curve shape indicators and the wavelength of maximum linear polarization (see Section 4) provide strong support for the reality of both small and large variations in the shape of the visual and IR extinction curve.

1.2 ULTRAVIOLET EXTINCTION   A number of satellites have provided extensive UV extinction measurements. These include: OAO-2 (Bless & Savage 1972, Savage 1975, Code et al. 1976), *Copernicus* (York et al. 1973), TD-1 (Nandy et al. 1975, 1976), and ANS (van Duinen, Wu & Kester 1976). In Figure 1, some of the average extinction curves from these satellites are plotted along with an abnormal curve for σ Sco. The different *average* UV extinction curves agree quite well. The small differences can probably be explained by bandpass effects and by true differences in the extinction between the different star samples used to obtain the averages. For example, the Bless & Savage (1972) average curve is probably lower in the region $\lambda^{-1} = 5$ to $8~\mu m^{-1}$, since the group of stars used to obtain the average contained a significant percentage from the Scorpius region, a region that exhibits low far UV extinction. The slight shift in the position of the extinction bump between the Code et al. (1976) curve and the rest of the curves may be the result of bandpass effects, since the Code et al. results are based on a combination of OAO-2 10 to 20 Å resolution scanner observations and broadband photometry observations of 70 stars.

The most remarkable aspect of the UV portion of the extinction curve is the broad bump centered near $4.6~\mu m^{-1}$ or 2175 Å. It appears that the feature is roughly symmetrical with a full width at half strength of about $1~\mu m^{-1}$ or 480 Å (Savage 1975). However, these estimates depend on the base line assumed. With a different choice, Nandy et al. (1975) obtain a full width at half strength of 360 Å. The position of peak absorption is remarkably constant although there may be some regions where the feature is shifted by about ± 50 Å from its normal position of 2175 Å (Savage 1975). The feature strength correlates well with $E(B–V)$ (Savage 1975, Nandy et al. 1976, Dorschner, Friedemann & Gürtler 1977). However, a few exceptions do exist. The most notable cases are θ^1 and NU Orionis, where the bump is weaker than normal (Bless & Savage 1972, Savage 1975, Savage & Bohlin 1978), and HD 156385 and HD 192163, where the bump is about one magnitude stronger than expected on the basis of estimates of $E(B–V)$ (Willis & Wilson 1975, 1977). The latter two objects are both Wolf-Rayet stars; HD 156385 is classified WC7, while HD 192163 is classified WN6. Circumstellar 2175 Å absorption also appears to exist in the spectrum of HD 45677 (Savage et al. 1978), a peculiar Be star with an enormous infrared excess due to circumstellar dust. It is noteworthy that the IR excess for this object resembles the emission from the dust shells surrounding carbon stars and WC9 stars. The 2175 Å feature has also been detected in the Large Magellanic Cloud (Koornneef 1978a) and possibly in a number of additional external galaxies and in quasars (Baldwin 1977, Wu 1977a,b). The feature is, thus, of universal significance, and possible explanations for its existence are given in Section 8. At present, small graphite particles must be considered the most plausible cause.

At shorter UV wavelengths the extinction curve exhibits a minimum followed by a rapid extinction rise. A broad weak local enhancement in extinction appears in the TD-1 extinction curves near 6.3 μm^{-1}, but the reality of this feature is not established (Koornneef 1978b), though we note that it is quite pronounced in the TD-1 average curves for the Orion-Monoceros region given in Nandy et al. (1976). Attempts to find diffuse features between 3600 and 1100 Å are discussed in Section 7.

In the far UV, variations in extinction from object to object can be large. An extreme case, that of σ Sco, is illustrated in Figure 1. The curve shown is an OAO-2 measurement from Bless & Savage (1972). This result has been confirmed with the spectrometer aboard the *Copernicus* satellite (Snow & York 1975). This confirmation, with a telescope having an entrance aperture so small that nebular contamination is unlikely, is significant since there was concern with the OAO-2 measurements that nebular contamination may have produced some of the abnormal UV results.

Recently, the International Ultraviolet Explorer (IUE) satellite was used to obtain extinction curves toward $\theta^1 C$ and $\theta^1 D$ Orionis (Savage & Bohlin 1978) and these new results confirm the general character of the OAO-2 curve of Bless & Savage (1972) for θ Ori (both θ^1 and θ^2) that also exhibits an abnormally low far-UV extinction. Thus, there is little doubt that large far-UV extinction variations occur for some localized galactic regions. The most convincing cases are for regions of reflection and/or emission nebulosity where the dust is situated close to hot luminous stars. Observed variations in the extinction curves, particularly for highly disturbed regions, provide significant clues about the origin of the UV extinction. In order to explain the observed shape differences it may be necessary to invoke multicomponent grain models (Bless & Savage 1972).

Investigations of the more subtle regional variations in UV extinction by different observers have lead to contradictory results. Nandy et al. (1976) divided TD-1 UV extinction observations of several hundred stars into nine groups according to their galactic positions, derived mean extinction curves for each group, and claimed that none of the curves deviated significantly from the grand average TD-1 curve illustrated in Figure 1. The maximum deviations found by Nandy et al. (1976) for these *mean* regional extinction curves, which they ascribe to errors, are illustrated with the vertical bars on some of the TD-1 data points in Figure 1. However, Koornneef (1978b) has re-analyzed the TD-1 data and concludes that definite regional extinction variations do exist.

Diffuse nebulae possibly provide information regarding the absorption properties of dust in the very far UV H- and He-ionizing portion of the spectrum (Mezger, Smith & Churchwell 1974, Panagia & Smith 1978). This suggestion is, however, quite controversial (see Brown, Lockman & Knapp 1978). Fairly elementary considerations (basically, simple photon counting) show that the He^+ and H^+ zones surrounding a hot star should be coincident if the central star is hotter than about 33,000 K (08 V). However, in Sgr B2 (Thum et al. 1978, Chaisson, Lichten & Rodriguez 1978) and in Orion (Peimbert & Torres-Peimbert 1977) the He^+/H^+ ratio is observed to be smaller than the cosmic abundance ratio (~ 0.1), indicating the ionized zones are not the same size. A reasonable explanation is that the absorption (not extinction) cross-section of the dust is about twice as large for helium-ionizing photons ($h\nu > 24.6$ eV) as for hydrogen-ionizing (12.6 eV $< h\nu < 24.6$ eV), so that the He^+ zone is reduced in size, relative to H^+, by dust absorbing its ionizing photons. Furthermore, nebular models (Balick 1975, Sarazin 1976) predict lower abundances for low stages of ionization (O^+, S^+, N^+, etc.) than are observed. The selective absorption by dust of radiation with $h\nu > 25$ eV over less energetic photons will explain this discrepancy as well. Finally, one can observe many giant H II regions in which there must be several O stars. There is a large spread in the He^+/H^+ ratios, as deduced from radio recombination lines (Panagia & Smith 1978). Emerson & Jennings (1978) show that those nebulae with the lowest He^+/H^+ ratio have the largest fraction of their ionizing photons absorbed by dust, although there are considerable observational uncertainties which make the correlation controversial. A serious objection to dust absorbing He-ionizing photons selectively is that no known substance has this property (Huffman 1977); in fact, the measured optical constants of graphite, silicates, etc., indicate that the absorption should peak at some 15 eV and then decline beyond the helium ionization edge.

1.3 AN AVERAGE EXTINCTION CURVE   Table 2 provides an average extinction curve extending from $\lambda^{-1} = 0$ to 10 μm^{-1}. This curve should be useful in correcting astronomical data for the effects of extinction. The curve represents a composite of the individual curves illustrated in Figure 1. However, an attempt was made to smoothly join the various sections. For the various regions the following measurements were adopted: for $\lambda^{-1} = 0$ to 0.5, the results of Sneden et al. (1978) and Schultz & Wiemer (1975) were averaged and $R = A_V/E(B-V)$ was assumed to be 3.1 which is consistent with a simple extrapolation of these data; for $\lambda^{-1} = 0.5$ to 2.3 the Schultz & Wiemer (1975) results were used; for $\lambda^{-1} = 2.3$ to 7 we use the Nandy et al. (1976) TD-1 data; for $\lambda^{-1} = 7$ to 9 the Code et al. (1976) OAO-2 results are employed; and for $\lambda^{-1} = 9$ to 10 the York et al. (1973) measurements for ξ Per, ζ Per, and α Cam were averaged. Anyone using this curve should always bear in mind that regional differences may be present. Also, objects may exist for which this average curve is completely inappropriate. For example, in the case of σ Sco at $\lambda^{-1} = 8$ μm^{-1} an observed value of $E(\lambda-V)/E(B-V) = 2.8$ is more realistic. But such large deviations from the average curve seem rare.

Table 2   An average interstellar extinction curve

	$\lambda(\mu m)$	$\lambda^{-1}(\mu m^{-1})$	$E(\lambda-V)/E(B-V)$	$A_\lambda/E(B-V)$
	∞	0	-3.10	0.00
L	3.4	0.29	-2.94	0.16
K	2.2	0.45	-2.72	0.38
J	1.25	0.80	-2.23	0.87
I	0.90	1.11	-1.60	1.50
R	0.70	1.43	-0.78	2.32
V	0.55	1.82	0	3.10
B	0.44	2.27	1.00	4.10
	0.40	2.50	1.30	4.40
	0.344	2.91	1.80	4.90
	0.274	3.65	3.10	6.20
	0.250	4.00	4.19	7.29
	0.240	4.17	4.90	8.00
	0.230	4.35	5.77	8.87
	0.219	4.57	6.57	9.67
	0.210	4.76	6.23	9.33
	0.200	5.00	5.52	8.62
	0.190	5.26	4.90	8.00
	0.180	5.56	4.65	7.75
	0.170	5.88	4.77	7.87
	0.160	6.25	5.02	8.12
	0.149	6.71	5.05	8.15
	0.139	7.18	5.39	8.49
	0.125	8.00	6.55	9.65
	0.118	8.50	7.45	10.55
	0.111	9.00	8.45	11.55
	0.105	9.50	9.80	12.90
	0.100	10.00	11.30	14.40

2   Distribution of Dust and the Dust-to-Gas Ratio

Dust is primarily confined to the galactic plane with an effective thickness (two scale heights) of about 200 pc. The average reddening per unit distance determined for stars in the plane at 1 kpc is 0.61 mag kpc^{-1} (Spitzer 1978). For $R = A_V/E(B-V) = 3.1$ this implies an average total extinction at the visual band of 1.9 mag kpc^{-1}. However, the distribution of the obscuring dust is very patchy. There are directions where the reddening per unit distance deviates by large factors (5 to 10 times) above or below the average. The irregular distribution of absorbing matter is clearly delineated in the extinction maps of FitzGerald (1968) and in the extinction contour plots of Lucke (1978). Lucke has updated the work of FitzGerald by determining the color excesses and photometric distances for ~ 4000 O and B stars using data from a number of recent photometric and spectroscopic catalogues. These results were then used to construct a number of contour plots of extinction that provide a three-dimensional picture of the distribution of large dust complexes in the local region of the galaxy. Perpendicular to the plane, a cosecant distribution, which is often assumed for extragalactic studies, is a poor representation of the absorption. The paper of Heiles & Jenkins (1976) is particularly useful for visualizing the distribution of interstellar matter. In this work, photographic representations of the galactic distribution of HI are provided for $|b| \geq 10°$. In addition, plots on a similar scale are given for interstellar linear polarization, the dark nebulae of Lynds (1962), the bright nebulae of Lynds (1965), and a photographic representation of the Shane & Wirtanen (1967) galaxy counts (which provides an estimate of the dust column density integrated to the edges of the galaxy). Only a few results have been reported so far on work in progress toward understanding the very local ($r < 200$ pc) distribution of dust (Strömgren 1972). For the very local material it is necessary to obtain precise estimates of $E(B-V)$ and to include A and F stars in the analysis in order to provide a reasonable sample of objects.

Extinction measurements are valuable in determining representative properties of interstellar clouds. Roughly speaking, the statistics of extinction imply the existance of ~ 6 "standard clouds" per kpc with $E(B-V)$/cloud of about 0.06, and ~ 0.8 "large clouds" per kpc with $E(B-V)$/cloud of about 0.29 (Spitzer 1978). Certainly the medium is more complex but the cloud concept provides a useful starting point for visualizing the patchy distribution of matter.

An improvement in our understanding of the association of interstellar dust and gas has been obtained through an extensive survey of interstellar atomic and molecular hydrogen with the *Copernicus* satellite (Savage et al. 1977, Bohlin, Savage & Drake 1978). These measurements are superior to the OAO-2 measurements reported by Jenkins & Savage (1974) because the high resolution of *Copernicus* permits a clear separation of the interstellar absorption lines from troublesome stellar lines. Furthermore, only *Copernicus* has the far-UV wavelength coverage that is required to observe molecular hydrogen in absorption against hot stars. Atomic hydrogen, N(HI), and the total hydrogen, $N(HI + H_2) = N(HI) + 2N(H_2)$, column densities are correlated with $E(B–V)$ in Figures 2a and b. These figures were produced by plotting linearly the data for 100 stars in Table 1 of Bohlin, Savage & Drake (1978). Be stars are denoted with the open symbols. The point for ρ Oph, which has an upward arrow attached, should be moved well off the figures to $N(HI) = 65 \times 10^{20}$ atoms cm^{-2} and $N(HI + H_2) = 72 \times 10^{20}$ atoms cm^{-2}. The correlation between dust and total hydrogen is significantly better than the correlation with atomic hydrogen. From the data shown in Figure 2b, Bohlin, Savage & Drake (1978) derive a weighted average total gas-to-color-excess ratio of

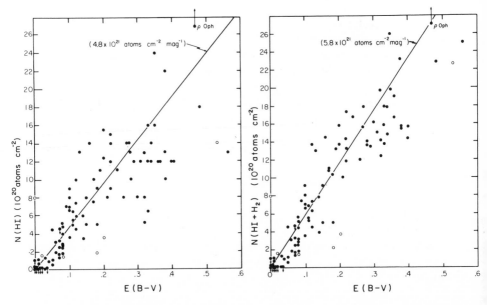

Figure 2 Correlations between gas column densities and interstellar reddening for 100 stars from the *Copernicus* atomic and molecular hydrogen survey (Savage et al. 1977, Bohlin, Savage & Drake 1978): (a) shows the atomic hydrogen column density, N(HI), versus $E(B–V)$, (b) shows the total hydrogen column density, $N(HI + H_2) = N(HI) + 2N(H_2)$, versus $E(B–V)$. Be stars are denoted with the open symbols. The solid line in (a) gives the average *atomic* hydrogen to $E(B–V)$ ratio 4.8×10^{21} atoms cm^{-2} mag^{-1}. In (b) the solid line gives the average *total* hydrogen to $E(B–V)$ ratio of 5.8×10^{21} atoms cm^{-2} mag^{-1}. The point for ρ Oph in (a) and (b) should be moved upward by about a factor of 2.7.

$\langle N(HI+H_2)/E(B-V)\rangle = \Sigma N(HI+H_2)/\Sigma E(B-V) = 5.8 \times 10^{21}$ atoms cm^{-2} mag^{-1}. This average result is illustrated with the solid line. A correction for ionization in the H II regions surrounding the early-type stars used as sources will tend to increase this number somewhat. Jenkins (1976) has estimated that the correction might amount to about 4%. A comparison of Figures 2a and b shows that the effect of molecule formation on the estimate of the total gas column density begins to become important for $E(B-V) \gtrsim 0.1$. Recent estimates of the *atomic* hydrogen-to-color-excess ratio using 21-cm data are $\langle N(HI)/E(B-V)\rangle = 5.0 \times 10^{21}$ atoms cm^{-2} mag^{-1} by Burstein & Heiles (1978) and 5.1×10^{21} atoms cm^{-2} mag^{-1} by Knapp & Kerr (1974). These values are very similar to the value obtained from the *Copernicus* results in Figure 2a for HI alone of 4.8×10^{21} cm^{-2} mag^{-1} (Bohlin, Savage & Drake 1978). Perhaps ionization could explain the small difference. For stars with accurate $E(B-V)$ the scatter in Figure 2b about the average relation is generally less than a factor of 1.5. An important exception is the dark cloud star ρ Oph with $N(HI+H_2)/E(B-V) = 15.4 \times 10^{21}$ atoms cm^{-2} mag^{-1}. A reduction in the visual reddening efficiency for the larger than normal size grains in the ρ Oph region (Carrasco, Strom & Strom 1973, Whittet & van Breda 1975) possibly explains this result. A search should be made for other objects that might exhibit an anomalous gas-to-color-excess ratio as these observations may provide useful information about the growth of grains in dense regions. The gas-to-color-excess ratio is well established, at least for the local region of the Galaxy. However, the value, 5.8×10^{21} atoms cm^{-2} mag^{-1}, should be used with caution for dense clouds. The point for ρ Oph suggests that this ratio might increase significantly for dust in dense clouds.

A total gas-to-color-excess ratio of 5.8×10^{21} atoms cm^{-2} mag^{-1} combined with current grain theories such as the theory of Mathis, Rumpl & Nordsieck (1977) implies that approximately 0.7% by mass of the interstellar material in diffuse clouds is in the form of dust. However, this number should be considered a lower limit since very large grains could contribute significantly to the grain mass but not to the extinction.

3   Light Scattering by Grains

In this section, we consider direct observations of the scattered light from dust grains in both reflection nebulae and in the diffuse galactic light. The procedure is simple: one knows (or assumes) the geometry of the light sources and of the scattering dust. The geometry is idealized into a simple form (plane-parallel uniform slabs, spheres, etc.) for which radiative transfer calculations are practical. One assumes the properties (albedo and phase function) for individual scattering from single grains, calculates the expected surface brightness from the dust distribution, and, finally, compares it to observations to determine the actual properties.

It has become traditional to calculate with the Henyey-Greenstein (1941) phase function, which is merely a convenient analytic function with two parameters, a, the albedo, and g, the average value of the cosine of the angle of scattering. Isotropic (or Rayleigh) scattering corresponds to $g = 0$, complete forward scattering to $g = 1$. The values of a and g are assumed to be arbitrary parameters to be varied until the predicted surface brightness matches the observed. Mie theory, for scattering from a sphere with a given size and index of refraction, provides analytic expressions for a and g. However, more than two parameters are required to characterize phase functions accurately. Therefore, the actual phase function of the sphere (also calculable from Mie theory) differs from the Henyey-Greenstein function of the same a and g. Besides showing complicated diffraction spikes, which are not important because integration over any reasonably broad size distribution washes them out, the actual pattern often shows a rather strong backscattering lobe not present in the Henyey-Greenstein function. This backscattering may be an artifact of the assumed spherical shape. On one hand, Zerull (1976) has experimentally measured the phase function for irregular particles. He finds fairly good agreement with Mie theory for $\theta = 180°$, but stronger scattering at intermediate angles. On the other hand, Chylek, Grams & Pinnick (1976) have also measured irregular particles and found the backscattering lobe to be virtually missing. At present, the situation is not clear. If the lobe is present, the phase function can be represented as the sum of two Henyey-Greenstein functions, one with $g < 0$ (Kattawar 1975).

We first consider the diffuse galactic light and the reflection by dark clouds, both of which are illuminated by the general stellar radiation field of the Galaxy. Next, we discuss reflection nebulae. We do not consider light scattered by dust inside H II regions (Isobe 1977). Such dust is known to be peculiar, as compared to the general interstellar medium, at least in the Orion nebula and M8.

3.1   THE DIFFUSE GALACTIC LIGHT   There have been many ground-based studies of the diffuse galactic light (DGL) (see Lillie & Witt 1976), all of which use a single Henyey-Greenstein function. Galactic models of varying complexity predict that interstellar dust is fairly efficient in scattering ($a \geq 0.4$, probably 0.7 ± 0.1). The values of a and g are not determined independently; a large value of g (~ 1) corresponds to a low value of a, and vice versa. The value of g is not well determined ($g = 0.7 \pm 0.2$), but must be large to avoid values of the albedo close to or exceeding unity. The data are not accurate enough to distinguish differences in a and g in the U, B, or V bandpasses. The physical reason why a low a must accompany a high g is that one sees readily detectable amounts of DGL only near the plane of the Galaxy, where there is a large amount of direct starlight as well. If g is large, most scattered photons have been deviated only slightly, and the scattered intensity is determined by the bright stellar background behind the dust. Conversely, with a lower g (i.e. a more isotropic phase function), the dust scattering depends more on the average value of the sky brightness, which is much lower than the value in the Galactic plane. Consequently, one needs a higher albedo to provide the same (observed) value of the DGL.

A basic problem in the visible spectral region is the faintness of the DGL in comparison to other sources of sky brightness. Averaged over the entire sky, at 5300 Å the zodiacal light, direct starlight, and DGL have brightness of 200, 100, and 35 S_{10}'s (Roach & Megill 1961), where one S_{10} is the surface brightness corresponding to one tenth-magnitude AO V star per square degree. In general, the brightness of the stars is uncertain by at least 20%. The uncertainty in the magnitude of the zodiacal light is large in comparison to the probable value of the DGL near the Galactic pole. At the Galactic equator, the the brightness of the DGL is not badly determined; about 70 S_{10} in Cygnus (along a spiral arm) at 5300 Å, and some 170 S_{10} for stars (Witt 1968). The contribution of stars within a single small area of the sky in which observations are made can be assessed by careful photometry. However, the illumination of the dust seen in that small area is from the irregularly fluctuating distribution of stars along the entire line of sight. The incident illumination is especially uncertain for dust clouds at some distance above the plane of the Galaxy, where the radiation field is expected to be quite different from the local one. Furthermore, the various dust clouds have a lower effective a and g than the individual particles in them (Mattila 1971).

In the UV spectral regions, observations are available from OAO-2, with a number of channels in the range $4250 \geq \lambda \geq 1550$ Å. With the assumption that $g = 0.7$, Lillie & Witt find that $a = 0.6$ for $\lambda = 1550$ Å, $a = 0.4 \pm 0.1$ for $1800 \leq \lambda \leq 2200$, and a increases to 0.7 ± 0.1 at $\lambda \geq 3300$. Morgan, Nandy & Thompson (1976), using another model for the radiation field, find consistent results; $a(2350$ Å$) \approx 0.7a(2740$ Å$)$, if g is constant over that range. However, there is appreciable additional uncertainty in a (perhaps several tenths) arising from the possibility of g's changing with wavelength.

The data do not require a dramatic change in either a or g across the 2200 Å bump; a seems to decline smoothly across the bump as λ decreases.

Hence, both the absorption and scattering cross-sections apparently reach a maximum there, and the bump cannot be ascribed unambiguously to either process.

The problems with UV studies of the DGL are rather different from the visual. On one hand, since the sun is comparatively faint in the UV, one has far less zodiacal light to contend with. On the other, the stellar illumination field was not known when the studies were made, and there are significant discrepancies among various observations and predictions for that field. For instance, the total radiation field at 1530 Å (DGL and stars) has been observed by Henry et al. (1977) for the northern hemisphere sky. They find very good agreement with Lillie & Witt at the Galactic equator, but obtain a mean brightness of about 60 S_{10} averaged over the sky. Lillie and Witt based their DGL analysis on a model of the stellar radiation field of Witt & Johnson (1973), which predicted a mean brightness of about 90 S_{10}. This difference indicates the difficulty of predicting the high-latitude stellar radiation field upon which the value of the DGL depends. Furthermore, the illumination is described far less well by the plane-parallel slab models for the star and dust. In fact, most of the local radiation at 1500 Å comes from rather bright stars, $m_V \lesssim 4$ (Henry 1977), which have an irregular distribution along the Gould belt. The clumping of the dust into clouds is also a much more severe problem in the UV than in the optical range.

3.2 SCATTERING FROM DUST CLOUDS   Another means of determining the optical properties of dust is from the surface brightness of dark nebulae that are reflecting radiation from the Galaxy. Mattila (1970) determined the absolute surface brightnesses of a low-latitude and of a high-latitude dark cloud. The predicted brightness of the cloud depends on the (a,g) values assumed for the grains in it, the optical depth of the cloud, and the density distribution assumed within the cloud. The values of a and g that explain the brightness of a cloud in Taurus ($b = 0°$) are similar to those from the DCL, with a increasing if g decreases. The interesting feature is that for clouds at high latitude, acceptable values of a and g increase *together*, for reasons discussed by Mattila (1970). Hence, if one assumes that dust in both high- and low-latitude clouds is intrinsically the same, the acceptable solutions cross at a steep angle in the (a,g) plane, and a and g are well determined. For 4300 Å, Mattila obtains $a = 0.65$, $g = 0.85$, similar to the DGL values. There are, however, difficulties with the method; the results are sensitive to the density distribution assumed within the cloud, which Mattila took to be uniform. Furthermore, the strong backscattering lobe of the actual dust phase function, not represented in a single Henyey-Greenstein function, possibly makes an appreciable difference.

FitzGerald, Stephens & Witt (1976) have used a different method to analyze scattering from dust clouds. They determine the brightness profile across the face of the "Thumbprint Nebula" relative to the adjacent sky brightness, so the photometry is differential. They compare the observations with predictions of a model with an r^{-2} density distribution. The nebula is 140 pc from the Galactic plane, so they rely upon a plane-parallel model of the Galaxy to provide the incident radiation field. They find that g is well determined for visible wavelengths at $g = 0.7$. The r^{-2} density distribution is suggested by star counts of the outer regions, but must be assumed within the regions for which it can not be verified. More seriously, the central optical absorption is large (17 mag at B), and the extinction appears to have $A_V/E(B-V) = 5.7 \pm 1.1$ instead of the normal value of 3.1. This large value is apparently characteristic of thick, dense clouds. Hence, one cannot confidently assume that the derived value of g applies to the normal interstellar medium.

Two reflection nebulae have recently received attention, the Merope Nebulosity in the Pleiades (Andriesse, Piersma & Witt 1977, Witt 1977) and the Orion Reflection Nebulosity (Carruthers & Opal 1977, Witt & Lillie 1978), which is the large, diffuse reflection extending over the entire constellation of Orion, containing Barnard's Loop and illuminated by the entire Orion cluster of OB stars. Unfortunately, in neither nebula is the geometry of the stars and dust very clear (Jura 1977). The wavelength variation of the scattered light, relative to the star(s), is consistent with a large contribution from an optically thick slab of dust behind the star(s), so that there is a large contribution from backscattered radiation. While no quantitative models have been fitted to the data, Andriesse, Piersma & Witt (1977), Witt (1977), and Witt & Lillie (1978) believe that the data indicate that one needs an almost isotropic ($g \leq 0.25$) phase function at 1550 Å, in marked contrast to Lillie & Witt's (1976) result of $g = 0.6$ for the DGL. This is a very important result, if true. If g is substantially lower than 0.7 at 1550 Å, which the DGL analysis of Lillie & Witt (1976) assumed, then the albedo must be appreciably increased for the reason stated at the beginning of Section 3.1. Furthermore, Lillie and Witt used a theoretical model of the radiation field that is now known to be too large, so the albedo must be increased further to compensate for the lower illumination that is actually present. Hence 0.6, the albedo claimed by Witt and Lillie at 1550 Å, is a strong lower limit to what would be necessary if g were ≤ 0.25. However, for classical scattering, i.e. in the absence of quantum mechanical effects, there is no distribution of particle sizes that can simultaneously provide both low g and high a in the UV, since low g is only obtained in the Rayleigh limit (size $\ll \lambda/2\pi$) where the albedo must be low. Thus, if $g \leq 0.25$ at 1550 Å, the scattering must be from some process other than classical scattering, such as that suggested by Platt (1956). We feel that it is important to determine to what extent the geometrical uncertainties in the reflection nebulae affect the interpretation of the observations.

In summary, in the visual region of the spectrum interstellar particles have $a \sim 0.7$ and $g \sim 0.7$. The albedo drops for $\lambda < 3000$ Å to about $a \sim 0.3$–0.4, if g remains at about 0.7. There is no evidence for dramatic changes in a or g in the vicinity of the extinction bump at 2200 Å. The phase parameter is not well determined, and variations in it would affect the derived albedo substantially. The situation for 1550 Å is quite unclear. The subject of the analysis of scattered light is controversial at present.

4   Interstellar Polarization

In this section, we discuss broad-band linear polarization, the linear polarization of narrower spectral features, and conclude with circular polarization.

4.1   LINEAR POLARIZATION   The linear polarization of stars in the $UBVR$ system is discussed by Coyne, Gehrels & Serkowski (1974) and Serkowski, Mathewson & Ford (1975). After all Wolf-Rayet and emission-line stars of luminosity classes III–V are eliminated, 364 stars remain, distributed over the entire sky. The polarization for each star reaches a maximum value, P_{max}, at some wavelength, λ_{max}, and there is a large spread in both quantities among the stars. However, a remarkable fact emerges: for all stars, the ratio $P(\lambda)/P_{max}$ is adequately fitted by a function of λ/λ_{max}: $P(\lambda)/P_{max} = \exp\left[-1.15 \ln^2 (\lambda/\lambda_{max})\right]$. The existence of such a scaling law implies that the optical constants of the polarizing material do not change with wavelength in the visual and near IR (Martin 1974), an important result that militates against graphite as the carrier of polarization, but allows dielectrics (silicates, ices, SiC, etc.). However, in the K (2.2 μm) band, Dyck & Jones (1978) show that the polarization is significantly higher than the expression would predict.

Serkowski, Mathewson & Ford (1975) find a good correlation of λ_{max} with the color ratio $E(V-K)/E(B-V)$, which was further tightened by improved IR photometry (Whittet & van Breda 1978). Table 2 and Figure 1 show that the quantity R, or $A_V/E(B-V)$, can be estimated from this color ratio by a short extrapolation from the K band (0.45 μm^{-1}) to infinite wavelength, so there is also a tight correlation of

λ_{max} with R:

$$R = (5.8 \pm 0.4) \, \lambda_{max}(\mu m),$$

where the constant of proportionality has been changed slightly from Whittet and van Breda's to reflect the extinction law of Table 2. This relation is very convenient to use for investigations of interstellar extinction, since one can substitute optical measurements of polarization for IR photometry. One must be cautious if light from the object may have intrinsic polarization (e.g. a Be star).

The average value of λ_{max} over the entire sky is 0.545 μm. However, as shown in Figure 6 of Serkowski, Mathewson & Ford (1975), there are regions (Orion, Scorpius, and Ophiuchus) where λ_{max} is significantly larger than average (up to 0.80 μm) and others (e.g. Cassiopeia and Cygnus) where it is smaller (down to 0.41 μm in the Cyg OB2 association). The regions of large λ_{max} are the denser clouds, and the large λ_{max} implies a larger than average particle size. The nearby stars also show a larger λ_{max} than average, presumably a selection effect because they must lie behind clouds of higher than average density in order to show measurable polarization.

The values of P_{max} are poorly correlated with $E(B-V)$, except that $P_{max}(\%)/E(B-V) \lesssim 9$. The upper limit of $P_{max}/E(B-V)$ presumably arises in the most favorable case of large and uniform alignment along the line of sight, and puts very severe constraints on grain alignment mechanisms.

The above discussion refers to polarization in the ordinary interstellar medium. In IR sources associated with molecular clouds, extremely strong near-IR ($\lambda > 1.6$ μm) polarization has been found, over 15% in some cases (Dyck & Beichman 1974, Dyck & Capps 1978). These objects, obscured so heavily by dust that their polarization in the optical is unknown, are in regions of recent star formation. There is a strong inverse correlation between physical size of the objects and the strength of polarization.

Linear polarization across spectral features such as the diffuse interstellar bands (see Section 7), the $\lambda2200$ bump, and the 9.7 μm silicate feature should increase if the carrier for the feature is also the aligned particle providing the continuum polarization. Measurements at $\lambda2200$ for two stars (Gehrels 1974) fit onto the standard law for P/P_{max}, indicating that the carrier of the $\lambda2200$ feature is not aligned. Clearly, higher spectral resolution UV polarimetry would be very interesting. There are two objects in which strong polarization in the 9.7 μm silicate feature has been observed, the Galactic center (Capps & Knacke 1976 and Knacke & Capps 1977), where $P(9.7 \mu m) \approx 6\%$, and the BN source associated with the Kleinmann-Low nebula in Orion, where $P(9.7~\mu m) \approx 15\%$ (Dyck & Beichman 1974). Each has a sharp increase in $P(\lambda)$ at the wavelength of the feature. This is a direct detection of the alignment of the silicate grains, and suggests that they are not responsible for the $\lambda 2200$ bump. However, it should be emphasized that the 9.7 μm and 2200 Å polarizations have been measured in quite different types of objects. In BN, the grains are known to be different, since the 3.07 μm ice absorption is seen. Possibly, the alignment mechanisms are different. It would be very interesting to detect the 2200 Å and 9.7 μm polarizations in the same object.

The polarization of the 9.7 μm feature also shows that the continuum polarization of the KL nebula and of BN is caused by absorption from aligned grains, rather than by emission. Aligned grains emit radiation with the direction of polarization along the direction in which the absorption is greatest. On the other hand, radiation suffering absorption by the same grains would be polarized along the direction in which the absorption is least, or perpendicular to the direction for emission. There is no doubt that the 9.7 μm feature is in absorption, and the direction of its polarization is the same as that of the continuum, strongly suggesting that the continuum polarization is from absorption by the grains. The polarization direction for several regions in the Kleinmann-Low nebula is the same, making it unlikely that the polarization arises from reflection by dust clouds.

Martin (1975) shows how the wavelength dependence of the polarization across the 9.7 μm feature can be used to obtain an upper limit to the band strength (the extinction cross-section at band center, per volume of grain). A low value is indicated, about two or three times lower in both BN and the Galactic center than in terrestrial or lunar rocks, and 15 times smaller than for pure minerals. This low strength is characteristic of disordered silicates (Day 1976, Krätschmer & Huffman 1979), and presumably explains why the emission feature is so broad near the Trapezium. A very high degree of grain alignment is required to explain the observations; just how much depends on the optical depth at 9.7 μm, which is not directly observable but must be inferred using assumptions regarding the geometry of the source and dust.

4.2   CIRCULAR POLARIZATION   Linearly polarized light, passing through aligned grains, can become circularly polarized because the grains introduce a phase shift in one direction relative to another. The birefringence, or amount of phase shift per volume of grain, can be calculated for grains of various shapes if the index of refraction $m(= n-ik)$ of their material is known. Martin (1972, 1974) has discussed the interpretation of circular polarization data. The most interesting feature is that its wavelength dependence is qualitatively different for a dielectric (a material with a small value of k), than for a metal (with k comparable to n). For a dielectric, such as a silicate, the circular polarization changes sign at some wavelength, λ_c. Furthermore, if $k = 0$, then $\lambda_c = \lambda_{max}$ (the wavelength at which the linear polarization is a maximum). As k increases for a given n, λ_c becomes progressively larger than λ_{max}. For metals with a wavelength-independent index, the circular polarization need not change sign; if it does, $\lambda_c \ll \lambda_{max}$. Thus, the difference of λ_c and λ_{max} is a measure of the value of k. Unfortunately, the value of k corresponding to a given value of $\lambda_c-\lambda_{max}$ depends sensitively on the value of n as well. Furthermore, if n and k are both changing with wavelength it is possible that $\lambda_c \approx \lambda_{max}$, even if k is not small. Magnetite is an example of such a substance (Shapiro 1975).

Martin & Angel (1976) observed λ_c for a variety of objects with both small and large values of λ_{max} and found $\lambda_c = \lambda_{max}$ in all cases. Graphite is definitely excluded as the carrier of polarization, and a very small value of k (as in a good dielectric such as a silicate) is indicated. Further analysis can yield information regarding the changes in grain alignment along the line of sight and provide information about the Galactic magnetic field (Martin & Campbell 1976).

5   Heavy Element Depletion

Studies of the depletion of interstellar elements from the gas phase provide an indirect but powerful method for studying the interstellar dust. The term depletion refers to the factor by which the gas phase column density of an element is below that expected on the basis of a measured total hydrogen column density and the assumption of cosmic abundances. Generally, solar abundances are used for the reference abundance since only the sun has a reliable set of abundances covering a wide range of elements. Any enrichment of the interstellar medium since the formation of the sun will introduce errors in these reference numbers. However, the solar abundance scale, with only a few exceptions, appears to be very similar to the abundances found in O and B stars and in gaseous nebulae (see Salpeter 1977 for a discussion and references). The missing gas assumed to be in the solid phase since observed molecular column densities can only explain a very small part of the measured depletions. The study of depletion has been greatly expanded by the unique observing capabilities of the Copernicus ultraviolet telescope. Since Spitzer & Jenkins (1975) and Salpeter (1977) reviewed this subject recently, we will emphasize current developments.

The accuracy of column densities of interstellar species determined from absorption line measurements depends critically on curve-of-growth effects. As a general rule, column densities for species with few very weak or very strong (damped) lines have errors up to a factor of two, while column densities for species producing lines on the flat part of the curve of growth can easily have errors amounting to factors of 10 or more (Nachman & Hobbs 1973). Therefore, in evaluating the significance of depletion results, one must carefully examine the positioning on the curve of growth of the line or lines used in the analysis. Often the errors quoted in depletion papers do not include estimates of the large systematic errors that would occur if the adopted curve of growth is inappropriate for the species being analyzed. An additional uncertainty of perhaps a factor of two is introduced when one adopts the solar reference abundance. Finally, the most extensive data have been obtained at a velocity resolution of 13 km s^{-1}. Therefore, the measurements are often an average over a number of unresolved cloud components.

5.1 DETAILED STUDIES OF A FEW CLOUDS   The most extensive studies of depletion of many elements in a limited number of clouds are for ζ Oph and ζ Pup (Morton 1975, 1978), for γ Ara (Morton & Hu 1975), and for ζ Per and o Per (Snow 1976, 1977). Figures 3a and b illustrate results for lines of sight with moderate [ζ Oph, $E(B-V) = 0.32$] and low [ζ Pup, $E(B-V) = 0.04$] reddening. Depletion is plotted versus condensation temperature, which is defined in the caption. For ζ Oph the highly refractory elements such as Fe, Ni, Ti, Ca, and Al exhibit large depletions (factors of 50 to 10^4). Na, K, Li, Mn, P, Si, and Mg exhibit moderate amounts of depletion (factors of 10 to 30), while a few elements (S and Zn) exhibit practically no depletion. For the very important C, N, and O group, the depletion appears to lie between 1 and 5. The oxygen depletion is of crucial importance for interstellar grain theories since oxygen is the most abundant element after H and He. For ζ Oph, estimates of the oxygen depletion have changed considerably over the past few years, as illustrated in Figure 3a. The 1975 results are from Morton (1975) and were based on two strong lines. Zeippen, Seaton & Morton (1977) review the ζ Oph oxygen depletion and on the basis of measurements of the weak intersystem line at 1355 Å derive a depletion ranging from about a factor of two to a factor of three. However, recently de Boer (1979) reanalyzed the Zeippen, Seaton & Morton (1977) data and concluded from curve-of-growth considerations and a profile fit that the oxygen toward ζ Oph is only depleted by about 25%. This revised depletion is consistent with the oxygen's only being tied up in mineral grains such as enstatite, olivine, and magnetite and does not require the presence of ice coatings, consistent with the absence of the 3.07 μm "ice" feature toward stars behind diffuse clouds (see Section 1.1). For nitrogen, Lugger et al. (1978) present results for ten lines of sight including ζ Oph, based on the analysis of NI lines ranging in f value from 10^{-6} to 10^{-1}. Since weak lines are included in the analysis, the results should be relatively reliable (factor of two errors) and indicate an average nitrogen depletion of a factor of two. This result, if correct, would appear to require the presence of solid NH_3, which is inconsistent with the missing 3.07 μm feature, although ultraviolet and cosmic ray processing of ice coatings has often been advocated as an explanation of the missing "ice" feature in diffuse clouds (Greenberg 1976, and references therein). However, considering the errors associated with abundance measurements, the result is also consistent with no nitrogen depletion. Unfortunately, the important carbon abundance is still very uncertain since only strong lines of C II have thus far been detected.

The correlation for reddened stars between depletion and condensation temperature was first pointed out by Field (1974). This correlation

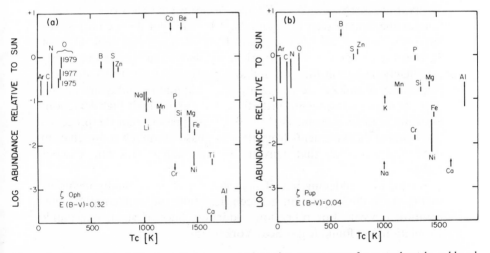

Figure 3 Elemental depletion is plotted versus condensation temperature for a moderately reddened line of sight in (*a*) and a lightly reddened line of sight in (*b*). The data are from Morton (1975) for ζ Oph in (*a*) and Morton (1978) for ζ Pup in (*b*). These results have been supplemented by more recent estimates of the oxygen abundance from Zeippen, Seaton & Morton (1977) and de Boer (1979). The nitrogen depletion plotted is from Lugger et al. (1978). The reference abundances are from Withbroe (1971) with the exception of the boron abundance which is from Boesgaard et al. (1974). Error bars are as given in the various publications. The condensation temperature is defined as the temperature at which half of a particular atom has condensed out of a gas of solar composition as it is slowly cooled but maintained in thermal and chemical equilibrium. If grains are formed in dense expanding circumstellar gas clouds under conditions of thermal and chemical equilibrium, then one might expect to see a correlation between depletion and condensation temperature (Field 1974).

possibly suggests that grain formation occurs under near equilibrium conditions in the dense environments in the vicinity of stars. However, Jura & York (1978) noted that recent equilibrium calculations suggest that P and Fe have similar condensation temperatures (1290 K versus 1335 K) but usually have very different depletions (2 to 3 for P and 30 to 100 for Fe). Also, Snow (1975) finds that with the exception of Li, Na, and K an equally good correlation exists between depletion and the first ionization potential of an element, which is perhaps related to the probability of an atom sticking to a grain's surface. Snow suggests that the noncorrelation for Li, Na, and K is possibly related to the fact that these are the only elements that form chemically saturated molecules with hydrogen; Watson & Salpeter (1972) have argued that whenever such a molecule is formed on a grain it is immediately ejected. Thus, Snow's hypothesis is that the depletion pattern is at least partly explained by the accretion of atoms by grains under nonequilibrium conditions in diffuse clouds. With such an hypothesis one would expect the depletion to increase in the densest clouds, an assertion possibly supported by observations (see Section 5.2). Whether or not we have a precise explanation for the pattern seen in Figure 3a, it is clear that these data provide important indirect information about the dust. The large depletions for Si, Mg, Fe, Cr, Ti, Ca, Ni, and Al imply that these species are nearly completely tied up in the dust and any grain theory must include them. In contrast, S and Zn are not greatly depleted and probably are an unimportant grain constituent. On the basis of the exceedingly high depletions for some of these elements (10^3 to 10^4), it is apparent that the processes that remove these elements from the gas phase are much more efficient than those processes that convert the same solids back into gaseous matter.

For the low reddening line of sight less depletion is usually recorded. In Figure 3b the correlation between depletion and condensation temperature is weak. For other low reddening lines of sight, reasonable correlations are found (e.g. λ Sco; York 1975).

5.2 STUDIES OF A LIMITED NUMBER OF ELEMENTS TOWARD MANY STARS
Figures 3a and b indicate a complex behavior in the depletion pattern from region to region in the Galaxy. Studies of this behavior can potentially provide valuable clues about the formation and destruction of dust grains. The classic Routly-Spitzer effect illustrates the kind of insights that can be gained from these studies. Routly & Spitzer (1952) noted that the Na I to Ca II column density ratio varied significantly with cloud velocity in the sense that N(Na I)/N(Ca II) is much smaller in the high velocity clouds. For a recent discussion of this effect see Siluk & Silk

(1974). It now seems well established that this abundance variation is the result of grain destruction in the high velocity clouds (Shull, York & Hobbs 1977). In this particular case Na is relatively undepleted while Ca is highly depleted and grain destruction of Ca-bearing grains returns a considerable amount of Ca to the gas phase and hence a smaller ratio of N(Na I) to N(Ca II) is observed. Similar effects have also been noted for Fe and Si when compared to S (Jenkins, Silk & Wallerstein 1976, Shull, York & Hobbs 1977). Thus, depletion studies can potentially provide information on the ease with which grains can be destroyed by interstellar shocks. For theoretical discussions of the destruction process see Shull (1977) and Barlow & Silk (1977).

Extensive survey measurements of elemental depletion toward many stars have been reported by Hobbs (1974, 1978) for Na, Ca, and K, by Stokes (1978) for Ti, and by Savage & Bohlin (1979) for Fe. Ca, Ti, and Fe exhibit large and variable depletions while Na and K exhibit at most a modest depletion with very little variability. The variability observed for Ca, Ti, and Fe amounts to about a factor of 100. However, the depletion of these three elements is reasonably well correlated (Stokes 1978, Savage & Bohlin 1979). From this one can probably conclude that grains bearing Ca, Ti, and Fe have roughly similar formation and/or destruction properties. However, detailed profile comparisons between Ca II and Ti II demonstrate that a one-to-one relation between the depletion variations is not present (Stokes 1978); hence, the Ca and Ti may be located in different grains. While the data are more uncertain, the measurements of de Boer & Lamers (1978) indicate that some of these comments for Ca, Ti, and Fe can also be extended to include Mn.

The question of whether the observed depletion depends on cloud parameters is of vital importance for establishing if accretion is a process that operates in diffuse clouds. Infrared measurements of the 3.07 μm "ice" feature in dense clouds show that accretion must be important in dense clouds and it is reasonable to expect similar processes to occur in the lower density diffuse clouds. The observational results on the dependence of depletion on cloud density are contradictory. A major difficulty is in estimating cloud density for lines of sight that contain a very inhomogeneous distribution of matter. Stokes (1978) used the column densities of Ca II and Ti II to estimate electron densities in 110 clouds, found that the electron densities were uncorrelated with depletion, and concluded that depletion is not a simple or unique function of local density. In contrast, Savage & Bohlin (1979) found that the interstellar Fe depletion correlates with several crude indicators of cloud density in the sense that the largest depletions are recorded in the densest clouds. The density indicators were average line-of-sight hydrogen density, the reddening per unit distance, and the fractional abundance of line-of-sight hydrogen in the molecular form. While the first two indicators are very approximate, an excellent correlation did hold up over nearly a three decade range in $n(HI + H_2)$ and $E(B-V)/r$. Unfortunately, the observed correlations do not necessarily imply that accretion is actually occurring in the denser regions since the correlations could equally well be explained by a destruction process that is more effective at low density.

6   Thermal Emission from Grains

Interstellar dust absorbs incident radiation and re-emits the energy in the IR. The intensity of the emitted 5–400 μm radiation provides clues as to the nature, temperature, and distribution of the dust, but the observational data require a considerably more complicated analysis than was originally believed. For instance, there is often a considerable variation in the radiation field incident upon the dust along the line of sight. Hence, some of the dust may be in a hot circumstellar shell, quite close to a star, while the rest may be in a cooler surrounding region. Furthermore, there is likely to be a wide range of grain sizes and a number of grain materials at any given point. Because of the dependence of the absorption and emission efficiencies on size, each grain size will assume a different temperature. A possible additional complication is that for very small grains (size < 0.01 μm, say), the absorption of a single photon is enough to cause a large fluctuation in the temperature of the grain (Greenberg & Hong 1974a, Purcell 1976), so that the equilibrium temperature appropriate to the size has little significance. However, Drapatz & Michel (1977) find that defects in the grain structure will increase the specific heat and greatly reduce these temperature excursions. Without defects, a 0.01 μm silicate grain in the interstellar radiation field heats to about 20 K after the absorption of a photon, and quickly cools to 4 K. Its steady-state temperature, obtained from balancing average energy input and radiation loss rates, is about 12 K. A 0.1 μm silicate grain will have smaller fluctuations and be at about 8 K. In H I regions, graphite grains of size 0.05 μm are at about 35 K. Inside H II regions, graphite grains are at about 200 K and 0.1 μm silicate grains are at about 80 K. Rather elaborate models, based on assumptions about the sizes and materials in the grains, have been necessary to interpret the observations of both molecular clouds and H II regions (Scoville & Kwan 1976, Aannestad 1978).

6.1   EMISSION FROM GASEOUS NEBULAE   There is abundant evidence that dust is intimately mixed with the ionized gas in at least some giant, ordinary, and compact H II regions (see Panagia 1977 and Mezger &

Wink 1977 for reviews). This evidence is partly based on the spatial coincidence of the free-free radio continuum and IR radiation. Furthermore, there is so much radiation in the far IR ($\lambda > 25$ μm) that some energy must be supplied to the dust by hydrogen-ionizing radiation from the star (Panagia 1977, Emerson & Jennings 1978). The IR spectra of many diverse objects are very similar (Thronson & Harper 1979). The intensity I_v peaks at about 70 μm but is broader than emission from dust with any reasonable optical properties at a single temperature. The peak intensity is lower than the Planck function intensity by substantial factors (~ 50) for optical H II regions, such as Orion and M 16, indicating that the sources are optically thin. For all sources, the emission is far brighter than the free-free emission would be, as extrapolated from the higher radio frequencies. The spectra of many objects decline in a similar way in the far-IR ($\lambda > 100$ μm) region, indicating that the grains' absorption efficiency (i.e. actual absorption cross section, divided by the geometrical cross section) decreases at least as fast as λ^{-1} or λ^{-2}. This decline is consistent either with disordered silicates (Day 1976), which are necessary to explain the shape of the 9.7 μm feature, or with graphite. It is likely that the decline of the far-IR spectrum is also consistent with emission from amorphous carbon. IC 418, an otherwise fairly typical planetary nebula, has an unusually narrow far-IR spectrum, which drops as λ^{-3} at long wavelengths (Moseley 1979).

Models (e.g. Natta & Panagia 1976) show that it is not possible to derive the total mass of the grains unless far-IR ($\lambda > 25$ μm) observations are available. Large grains, containing most of the mass, absorb only a small fraction of the stellar energy and radiate it only in the far-IR because they are cold. Thus, one must view claims of dust deficiencies in H II regions with caution if the observations do not extend to ~ 100 μm.

The rate of formation of planetary nebulae indicates that they are a significant source of mass input into the interstellar medium. Furthermore, their spectra show continuous IR radiation from heated dust, as well as some narrower emission features, which probably arise from the dust. Dust in planetary nebulae has been reviewed by Balick (1978) and Mathis (1978). New observations have been reported by Moseley (1979), Willner et al. (1978), and McCarthy, Forrest & Houck (1978). The continua peak at shorter wavelengths (~ 30 μm) than in H II regions, indicating warmer grains. As for H II regions, at long wavelengths the absorption efficiency usually drops as λ^{-1} to λ^{-2}.

The 9.7 μm silicate band is not seen in any planetary nebula. Since the C/O abundance ratio in planetaries is significantly greater than unity (Torres-Peimbert & Peimbert 1977, Shields 1978), one would expect the condensates to be SiC and graphite, or possibly amorphous carbon, rather than silicates. Thus, the lack of the 9.7 μm band in planetaries further strengthens its identification with silicates. The broad far-IR emissivity of the planetary is, therefore, presumably caused by graphite or amorphous carbon.

There are several unidentified IR features found in a wide variety of objects, including several H II regions, several planetary nebulae, the galaxies M82 and NGC 253, and some stars with dusty shells (see Merrill 1977 for a review and references). The features occur at 3.3, 6.2, 7.7, 8.7, and 11.3 μm. The 11.3 μm feature is usually narrower than the 11.2 μm emission indentified with SiC in carbon stars (Treffers & Cohen 1974). The identification of the 11.3 μm feature with carbonates is appealing, but can be ruled out by the lack of the 7 μm and 45 μm features, which should be even stronger but are not observed (McCarthy, Forrest & Houck 1978, Russell, Soifer & Willner 1978). The 3.3 μm feature, usually with a wing at 3.4 μm, is apparently continuous at a resolution of $\lambda/\Delta\lambda \sim 3000$ (H. Smith, quoted in an extensive discussion of the feature by Russell, Soifer & Merrill 1977) and is apparently not caused by molecular bands. However, the profile does appear to be different in the planetary nebulae NGC 7027, NGC 6572, and IC 418 (Willner et al. 1978), and is uncorrelated with the excitation of the nebula. IC 418 and NGC 6572 have 11.2 μm features similar to those attributed to SiC in carbon stars, but even broader; possibly the SiC grains are very large.

The Galactic plane has been detected in the 100 μm spectral region (Rouan et al. 1977, Low et al. 1977, Serra et al. 1978). There is diffuse radiation not associated with specific sources. The emission may come mainly from dust in the vicinity of hot young stars. The most direct implication is that the rate of star formation per mass of gas is larger in the inner regions of the Galaxy than it is near the sun.

6.2 EMISSION FROM STARS AND NOVAE   In this section we briefly consider stars that are expelling dust into the interstellar medium. Cool stars are discussed in an excellent review by Merrill (1977), and many of the observations are in Forrest, Gillett & Stein (1975) and Merrill & Stein (1976a,b,c). Briefly, M stars (O/C > 1) show the 9.7 μm silicate band in either emission or absorption, depending upon the extent of the circumstellar dust shell. When a large emission feature is observed, the profile is very similar to the emission found in the Orion Nebula near the Trapezium. The 18 μm silicate band is also in absorption in some cases (Forrest et al. 1978), but is partially filled in by emission. Models (Jones & Merrill 1976) are required for a detailed interpretation, but there is little doubt that M stars are injecting silicate grains into the interstellar medium.

By contrast, carbon stars do not show the 9.7 μm emission feature, but rather 10–12 μm emission of SiC (Treffers & Cohen 1974). This emission is much wider than the unidentified 11.3 μm feature of planetary nebulae. Carbon stars also show a featureless component with a broad far-IR emissivity, presumably from graphite or possibly amorphous carbon.

Most early-type stars do not show IR excesses that clearly indicate grain emission. A number of exceptions (Allen 1973) include several Herbig Ae/Be stars and some peculiar Ae, Be objects. An extreme case is HD 45677, which exhibits a featureless IR emission spectrum resembling that found for carbon stars. The measurements of the 2200 Å feature for this star were discussed in Section 1.2. A number of WC7–WC9 stars (Gehrz & Hackwell 1974, Cohen, Barlow & Kuhi 1975) show the broad,

Table 3 Spectroscopic features produced by interstellar and/or circumstellar dust

Feature	Where seen[a]	Where not seen[a,b]	Common explanation
Rapid extinction rise for $\lambda < 1300$ Å	DC	—	small interstellar grains
2200 Å feature	DC,CS	—	graphite
Diffuse interstellar features[c]	DC	CS	—
Very broad structure in the visual	DC	—	magnetite?
3.07 μm	MC	DC,CSC,CSO,DN	H_2O and/or NH_3 ice
9.7 μm (sometimes with 18 μm)	DC,MC,CSO,DN,C	N,CSC,PN	silicates
11.2 μm	CSC	—	silicon carbide
Unidentified IR features at 3.3, 3.4, 6.2, 7.7, 8.7, and 11.3 μm[c]	CS,DN	DC	—

[a] C   comets
CS   circumstellar shells (composition unknown)
CSC circumstellar shells (carbon-rich)
CSO circumstellar shells (oxygen-rich)
DN   diffuse nebulae
DC   diffuse clouds
MC   molecular clouds
N   novae (presumably carbon-rich)
PN   planetary nebulae (carbon-rich)
[b] Entries in this column imply that the feature or features have been looked for but not detected.
[c] Some of these may have a molecular origin.

featureless emission expected from graphite, with no trace of silicate or silicon carbide features.

Some novae also form dust grains. References are given and the subject is reviewed by Gallagher & Starrfield (1978). The grain formation is related to the speed class, with the slow novae being favored. There is no 9.7 μm silicate emission. Theoretical models for novae suggest that $C/O > 1$ in order to produce the explosion. The implication, then, is that novae are producing carbon grains.

Table 3 summarizes the objects in which various features are found, as well as the objects in which they have not been found after a careful search. The table points out that the 9.7 μm silicate feature has been seen in several comets (see Ney 1977 for a review). Snow (1973) notes that the diffuse interstellar features are not formed in circumstellar dust shells. References to the other entries are contained in the text.

7   Diffuse Interstellar Features

The diffuse interstellar features (sometimes called diffuse interstellar bands) represent one of the longest standing unsolved problems in all of spectroscopy. This is unfortunate since these features probably contain a multitude of clues about the composition and physical nature of interstellar dust. While it is not established that these features have a solid-state origin, it is worth noting that the diffuse features are seen toward stars behind diffuse interstellar clouds, and thus far only simple diatomic molecules have been detected along such lines of sight. Therefore, explanations that involve polyatomic molecules would appear untenable. However, for arguments favoring a molecular origin see Smith, Snow & York (1977).

The diffuse features are numerous. Herbig (1975), in his extensive and important investigation covering the wavelength region 4400–6850 Å, detected 39 features of certain or probable interstellar origin with full widths at half intensity ranging from about 1 Å to about 20 Å. Recently, the study of these features has been extended to longer and shorter wavelengths. Sanner, Snell & van den Bout (1978) present results for several additional features between 6500 and 8900 Å. Snow, York & Resnick (1977) possibly detected an ultraviolet feature at 1416 Å in a search between 1114 and 1450 Å. However, Savage (1975) finds no evidence for diffuse features between 2000 and 3600 Å with the exception of the 2200 Å feature.

A compilation of diffuse feature measurements is provided by Snow, York & Welty (1977), where measurements for four features are collected and reduced statistically to a common measurement system. Much of the effort in studying the diffuse features has gone into investigating the correlations of diffuse feature strengths with various interstellar quantities such as $E(B-V)$ and other reddening indicators. The principal results of these correlation studies are summarized in articles by Herbig (1975), Smith, Snow & York (1977), Schmidt (1978), and Sneden et al. (1978). Correlation studies may have been overemphasized and studies that might provide more useful *direct* physical information about the features have been neglected.

Van de Hulst (1957) first discussed the asymmetric profiles expected for the diffuse features if they are produced by scattering and absorption in interstellar grains. The degree of asymmetry depends on the particle optical constants and size. Therefore, profile measurements can potentially provide diagnostic information on the sizes of the diffuse feature carriers. For more recent theoretical discussions of diffuse feature profiles, see Bromage (1972), Savage (1976), and Purcell & Shapiro (1977). The early attempts to measure diffuse feature profiles concentrated on the broad features in the blue and the results were often contradictory. The most modern measurements for $\lambda 4430$ yield symmetric profiles (Danks & Lambert 1975, Martin & Angel 1975). For the broad features, reliable profile measurements are difficult to make because of the problem of establishing an accurate continuum level over a large wavelength span. This problem is further aggravated by the fact that numerous stellar lines occur in the general vicinity of the strongest broad diffuse features such as the feature near 4430 Å. Higher precision profile measurements can be obtained for the narrower features in the red. This is because these features are deeper and the continuum placement is easier. However, for the very narrow features, Doppler smearing due to the existence of multiple clouds can present a problem. High resolution profiles for features at 5780, 5797, 6379, and 6614 Å have been reported by Savage (1976), Danks & Lambert (1976), and Welter & Savage (1977). Asymmetries were found for the features at 5780, 5797, and 6614 Å. However, the feature at 6379 Å was symmetrical. From a detailed analysis of the observed profiles for the features at 5780, 6379, and 6614 Å, Savage (1976) and Welter & Savage (1977) conclude that a solid-state origin is the most likely explanation for these three features. It is also noted that the processes of autoionization and preionization, at least in the simple cases, could be ruled out since the profiles do not appear to have the required broad Lorentzian wings. An extension of this profile work using detectors capable of producing spectra with high signal to noise appears desirable. Perhaps new studies could provide a convincing demonstration of intrinsic profile variability from star to star which could more readily be explained by theories involving interstellar solids than by molecular absorption.

Studies of the polarization characteristics across the diffuse features are also an important way of establishing their origin (Martin & Angel 1974, Greenberg & Hong 1974b). High quality measurements now exist for the features at 4430, 5780, and 6284 Å (Martin & Angel 1974, 1975, Fahlman & Walker 1975) and the results all imply no polarization variations across the bands. These studies imply that the three features are not formed in those grains that produce the visible interstellar polarization. While this seems to favor the molecular explanation, the absence of a large polarization change across the 2200 Å feature (see Section 4) implies that nonpolarizing grain components do exist in the interstellar medium.

The ultimate identification of the diffuse features will probably come through laboratory investigations of the spectra and solids at low temperature. Unfortunately, very little laboratory work on plausible solids is currently being pursued.

8   Composition of Grains

In this section, we first discuss a possible picture of the composition and origin of interstellar grains, somewhat along the lines of Field (1978). We then discuss some of the problems associated with this picture and some alternative ideas. The growth of grains has been reviewed by Salpeter (1977) and will not be considered here.

As we discussed in Section 6, it is widely accepted that silicates are being injected by oxygen-rich M stars into the interstellar medium. Carbon-rich objects (primarily planetary nebulae, but also carbon stars and presumably novae) are also losing mass and injecting grains into the interstellar medium, probably both SiC and graphite or amorphous carbon. Small uncoated graphite grains have an extinction bump near 2200 Å, although the exact wavelength at which it occurs depends on the condition of the material and upon the shape and size of the particles. Graphite can also be abundant enough to produce the bump, while silicates can not if the absorption is classical. If this picture is right, the lack of polarization of the bump (which needs further observational confirmation) indicates that small graphite particles do not polarize, probably because they are not aligned. Similarly, the wavelength dependences of the circular and linear optical polarization militate against moderate-sized graphite particles as producers of the polarization and suggest that a dielectric, such as a silicate, is responsible. Furthermore, the 9.7 μm silicate feature is highly polarized in some objects, clearly indicating that silicates can be aligned. A power-law particle size distribution ($n[a] \propto a^{-3.5}$) of un-coated graphite and silicates produces a very good fit to the observed extinction (Mathis, Rumpl & Nordsieck 1977), while other substances

(SiC, magnetite, or iron) in combination with graphite are not ruled out. There have been other similar suggestions of uncoated mixtures (e.g. Gilra 1971).

The presence of extensive coatings, particularly H_2O, in the diffuse interstellar clouds has important consequences for the properties of grains. As discussed in Section 6, the smallest grains (<0.01 μm, say) are probably too warm to acquire coatings. Larger particles, responsible for the visible-UV extinction, might become ice-coated. Shortly after the first UV depletion studies, the fundamental question was posed by Greenberg (1974): if interstellar nitrogen and oxygen abundances are substantially depleted relative to solar abundances, where have all their atoms gone? The only sufficiently abundant substance that can combine with the oxygen is hydrogen, so if oxygen is depleted, water ice coatings must exist on at least some grains. However, the 3.07 μm H_2O band is seen only in very dense molecular clouds, which are quite unlike the diffuse clouds in which the O and N gaseous abundance studies in the UV are made. To relieve this paradox, Greenberg (1976, and references therein) suggests that radiation converts the H_2O molecules into free radicals of OH, which have an absorption band at 2.7 μm. This is the wavelength of water vapor absorption and is, therefore, unobservable even from aircraft and balloon altitudes. However, there is another possible solution to the "mystery of the missing oxygen." As was discussed in Section 5.1, the most recent (de Boer 1979) studies of the oxygen depletion for ζ Oph allow, but do not require, an oxygen depletion consistent with silicates only.

Satellite measurements of the possible 2.7 μm OH feature would settle the question of whether extensive H_2O mantles, processed to free OH radicals, occur outside of molecular clouds. The uncertainties in the nitrogen depletion have already been discussed (Section 5.1) and are probably consistent with either little or most of the nitrogen's being in coatings.

Since we should confine ourselves to the observed properties of grains, we will only very briefly mention some of the many suggestions which have been made regarding the nature of interstellar particles and the origins of the $\lambda2200$ feature. Graphite is certainly not the only means of producing the $\lambda2200$ bump. The type of bonds occurring in graphite occurs in other long-chain carbon molecules, which might be formed on the surfaces, resulting in an oily, tarlike material (see Salpeter 1977, p. 289). Duley & Millar (1978), from consideration of the details of the depletions of various elements, feel that grains consist of diatomic oxides of magnesium, iron, silicon, etc., the surfaces of which contain unsaturated O^{-2} ions (Duley 1976) which produce the $\lambda2200$ bump. Andriesse (1977) suggests that the bump arises from the plasma oscillations of electrons in very small dielectric particles ($\lesssim 30$ Å, or $\lesssim 1000$ atoms in each), while in this theory the 9.7 μm feature arises from the plasma oscillations of the ions in particles of sizes $\lesssim 0.3\,\mu$m. The extinction at other wavelengths arises from off-resonance absorption of both ions and electrons. There have been many other suggestions. A few include lattice defects in small crystals (Drapatz & Michel 1976), and small silicate grains, which would apparently require an abundance of silicon greater than the solar value (Huffman & Stapp 1971).

ACKNOWLEDGMENTS

We are grateful to a number of colleagues for helpful comments about the manuscript, especially R. C. Bohlin, K. S. de Boer, D. A. Harper, J. Koornneef, K. M. Merrill, and A. N. Witt. John S. Mathis acknowledges support from NSF and Blair D. Savage from NASA.

Literature Cited

Aannestad, P. A. 1978. *Ap. J.* 220:538
Aannestad, P. A., Purcell, E. M. 1973. *Ann. Rev. Astron. Astrophys.* 11:309
Allen, D. A. 1973. *MNRAS* 161:145
Andriesse, C. D. 1977. *Vistas Astron.* 21:107
Andriesse, C. D., Piersma, T. R., Witt, A. N. 1977. *Astron. Astrophys.* 54:841
Baldwin, J. A. 1977. *MNRAS* 178:67
Balick, B. 1975. *Ap. J.* 201:705
Balick, B. 1978. In *Planetary Nebulae, Theory and Observations. IAU Symp. 76,* ed. Y. Terzian, p. 275. Dordrecht: Reidel
Barlow, M. J., Silk, J. 1977. *Ap. J. Lett.* 211:L83
Becklin, E. E., Matthews, K., Neugebauer, G., Willner, S. P. 1978. *Ap. J.* 220:831
Becklin, E. E., Neugebauer, G. 1967. *Ap. J.* 147:799
Bless, R. C., Savage, B. D. 1972. *Ap. J.* 171:293
Boesgaard, A. M., Praderie, F., Leckrone, D. S., Faraggiana, R., Hack, M. 1974. *Ap. J. Lett.* 194:L143
Bohlin, R. C., Savage, B. D., Drake, J. F. 1978. *Ap. J.* 224:132
Bromage, G. E. 1972. *Astrophys. Space Sci.* 15:426
Brown, R. L., Lockman, F. J., Knapp, G. R. 1978. *Ann. Rev. Astron. Astrophys.* 16:445
Burstein, D., Heiles, C. 1978. *Ap. J.* 225:40
Capps, R. W., Knacke, R. F. 1976. *Ap. J.* 210:76
Carrasco, L., Strom, S. E., Strom, K. M. 1973. *Ap. J.* 182:95
Carruthers, G. R., Opal, C. B. 1977. *Ap. J. Lett.* 212:L27

Chaisson, E. J., Lichten, S. M., Rodriguez, L. F. 1978. *Ap. J.* 221:810
Chylek, P., Grams, G. W., Pinnick, R. G. 1976. *Science* 193:480
Code, A. D., Davis, J., Bless, R. C., Hanbury Brown, R. 1976. *Ap. J.* 203:417
Cohen, M., Barlow, M. J., Kuhi, L. V. 1975. *Astron. Astrophys.* 40:291
Coyne, G. V., Gehrels, T., Serkowski, K. 1974. *Astron. J.* 79:581
Danks, A. C., Lambert, D. L. 1975. *Astron. Astrophys.* 41:455
Danks, A. C., Lambert, D. L. 1976. *MNRAS* 174:571
Day, K. L. 1976. *Ap. J.* 210:614
de Boer, K. S. 1979. *Ap. J.* 229:132
de Boer, K. S., Lamers, H. J. G. L. M. 1978. *Astron. Astrophys.* 69:327
de Jong, T. 1976. *Mem. Soc. Astron. Italiana* 45:189
Dorschner, J., Friedemann, C., Gürtler, J. 1977. *Astron. Astrophys.* 58:201
Drapatz, S., Michel, K. W. 1976. *Mitt. Astron. Ges.* 40:187
Drapatz, S., Michel, K. W. 1977. *Astron. Astrophys.* 56:353
Duley, W. W. 1976. *Astrophys. Space Sci.* 45:253
Duley, W. W., Millar, T. J. 1978. *Ap. J.* 220:124
Dyck, H. M., Beichman, C. A. 1974. *Ap. J.* 194:57
Dyck, H. M., Capps, R. W. 1978. *Ap. J. Lett.* 220:L49
Dyck, H. M., Jones, T. J. 1978. *Astron. J.* 83:594

Egan, W. G., Hilgeman, T. 1978. *Nature* 273:369

Elsässer, H., Fechtig, H., eds. 1976. *Interplanetary Dust and Zodiacal Light, IAU Colloq. 31.* Berlin: Springer. 493 pp.

Emerson, J. P., Jennings, R. E. 1978. *Astron. Astrophys.* 69:129

Fahlman, G. G., Walker, G. A. H. 1975. *Ap. J.* 200:22

Field, G. B. 1974. *Ap. J.* 187:453

Field, G. B. 1978. In *Planetary Nebulae, Theory and Observations, IAU Symp.* 76, ed. Y. Terzian, p. 367. Dordrecht: Reidel

Field, G. B., Cameron, A. G. W., eds. 1975. *The Dusty Universe.* New York: Neale Watson. 323 pp.

FitzGerald, M. P. 1968. *Astron. J.* 73:983

FitzGerald, M. P., Stephens, T. C., Witt, A. N. 1976. *Ap. J.* 208:709

Forrest, W. J., Gillett, F. C., Houck, J. R., McCarthy, J. F., Merrill, K. M., Pipher, J. L., Puetter, R. C., Russell, R. W., Soifer, B. T., Willner, S. P. 1978. *Ap. J.* 219:114

Forrest, W. J., Gillett, F. C., Stein, W. A. 1975. *Ap. J.* 195:423

Forrest, W. J., Houck, J. R., Reed, R. A. 1976. *Ap. J. Lett.* 208:L133

Forrest, W. J., Soifer, B. T. 1976. *Ap. J. Lett.* 208:L129

Frogel, J. A., Persson, S. E., Aaronson, M. 1977. *Ap. J.* 213:723

Gallagher, J. S., Starrfield, S. 1978. *Ann. Rev. Astron. Astrophys.* 16:171

Gehrels, T. 1974. *Astron. J.* 79:590

Gehrz, R. D., Hackwell, J. A. 1974. *Ap. J.* 194:619

Gillett, F. C., Forrest, W. J. 1973. *Ap. J.* 179:483

Gillett, F. C., Forrest, W. J., Merrill, K. M., Capps, R. W., Soifer, B. T. 1975a. *Ap. J.* 200:609

Gillett, F. C., Jones, T. W., Merrill, K. M., Stein, W. A. 1975b. *Astron. Astrophys.* 45:77

Gilra, D. P. 1971. *Nature* 220:237

Greenberg, J. M. 1974. *Ap. J. Lett.* 189:L81

Greenberg, J. M. 1976. *Astrophys. Space Sci.* 39:9

Greenberg, J. M. 1978. In *Cosmic Dust*, ed. J. A. M. McDonnell, p. 187. New York: Wiley

Greenberg, J. M., Hong, S. S. 1974a. In *Galactic Radio Astronomy, IAU Symp. 60,* ed. F. J. Kerr, S. C. Simonson III, p. 155. Dordrecht: Reidel

Greenberg, J. M., Hong, S. S. 1974b. In *Planets, Stars and Nebulae Studied with Photopolarimetry*, ed. T. Gehrels. p. 916. Tucson: Univ. Arizona Press

Greenberg, J. M., van de Hulst, H. C., eds. 1973. *Interstellar Dust and Related Topics, IAU Symp. 52.* Dordrecht: Reidel

Hayes, D. S., Greenberg, J. M., Mavko, G. E., Radick, R. R., Rex, K. H. 1973. In *Interstellar Dust and Related Topics, IAU Symp. 52*, ed. J. M. Greenberg, H. C. van de Hulst, p. 83. Dordrecht: Reidel

Heiles, C., Jenkins, E. B. 1976. *Astron. Astrophys.* 46:333

Henry, R. C. 1977. *Ap. J. Suppl.* 33:451

Henry, R. C., Swandic, J. R., Schulman, S. D., Fritz, G. 1977. *Ap. J.* 212:707

Henyey, L. G., Greenstein, J. L. 1941. *Ap. J.* 93:70

Herbig, G. H. 1975. *Ap. J.* 196:129

Hobbs, L. M. 1974. *Ap. J.* 191:381

Hobbs, L. M. 1978. *Ap. J. Suppl.* 38:129

Hoyle, F., Wickramasinghe, N. C. 1977. *Nature* 268:610

Hudson, H. S., Soifer, B. T. 1976. *Ap. J.* 206:100

Huffman, D. R. 1977. *Adv. Phys.* 26:129

Huffman, D. R., Stapp, J. L. 1971. *Nature Phys. Sci.* 229:45

Isobe, S. 1977. In *Topics in Interstellar Matter*, ed. H. van Woerden, p. 61. Dordrecht: Reidel

Jenkins, E. B. 1976. In *The Structure and Content of the Galaxy and Galactic Gamma Rays*, ed. C. E. Fichtel, F. W. Stecker, p. 239. Greenbelt: NASA

Jenkins, E. B., Savage, B. D. 1974. *Ap. J.* 187:243

Jenkins, E. B., Silk, J., Wallerstein, G. 1976. *Ap. J. Suppl.* 32:681

Johnson, H. L. 1968. In *Nebulae and Interstellar Matter*, ed. B. M. Middlehurst, L. H. Aller, p. 167. Chicago: Univ. Chicago Press

Johnson, H. L. 1977. *Rev. Mex. Astron. Astrof.* 2:175

Jones, T. W., Merrill, K. M. 1976. *Ap. J.* 209:509

Jura, M. 1977. *Ap. J.* 218:749

Jura, M., York, D. G. 1978. *Ap. J.* 219:861

Kattawar, G. W. 1975. *J. Quant. Spectrosc. Radiat. Transfer* 15:839

Knacke, R. F., Capps, R. W. 1977. *Ap. J.* 216:271

Knapp, G. R., Kerr, F. J. 1974. *Astron. Astrophys.* 35:361

Koornneef, J. 1978a. *Astron. Astrophys.* 64:179

Koornneef, J. 1978b. *Astron. Astrophys.* 68:139

Krätschmer, W., Huffman, D. R. 1979. Preprint

Lillie, C. F., Witt, A. N. 1976. *Ap. J.* 208:64

Low, F. J., Kurtz, R. F., Poteet, W. M., Nishimura, T. 1977. *Ap. J. Lett.* 214:L115

Lucke, P. B. 1978. *Astron. Astrophys.* 64:367

Lugger, P. M., York, D. G., Blanchard, T., Morton, D. C. 1978. *Ap. J.* 224:1059

Lynds, B. T. 1962. *Ap. J. Suppl.* 7:1

Lynds, B. T. 1965. *Ap. J. Suppl.* 12:163
Martin, P. G. 1972. *MNRAS* 159:179
Martin, P. G. 1974. *Ap. J.* 187:461
Martin, P. G. 1975. *Ap. J.* 201:373
Martin, P. G., Angel, J. R. P. 1974. *Ap. J.* 188:517
Martin, P. G., Angel, J. R. P. 1975. *Ap. J.* 195:379
Martin, P. G., Angel, J. R. P. 1976. *Ap. J.* 207:126
Martin, P. G., Campbell, B. 1976. *Ap. J.* 208:727
Mathis, J. S. 1978. In *Planetary Nebulae, Theory and Observations, IAU Symp. 76*, ed. Y. Terzian, p. 281. Dordrecht: Reidel
Mathis, J. S., Rumpl, W., Nordsieck, K. H. 1977. *Ap. J.* 217:425
Mattila, K. 1970. *Astron. Astrophys.* 9:53
Mattila, K. 1971. *Astron. Astrophys.* 15:292
McCarthy, J. F., Forrest, W. J., Houck, J. R. 1978. *Ap. J.* 224:109
McDonnell, J. A. M., ed. 1978. *Cosmic Dust.* New York: Wiley. 693 pp.
McMillan, R. S. 1978. *Ap. J.* 225:417
Merrill, K. M. 1977. In *The Interaction of Variable Stars with Their Environment, IAU Colloq. 42*, ed. R. Kippenhahn, J. Rahe, W. Strohmeier, p. 446. Bamberg: Remeis-Sternwarte
Merrill, K. M., Russell, R. W., Soifer, B. T. 1976. *Ap. J.* 207:763
Merrill, K. M., Stein, W. A. 1976a. *Publ. Astron. Soc. Pac.* 88:285
Merrill, K. M., Stein, W. A. 1976b. *Publ. Astron. Soc. Pac.* 88:294
Merrill, K. M., Stein, W. A. 1976c. *Publ. Astron. Soc. Pac.* 88:874
Mezger, P. G., Smith, L. F., Churchwell, E. B. 1974. *Astron. Astrophys.* 32:269
Mezger, P. G., Wink, J. E. 1977. *Infrared and Submillimeter Astronomy*, ed. G. G. Fazio, p. 55. Dordrecht: Reidel
Morgan, D. H., Nandy, K., Thompson, G. I. 1976. *MNRAS* 177:531
Morton, D. C. 1975. *Ap. J.* 197:85
Morton, D. C. 1978. *Ap. J.* 222:863
Morton, D. C., Hu, E. M. 1975. *Ap. J.* 202:638
Moseley, H. 1979. *Ap. J.* In press
Nachman, P., Hobbs, L. M. 1973. *Ap. J.* 182:481
Nandy, K., Thompson, G. I., Jamar, C., Monfils, A., Wilson, R. 1975. *Astron. Astrophys.* 44:195
Nandy, K., Thompson, G. I., Jamar, C., Monfils, A., Wilson, R. 1976. *Astron. Astrophys.* 51:63
Natta, A., Panagia, N. 1976. *Astron. Astrophys.* 50:191
Neugebauer, G., Becklin, E. E., Matthews, K., Wynn-Williams, C. G. 1978. *Ap. J.* 220:149

Ney, E. P. 1977. *Science* 195:541
Ney, E. P., Strecker, D. W., Gehrz, R. D. 1973. *Ap. J.* 180:807
Oort, J. H. 1977. *Ann. Rev. Astron. Astrophys.* 15:295
Panagia, N. 1977. In *Infrared and Submillimeter Astronomy*, ed. G. G. Fazio, p. 43. Dordrecht: Reidel
Panagia, N., Smith, L. F. 1978. *Astron. Astrophys.* 62:277
Peimbert, M., Torres-Peimbert, S. 1977. *MNRAS* 179:217
Penston, M. V., Hunter, J. K., O'Neill, A. 1975. *MNRAS* 171:219
Persson, S., Frogel, J. A., Aaronson, M. 1976. *Ap. J.* 208:753
Platt, J. R. 1956. *Ap. J.* 123:486
Purcell, E. M. 1976. *Ap. J.* 206:685
Purcell, E. M., Shapiro, P. R. 1977. *Ap. J.* 214:92
Roach, F. E., Megill, L. R. 1961. *Ap. J.* 133:228
Rouan, D., Lena, P. J., Puget, J. L., de Boer, K. S., Wijnbergen, J. J. 1977. *Ap. J. Lett.* 213:L35
Routly, P. M., Spitzer, L. 1952. *Ap. J.* 115:227
Russell, R. W., Soifer, B. T., Merrill, K. M. 1977. *Ap. J.* 213:66
Russell, R. W., Soifer, B. T., Willner, S. P. 1978. *Ap. J.* 220:568
Salpeter, E. E. 1977. *Ann. Rev. Astron. Astrophys.* 15:267
Sanner, F., Snell, R., van den Bout, P. 1978. *Ap. J.* 226:460
Sarazin, C. L. 1976. *Ap. J.* 208:323
Savage, B. D. 1975. *Ap. J.* 199:92
Savage, B. D. 1976. *Ap. J.* 205:122
Savage, B. D., Bohlin, R. C. 1978. *Bull. Am. Astron. Soc.* 10:445
Savage, B. D., Bohlin, R. C. 1979. *Ap. J.* 229:136
Savage, B. D., Bohlin, R. C., Drake, J. F., Budich, W. 1977. *Ap. J.* 216:291
Savage, B. D., Wesselius, P. R., Swings, J. P., Thé, P. S. 1978. *Ap. J.* 224:149
Schalén, C. 1975. *Astron. Astrophys.* 42:251
Schiffer, F. H., Mathis, J. S. 1974. *Ap. J.* 194:597
Schild, R. E. 1977. *Astron. J.* 82:337
Schmidt, E. G. 1978. *Ap. J.* 223:458
Schultz, G. V., Wiemer, W. 1975. *Astron. Astrophys.* 43:133
Scoville, N. Z., Kwan, J. 1976. *Ap. J.* 206:718
Serkowski, K., Mathewson, D. S., Ford, V. L. 1975. *Ap. J.* 196:261
Serra, G., Puget, J. L., Ryter, C. E., Wijnbergen, J. J. 1978. *Ap. J. Lett.* 222:L21
Shane, C. D., Wirtanen, C. A. 1967. *Lick Obs. Publ.* 22:1
Shapiro, P. R. 1975. *Ap. J.* 201:151
Shields, G. A. 1978. *Ap. J.* 219:559

Shull, J. M. 1977. *Ap. J.* 215:805
Shull, J. M., York, D. G., Hobbs, L. M. 1977. *Ap. J. Lett.* 211:L139
Siluk, R. S., Silk, J. 1974. *Ap. J.* 192:51
Smith, W. H., Snow, T. P., York, D. G. 1977. *Ap. J.* 218:124
Sneden, C., Gehrz, R. D., Hackwell, J. A., York, D. G., Snow, T. P. 1978. *Ap. J.* 223:168
Snow, T. P. 1973. *Publ. Astron. Soc. Pac.* 85:590
Snow, T. P. 1975. *Ap. J. Lett.* 202:L87
Snow, T. P. 1976. *Ap. J.* 204:759
Snow, T. P. 1977. *Ap. J.* 216:724
Snow, T. P., York, D. G. 1975. *Astrophys. Space Sci.* 34:19
Snow, T. P., York, D. G., Resnick, M. 1977. *Publ. Astron. Soc. Pac.* 89:758
Snow, T. P., York, D. G., Welty, D. E. 1977. *Astron. J.* 82:113
Soifer, B. T., Russell, R. W., Merrill, K. M. 1976. *Ap. J. Lett.* 207:L83
Spitzer, L. 1978. *Physical Processes in the Interstellar Medium.* New York: Wiley. 318 pp.
Spitzer, L., Jenkins, E. B. 1975. *Ann. Rev. Astron. Astrophys.* 13:133
Stein, W. A. 1975. *Publ. Astron. Soc. Pac.* 87:5
Stokes, G. M. 1978. *Ap. J. Suppl.* 36:115
Strömgren, B. 1972. *Q. J. R. Astron. Soc.* 13:153
Torres-Peimbert, S., Peimbert, M. 1977. *Rev. Mex. Astron. Astrof.* 2:181
Thronson, H., Harper, D. A. 1979. *Ap. J.* In press
Thum, C., Mezger, P. G., Pankonin, V., Schraml, J. 1978. *Astron. Astrophys.* 64:L17
Treffers, R., Cohen, M. 1974. *Ap. J.* 188:545
van de Hulst, H. C. 1957. *Light Scattering by Small Particles.* New York: Wiley. 470 pp.
van Duinen, J. R., Wu, C.-C., Kester, D. 1976. *Department of Space Research Groningen Internal Note,* ROG NR 76-4
Watson, W. D., Salpeter, E. E. 1972. *Ap. J.* 174:321
Welter, G. L., Savage, B. D. 1977. *Ap. J.* 215:788

Wesson, P. S. 1974. *Space Sci. Rev.* 15:469
Whiteoak, J. B. 1966. *Ap. J.* 144:305
Whittet, D. C. B., van Breda, I. G. 1975. *Astrophys. Space Sci.* 38:L3
Whittet, D. C. B., van Breda, I. G. 1978. *Astron. Astrophys.* 66:57
Whittet, D. C. B., van Breda, I. G., Glass, I. S. 1976. *MNRAS* 177:625
Wickramasinghe, N. C., Morgan, D. H., eds. 1976. *Solid State Astrophysics, Astrophys. Space Sci. Library Vol. 55.* Dordrecht: Reidel. 314 pp. Also in *Astrophys. Space Sci.* 34:1 (1975)
Wickramasinghe, N. C., Nandy, K. 1972. *Rep. Prog. Phys.* 35:157
Willis, A. J., Wilson, R. 1975. *Astron. Astrophys.* 44:205
Willis, A. J., Wilson, R. 1977. *Astron. Astrophys.* 59:133
Willner, S. P. 1976. *Ap. J.* 206:728
Willner, S. P. 1977. *Ap. J.* 214:706
Willner, S. P., Jones, B., Puetter, R. C., Russell, R. W., Soifer, B. T. 1978. Preprint
Withbroe, G. L. 1971. In *The Menzel Symposium, NBS Spec. Pub. 353,* ed. K. B. Gebbie. Washington: NBS
Witt, A. N. 1968. *Ap. J.* 152:59
Witt, A. N. 1977. *Publ. Astron. Soc. Pac.* 89:750
Witt, A. N., Johnson, M. W. 1973. *Ap. J.* 181:363
Witt, A. N., Lillie, C. F. 1978. *Ap. J.* 222:909
Wu, C.-C. 1977a. *Ap. J. Lett.* 217:L117
Wu, C.-C. 1977b. *Bull. Am. Astron. Soc.* 9:296
Wynn-Williams, C. G., Becklin, E. E. 1974. *Publ. Astron. Soc. Pac.* 86:5
York, D. G. 1975. *Ap. J. Lett.* 196:L103
York, D. G., Drake, J. F., Jenkins, E. B., Morton, D. C., Rogerson, J. B., Spitzer, L. 1973. *Ap. J. Lett.* 182:L1
Zeippen, C. J., Seaton, M. J., Morton, D. C. 1977. *MNRAS* 181:527
Zerull, R. 1976. In *Interplanetary Dust and Zodiacal Light, IAU Colloq. 31,* ed. H. Elsasser, H. Fechtig, p. 130. Berlin: Springer

Ann. Rev. Astron. Astrophys. 1979. 17:113–34
Copyright © 1979 by Annual Reviews Inc. All rights reserved

COMPUTER IMAGE PROCESSING

✕2147

Ronald N. Bracewell

Radio Astronomy Institute, Stanford University, Stanford, California 94305

INTRODUCTION

With the advent of modern electronic control systems and modern computers have come big changes in the handling of astronomical observations. At the time when *Ground-Based Astronomy: A Ten Year Program* (Whitford 1964) was prepared the coming revolution in automatic observing was already apparent but at that time it could be said that "only a few optical telescopes are even partially automated, either in their basic operation or for final data reduction." Development of new automatic optical instruments was seen as urgent especially in contrast to the situation in radio astronomy. Automation of observatories and instruments then lagged behind availability of computers. By 1972 with the appearance of *Astronomy and Astrophysics for the 1970's* (Greenstein 1972) the revolution was taken for granted. There was no chapter on automation. Instead the astronomical community was asking for a Very Large Array, an extraordinarily ambitious automatic image-forming instrument with arcsecond resolution, a 400- to 600-inch telescope with advanced television sensors, automatic controls for setting and guiding and computers for immediate data reduction, and a continuing list of instruments for other fields of astrophysics all reflecting advances in recent technologies—electronic cameras, solid state electronics, low temperature devices, X-ray imaging and others.

Floods of data created by these developments required unprecedented attention to information processing as witnessed by a previous contribution to these Annual Reviews (Clark 1970). Astronomy has long been familiar with the handling of great masses of data as in the distinguished catalogues of stars and galaxies and in the compilation of solar and stellar spectra. And the performance of intricate numerical analysis on a large scale is also part of the tradition of astronomy as in the working

113

0066-4146/79/0915-0113$01.00

out of the theory of the moon and planets and in the preparation of ephemerides. The change that occurred was in the rate of creation of new observations and the need to reduce them promptly.

Because information processing in astronomy has now become too broad a field to review as a whole, this chapter restricts itself to computer image processing, and in particular to radio images, Fourier synthesis, restoration, speckle interferometry, and X-ray imaging.

RADIO IMAGE FORMATION

When sizable parabolic radar antennas came into use in the 1940s it was apparent that an electromagnetic image forms in the focal plane and that the information available there is mostly wasted. The method of collecting the echo power was, and still is, to pipe the power in one picture element away to a receiver. Then by scanning the antenna one might consecutively build up an image that in principle could be extracted all at once, as in photography. It is as though an optical telescope equipped with one photosensitive cell were to be used to construct the image of an extended field. Simultaneous extraction, however, was not achieved easily.

Image formation, or mapping, by simple scanning by a paraboloid as introduced by Reber (1944) is still practiced today but pencil beam arrays, which were introduced (Mills et al. 1958) for higher resolution, soon led to automation. At microwavelengths, resolution better than 1' was soon achieved (Swarup et al. 1963) by the introduction of a technique for internal monitoring of the phase paths to be traversed by the faint signals (Bracewell & Swarup 1961, Swarup & Yang 1961). Descendants of this innovation include closed servo loops for stabilizing the phase paths and transmission systems of high stability requiring only occasional monitoring. With the tendency of high resolution to produce more data came the first telescope that automatically produced data in publishable form. Daily digital solar images for the eleven years 1962–1973 published in the traditional way, but available also in computer readable punched card or magnetic tape format (Graf & Bracewell 1975), combined sophisticated developments in high-resolution antennas with currently available computation.

A further step toward simultaneous imaging was the introduction of earth-rotation synthesis in 1953 (Christiansen & Warburton 1955). When transportable antenna elements were used, earth-rotation synthesis was called supersynthesis (Ryle 1962). As first practiced, this technique economized on capital construction costs by substituting increased observing time. In a sense, then, this was a step away from rapid imaging but the drive for high-resolution images under cost constraint then led to the fast interferometer incorporating the minimum redundancy array (Bracewell 1966, Bracewell et al. 1973). This instrument proves to be the simultaneous image-forming radio-telescope, its multiple receivers performing the function of multiple detectors that could be imagined in the focal plane of a paraboloid but cannot readily be fitted in there.

Suppose that each element of a two-dimensional array of antennas closely packed on an area A delivers its signal to a signal divider with N^2 identical outputs, and that N^2 different outputs, one from each antenna, are combined at a central point having suffered equal time delays to reach that point. The resultant delivered is the same as would be found at the focus of a great paraboloid occupying the area A because the geometrical property of the paraboloid is to introduce equal time delays between passage through the aperture plane and arrival at the focus. The delays are free-space delays rather than transmission-line delays. The field that would be found at an off-axis picture element in the focal plane is the same as would be delivered by a signal combiner accepting signals from each antenna through transmission lines of unequal time delays. The time delays would be such as to cancel the time delay to each antenna associated with the tilted plane wavefront arriving from the direction in space corresponding to the off-axis element. Thus, the set of N^2 combinations that can be formed provide simultaneously the focal plane radio image that otherwise has to be explored sequentially.

The minimum redundancy principle now permits the omission of all pairs of elements having the same vector spacing because the complex visibility (Bracewell 1958a) measured with a two-antenna interferometer is dependent, in the absence of parallax, only on the vector spacing of the elements. When redundant elements are removed, leaving a total of M, the combination signals no longer constitute an image but they contain all the information that the image would contain and the image can be computed. Prior to the introduction of the concept of complex visibility the term visibility (of an interferogram) meant a contrast ratio, as it still does in optical interferometry. The spatial phase of an astronomical interferogram was (and optically still is) difficult to measure (Labeyrie 1976) but the generalization to include phase permits the Fourier transformation relationship with the image to be enunciated and explains how the transmission system may be reduced to the state where each antenna signal is divided $M-1$ ways, transmitted to a central point, and fed to $M(M-1)/2$ multipliers. As the earth rotates the product signal rises and falls sinusoidally with an amplitude and phase representing the desired complex visibility. Zernicke's complex degree of coherence, which is a property of the field regardless of the existence of interference fringes, can be deduced from the complex visibility of an interference pattern given the antenna gains, transmission-line losses, line lengths, and other parameters of the instrument whose interaction with the field generates the observable interference. Line dispersion, which affects visibility but not coherence, can normally be made negligible by delay compensation appropriate to the bandwidth and source direction.

If now we go further and do not require the array to exhibit minimum redundancy at any one time, the sensitivity diagram in the Fourier transform plane (Bracewell 1961) accumulates as the earth rotates (Swenson & Mathur 1968). Various configurations of elements for

Table 1 Earth rotation aperture synthesis telescopes

Location	Elements	Element beam	Frequency (MHz)	Max baseline (meters)	Synthesized beam	References[a]
Cambridge,	2	100'	1400	730	47"	1
England	3	175'	400	1550	80"	2
		50'	1400		23"	
		14'	5000		7"	
	8	6'	5000	4560	2"	3
Greenbank,	3	39'	1400	2700	16"	4
West Virginia[b]		20'	2700		8"	
		7'	8100		3"	
Stanford, California	5	7'	11000	206	19"	5
Fleurs, Australia[c]	68	100'	1400	800	40"	6
Big Pine, California[d]	3	77'	600	1080	16"	
		33'	1400		7"	
		27'	1700		6"	
		9'	5000		2"	
Penticton, British						
Columbia	4	105'	1400	600	60"	7
Ooty, India	4	—	330	3500	48" × 330"	8
Hat Creek,						
California	2	11'	20000	300	9"	9
Westerbork, The	12	83'	600	1600	55"	10
Netherlands		36'	1400		23"	
		11'	5000		7"	

[a] Key to references: 1. Baldwin et al. (1970). 2. Ryle (1962). 3. Ryle (1972). 4. Hogg et al. (1969). 5. Bracewell et al. (1973). 6. Christiansen (1973). 7. Roger et al. (1973). 8. Swarup & Bagri (1973). 9. Hills et al. (1973). 10. Baars et al. (1973).
[b] Recently a 14-meter dish has been included in the system which can be used to provide baselines of 11 and 35 km. The maximum resolution of this system is about 1.2 at 1400 MHz.
[c] This instrument consists of both an east–west and a north–south synthesis telescope, each containing 34 elements.
[d] This instrument has both east–west and north–south synthesis capabilities, and has an additional 40-meter dish.

implementing earth-rotation synthesis are presented in Table 1 (Brouw 1975).

To this table should now be added the Very Large Array under construction by the National Radio Astronomy Observatory in New Mexico (Heeschen 1975). This ambitious instrument, which was proposed more than a decade ago (National Radio Astronomy Observatory 1967), is already in partial use pending completion in 1981. Ultimately it will comprise 27 paraboloids, each 25 m in diameter and transportable along three radiating arms 21 km long. At the highest frequency of operation, about 15 GHz, the resolution will be approximately one arcsecond. Consequently the data handling will be prodigious.

Mathematical discussions of aperture synthesis with references to earlier sources have been given by Brouw (1975) and Cole (1977). Aperture synthesis was first conceived by McCready, Pawsey & Payne-Scott (1946) who asserted that "it is possible in principle to determine the actual form of the distribution in a complex case by Fourier synthesis using information derived from a large number of components." They gave the cosine integral for a symmetrical one-dimensional source but their formula does not apply to an arbitrary one-dimensional distribution as they thought. Interferometer phase measurement, as ultimately placed on a firm basis by internal phase calibration, remained elusive for many years, but by the time it was achieved the notion of complex visibility was available and permitted the two-dimensional formulation of the Fourier transform relation in its general form, which does apply to arbitrary distributions. However, there is a discrepancy between the rigorous Fourier transform relation (Bracewell 1958a) and what is needed in practice in earth-rotation synthesis because the measured complex visibilities do not lie in the (u,v)-plane, or Fourier domain, in terms of which the basic relation is stated. In practice the ensemble of vector displacements between pairs of antennas thereby occupies a three-dimensional volume rather than occupying a plane. To invert the data thus involves first estimating what the complex visibilities would have been on some reference plane passing through the volume. To understand the problem we make use of the known property that the delay-dependent correlation function $\hat{\Gamma}(x_A, x_B, \tau)$ obeys Maxwell's wave equation (Wolf 1955) and the same is true of the complex degree of coherence, which has been shown to be equal to the complex visibility $\mathcal{V}_{AB}$ measured by an interferometer, under certain conditions. Thus if the complex visibility were given on a reference plane, the visibility at other points could be computed by solving Maxwell's equations. For example, the visibility would propagate in such a way that on a parallel plane the visibility would be the Fresnel diffraction pattern of the given pattern. As the speed of propagation of features in a Fresnel diffraction field is not the same as the speed of light (and in fact is greater) it is not precisely correct to assume that

$$\mathscr{V}_{AB} = \mathscr{V}_{AC} \exp{(i\,2\pi\,d/\lambda)}$$

where B and C are adjacent points on the same ray separated by distance d and λ is the free-space wavelength. The equation also implies that $|\mathscr{V}_{AB}| = |\mathscr{V}_{AC}|$ which also can only be an approximation. As precision becomes of more importance in radio astronomical image construction it will be possible to handle this difficulty rigorously and to assess the adequacy of approximate methods. One approach would be to use the above formula, as in current practice, to correct the visibility phases at points distributed through a volume to the corresponding projected points on a reference plane, then carry out the Fresnel diffraction computation for the distributed points, determine the discrepancies, and make a first-order correction by a further application of the approximate formula. This topic has been discussed in terms of three-dimensional Fourier transforms, erroneously in some cases, by a number of authors but requires further work along the lines indicated above.

A quite unique synthesis instrument, not dependent on rotation synthesis, was introduced by Wild (1961, 1965) and permits sun mapping at 80 MHz in one second or less. Because of the high intensity of solar emission, progressive adjustment of the phase paths to the antennas can be carried out in a short time and still result in an adequate signal-to-noise ratio. A 3.8′ beam can be formed (Wild 1967) and scanned over the sun, without the need for visibility measurement and subsequent Fourier transformation, and movies can be obtained showing transient phenomena as they develop (Wild & Smerd 1972). Wild's instrument is an intimately computerized one. Not only is the ouput image computed but the antenna phase paths are computed and computer-controlled and the beam is constructed by computer combination of diffraction patterns formed sequentially by azimuthal alteration of the antenna phases in various ways. Experience with the data processing (Labrum et al. 1975) has shown that "preliminary data reduction by on-line computers will become more and more necessary to avoid the problem of storing, and subsequently processing, unmanageable quantities of raw data."

IMPLEMENTATION OF FOURIER SYNTHESIS

The Fourier Transform Relation

For a vector displacement (u,v) in a reference plane (which may for example be taken perpendicular to the nominal direction of the source whose image is desired) the complex visibility $\mathcal{V}(u,v)$ due to a source brightness distribution $B(l,m)$ where l and m are direction cosines is given by

$$\mathcal{V}(u,v) = \int \int \exp\left[-i\,2\pi(lu+mv)\right] B(l,m)dl\,dm \Big/ \int \int B(l,m)dl\,dm$$

under certain conditions. The quantities u and v are distances measured in wavelengths and therefore dimensionless. In practice, as we have seen, we have to deal with vector displacements (u,v,w) in three-dimensional space and the measurements $\mathcal{V}(u,v,w)$ obtained will be distributed through a volume in a systematic way, but by no means tending to fill the volume. Suppose they are distributed on a sheet so that w is a single-valued function of u and v. The values will be given by

$$\mathcal{V}(u,v,w) = \int \int \exp\left[-i\,2\pi(lu+mv+nw)\right] B(l,m)dl\,dm$$

where n is the third direction cosine, $n = (1-l^2-m^2)^{1/2}$. If the factor $\exp(-i\,2\pi\,nw)$ does not vary significantly over the cone of directions occupied by the source we may take the factor outside the integral and invert to obtain

$$B(l,m) = \int \int \exp\left[i\,2\pi(lu+mv)\right] \mathcal{V}_0(u,v)du\,dv$$

where $\mathcal{V}_0(u,v) = \mathcal{V}(u,v,w)\exp(i\,2\pi\,nw)$. However, the larger high-resolution arrays are encountering significant image distortion because $\exp(-i\,2\pi\,nw)$ does vary over the synthesized field.

In addition to this there are other problems with the inversion of the data. To discuss these it is sufficient to work in the (u,v)-plane as if an array at the north pole were looking at a source near the celestial pole. In this case

$$B(l,m) = \int \int \exp\left[i\,2\pi(lu+mv)\right] \mathcal{V}(u,v)du\,dv.$$

It is understood that the integral cannot be extended beyond a finite boundary in the (u,v)-plane and for that reason the image that is formed is likely to be deficient in the higher spatial frequencies associated with the larger vector spacings that have not been reached. As the hope of getting something for nothing springs eternal, it is not surprising to learn that extrapolation is being vigorously practiced and that controversy has resulted as discussed in a later section.

Area Density Compensation

The next important problem is to evaluate an integral over a region where the integrand is known only semidiscretely on a set of loci. The fact that one coordinate is continuous as a result of the continuous flow of time is of no great significance because discretization in time is necessary in order to form the discrete data sets required by the digital computer. If the data came on the intersections of a rectangular lattice the situation would be favorable for computing. As is well known, the fast Fourier transform algorithm, which has proved most efficient in the handling of large images, accepts and indeed demands a rectangular array as input. But the measurements are distributed irregularly on a set of points (u_j,v_j). Introduce a *sampling function* $S(u,v)$ comprising J spikes at the sampling points as defined by

$$S(u,v) = \sum_{j}^{J/2} \left[{}^2\delta(u - u_j, v - v_j) + {}^2\delta(u + u_j, v + v_j) \right].$$

The two-dimensional delta function ${}^2\delta(u,v)$ is defined in terms of the Dirac delta function by $\delta(u)\delta(v)$. The available information is equivalent to that contained in the product of $S(u,v)$ with $\mathscr{V}(u,v)$ because values of $\mathscr{V}$ between the sampling points are lost and the values at the sampling points are retained. The product $S(u,v)\,\mathscr{V}(u,v)$ (regarded as a generalized function) possesses a Fourier transform which in some sense is an estimate of $B(l,m)$ and is given, using a double asterisk for two-dimensional con-volution, by

$$s(l,m) ** B(l,m),$$

where $s(l,m)$ is the two-dimensional Fourier transform of the sampling function $S(u,v)$. If the sampling points were distributed on a square grid it would only remain to smooth out the effects of the sharp cutoff at the boundary of the occupied area of the (u,v)-plane. But the irregular pattern of sampling points will result in a variable sampling density that will distort the image if uncompensated. Thompson & Bracewell (1974) describe a *principal transfer function* $S(u,v)/\rho_d(u,v)$ that incorporates such correction and depends on a concept of area density $\rho_d(u,v)$ of sampling points. The Fourier transform of the principal transfer function is the *principal response pattern* $p(l,m)$; it is what we would get as an image if a point source in the direction $(l,m) = (0,0)$ were processed in this way. Ordinarily one would like to normalize so that $S(0,0)/\rho_d(0,0) = 1$, but here we leave the normalization open to allow the possibility that $S(0,0)$ may be zero as will happen if there is no sample at $(u,v) = (0,0)$.

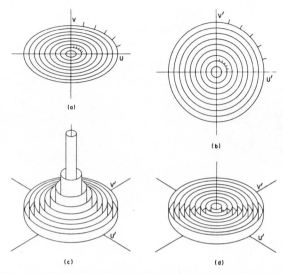

Figure 1 (*a*) The sensitive loci on the (*u*,*v*)-plane for a five-element east-west array. The interval between time marks is one hour. (*b*) The same loci on the (*u′*,*v′*)-plane, where they become circles and the time marks become equally spaced. (*c*) The transfer function $S(u,v)$ which arises from simple superposition of the data. (*d*) The principal transfer function showing compensation for area density. Reproduced from *Ap. J.* 182:80.

Even if there is a sample at (0,0) ordinary normalization still breaks down.

The principal transfer function and principal response pattern have been discussed in a basic case that applies to most synthesis instruments (Bracewell & Thompson 1973). Figure 1 shows how the sampling points on the (*u*,*v*)-plane merge to form concentric ellipses for a five-element minimum redundancy east-west array and how a simple coordinate transformation converts the loci to concentric circles. Figure 1*c* shows the sampling function $S(u,v)$ where in this case the two-dimensional delta functions are shown merged into ring deltas. The excess and undesirable sampling density on the inner circles is apparent. Figure 1*d* illustrates the compensating effect of a choice of $\rho_a(u,v)$ appropriate to this case. At this point no moderating of the boundary cutoff has been carried out; the sensitivity over the plane has merely been compensated so as to level out the transfer function to the degree that the geometry of the loci permits.

The Ringlobe Phenomenon

Associated with the transfer function of Figure 1*d* is the principal response pattern $p(l,m)$ of Figure 2. The sharp central beam is surrounded

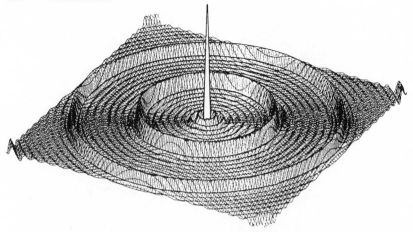

Figure 2   The principal response pattern of the five-element east-west rotation-synthesis array for which the principal transfer function is shown in Figure 1*d*. Reproduced from *Ap. J.* 182:82.

by the conspicuous ringlobes that have become familiar in rotation synthesis images. Performing the transformation, which because of the circular symmetry of the primed coordinates reduces to a Hankel transformation, we have

$$p(l',m') = \sum_{n=1}^{N} 2\pi n Q' J_0[2\pi n Q'(l'^2 + m'^2)^{1/2}]$$

where the number of circles is N and their spacing is Q'. It is possible to decompose this expression into a main beam

$$P_{\text{main}}(l',m') = 4(N+1/2)^2 \, Q' \text{ jinc } [2(N+1/2)Q'(l'^2+m'^2)^{1/2}],$$

where jinc $\xi = [J_1(\pi\xi)]/2\xi$, which has precisely the form of the Airy diffraction pattern of a circular aperture of radius $(N+1/2)Q'$, plus an infinite series of ringlobes whose radii are integral multiples of $1/Q'$. Rather curiously, the ringlobes have an asymmetrical profile which, if scanned across a source, removes the spatial frequencies that are beyond cutoff and takes the half-order derivative of what is left. This results in a net bias of background brightness on the inner side of the ringlobe, an effect which is very noticeable in optical diffraction patterns obtained experimentally (Tichenor & Bracewell 1973) and is beginning to be taken into account in radio astronomy (van de Hulst 1978).

Interpolation

If the fast Fourier transform is to be utilized it is necessary to interpolate the complex visibility values onto a rectangular array, a procedure that is likely to require more computer time than the Fourier transformation and therefore warrants careful attention. Interpolation is a well-studied branch of mathematics but several programmers have abandoned classical interpolation procedures in favor of special methods. The study of the artifacts produced is still under way.

Cell summing Mathur (1969) assigns to each array point a value equal to the sum of the visibilities measured at the sampling points contained within the cell centered on the array point. Unjustified emphasis is thus given at a point where two measurements happen to have been made in the surrounding cell as compared with a point where only one measurement was made. No compensation is made for area density.

To overcome this objection, *cell averaging* was introduced (Hogg et al. 1969) for use with the Green Bank variable baseline interferometer and was contemplated for the Very Large Array (National Radio Astronomy Observatory 1967). Cell averaging also possesses objectionable features. It is not an interpolation procedure, because the cell average, regarded as a function of continuous u and v, is a surface that does not pass through the sample values. The cell average is not a smooth function in the sense that Lagrange and spline interpolation strive to be smooth because it can dip sharply to zero wherever there is room for an empty cell between samples. If there is one sample in the cell, it is shifted from its own location to the cell center thus producing an error proportional to the visibility gradient.

These are serious objections. If the cell size is reduced the gradient errors are reduced but the greater number of empty cells will result in spurious detail being introduced by the more frequent dips to zero. There is also a curious distortion apparent in published maps caused by an unforeseen dependence on the orientation of the sampling loci with respect to the array axes.

Gaussian convolution Brouw (1971, 1975) assigns Gaussian weights to samples in the neighborhood of an array point, the less weight the farther away the sample. Experience indicates that a suitable choice of the width of the Gaussian function and including the samples within the area of about 15 cells allows the interpolation time to be made comparable with the transformation time. After Fourier transformation a factor is applied that ranges from unity at the center to four at the edge of the synthesized field. The results of this procedure have been satisfactory, judging by the appearance of the images, but there should be scope for improved precision for the following reasons. If the number of sampling points falling in the 15-cell area is above average the value assigned at the array point is incorrectly emphasized. Gaussian gridding is not a true interpolation procedure because, at a sampling point, it does not assign the value that was measured there. A possible remedy has been discussed (Bracewell 1979).

Another interpolation procedure that has proved satisfactory in use is described by Kenderdine (1974). It is clear that implementation of the Fourier transform relation in earth-rotation synthesis has developed empirically and that new approaches are going to be tested empirically. One such approach would be to start with a standard interpolation procedure that does at least put a smooth surface through the sample values and to cut corners progressively on the computing until there is a balance between image fidelity and computer economy.

When the interpolation has been done, suitable (u,v)-plane tapering will be introduced by the user to smooth out the effects of the boundary cutoff to the extent deemed appropriate.

RESTORATION

Map Cleaning and Maximum Entropy

If a tapering factor is introduced certain undesirable effects occur. Compact sources are presented with increased width and if there are double sources there will be a loss of resolution. The peak brightness, whose absolute value may be of astrophysical significance, will be biased toward a low value. On the other hand, tapering can suppress the spurious oscillations associated with the principal solution of antenna smoothing integral equation. If such oscillations occur against a faint background the resultant image brightness can go negative. An error that produces negative brightness has often been felt to be more serious than an error of the same amount in the presence of a brighter background where the net brightness does not go negative. And yet, the error in peak brightness that is introduced by tapering may be of more astrophysical importance than an equal error in a dark area where negative brightness is produced. Choice and use of tapering factors are clearly delicate matters dependent upon the kind of information that is to be extracted.

Tapering can be regarded as a sacrifice of information. Extrapolation of the visibility to (u,v)-values beyond the boundary reached by observation would seem to be the antithesis of tapering. It can be argued that what was not observed cannot be guessed and yet a case can be made for extrapolation as an alternative method for smoothing out the cliff at the boundary. For example, when radio astronomical sources were only partly resolved one could model the source as say two elliptical components at such and such an orientation and spacing and determine the model parameters by fitting to the visibility data. In effect, one had extrapolated and, if the source model was good, the results could be satisfactory. Prior information about the source could lead to a good model and could be incorporated in this way. This is an established and accepted approach and while it need not be thought of as extrapolation we can take the Fourier transform of the proposed image and see exactly what extrapolation is implied.

Now the method CLEAN due to Högbom (1974), which frees compact features from grating lobes and sidelobes and separates features that were previously confused, is not presented as extrapolation but in fact it does extrapolate. So also does the method of maximum entropy (Smyllie et al. 1973, Ables 1974).

Unlike the evaluation of model parameters, CLEAN is an automatic procedure not requiring the injection of prior source information going beyond the visibilities on which the image is based. Consequently there has been some doubt as to the significance of the improved maps especially where one is not dealing with widely spaced discrete sources on a dark field. It would seem that this question could be settled empirically if mathematical analysis is too difficult but there is a logical flaw in those computer experiments where an artificial map is degraded by computer then subjected to CLEAN. It would be better if CLEAN were applied by one investigator to a degraded map supplied by an adversary interested in defeating CLEAN and if the original artificial map were unseen by the other party. As things are, we have innumerable examples of excellent restorations carried out by a person who knew the answer and had the opportunity of selecting the cases to be exhibited. CLEAN has come into general radio astronomical use, gives pleasing results, and, as the limitations come to be better understood, is likely to see wider application.

Maximum entropy methods have been slower to come into use partly because of greater difficulty of finding implementations in two dimensions but an elegant development has occurred (Wernecke & D'Addario 1977, Gull & Daniel 1978) which introduces a Lagrange multiplier technique to produce a smooth image that is compatible with rather than identical to the data. In addition a certain mystery surrounds the logarithmic integrand which has been chosen one way by some authors (Frieden 1972) and differently by others (Ponsonby 1973). The two definitions are

$$\int \int \log B(l,m) dl\, dm \quad \text{and} \quad -\int \int B(l,m) \log B(l,m) dl\, dm.$$

Both definitions place heavy emphasis on brightness values near zero, where the absolute value of the logarithm is large. Where the image brightness falls to zero, as it may at the outer edges or even in the interior of the field, the logarithm cannot be taken. Therefore, an arbitrary boundary has to be established at the five percent or some other contour. But of course this contour is not known in advance. And a lot of entropy lies around the low brightness fringe so the choice of the boundary is not a trifle. Likewise, if the map to be improved contains observational errors, as it must, it is quite legitimate for negative brightness to occur where the true brightness is comparable with the error level. To treat such negative values as zero would be philosophically bothersome because of the bias introduced. Maximum entropy algorithms in action, however, must actually assign an arbitrary positive value in places where negatives occur because they cannot take the logarithm of zero. These considerations tend to detract from the elegance of the logarithm and in a way have higher priority than the discussions of when to use log B and when B log B (Ponsonby 1979). A recent international meeting concerned with astronomical image construction received contributions on maximum entropy relating the method to Bayesian estimation (Spencer & Cornwell 1979), showing that in general the algorithm cannot be regarded as "safe" (Dainty et al. 1979, Bhandari 1979) and discussing other methods (Baker 1979, Subramanya 1979).

Defective Data

Because of interruption of observations, sectors of the (u,v)-plane may be missing. The consequences of such a defect have been examined in detail. One approach is to use the method of radial interpolation described by Thompson & Bracewell (1974), which is rigorous, and then to interpolate azimuthally across the missing sector by standard methods. The result will be faithful across the narrow end of the gap but at the wide end there will be a loss of those components with high azimuthal spatial frequencies. Wernecke & D'Addario (1977) tested maximum entropy on a special case with encouraging results but a comparison with interpolation has not been made.

Missing regions also arise with Very Long Baseline Interferometry but in this situation there is also another important defect, namely incomplete phase information. Building on an idea due to Jennison (1958) methods of constraining the image by use of the closure phase have been introduced (Wilkinson & Readhead 1979, Cotton & Wittels 1979) and *ad hoc* attacks on obvious systematic errors have often been made as in a recent example of Hamaker (1979).

Restoration in the Presence of Noise

When the importance of the error spectrum in restoration was first recognized (Bracewell 1958b) the determination of the error spectrum and even more the cross-correlation with signal, which was also shown to be significant, was beyond the computational capacity of the time. Now, however, it would be perfectly feasible to determine the statistics of errors under operating conditions. In some cases it would suffice to record outputs in the absence of a source. In other cases where the presence of a source contributes statistical fluctuations one could observe the same source repeatedly and subtract the mean to obtain an ensemble of noise images. The spatial autocorrelation function of the noise image, or equivalently its two-dimensional spectrum, is one of the quantities that appears in the simple theory. The cross-correlation between the noise image and the wanted image is not necessarily zero and also appears.

It is almost certain that error images, if obtained routinely for examination, would reveal systematic features going beyond the assumptions of randomness that are made in the theory and that diagnostic insight would be gained. As the precision permitted by other components of radio telescopes improves it is certain that more attention will be given to determination of the statistics of the random errors and to the systematic errors.

More work remains to be done on the theory, which is incomplete at present. For one thing, theoretical results applicable to stationary random processes with Gaussian distribution, that is, to infinite ensembles possessing statistics given in advance, do not apply directly here. Where signal and noise are given separately, as in Wiener filter theory, one obtains formulas that do not contain the observed quantity which we have available as our starting point. Our point of departure is one image, a combination of an unknown but definite "true image" or signal with one sample from an ensemble of noise images. Thus we do not start with ensemble statistics of either signal or noise.

SPECKLE INTERFEROMETRY

In a Michelson stellar interferometer, where for example a mask with two well-spaced holes is placed over the objective lens of a refracting telescope, it is well known that the observer can see interference fringes in the eyepiece. This was often done by double-star observers in order to gain a little extra resolution in the determination of position angles and, in modified form, led to the first stellar angular diameter measurements. Now, if a photographic plate is used instead of the eye, the interference pattern may disappear because time variation in the structure of the atmosphere causes the fringes to run back and forth under their envelope, which is the Airy disk of the hole. The fringe separation is inversely proportional to the hole spacing so that there will be about as many fringes under the envelope as there are hole diameters in the hole spacing.

A telescope without a mask exhibits in its focal plane a superposition of such phenomena, the role of the hole being played by patches of good wavefront whose dimensions are of the order of 10 cm. A patch of wavefront that is reasonably flat will produce in the focal plane a disturbance inversely proportional in extent to the patch size. In terms of the Airy disk of the telescope aperture there will be about as many Airy disks under the envelope of the disturbance due to a single patch as there are patches in the telescope aperture. Superposition of the disturbances from all the patches in the aperture results in a pattern of cancellation and reinforcement described as a speckle pattern. As the patch structure of the wavefront changes with time the speckle pattern shifts and if a time exposure is made on a photographic plate the blackening will be more or less uniform under the envelope. The size of the blackened area determines the resolution under the prevailing atmospheric seeing conditions and for large telescopes will be much larger than the Airy disk corresponding to the full aperture.

The idea behind speckle interferometry (Labeyrie 1970) was to extract the information that is in a brief exposure but is lost in a long exposure. It is true that a brief exposure is faint and that the image is a scramble of randomly arranged speckles. But if the source is a double star *each* speckle becomes a speckle-pair. Since the speckle size is connected with the Airy disk of the full aperture, the resolution now corresponds to the full aperture width rather than to the patch size as in a long exposure. If the intensity were high enough the high-resolution information might be discernible by eye but two special tricks enable high-resolution images to be extracted from an ensemble of photographs provided the exposures are short enough to catch the speckle pattern.

If the autocorrelation of a single speckle photograph is calculated the effect will be to enhance the signal-to-noise ratio by the square root of the number of speckles. This step provides just what is needed in order to superimpose numerous exposures which, if superimposed directly, would merely give the degraded image of a long exposure. Since autocorrelation functions are centrosymmetric $[f(x,y) = f(-x,-y)]$, they may be accumulated with an increase in signal-to-noise ratio equal to the square root of the number of correlograms.

These two steps could be performed by analogue methods but are greatly facilitated by digital computation. However, it is not necessary to follow the procedure described. An alternative is to compute the two-dimensional Fourier transform of the speckle image, accumulate the squared moduli of the transforms, and finally to retransform. This is precisely equivalent and, in the digital case, the choice is entirely one of computational ease.

Optical processing of image-tube speckle photographs taken with the

Hale telescope is described by Gezari et al. (1972). About 125 exposures as short as one millisecond are subjected to optical Fourier transformation and accumulated by successive exposures on a single photographic plate. Parallel processing of an unresolved reference star permits correction for the prevailing seeing. Later reports (Bonneau & Labeyrie 1973, Labeyrie et al. 1974) describe experience with more advanced data processing again with the 200-inch telescope. A system using a vidicon television camera at the prime focus and connected directly to a computer is described by Labeyrie (1974; see Figure 3) and Lynds et al. (1976).

Theoretical discussions have been given by Liu & Lohmann (1973) and by Dainty (1976a, 1976b) together with pictures of the results obtained. Dependence of the modulation transfer function (Goodman 1968) on exposure time and spectral band have been measured by Karo & Schneiderman (1978), who, with Korff (1973), also discuss the effect of the scale of turbulence r_0.

As described above, speckle interferometry can give an improvement in resolution of about 50 times with a 5-m telescope but the final image is an autocorrelation function which is of necessity centrosymmetric. A double star would appear as three spots of light. As far as double-star separations are concerned, however, it is sufficient to measure the fringe spacing of the accumulated transform without taking the last step of retransforming.

When more complex source distributions are considered it is clearly going to be more difficult to extract the true configuration from the autocorrelation, although one might bear in mind the success of X-ray crystallography in doing just that. Seeing that each speckle photograph is a random replication of true images rather than of correlograms it appears that the loss of information lies in the processing and is not inherent. Knox & Thompson (1974) have described a method that would preserve the phase of the Fourier transform. Bates et al. (1973) have shown how to regard a speckle photograph as a hologram where the light from one unresolved component of a source system such as a multiple star acts as a reference. Experimental results from the optical bench (Gough & Bates 1974) support this view which can be used to understand other second-order interferometers such as the Brown and Twiss intensity interferometer and the Knox-Thompson procedure (Bates & Gough 1975). Whether good images of complex sources (Bates et al. 1978, Bates & Milner 1979) can be made depends on the source brightness, how many exposures have to be made, and limitations of computers. The interplay between these factors is shown in a paper by Bates (1977) on two-level quantization of speckle images. On the outcome of such analysis will depend the future of imaginative schemes to link arrays of telescopes by light beams with a view to obtaining angular resolutions set by the array dimensions as in radio astronomy (Labeyrie 1976).

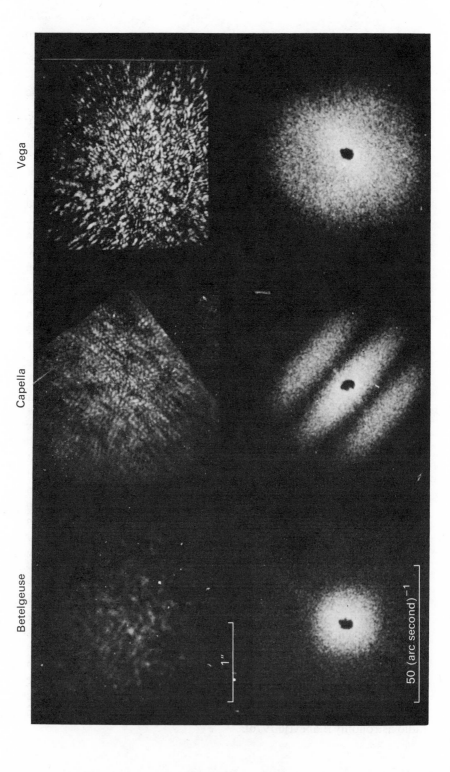

Betelgeuse   Capella   Vega

1"

50 (arc second)$^{-1}$

X-RAY IMAGING

Grazing Optics

X-ray astronomy (Giacconi & Gursky 1974) has developed explosively since the discovery in 1962 of the first nonsolar X-ray source. Imaging by X-rays can be performed with principles of conventional reflection optics but grazing incidence is required, otherwise the mirrors will become transparent. Angles of incidence of about 89° were used in the major installation in Skylab (Vaiana et al. 1977). The images are recorded on film and scanned by a microdensitometer with due attention to calibration wedges and recorded digitally on magnetic tape. Restoration can be applied, for example by a two-dimensional Fourier transform technique (Krieger 1977), to correct for a point spread function whose spread is due to scattering from irregularities of the mirror surfaces and is wavelength dependent. As the situation in this field of astronomical imaging does not yet warrant the sophistication of radioastronomical imaging, the subject of restoration has not yet generated vigorous controversy.

Coded Apertures

A quite different imaging scheme is presented by the pinhole camera where a small aperture in an opaque plate is placed in front of film. With a source of low intensity, exposure time can be conserved by opening up the aperture but then resolution is lost on the image. A way out of this is to use many pinholes to form many spatially displaced images on the film. To some extent the various images will interfere with one another but decoding methods can be imagined along the lines discussed in con-

---

← *Figure 3*   (*top row*) Direct highly magnified photographs of the stars Betelgeuse, Capella, and Vega taken with the 200-inch (5 m) Hale telescope using a focal ratio of f/200, exposure time of 0.010 s, and $\Delta\lambda = 200$ Å at $\lambda = 5000$ Å. The seeing during the observations was about 2″. Vega is not resolved, and the granular detail or "speckle" in the image is caused by atmospheric turbulence and the telescope optics. Betelgeuse is resolved by the telescope, with enlarged speckle arising from the convolution of its true image (diameter about 0″.06) with a point spread function which is illustrated by the image of an unresolved star as Vega. In the photograph of Capella a doubling of the image detail is clearly evident due to its binary structure. Each pair of speckles can be thought of as a distorted, diffraction limited image of the two stars in Capella. The separation of the two components is about $\rho = 0″.055$, and the stellar discs are not resolved. (*bottom row*) Co-added optical Fourier transforms of the stellar images in the top row. The transform of Vega is typical of an unresolved star. Betelgeuse is well resolved, as indicated by an absence of high spatial frequencies in its Fourier transform. The transform of Capella shows a fringe pattern which defines the position angle and separation of the binary system. The optical transform and data reduction techniques are described by Gezari et al. (1972). Photograph courtesy of A. Labeyrie.

nection with speckle interferometry. Naturally some pseudo-random arrangement of pinholes will be selected whose properties are satisfactory. A particular kind of coded aperture that is of special interest consists of a pattern of transparency and opacity arranged as with a Fresnel zone plate (Elwert & Feitzinger 1970, Barrett & Horrigan 1973). Each point source of X rays then throws a shadow on the film that is a replica of the zone plate but displaced in accordance with the direction of the source and with appropriate intensity. When the shadowgram is developed it may be illuminated with a source of coherent light. Each zone plate now acts as a lens and produces an image resembling the source distribution. Imperfections result from the fact that the intersection of two transparent zones cannot possess a transmittance greater than unity. While this is an analogue procedure not susceptible to digital interpretation, the analysis of the optical system is important in showing the possibilities and limitations of possible algorithms that could be applied directly to digital readouts from a matrix of X-ray detectors occupying the shadowgraph plane.

GENERAL DIGITAL IMAGE PROCESSING

Photographs and television pictures telemetered to earth from the moon and the subsequent color photographs from Mars first drew wide attention to the quality of images that have been handled in digital form. A good deal has been written on many aspects of digital imaging, which is of course not specific to astronomy. Therefore it will suffice here to mention Huang (1975) and the special issue of the *IEEE Proceedings* for July 1972 as sources of information and of further references. Among the journals reporting new developments are *Computer Graphics and Image Processing*, and *IEEE Transactions on Computers*.

For examples of images of external galaxies that have been treated digitally at the European Southern Observatory, see Middelburg (1977).

Literature Cited

Ables, J. G. 1974. *Astron. Astrophys. Suppl.* 15:383–93
Baars, J. W. M., Van der Brugge, J. F., Casse, J. L., Hamaker, J. P., Sondaar, L. H., Visser, J. J., Wellington, K. J. 1973. *Proc. IEEE* 61:1258–66
Baker, P. L. 1979. *Formation of Images from Spatial Coherence Functions in Astronomy, IAU Colloq. No. 49.* Groningen, August 1978. In press
Baldwin, J. E., Jennings, J. E., Shakeshaft, J. R., Warner, P. J., Wilson, D. M. A.,

Wright, M. C. H. 1970. *MNRAS* 150:253–70
Barrett, H. H., Horrigan, F. A. 1973. *Appl. Opt.* 12:2686–702
Bates, R. H. T. 1977. *MNRAS* 181:365–74
Bates, R. H. T., Gough, P. T. 1975. *IEEE Trans. Computers* C-24:449–56
Bates, R. H. T., Gough, P. T., Napier, P. J. 1973. *Astron. Astrophys.* 22:319–20
Bates, R. H. T., Milner, M. O. 1979. See Baker 1979
Bates, R. H. T., Milner, M. O., Lund, G. I.,

Seagar, A. D. 1978. *Opt. Commun.* 26 : 22–26

Bhandari, R. 1979. See Baker 1979

Bonneau, D., Labeyrie, A. 1973. *Ap. J.* 181 : L1–L4

Bracewell, R. N. 1958a. *Proc. IRE* 46 : 97–105

Bracewell, R. N. 1958b. *Proc. IRE* 46 : 106–11

Bracewell, R. N. 1961. *IRE Trans. Antennas & Propagation* AP-9 : 59–67

Bracewell, R. N. 1966. *Progress in Scientific Radio*, pp. 242–44. *Publication 1468*. Washington, DC: Natl. Acad. Sci., Natl. Res. Council

Bracewell, R. N. 1979. Image Reconstruction in Radio Astronomy. In *Image Reconstruction from Projections : Implementation and Applications*, ed. G. Herman. Berlin: Springer. In press

Bracewell, R. N., Colvin, R. S., D'Addario, L. R., Grebenkemper, C. J., Price, K. M., Thompson, A. R. 1973. *Proc. IEEE* 61 : 1249–57

Bracewell, R. N., Swarup, G. 1961. *IRE Trans. Antennas & Propagation* AP-9 : 22–30

Bracewell, R. N., Thompson, A. R. 1973. *Ap. J.* 182 : 77–94

Brouw, W. N. 1971. PhD thesis. Univ. Leiden

Brouw, W. N. 1975. *Methods Comput. Phys.* 14 : 131–75

Christiansen, W. N. 1973. *Proc. Inst. Radio Electron. Eng. Aust.* 34 : 302–8

Christiansen, W. N., Warburton, J. A. 1955. *Aust. J. Phys.* 8 : 474–86

Clark, B. G. 1970. *Ann. Rev. Astron. Astrophys.* 8 : 115–38

Cole, T. W. 1977. *Prog. Opt.* 15 : 187–244

Cotton, W. D., Wittels, J. L. 1979. See Baker 1979

Dainty, J. C. 1976a. *Laser Speckle and Related Phenomena*. Berlin: Springer

Dainty, J. C. 1976b. *Prog. Opt.* 14 : 1–46

Dainty, J. C., Fiddy, M. A., Greenaway, A. H. 1979. See Baker 1979

Elwert, G., Feitzinger, J. V. 1970. *Optik* 31 : 390–91

Frieden, B. R. 1972. *J. Opt. Soc. Am.* 62 : 511–18

Gezari, D. Y., Labeyrie, A., Stachnik, R. V. 1972. *Ap. J.* 173 : L1–L5

Giacconi, R., Gursky, H., eds. 1974. *X-Ray Astronomy*. Dordrecht: Reidel

Goodman, J. W. 1968. *Introduction to Fourier Optics*. New York: McGraw-Hill

Gough, P. T., Bates, R. H. 1974. *Optica Acta* 21 : 243–54

Graf, W., Bracewell, R. N. 1975. *Synoptic Maps of Solar 9.1 cm Microwave Emission from June 1962 to August 1973. Report UAG-44.* Boulder, Colorado: World Data Center A for Solar-Terrestrial Physics

Greenstein, J. L. 1972. *Astronomy and Astrophysics for the 1970's.* Washington, DC: Natl. Acad. Sci. 136 pp.

Gull, S. F., Daniel, G. J. 1978. *Nature* 272 : 686–90

Hamaker, J. P. 1979. See Baker 1979

Heeschen, D. S. 1975. *Sky Telesc.* 49 : 334–51

Hills, R. E., Janssen, M. A., Thornton, D. D., Welch, W. J. 1973. *Proc. IEEE* 61 : 1278–82

Högbom, J. A. 1974. *Astron. Astrophys. Suppl.* 15 : 417–26

Hogg, D. E., Macdonald, G. H., Conway, R. G., Wade, C. M. 1969. *Astron. J.* 74 : 1206–13

Huang, T. S., ed. 1975. *Picture Processing and Digital Filtering. Top. Appl. Phys., Vol. 6.* Berlin: Springer

Jennison, R. C. 1958. *MNRAS* 118 : 276–84

Karo, D. P., Schneiderman, A. M. 1978. *J. Opt. Soc. Am.* 68 : 480–85

Kenderdine, S. 1974. *Astron. Astrophys. Suppl.* 15 : 413–15

Knox, K. T., Thompson, B. J. 1974. *Ap. J.* 193 : L45–L48

Korff, D. 1973. *J. Opt. Soc. Am.* 63 : 971–80

Krieger, A. S. 1977. *X-Ray Imaging.* Soc. Photo-Optical Instrum. Eng., Vol. 106, pp. 24–33

Labeyrie, A. 1970. *Astron. Astrophys.* 6 : 85–87

Labeyrie, A. 1974. *Nouv. Rev. Opt.* 5 : 141–51

Labeyrie, A. 1976. *Prog. Opt.* 14 : 47–87

Labeyrie, A., Bonneau, D., Stachnik, R. V., Gezari, D. Y. 1974. *Ap. J.* 194 : L147–L151

Labrum, N. R., McLean, D. J., Wild, J. P. 1975. *Methods Comput. Phys.* 14 : 1–53

Liu, C. Y. C., Lohmann, A. W. 1973. *Opt. Commun.* 8 : 372–77

Lynds, C. R., Worden, S. P., Harvey, J. W. 1976. *Ap. J.* 207 : 174–80

Mathur, N. C. 1969. *Radio Sci.* 4 : 235–44

McCready, L. L., Pawsey, J. L., Payne-Scott, R. 1946. *Proc. R. Soc. London Ser. A* 190 : 357–74

Middleburg, F. 1977. *The Messenger (El Mensajero)* 10 : 16–19

Mills, B. Y., Little, A. G., Sheridan, K. V., Slee, O. B. 1958. *Proc. IRE* 46 : 67–84

National Radio Astronomy Observatory. 1967. *Proposal for a Very Large Array Radio Telescope.* Green Bank, West Virginia: NRAO

Ponsonby, J. E. B. 1973. *MNRAS* 163 : 369–80

Ponsonby, J. E. B. 1979. See Baker 1979

Reber, G. 1944. *Ap. J.* 100 : 279–87

Roger, R. S., Costain, C. H., Lacey, J. D., Landecker, T. L., Bowers, F. K. 1973. *Proc. IEEE* 61 : 1270–76

Ryle, M. 1962. *Nature* 194 : 517–18

Ryle, M. 1972. *Nature* 239 : 435–38

134   BRACEWELL

Smyllie, D. E., Clarke, G. K. C., Ulrych, T. J. 1973. Analysis of the Irregularities in the Earth's Rotation. In *Methods Comput. Phys.* 13:391–430

Spencer, R. E., Cornwell, T. J. See Baker 1979

Subramanya, C. R. 1979. See Baker 1979

Swarup, G., Bagri, D. C. 1973. *Proc. IEEE* 61:1285–87

Swarup, G., Thompson, A. R., Bracewell, R. N. 1963. *Ap. J.* 1:305–9

Swarup, G., Yang, K. S. 1961. *IRE Trans. Antennas & Propagation* AP-9:75–81

Swenson, G. W. Jr., Mathur, N. C. 1968. *Proc. IEEE* 56:2114–30

Tichenor, D., Bracewell, R. N. 1973. *J. Opt. Soc. Am.* 63:1620–22

Thompson, A. R., Bracewell, R. N. 1974. *Astron. J.* 79:11–24

Vaiana, G. S., Van Speybroeck, L., Zombeck, M. V., Krieger, A. S., Silk, J. K.,

Timothy, A. F. 1977. *Space Sci. Instrum.* 3:19–76

van de Hulst, J. M. 1978. PhD thesis. Univ. Groningen

Wernecke, S. J., D'Addario, L. R. 1977. *IEEE Trans. Computers* C-26:351–64

Whitford, A. E. 1964. *Ground-Based Astronomy: A Ten-Year Program.* Washington, DC: Natl. Acad. Sci. 105 pp.

Wild, J. P. 1961. *Proc. R. Soc. London Ser. A* 262:84–99

Wild, J. P. 1965. *Proc. R. Soc. London Ser. A* 286:499–509

Wild, J. P. 1967. *Proc. Inst. Radio Electron. Eng. Aust.* 28:277–384

Wild, J. P., Smerd, S. F. 1972. *Ann. Rev. Astron. Astrophys.* 10:159–96

Wilkinson, P. N., Readhead, A. C. S. 1979. See Baker 1979

Wolf, E. 1955. *Proc. R. Soc. London Ser. A* 230:246–65

Ann. Rev. Astron. Astrophys. 1979. 17:135–87
Copyright © 1979 by Annual Reviews Inc. All rights reserved

MASSES AND MASS-TO-LIGHT RATIOS OF GALAXIES[1]

S. M. Faber[2]

Lick Observatory, Board of Studies in Astronomy and Astrophysics, University of California, Santa Cruz, California 95064

J. S. Gallagher

Department of Astronomy, University of Illinois, Urbana, Illinois 61801

1 INTRODUCTION

Is there more to a galaxy than meets the eye (or can be seen on a photograph)? Many decades ago, Zwicky (1933) and Smith (1936) showed that if the Virgo cluster of galaxies is bound, the total mass must considerably exceed the sum of the masses of the individual member galaxies; i.e. there appeared to be "missing mass" in the cluster. As more data became available, the discrepancy persisted between masses of individual galaxies determined from optical rotation curves and the larger average galaxy mass needed to bind groups and clusters (e.g. Neyman, Page & Scott 1961).

Recently, however, new information has pointed toward larger total masses for individual galaxies, thus decreasing the traditional discrepancy between various methods of mass measurement. Arguing that thin self-gravitating stellar disks are unstable against bar-like modes, Ostriker & Peebles (1973) suggested that the disks of normal spiral galaxies must be imbedded in optically undetected, stabilizing massive halos. Ostriker, Peebles & Yahil (1974) and Einasto, Kaasik & Saar (1974) collected observational evidence in support of the existence of such halos (although Burbidge 1975 used similar data to reach the opposite conclusion). At nearly the same time, high-resolution 21-cm observations of nearby galaxies were showing that H I often extends well beyond the optical

[1] Supported in part by the Alfred P. Sloan Foundation and National Science Foundation Grant No. AST 77-18270.
[2] Alfred P. Sloan Research Fellow.

135

0066-4146/79/0915-0135$01.00

boundaries of galaxies and that rotation velocities are constant at large galactocentric distances. Simply interpreted, these measurements implied the presence of substantial mass outside the optically visible dimensions of galaxies.

In this review, then, we are especially concerned with the current status of the "missing mass" problem: has it been resolved by new data, or does it linger on essentially unchanged in magnitude? To answer this question we rely on mass-to-light ratios as our primary tool, since they provide a direct intercomparison of galaxy masses measured for many different samples using varied techniques. We further assume that all Doppler shifts are caused by actual velocities of recession, though this view is not universally held (e.g. Arp 1974). Finally, in discussing the possible presence of invisible mass in galaxies, we do not wish to assume any model for its structure or spatial distribution. For this reason, we choose the neutral term "massive envelope" to describe the unseen mass. Although "massive halo" is often used in a similar context, it connotes a more or less smooth and spherical mass distribution, a possibly misleading notion since the present data contain little actual information on the spatial structure of any extended components.

In comparing mass-to-light ratios from different sources, we use a standard system of M/L_B. We define total magnitudes of galaxies on the B_T system of the *Second Reference Catalogue of Bright Galaxies* (de Vaucouleurs, de Vaucouleurs & Corwin 1976, hereafter RC2). We have corrected the B_T magnitudes for internal extinction using the precepts of the RC2, which are nearly independent of type. The resultant luminosities are "face-on" values only; an additional 10–20% increase would be necessary to produce totally absorption-free magnitudes, which, strictly speaking, are those with which the local M/L_B for the solar neighborhood should be compared (Section 2.1). The galactic extinction assumed is $A_B = 0.133(\csc|b|-1)$, close to Sandage's (1973) formulation and in reasonable agreement with the more recent results of Burstein & Heiles (1978). The value of the solar absolute magnitude is here taken to be $+5.48$ in B (Allen 1973) or $+5.37$ in the photographic system used by Holmberg (Stebbins & Kron 1957). Finally, $H_0 = 50$ km s^{-1} Mpc^{-1} is used consistently throughout.

All values of M/L_B used in this paper are corrected to this standard system. Failure to adopt a standard system of M/L_B can easily lead to errors of a factor of two or three in comparisons among mass-to-light ratios from various authors. The standard system adopted here is sensitive to the adopted magnitude and extinction corrections. The resulting uncertainties in the overall scale are ±30–40%.

2 THE MILKY WAY

Because of the sun's location in the central plane of the galactic disk, the large-scale structure and global mass of the Milky Way are more difficult to determine than those of a nearby external galaxy. However, our position within our own galaxy gives us a bird's eye view of the mass in the solar neighborhood plus a chance to study the dynamics and mass distributions of various subpopulations in the Milky Way.

2.1 *The Solar Neighborhood, a Benchmark in M/L*

Because most of the mass resides in intrinsically faint stars, the stellar mass density can be directly determined only in the immediate neighborhood of the sun. In practice this is accomplished by combining the luminosity function, $\Phi(M)$, with the mass-luminosity relationship for each stellar group to find the stellar density, ρ_s. The faint-star luminosity function now appears to be well determined for $M_V \lesssim +15$ (Wielen 1974, Luyten 1968, 1974). From Gliese's (1969) data, Wielen derives $\rho_s = 0.046 \; M_\odot \; \text{pc}^{-3}$ while Luyten (1968) gives $\rho_s = 0.064 \; M_\odot \; \text{pc}^{-3}$.

The greatest uncertainty in ρ_s arises from the difficulty of properly applying the mass-luminosity relationship to the observed sample. The mass-luminosity relation (Veeder 1974) and observational data (van de Kamp 1971) are most reliable for main sequence stars, which Wielen finds amount to $0.038 \; M_\odot \; \text{pc}^{-3}$. Sources of error include confusion between low-mass stars and more massive, cooled degenerate stars (e.g. Hintzen & Strittmatter 1974) and the interpretation of the drop in $\Phi(M)$ for $M_V \gtrsim +15$. This could be due to a real absence of very low-mass objects. On the other hand, stars with $M \lesssim 0.08 \; M_\odot$ are not able to support stable H burning (Graboske & Grossman 1971, Straka 1971) and so might cool sufficiently rapidly to produce an apparent deficiency of low-luminosity stars (Kumar 1969, Greenstein, Neugebauer & Becklin 1970, Hoxie 1970). Joeveer & Einasto (1976) have estimated that this effect requires increasing the contribution of low-mass stars by about $0.02 \; M_\odot \; \text{pc}^{-3}$.

The other major mass contribution resides in white dwarfs. Luyten (1975) finds $n_{\text{WD}} \simeq 0.006 \; \text{pc}^{-3}$, while Sion & Liebert (1977) obtain $n_{\text{WD}} \gtrsim 0.01 \; \text{pc}^{-3}$. For a mean white dwarf mass of $0.7 \; M_\odot$ (Wegner 1974, Greenstein et al. 1977), $\rho_{\text{WD}} \gtrsim 0.004\text{--}0.007 \; M_\odot \; \text{pc}^{-3}$, compared to the theoretical prediction of $0.012\text{--}0.03 \; M_\odot \; \text{pc}^{-3}$ (Hills 1978), which depends on the age of the disk (Hills took 1.2×10^{10} yr). Hills also shows that the mass in neutron stars is probably negligible.

The stellar mass density near the sun thus most likely lies in the range $0.05 < \rho_s < 0.09 \, M_\odot \, \text{pc}^{-3}$. To this must be added the mass in insterstellar matter, which from Savage et al.'s (1977) measurement of the mean density of hydrogen is $\rho_{ISM} \simeq 0.03 \, M_\odot \, \text{pc}^{-3}$. The total density, ρ, then lies between 0.08 and 0.12 $M_\odot \, \text{pc}^{-3}$, consistent with the estimate of $0.09 \pm 0.02 \, M_\odot \, \text{pc}^{-3}$ found by Joeveer & Einasto (1976).

The local mass density can also be measured by observing the z density and velocity dispersion for a homogeneous stellar population (Oort 1965). Although straightforward in principle, the accurate measurement of the galactic acceleration gradient perpendicular to the plane has proved elusive. Different determinations are in conflict with each other and in some cases yield nonphysical results (Dessureau & Upgren 1975, Joeveer & Einasto 1976, King 1977). The most likely value of the density found by this method, $\rho_{dyn} \approx 0.14 \, M_\odot \, \text{pc}^{-3}$ (Jones 1976), must be considered uncertain. Thus at present there is no compelling evidence for significant undiscovered mass in the immediate solar vicinity. This result is consistent with a model mass-distribution for the galaxy computed by Ostriker & Caldwell (1979); the model has much unseen mass in an extended halo but very little in the neighborhood of the sun.

To compute the local mass-to-light ratio, we need the local luminosity density, $\mathscr{L}$. This quantity follows directly from $\Phi(M)$, which must now be based on a large volume since rare stars make a significant contribution to the luminosity. The $\Phi(M)$ of Starikova (1960) and McCuskey (1966, Table 8) respectively give $\mathscr{L}_V = 0.049$ and $\mathscr{L}_V = 0.063 \, L_\odot \, \text{pc}^{-3}$. A recent study by F. Malagnini (private communication) suggests that the results of Starikova may be preferable, so we adopt $\mathscr{L}_V = 0.055 \pm 0.01 \, L_\odot \, \text{pc}^{-3}$. Since Malagnini finds $B-V = 0.62$ for the solar neighborhood, $\mathscr{L}_V \cong \mathscr{L}_B$. For $\rho = 0.09 \pm 0.02$, $M/L_B = 1.1$–2.4. Using $\rho_{dyn} = 0.15$, we find $M/L_B = $ 2.3–3.3. If the light contribution from younger stars (spectral type earlier than G2 on the main sequence) were removed, then the local M/L would be approximately doubled.

2.2 Mass of the Milky Way

Historically the mass of the Milky Way has been determined from the rotation curve. This involves two distinct but interrelated observational problems: finding the shape of the rotation curve interior and exterior to the solar radius R_0, usually from 21-cm H I studies (Kerr & Westerhout 1965, Burton 1974), and setting the scale of the rotation curve by estimating the circular velocity at the sun, V_0. The latter measurement is difficult due to lack of a suitable inertial reference frame. One approach is to use extreme Pop II objects as a reference; this technique yields a lower limit, since the amount of rotation of the Pop II spheroid is unknown.

Using this method, Oort (1965) showed that $V_0 \gtrsim 190 \pm 30$ km sec^{-1}, in good agreement with Hartwick & Sargent's (1978) value of 220 km sec^{-1} based on velocities of globular clusters and dwarf spheroidal galaxies. On the other hand, a best-fit solution of 300 km sec^{-1} is obtained from the dynamics of the Local Group (see Section 6.4). Between these two extremes is the officially adopted I.A.U. value of 250 km sec^{-1} for a solar radius of 10 kpc.

A fresh attack on the determination of V_0 has been made recently by Gunn, Knapp & Tremaine (1979). They combine observations of H I interior to the sun with the requirement that the rotation curve join smoothly to their suggested flat rotation curve exterior to the sun. Their reasoning is too complex to detail here, but their preferred value of V_0 is 220 km sec^{-1}. In our opinion the uncertainties are large, but the method does minimally require $V_0 \lesssim 260$, in contrast to $V_0 = 300$ found from Local Group dynamics.

Once V_0 and R_0 are known, the mass in the Milky Way interior to the sun can be obtained by a variety of modelling techniques (see, e.g.,

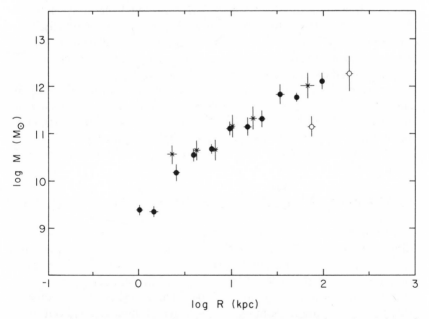

Figure 1 Mass of the Milky Way interior to radius R determined from observations of globular clusters and dwarf spheroidal galaxies; all data have been averaged in radial bins. The vertical bars are the standard error of the mean of each bin. Filled dots are mass measurements from globular cluster tidal radii, stars refer to dynamical mass determinations, and open circles are masses derived from tidal radii of dwarf spheroidal galaxies.

Schmidt 1965). However, these results have lately diminished in significance in the face of mounting evidence for large amounts of nonluminous matter far beyond the sun's orbit. From stellar motions, Fitzgerald et al. (1978) found that the rotation curve stays flat outside the sun for several kiloparsecs. Hartwick & Sargent (1978) analyzed the distribution of radial velocities of globular clusters and nearby dwarf spheroidal galaxies, which they took to be bound to the Milky Way. The outermost sample tests the potential at an effective radius of about 60 kpc and gives an interior mass of 8×10^{11} $M_\odot$ for an isotropic distribution of velocity components.

Finally, using the globular cluster system as a probe of the galactic potential and tidal fields, Webbink (in preparation) has mapped the mass distribution out to a radius of ~ 100 kpc. Independent estimates of galactic mass were obtained from the tidal radii and radial velocities of 126 globular clusters and 7 dwarf spheroidal companions of the galaxy. The deduced mass distribution of the Milky Way based on radial bin averages is shown in Figure 1. The present tidally-limited radius of the galaxy due to M31 is ~ 200 kpc. Within this assumed radius, Webbink derives a total mass of $1.4 \pm 0.3 \times 10^{12}$ $M_\odot$ from tidal effects on globular clusters and $1.4 \pm 0.8 \times 10^{12}$ $M_\odot$ from globular cluster radial velocities. The consistency between Webbink's two completely independent determinations of the mass distribution is strong empirical evidence for the existence of a dark envelope around the Milky Way.

To obtain M/L_B for the galaxy, we use Sandage & Tammann's (1976) calibration of L_B versus rotation velocity to estimate the luminosity of the galaxy. The result is 2.0×10^{10} $L_\odot$. With Webbink's mass estimate, we obtain $M/L_B \lesssim 70 \pm 20$ on our mass-to-light system. This value is an upper limit, since the mass may not extend as far as the assumed tidal cutoff of 200 kpc.

3   MASS-TO-LIGHT RATIOS OF SPIRAL GALAXIES

The rotation of spiral nebulae was first noticed by Wolf (1914) and Slipher (1914) fully a decade before astronomers discovered the true nature of galaxies. Pease (1916, 1918) made the first measurements of what we now call the stellar "rotation curve" in the nuclear regions of M31 and the Sombrero galaxy (M104). These observations required truly heroic dedication; both exposures lasted 80 hours spread out over a period of 3 months!

From these modest beginnings, the study of rotation curves has since matured to become our single most powerful tool for determining mass distributions inside galaxies. For many decades, optical spectroscopists dominated the field by measuring velocities of H II regions. Lately, it has been discovered that neutral atomic hydrogen extends outward past the optically bright regions of most spiral galaxies, and the 21-cm line is now being fully exploited to follow the dynamics to radii far beyond the last observable H II region.

The fundamental theory of inferring mass distributions from rotation curves has been discussed by Burbidge & Burbidge (1975), de Vaucouleurs & Freeman (1973), Freeman (1975), and Schmidt (1965), among others. Brosche, Einasto & Rümmel (1974) give a bibliography of dynamical measurements for all galaxies complete through May 1973, and van der Kruit (1978) reviews recent observational results.

3.1 *Observed Rotation Curves*

Roberts (1975a) has dramatically illustrated the difficulty of using the older optical rotation curves to probe the outer mass distributions of spiral galaxies. A convenient measure of the optical extent of a galaxy is the Holmberg radius (Holmberg 1958), which is the major-axis radius at a surface brightness of 26.5 photographic mag arcsec^{-2}. Roberts found that the median extent of optical rotation curves published up to 1975 was only 0.3 Holmberg radii. These curves typically showed a steep rise in rotational velocity near the nucleus, then a short section of leveling off [e.g. NGC 157 (Burbidge, Burbidge & Prendergast 1961)]. There the data ended, usually because the surface brightness of the galaxy was so low that further measurements were impossible.

Until recently it was customary to assume that in such cases the turnover of the rotation curve had been reached and that the rotational velocity declined smoothly past the turnover radius, eventually to reach the Keplerian falloff, $V \propto R^{-1/2}$, at large R. Considerable theoretical machinery was constructed to model the curve and deduce the mass distribution and total mass from the measured points. Burbidge & Burbidge (1975) give a comprehensive description of these techniques; Bosma (1978) provides a useful additional discussion of Toomre (1963) disk models. A simple and commonly used approximation is based on Brandt's (1960) parametrization of the rotation curve:

$$V_{\rm rot}(R) = \frac{V_{\rm max}(R/R_{\rm max})}{\left(1/3 + 2/3\left(\dfrac{R}{R_{\rm max}}\right)^n\right)^{3/2n}} \tag{1}$$

where $V_{\rm max}$ is the maximum rotational velocity, $R_{\rm max}$ is the radius at which the maximum rotational velocity occurs, and n is a shape parameter which determines how rapidly the curve reaches a Keplerian falloff. The value of n is determined by fitting the curve up to the last

Table 1   Galaxies with extended rotation curves

Object	Type[a]	Distance[b]	$L_B{}^c$ ($10^9\ L_\odot$)	Corrected Holmberg radius[d]	Rotation velocity[e] (km s^{-1})	Mass[f] ($10^9\ M_\odot$)	M/L_B	Source[g]
N3626	RSA(rs)0$^+$	21.1	16.2	11.3	233	142	8.8	1
N4324	SA(r)0$^+$	23.0	10.1	9.0	163	55	5.5	1
N3593	SA(s)0/a:	11.8	5.9	10.8	108	29	4.9	1
N3623	SAB(rs)a	11.8	27.0	15.8	212	94	3.5	1
N4378	RSA(s)a	48.6	45.0	30	280	540	11.9	2
N4594	SA(s)a	18.2	10.5*	21	350	590	5.6	3,4
M81	SA(s)ab	3.6	20.3	15.8	217	172	8.5	2
N4151	PSAB(rs)ab	19.6	23.0	19.2	146	95	4.1	5,1
N4698	SA(s)ab	23.0	28.0	19.1	246	270	9.6	1
N4736	RSA(r)ab	6.8	22.0	14.2	180	107	4.9	6
N4826	RSA(rs)ab	6.8	15.6	10.7	169	71	4.5	1
M31	SA(s)b	0.69	20.0	15.6	233	196	7.6	7,8,9
N891	SA b	14.4	31.0	21.7	225	255	8.2	10
N2590	Sb	95.9	105.0	34	265	560	5.4	2
N2841	SA(r)b	12.0	28.0	16.9	273	290	10.6	10
N3627	SAB(s)b	11.8	36.0	20	195	176	4.9	1
N4501	SA(rs)b	23.0	79.0	27	295	550	7.0	1
N4565	SA b	18.4	71.0	36	250	520	7.3	1,10
N5383	SB(rs)b	47.0	55.0	29	215	300	5.5	10
M51	SA(s)bcp	12.0	66.0	23	203	220	3.3	11,12
N801	Sbc-c	119.0	240.0	51	220	570	2.4	2
N1620	SA(rs)bc	68.5	92.0	32	240	430	4.7	2
N3145	SB(rs)bc	68.3	105.0	34	250	500	4.7	2
N4258	SAB(s)bc	6.8	25.0	20	200	180	7.3	13
N4527	SAB(s)bc	23.0	35.0	20	186	160	4.5	1
N4536	SAB(rs)bc	23.0	44.0	25	188	210	4.8	1
N5055	SA(rs)bc	12.0	50.0	25	199	230	4.5	10
N7331	SA(s)bc	21.0	85.0	33	227	390	4.6	1,10
N7541	SB(rs)bcp	57.5	70.0	30	230	370	5.2	2
N7664	Sbc-c	74.2	74.0	40	200	370	5.0	2
N2998	SAB(rs)c	95.6	149.0	43	210	440	3.0	2
N3198	SB(rs)c	13.5	15.9	19.5	140	89	5.6	10
N3672	SA(s)c	33.1	48.0	22	180	165	3.4	2
N5033	SA(s)c	21.0	48.0	33	209	330	6.8	10
N5907	SA c	17.2	36.0	25	235	320	9.0	10
M33	SA(s)cd	0.72	3.3	7.8	90	14.6	4.5	14,15
M83	SAB(s)cd	6.3	37.0	12.3	170	83	2.3	16
M101	SAB(rs)cd	8.0	53.0	32	194	280	5.3	17,18
N672	SB(s)cd	10.8	7.7	14.5	129	56	7.2	1
N2403	SAB(s)cd	3.6	7.6	10.7	130	42	5.5	19
N4244	SA(s)cd	6.8	8.1	11.5	106	30	3.7	10
N4517	SA(s)cd:	23.0	53.0	29	158	169	3.2	1
N4559	SAB(rs)cd	18.4	50.0	26	132	106	2.1	1
N925	SAB(s)d	14.4	25.0	26	122	89	3.5	1
N4631	SB(s)d	14.0	63.0	27	150	143	2.3	1,20
N4236	SB(s)dm	3.6	2.9	10.8	72	12.8	4.4	19

Table 1 (*continued*)

Object	Type[a]	Distance[b]	$L_B{}^c$ ($10^9 L_\odot$)	Corrected Holmberg radius[d]	Rotation velocity[e] (km s^{-1})	Mass[f] ($10^9 M_\odot$)	M/L_B	Source[g]
SMC	SB(s)m	0.065	0.67	2.9	40	1.05	1.6	21
N3109	SB(s)m	2.6	1.57	5.2	40	1.93	1.2	22
N4656	SB(s)mp	14.0	26.0	21	84	40	1.4	1
I1727	SB(s)m	10.8	3.9	13.5	74	17.0	4.3	1
N6822	IB(s)m	0.70	0.18	2.0	15:	0.10	0.58	21

[a] De Vaucouleurs type from RC2 (estimate when not available).
[b] Distance based on group membership (Sandage & Tammann 1975, de Vaucouleurs 1975) and mean group velocity. Radial velocity of galaxy used if not a group member.
[c] B luminosity based on B_T from RC2; corrections for internal absorption from RC2; galactic absorption correction $A_B = 0.133 \, [\csc(b)\text{-}1]$; $M_B(\odot) = 5.48$ mag.
[d] Radius on Holmberg's system corrected for inclination using diameter correction from RC2.
[e] Adopted rotation velocity at corrected Holmberg radius.
[f] Mass within Holmberg radius assuming spherical distribution.
[g] Sources for rotation velocity:

1. Krumm & Salpeter, private communication	12. Segalowitz 1976
2. Rubin, Ford & Thonnard 1978	13. van Albada & Shane 1976
3. Faber et al. 1977	14. Warner, Wright & Baldwin 1973
4. Schweizer 1978	15. Rogstad, Wright & Lockhart 1976
5. Bosma, Ekers & Lequeux 1977	16. Rogstad, Lockhart & Wright 1974
6. Bosma, van der Hulst & Sullivan 1977	17. Allen 1975
7. Emerson 1976	18. Rogstad & Shostak 1971
8. Newton & Emerson 1977	19. Shostak 1973
9. Roberts & Whitehurst 1975	20. Weliachew, Sancisi & Guélin 1978
10. Bosma 1978	21. Tully et al. 1978
11. Shane 1975	22. Huchtmeier 1975

measured point. If it is *assumed* that the velocity beyond this radius is adequately approximated by the Brandt model, the total mass of the galaxy, M_T, is $(3/2)^{3/n} V_{max}^2 R_{max}/G$.

In the context of extragalactic astronomy a decade ago, the Brandt model and its relatives were a logical way to model the outer regions of a galaxy. After all, the light was falling off rapidly at the last measured point, and in several galaxies the rotation curve also seemed to be falling appreciably as well [e.g. NGC 5055 (Burbidge, Burbidge & Prendergast 1960)]. However, workers at that time were well aware that a convenient extrapolation was being used which might not represent reality. "One does not know how much the tail wags the dog," cautioned Burbidge and Burbidge.

Radio 21-cm observations, which in many galaxies now extend well past the Holmberg radius, do not confirm this extrapolation. The radio rotation curves remain flat to the limit of observation, in some cases beyond 50 kpc, indicating much larger total masses than given by the Brandt formula. This result was strongly hinted at in early observations by Shostak & Rogstad (1973), Rogstad et al. (1973, 1974), and Seielstad & Wright (1973). Further evidence came from observations of M31 by

Roberts (Roberts 1975a, Roberts & Whitehurst 1975) and from an early compilation of rotation curves by Huchtmeier (1975). These initial results have been overwhelmingly confirmed by the more recent work of Bosma (1978), Krumm, and Salpeter (Salpeter 1978), and other references summarized in Table 1. At the same time, Rubin, Ford and co-workers (summarized by Rubin, Ford & Thonnard 1978) have pushed optical observations to greater radii by exploiting improvements in spectrograph and image-tube design. A montage of representative modern rotation curves collected by Bosma is shown in Figure 2.

There are now approximately 50 galaxies for which reliable rotation curves exist out to large radii (see Table 1). Very few are seen to turn over at all, and only three (M81, M51, and M101) show significant declines. All three of these galaxies, however, have nearby companions which may well perturb the outer H I. Furthermore, M81 has a large bulge which might produce a turnover in velocity because of its strong central condensation, while the H I in M101 shows strongly asymmetric motions on opposite sides of the major axis. In short, all three of these galaxies might well be atypical objects.

The reality of flat rotation curves has been questioned on several grounds. Doubts have been raised, for example, as to whether the H I is truly in circular motion. There are indeed good reasons to fear that within the inner regions of galaxies, ionized gas is not always in circular orbit. For example, marked asymmetries in the inner rotation curves amounting to ~ 100 km sec^{-1} exist in both M31 (Rubin & Ford 1971, de Harveng & Pellet 1975) and M81 (Goad 1976). Moreover, the emission-line rotation curves of bulge-dominated early-type spirals do not rise nearly as steeply as the light distributions suggest they should. The brightness profiles of such bulges near the nuclei are similar to those of elliptical galaxies (Kormendy 1977a, Burstein 1978, Kormendy & Bruzual 1978), and by analogy we would expect the rotation curve to rise to a sharp maximum within a few arc seconds, provided the mass-to-light ratio is uniform. In the three Sa galaxies with observed rotation curves, NGC 4378 (Rubin et al. 1978), NGC 4594 (Schweizer 1978), and NGC 681 (Burbidge, Burbidge & Prendergast 1965), this steep rise is not observed. NGC 4594 is an especially puzzling case; M/L_B in the nucleus based on the velocity dispersion is 12.6 (Williams 1977) yet is only 0.26 at 1 kpc according to the rotation curve (Schweizer 1978). Since the spectrum and colors (S. Faber, unpublished) give no hint of a significant change in the mass-to-light ratio of the stellar population, we seriously doubt that the ionized gas is in circular motion. Einasto (1972) has expressed a similar opinion that observed velocities in the ionized gas in the bulge of M31 are too low to reflect true rotation. Lack of adequate spatial resolution and noncircular motions due to bar-like distortions (Bosma 1978) are additional complications affecting rotation curves near nuclei.

For these reasons, rotation curves in the inner regions of galaxies might not be useful indicators of the mass distribution. We prefer to concentrate here on the outer regions, where noncircular motions and lack of angular resolution pose fewer problems.

Even at large radii, however, the interpretation of these observations

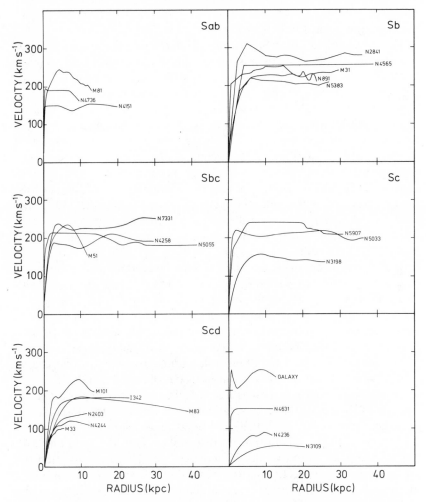

Figure 2   Rotation curves of 25 galaxies of various morphological types from Bosma (1978).

is a subtle matter. For example, sidelobes on radio telescopes could produce a fictitious flat rotation curve due to spillover from the bright H I at smaller radii. The initial results of Krumm and Salpeter (Salpeter 1978) were criticized for this reason by Sancisi (1978), but F. Briggs, N. Krumm, and E. Salpeter (in preparation) have since carefully calibrated the sidelobes of the Arecibo dish, and the final data should be free from this effect. Moreover, it is hard to see how sidelobes could affect the entire body of available data, since 21-cm measurements have been with many different instruments, single dishes as well as interferometers. The fact that flat rotation curves are also measured with optical techniques further strengthens this conclusion.

For all these reasons, it seems most unlikely that flat rotation curves are merely an artifact of observational errors. Even so, various dynamical arguments have been advanced which question the conventional identification with local circular velocity. For example, the outermost H I layer in many spirals is significantly warped out of the main plane of the galaxy (e.g. Rogstad, Lockhart & Wright 1974, Sancisi 1976). These warps might be accompanied by motions which mimic a flat rotation curve along the major axis if the inclination of the warp were properly arranged. However, for several of the galaxies in Table 1, warps have been fully modelled using the entire information available in two dimensions, and a flat rotation curve still persists. Further, the sheer bulk of the data is beginning to tell: it is hard to see how warps with random projection factors could conspire to produce a flat rotation curve in so many different galaxies.

It has also been suggested that H I at large radii might represent recent infall and not yet be in dynamical equilibrium. However, with few exceptions (e.g. M101) the velocities on opposite sides of the galaxy are reasonably symmetric, and circular motions (with a possible warp) satisfactorily fit the observations leaving relatively small residuals (Bosma 1978). Furthermore, because the flat portions in many galaxies extend over a large fraction of the observable radius, one would be forced to conclude that a large portion of the gas is out of equilibrium.

In summary, we feel that no generally valid alternative explanation has been put forward for these flat rotation curves and that the observations and their implications must therefore be taken very seriously.

For an assumed spherical mass distribution and V_{rot} constant with radius, the mass within radius R increases linearly with radius and the surface mass density declines as R^{-1}. Since the surface brightness of spirals declines exponentially (Freeman 1970, Schweizer 1976), this simple model predicts a strong increase in the local mass-to-light ratio projected on the sky as long as the rotation curve stays flat. Bosma (1978) finds that the precise form of this increase depends rather sensitively on the nature of the mass model assumed, whether spherical or disk. However, the increase in local M/L cannot be made to disappear completely by varying the model. Bosma (1978) and Roberts & Whitehurst (1975) have obtained local values of M/L_B of 100–300 in seven spirals at the outer limits of the observations, much larger than M/L_B for the stellar population in the solar neighborhood (Section 2.1).

Despite the complications mentioned above, mass determination using rotation curves is a relatively simple procedure. There are none of the statistical projection factors and group membership decisions which plague the analysis of binary and group motions. Based on present data and standard interpretations, it seems relatively certain that dark material is being detected.

The amount of extra mass actually implied by these rotation curves is itself relatively trivial; galaxy masses on average have perhaps doubled over the older optical estimates, not enough to satisfy the mass discrepancy for groups and clusters, as we shall see. Nevertheless, flat rotation curves have profound implications for the problem of missing mass, first because the detection of unseen matter is relatively secure and second because at least some of the missing mass seems to be associated with individual galaxies themselves.

3.2   Mass-to-Light Ratios

The discovery of flat rotation curves has thrown out any hopes we might have had of estimating the total masses of galaxies based on an extrapolation of their rotation curves. Such an extrapolation would necessarily involve an assumption as to how far the flat rotation curves continue, which is something one certainly would not want to guess at this time. The notion of total masses based on internal motions still appears frequently in today's literature and is one we would like to discourage; the derivation invariably involves assumptions which are unjustified, and the results can be misleading.

If we confine ourselves to what is actually measured, we are led inevitably to the concept of mass and mass-to-light ratio within a specified radius. Ideally one would like to use a radius related to some natural length-scale for the galaxy, for example, the e-folding length for exponential disks (α^{-1}). This is not realizable at present because α^{-1} is known for too few galaxies. As a practical necessity, we adopt an isophotal radius, even though this radius is systematically smaller for systems of low surface brightness, such as late-type spirals and irregulars. The Holmberg radius R_{HO} seems a good choice beca se it is comparable in size to the extent of presently available rotation curves and is easily derived by transformation from the large body of diameter data in the *Second Reference Catalogue.*

Table 1 collects information on galaxies available at present with published rotation curves which extend to at least 0.5 Holmberg radii and for which inclination corrections can be reliably estimated. The great majority have velocity curves extending nearly to R_{HO} or beyond. For estimating the mass within R_{HO}, elaborate techniques which exploit every bump and wiggle in the rotation curve seem to us unnecessary. For reasons discussed earlier, the inner sections of the measured curves may not contain useful information and furthermore do not strongly influence the total mass determination. Moreover, if one is looking for trends in mass and M/L with Hubble type, simplicity is a virtue. Insofar as possible, one must avoid the use of assumptions which vary with type since such procedures may themselves introduce spurious trends.

For all these reasons, it seems justified simply to assume that the mass is spherically distributed within R_{HO} and to calculate the mass at R_{HO} as $M_{HO} = R_{HO} V_{HO}^2/G$, where V_{HO} is the observed velocity of rotation at the

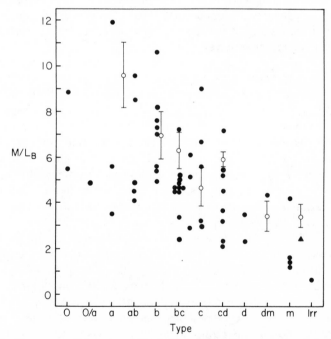

Figure 3 Blue mass-to-light ratios within the Holmberg radius versus morphological type. Black dots: individual galaxies in Table 1. Open circles: logarithmic means from single-dish measurements by Dickel & Rood (1978). Triangle: late-type DDO irregulars from Fisher & Tully (1975).

Holmberg radius. The assumption of a spherical distribution of material has some theoretical justification (see Section 8) but is basically unproven at this time. The use of highly flattened spheroids would yield masses roughly 35% smaller.

The resultant values of M/L_B are computed on the system of luminosities described in Section 1 and are plotted in Figure 3 versus morphological type. A trend is apparent in that later types seem to have lower M/L_B than early types. This conclusion is weak, however, because the number of late-type spirals in the sample is small.

We have therefore used single dish 21-cm profiles to verify this trend. Single-dish observations are useful here because the steep-sided H I profile yields a velocity width that is closely related to the velocity of rotation in the galaxy. Many workers have converted such velocity widths into total masses using the Brandt approximation (Roberts 1969, S. Peterson 1978, Balkowski 1973, Dickel & Rood 1978). We have assumed instead a flat rotation curve and calculated the mass within the Holmberg radius, as described above.

Our results are based on the extensive observations of Dickel and Rood for 121 disk galaxies. Comparison of 21-cm line widths with rotational velocities at R_{HO} for 16 galaxies in common with Table 1 indicates that $\langle V^2 \rangle$ of Dickel and Rood is 25% larger than $\langle V_{rot}^2 \rangle$ from Table 1. We applied this correction to the single-dish masses. The results for M/L_B (logarithmic means) based on this sample are shown as open circles in Figure 3, where the error bars reflect only the formal error in the mean, not the possible systematic errors. These data confirm the trend with type suggested by the individual rotation curves.

To further strengthen the data on the latest types, we used the indicative masses of Fisher & Tully (1975) for DDO irregular galaxies, adjusted to our system. The result is plotted as a triangle in Figure 3, and agrees with the low M/L_B for late types found from the other two methods. Recent data by Shostak (1978) (not plotted) also support the correlation in Figure 3 (see also Nordsieck 1973).

In summary, a trend in M/L_B within the Holmberg radius seems fairly well established. Whether or not a trend exists in *total* mass-to-light ratio cannot be determined from rotational velocities at the present time.

An estimate for S0's from Section 4 is also included. For completeness, M/L_V and M/L_K are also given, transformed from M/L_B using the data described in the notes. Interestingly, the trend in M/L with morphological type seen with B luminosities disappears when K magnitudes are used, and M/L_K is approximately constant. This is strong confirmation of the prediction of this effect made on quite different quasi-theoretical grounds by Aaronson, Huchra & Mould (1979).

The observed trend in M/L_B with type is generally what one would expect from variations in stellar content along the Hubble sequence. For example, Larson & Tinsley (1978) have found that M/L_B is well correlated with $B-V$ for model stellar populations over a wide range of ages. In their work, the mass refers to stars alone. To compare with their results, we must correct the observed M/L_B for mass due to gas. To do this, we have made the crude assumption that Robert's (1975b) global values of M_{HI}/L_B versus Hubble type apply also within the Holmberg radius and have neglected any contribution by molecular hydrogen.

The result is the column labeled M^*/L_B in Table 2, to be compared with Larson and Tinsley's model M^*/L_B, also in Table 2. Exact agreement is not expected, since the amount of material in evolved degenerate stars in the model is quite uncertain. Furthermore any matter in dark envelopes is not included in the Larson-Tinsley model. Nevertheless, within a scaling factor, the models appear to represent the total range of M^*/L_B rather well. The detailed agreement is not quite so good, however; the strong change in color of spirals from type Sa to Sd would suggest a noticeable decrease in M^*/L_B whereas the data show only a slight decrease.

Finally, we note that M/L_B for spirals within the Holmberg radius is ~ 4–6, not much greater than the local M/L_B for the solar neighborhood, which is ~ 1–3 (Section 2). This comparison indicates that unseen matter does not strongly dominate the mass within R_{HO}. This result is consistent with the Ostriker-Caldwell model of the Milky Way; their model contains only 25% dark matter within 20 kpc (roughly the Holmberg radius for our galaxy).

Table 2 Mass-to-light ratios within the Holmberg radius[a]

Type	M/L_B	M/L_V[b]	M/L_K[c]	M^*/L_B[d]	Model M^*/L_B[e]
S0[-]	10:[f]	7.6:	1.4:	10:	5.3
S0[+]-Sa	6.2 ± 1.1	5.4	1.1	6.1	3.5
Sab-bc	6.5 ± 0.5	6.1	1.2	6.3	2.8
Sbc-Sc	4.7 ± 0.4	5.0	1.1	4.4	1.6
Scd-Sd	3.9 ± 0.6	4.5	1.4	3.5	0.80
Sdm-Irr	1.7 ± 0.6	2.0	0.9	0.9	0.80

[a] R_{HO} corrected for inclination according to RC2.
[b] $(B-V)_\odot = 0.65$ (Allen 1973). Mean $B-V$ for galaxies from de Vaucouleurs & de Vaucouleurs (1972).
[c] $(V-K)_\odot = 1.42$ (Allen 1973). Mean $V-K$ for galaxies from Aaronson (1978).
[d] Includes mass of stars only in M^*/L_B. See text for details.
[e] Model estimates of M^*/L_B by Larson & Tinsley (1978).
[f] From Section 4.

4   MASS-TO-LIGHT RATIOS OF E AND S0 GALAXIES

Three basic methods can be used to determine the mass and mass-to-light ratio of a spheroidal stellar system. The first of these utilizes the global virial theorem, extensively discussed by Poveda (1958):

$$\tfrac{1}{2}d^2 I/dt^2 = 2T + \Omega, \tag{2}$$

where I is the moment of inertia, T the kinetic energy, and Ω the gravitational potential energy (Limber 1959). For a galaxy in equilibrium, the left-hand side vanishes. Let us assume further that the galaxy is spherical and nonrotating and that the kinetic energy of each star per unit mass is independent of mass. Then

$$M\langle V^2 \rangle + \Omega = 0, \tag{3}$$

where the potential energy Ω is given by

$$\Omega = -G \int_0^R \frac{M(r)dM}{r}; \tag{4}$$

M, R are the total mass and radius, $M(r)$ is the mass contained within a sphere of radius r, and $\langle V^2 \rangle$ is the mass-weighted average of the square of the space velocities of the stars relative to the center of mass of the galaxy.

Equations (3) and (4) involve theoretical parameters which are far removed from observed quantities. $\langle V^2 \rangle$ for example is usually estimated from the observed line-of-sight velocity dispersion (σ) in the nucleus by asssuming that σ^2 is constant throughout the galaxy, an assumption usually not supported by any observational data. The estimate of the total potential energy Ω is likewise subject to great uncertainty. It is generally assumed that the light distribution is an adequate tracer of the mass and that the luminosity profiles of most ellipticals are similar and are adequately described by empirical expressions, such as the $R^{1/4}$ law of de Vaucouleurs (1948) (see also Young 1976). The integral in (4) is then $\Omega = -0.33\, GM^2 R_e$, where R_e is the isophotal radius containing half the light (and mass). This approach assumes that the outer structure of the galaxy obeys de Vaucouleur's law. In fact, there seem to be significant departures from de Vaucouleur's law in the outer profiles of elliptical galaxies which correlate with environment (Kormendy 1977b, Strom & Strom 1978). Furthermore, the total light in the envelopes of some cD ellipticals shows no sign of converging to a finite value (Oemler

1976, Carter 1977), making the determination of R_e operationally impossible. Finally, if ellipticals contain appreciable amounts of dark material which is more extended than the luminous material, R_e as determined for the stars alone may have no connection with the true mass distribution of the galaxy.

The virial theorem thus leads to uncertain results basically because it treats the whole galaxy, including the poorly understood outer regions. To circumvent this difficulty, King (King & Minkowski 1972 and in preparation) has devised a second method to determine M/L; this method is based on stellar hydrodynamical equations applied to the core only. The observational data required include the central surface brightness, core radius (the point where surface brightness drops to $\frac{1}{2}$ of central value), and core line-of-sight velocity dispersion. From these one determines the core density and core mass-to-light ratio. Total mass is not derived. The method assumes only that the nuclear velocity distribution is Gaussian and isotropic with constant σ over the core region, in agreement with the properties of model star clusters whose cores closely resemble the nuclear regions of elliptical galaxies (King 1966). Young et al. (1978) and Sargent et al. (1978) have developed a similar formalism which is applicable to regions outside the core.

King's formula contains an explicit correction for rotational motion based on the observed ellipticity. However, several studies (e.g. Bertola & Capaccioli 1975, Illingworth 1977, C. Peterson 1978) have shown that even flattened ellipticals rotate very slowly and are almost completely pressure supported; rotational corrections should therefore be small. Binney (1976) and Miller (1978) have presented alternative models for elliptical galaxies having anisotropic velocity dispersions. These models imply a correction to our assumption of an isotropic velocity distribution in the core, but the effect should again be small.

The last method for determining M/L in E and S0 galaxies is the most straightforward: find a test particle in circular motion about the spheroidal component. This approach is applicable to the stellar disks of S0 galaxies and to gas in orbit about an elliptical [NGC 4278 is apparently such a galaxy (Knapp, Kerr & Williams 1978)]. For S0 disks seen directly edge-on, the observed rotational velocity must be increased by 30–40% to correct for stars at large spatial radii projected along the line-of-sight (Bertola & Cappaccioli 1977, 1978).

Burbidge & Burbidge (1975) summarized the results on M/L_B in early-type galaxies through 1969. Their mean value was 19.7, for a variety of objects and techniques. King & Minkowski (1972) reported values of 7–20 for luminous elliptical galaxies based on King's method applied to the cores and utilizing Minkowski's velocity dispersions.

Since this early work, the trend in M/L_B has been generally downward owing to two factors: remeasurements of velocity dispersions significantly smaller than earlier values, and the general adoption of core analyses in place of the global virial theorem. It is not easy, however, to summarize these recent results because there is still significant disagreement between various groups as to the correct measurement of σ. The two largest sets of data available are those of Faber & Jackson (1976) (FJ) and Sargent, Young and co-workers (SY) (Sargent et al. 1977, 1978, Young et al. 1978). Although it seemed initially that the values of FJ exceeded those of SY by 28%, new measurements (S. Faber, unpublished, Schechter & Gunn 1979) now make it seem likely that the two systems agree within 10%. In comparison to these results, however, the measurements of Williams (1977), Morton and co-workers (Morton & Chevalier 1972, 1973, Morton, Andereck & Bernard 1977), and de Vaucouleurs (1974) average about 35% smaller. This comparison is quite uncertain, however, because the number of objects in common is in all cases very small.

If these systematic differences in σ are taken into account, one finds that the agreement between investigators is good, with M/L_B typically 5–10. Sargent et al. (1977) found substantially higher M/L_B for elliptical galaxies but did so by applying the virial theorem to the entire galaxy assuming constant σ. Since σ was available only for the core, we believe that an analysis based only on the measured core quantities is preferable.

Determinations of velocity dispersions for normal elliptical galaxies have revealed two possible regularities. Both FJ and Sargent et al. (1977) found that σ increased with total galaxy luminosity approximately as $L_B^{1/4}$. For a sample of ellipticals with core radii and central surface brightness determined by I. R. King (unpublished), FJ derived power-law correlations between luminosity, core radius, and central surface brightness (see also Kormendy 1977b). Using these correlations plus the $L^{1/4}$ law for velocity dispersions, FJ found that M/L_B increased with luminosity as $L^{1/2}$. Schechter & Gunn (1979) found no such correlation, based on a published subset of King's data (King 1978). However, the published data were not corrected for seeing effects, unlike the unpublished list used by FJ. This difference apparently accounts for the discrepancy.

Using a method rather different from those above, Ford et al. (1977) have estimated the mass of M32 from motions of planetary nebulae far from the nucleus. Their result is $4.3 \times 10^8 \ M_\odot$, from which $M/L_B = 1.8$. This result would seem to be consistent with the possibility noted above that M/L_B is smaller in less luminous ellipticals.

The nucleus of M87 differs significantly from other ellipticals in having a bright central luminosity spike and a rapid decline in σ just outside the spike (Young et al. 1978, Sargent et al. 1978). A mass with very

Table 3 Mass-to-light ratios in early-type galaxies

Core values:

Object	M/L_B	Source
Mean luminous E	8.5	Faber & Jackson 1976
M31 bulge outside nucleus	8.5	Photometry from Light et al. 1974; $\sigma = 150$
NGC 4473, inside 5 kpc	5.4	Young et al. 1978

Other values:

Object	Type	Distance (Mpc)	R^a (kpc)	R/R_{HO}	V_{rot} [b]	M/L_B [c]	Source
NGC 128	S0p	90.0	17.4	0.4	250	7.2	Bertola & Cappaccioli 1977
NGC 3115	S0$^-$	9.5	4.6	0.3	350	12.3	Rubin et al. 1976
NGC 3636	RSA(rs)0$^+$	21.1	11.3	1.0	233	8.8	Table 1
NGC 4278	E	14.4	16.8	1.9	195	19.8	Knapp et al. 1978
NGC 4324	SA(r)0$^+$	23.0	9.0	1.0	163	5.5	Table 1
NGC 4762	SB(r)0^0	23.0	7.8	0.3	210	5.1	Bertola & Cappaccioli 1978

[a] Radial extent of observed rotation curve.
[b] Assumed velocity of rotation, corrected when necessary for projection effects.
[c] M/L_B within radius R, assuming mass spherically distributed.

high M/L, perhaps even a black hole, apparently exists in the middle of the core. Since very few elliptical galaxies have been studied with such high spatial resolution, it is not known whether such cases are common. Recent results for M/L_B among early-type galaxies (corrected to the M/L system of the preceding section) appear in Table 3. The first group contains objects for which nuclear values of M/L_B have been measured using King's method or a related treatment. The mean value obtained by Faber and Jackson is 8.5 for 10 galaxies, while the value of 8.5 for the inner bulge of M31 is new in this review. The second group consists of galaxies for which the circular rotation of test particles can be measured. The mass and M/L_B within radius R have been computed assuming a spherically symmetric mass distribution as in the preceding section.

Taken at face value, these data suggest that there is no gross increase in M/L_B from the core to the Holmberg radius. This conclusion is supported by the tendency of velocity dispersions to decrease away from the nucleus in M32 (Ford et al. 1977), NGC 3379 and NGC 4472 (FJ), and NGC 4486 (Sargent et al. 1978), leading to constant M/L in the inner regions. On the other hand, σ does not decline with radius in NGC 4473 (Young et al. 1978). Schechter and Gunn find that σ is basically constant in 12 more ellipticals, but their measurements extend to only a few core radii.

In summary, rotation curve data indicate that M/L_B within the Holmberg radius is approximately 10 for S0's. This number is entered in Table 2. No comparable estimate for E's can be given at this time owing to inadequate data on the velocity dispersions away from the nuclei.

Of great importance is the question whether a strong increase in M/L occurs beyond the Holmberg radius, as seems to be the case with spirals. The evidence on this point is fragmentary but highly suggestive of dark envelopes around early-type galaxies as well. NGC 4278 is the only galaxy in Table 1 for which rotation measurements extend beyond R_{HO}, and its M/L_B seems significantly higher than the others. Faber et al. (1977) found no decrease in the velocity dispersion in the halo of cD galaxy Abell 401 at a radius of 44 kpc, but the accuracy of the measurement was not high. Dressler & Rose (1979) have detected an actual increase in σ out to 100 kpc in the halo of the cD galaxy Abell 2029 with much better data, implying a strong increase in the local mass-to-light ratio. Finally, we mention the novel mass determination of M87 based on the assumption that the X-ray emission centered on M87 is due to thermal bremsstrahlung from isothermal gas in hydrostatic equilibrium within the potential well of the galaxy (Bahcall & Sarazin 1977, Mathews 1978). Mathews finds in this case that the total mass of M87 exceeds 10^{13} $M_\odot$ and the total M/L_B is several hundred.

All these data point strongly to the existence of dark matter around at least some elliptical galaxies. With the recent increased availability of efficient two-dimensional detectors for spectroscopy, additional information on the dynamics of the outer regions of elliptical galaxies should soon be forthcoming. Ultimately, radial velocities of globular clusters will be used to probe the structure of spheroidal systems at very large radii, but these observations seem to lie just beyond the capabilities of present equipment.

5   MASS-TO-LIGHT RATIOS OF BINARY GALAXIES

The derivation of masses of binary galaxies rests on several important assumptions. Let us first consider the simplest case of circular orbits. Assume that the galaxies are gravitationally bound, they interact as point masses, the distribution of orbital inclinations and phases is random, there is no intergalactic matter, and there is no dynamical interaction with matter outside the binary. A full discussion of this case is given by Page (1961), for example, but we present the rudiments here. Let M_T be the mass of each component (the two masses are assumed equal), R the spatial separation, and V the total orbital velocity. Then $M_T = V^2 R/2G$. Owing to projection effects, V and R are unobservable. One measures instead Δv, the radial velocity difference, and r_p, the projected separation. These are given by $\Delta v = V \cos \phi \cos \psi$, and $r_p = R \cos \phi$, where ϕ is the angle between the spatial separation R and the plane of the sky and ψ is the angle between the orbital velocity V and the plane determined by the two galaxies and the observer. Let us define an "indicative mass":

$$F(M_T) \equiv \frac{\Delta v^2 r_p}{2G} = M_T \cos^3 \phi \cos^2 \psi. \tag{5}$$

$F(M_T)$ is always less than or equal to the true mass according to the projection factor $\cos^3 \phi \cos^2 \psi \equiv F_p(\phi,\psi)$. To obtain the true mass in the simplest possible way, we can average over some appropriate statistical sample to derive a mean projection factor. Then, in principle, for a large enough statistical sample,

$$\langle M_T \rangle = \frac{\langle F(M_T) \rangle}{\langle F_p(\phi,\psi) \rangle}. \tag{6}$$

Page (1952) assumed that for a collection of circular orbits, the angles ϕ and ψ are randomly oriented, whence $\langle F_p \rangle = 3\pi/32 = 0.295$. This assumption is unrealistic because binary galaxies are selected and analyzed on the basis of their projected separation, r_p. Once r_p is specified, ϕ is no longer a random variable. The distribution of true spatial separations combines with the observed r_p to make certain values of ϕ more likely than others.

Depending upon the true distribution of spatial separations, $\langle F_p \rangle$ can therefore be a strong function of r_p. Page (1961) later amended his treatment to include this dependence. As a model for the spatial distribution function, he used

$$D(R) = K\left[1 - \left(\frac{R}{R_{max}}\right)^3\right],$$
(7)

an expression empirically derived by Holmberg (1954). This distribution produces a marked increase in $\langle F_p \rangle$ as r_p approaches R_{max}, the maximum spatial separation of binary galaxies. Such a variation in $\langle F_p \rangle$, if real, is quite inconvenient. It means we cannot compare indicative mass values $F(M_T)$ at different projected separations without first correcting them for trends in $\langle F_p \rangle$. Comparisons between independent data sets thus become much more cumbersome.

An important breakthrough in binary galaxy studies has been the realization that Equation (7) is not correct and that apparently $D(R)$ is a power law over the observable range 20 to 500 kpc. Turner (1976a,b) and S. Peterson (1978), both working from samples of binary galaxies chosen by means of well-determined selection criteria, find that $D(r) \propto R^{-\gamma}$, where $\gamma = 0.5 \pm 0.1$. Since a power law has no scale-length, $\langle F_p \rangle$ is independent of r_p and is given by (Peterson 1978):

$$\langle F_p \rangle = 4\, \frac{\Gamma^2\left(\dfrac{\gamma+4}{2}\right)}{\Gamma(\gamma+4)} \times \frac{\Gamma(\gamma+1)}{\Gamma^2\left(\dfrac{\gamma+1}{2}\right)}.$$
(8)

Fortunately, the sensitivity to γ is not great; over the range $\gamma = 0.0$ to $\gamma = 1.0$, $\langle F_p \rangle$ increases from 0.212 to 0.295. The fact that the projection factor is independent of radius for a power law makes the analysis much more transparent since radial trends in mass and M/L_B can now be determined immediately from trends in indicative mass.

The use of a mean correction factor $\langle F_p \rangle$ to rectify the observed values of $F(M_T)$ is unwise for a number of reasons. First, regardless of what criteria are employed to select the sample of binaries, spurious pairs will creep in. Such pairs tend to have large values of Δv^2 and hence of $F(M_T)$. Thus $\langle M_T \rangle$ for the sample is sensitive to the adopted cutoff in Δv. To avoid this difficulty, one should use a method based on the entire distribution $F(M_T)$ and which weights the observations more or less equally.

Second, the use of a simple mean ignores all information contained in the shape of the distribution $F(M_T)$ and, more particularly, in the bivariate probability density $p(\Delta v, r_p)$. As pointed out by Noerdlinger (1975), $p(\Delta v, r_p)$ is sensitive to the orbital eccentricities of galaxies in the sample. Turner (1976b) was able to rule out rather conclusively highly eccentric models for binary orbits on this basis.

Finally, the distribution of $F(M_T)$ is highly skewed, owing to the skewness of $F_p(\phi, \psi)$ itself. $F(M_T)$ peaks strongly at zero, and for circular orbits the median value of $F(M_T)$ is only 11% of the value of $\langle M_T \rangle$ ultimately derived. Peterson argues that such a skewed distribution is not well represented by its mean.

The use of a simple mean projection factor nevertheless provides a quick and moderately accurate way to compare data sets from various workers and to estimate the effects of different assumptions concerning the geometry of the orbits. From the work of Turner (1976b), who analyzed his data using both a simple mean and a more elaborate technique (see below), we estimate that use of the simple mean yields results accurate to $\pm 20\%$.

Because of the greater complexity of analyzing noncircular orbits, the general solution analogous to Equation (8) has not yet been derived. If the orbits are eccentric, the angles ϕ and ψ as defined in this section are not applicable. Nevertheless, a mean projection factor still exists, call it $\langle \eta \rangle$, which reduces to $\langle F_p \rangle$ in the circular case. Karachentsev (1970) and Noerdlinger (1975) have calculated values of $\langle \eta \rangle$ for collections of binaries with eccentric orbits. However, in their existing form these expressions share the same drawback as Page's original estimate: they do not take into account the bias introduced into the angle ϕ once r_p is specified. To treat the noncircular case, it would seem best at present to rely on Turner's model simulations of binaries with various eccentricities. These show that, for eccentric orbits, the projection factor is smaller and the derived masses therefore larger. If $\gamma = 0.5$, for example, the minimum value of $\langle \eta \rangle$ is 0.105 for purely radial orbits, compared to 0.261 for circular orbits. Qualitatively this effect is easy to understand: if the orbits are eccentric, one views them preferentially near apogalacticon, where the kinetic energies are lower than average. Therefore the observed $F(M_T)$ are systematically lower than in the circular case.

The success of the binary method clearly depends on being able to identify pairs that are real physical systems. The first discussion of this problem and the resulting catalog of doubles was prepared by Holmberg (1937). Other major lists having historical interest include Vorontsov-

Velyaminov's *Atlas of Interacting Galaxies* (1959), Arp's *Atlas of Peculiar Galaxies* (1966), and Karachentsev's comprehensive list of 603 pairs (Karachentsev 1972). Recent velocity measurements have been published by Karachentsev et al. (1976), Karachentsev (1978), Turner (1976a), and S. Peterson (1978).

In all, there have been four major studies of masses in binary galaxies. Page's work (Page 1961, 1962) was the standard reference for many years. Assuming Holmberg's relation (Equation 7) and circular orbits, he obtained $M/L_B = 0.7 \pm 0.9$ for spiral-spiral pairs and $M/L_B = 46 \pm 46$ for pairs containing ellipticals and/or S0's (corrected to $H_0 = 50$ km sec^{-1} Mpc^{-1}). The value for spirals is much smaller than the average of ~ 5 within the Holmberg radius derived in Section 2, while the value for early-type galaxies is significantly larger than our estimates for single galaxies presented in Section 3.

Turner (1976a,b) introduced a new standard of rigor into the study of binary galaxies by selecting a binary sample according to well-defined criteria and using, instead of the traditional mean projection correction $\langle \eta \rangle$, a rank-sum test that essentially compared the observed frequency distribution $p(\Delta v, r_p)$ with simulated versions of $p(\Delta v, r_p)$ for various orbital eccentricities. He was also able to model extended spherical halos surrounding the galaxies. Using the shape of the distribution $p(\Delta v, r_p)$, he showed that binary orbits cannot be highly eccentric and that if massive halos are present, they must have radii ≤ 100 kpc.

S. Peterson (1978) applied a rank-sum procedure to the distribution $F(M_T)$. His analysis was confined to the case of circular orbits, but his radial velocities, many of which came from 21-cm measurements, were more accurate than Turner's largely optical velocities. His selection criteria also differed from Turner's: his binaries were less isolated but extended to much larger radial separations because no outer angular cutoff was employed.

In Table 4 Turner's and Peterson's results for spiral-spiral pairs are compared. The original values of M/L_B given by these authors have been corrected to our system. For reference, Table 4 also includes median values of Δv and r_p for the Turner and Peterson samples. We have also reduced the results of Karachenstev (1977, 1978) to our system and included them in Table 4. Unfortunately only fragmentary accounts of Karachentsev's work were available to us, and we have had to combine results from several different papers. Details appear in the footnotes. We regret any inaccuracies introduced by this procedure.

Inspection of Table 4 shows that values of M/L_B derived by various authors differ substantially. Is this discrepancy due to differences in the data sets or to differences in statistical treatment? To answer this ques-

Table 4   Binary mass-to-light ratios in spiral-spiral pairs

	M/L_B	Median r_p[a]	Median Δv[b]
Turner 1976a,b:		50 kpc	105 km s^{-1}
$\varepsilon = 0.00$	17 ± 4		
$\varepsilon = 0.67$	20 ± 5		
$\varepsilon = 0.74$	24 ± 7		
Small halo[c]	33 ± 10		
S. Peterson 1978:		110 kpc	125 km s^{-1}
$\varepsilon = 0$; all objects	32 ± 11		
$\varepsilon = 0$; $r_p > 112$ kpc	37 ± 14		
Karachentsev:		27 kpc[d]	125 km s^{-1}[e]
$\varepsilon = 0$; $\eta = 0.295$	5.2 ± 2.3[d]		
$\varepsilon = 0$; $\eta = 0.261$	5.9 ± 2.7[f]		

[a] Projected separation.
[b] Only systems with $\Delta v \leq 750$ km s^{-1} included.
[c] M/L_B for total galaxy, including dark halo, is given. Radius of halo = 100 kpc.
[d] Karachentsev 1977.
[e] Karachentsev 1978.
[f] Transformed from line above using $\eta = 0.261$.

tion, it is helpful to consider just those values based on the assumption of point masses, circular orbits, and $\gamma = 0.5$. (We have provided an additional entry for Karachentsev's data which converts his results to this case.) We then obtain M/L_B (Turner) = 17 ± 4, M/L_B (Peterson) = 32 ± 11, and M/L_B (Karachentsev) = 5.9 ± 2.7, values which still differ by more than the observational errors. Therefore the differences must be due to the data themselves.

Histograms of Δv (uncorrected for observational errors) for all three samples are quite similar. The median values of Δv (for $\Delta v \leq 750$ km sec^{-1}) in Table 4 confirm this fact. Thus, the observed differences in the Δv distributions do not seem large enough to account for the discrepancy in M/L_B (see below).

On the other hand, the three samples differ markedly in their distribution of projected separation, r_p. Evidently Karachentsev's sample primarily includes rather close pairs, a conclusion supported by the high percentage of interacting galaxies in his sample (Karachentsev 1977). Turner's sample is intermediate, while Peterson's, chosen without any arbitrary cutoff in angular separation, contains many very wide binaries.

It seems to us that these results are consistent with an increase in M/L_B with radius, as would be expected from massive envelopes. This view is supported by the fact that M/L_B for Peterson's outermost pairs (with $r_p > 112$ kpc) is in good agreement with M/L_B for Turner's sample if massive halos having radii of 100 kpc are assumed. Peterson furthermore finds that M_T increases linearly with r_p out to ~ 100 kpc and then levels off, whereas Δv is constant out to ~ 100 kpc, and then begins to decline. Although of limited statistical significance (Peterson's radial bins overlap), this behavior is also consistent with the existence of massive envelopes having a limited extent. On the other hand, the same data also show that although M_T increases with r_p out to 100 kpc, M/L_B appears to be approximately constant from 20 kpc to 500 kpc, a result which Peterson takes as strong evidence *against* massive envelopes.

In our opinion the data are not yet strong enough to take any of these radial trends too seriously, and the global result must be given highest weight. Taken as a whole, the binary data of Turner and Peterson imply that $M/L_B \approx 35$ at large separations. As the average value for spirals within the Holmberg radius is only ~ 5, this result would seem to argue convincingly for additional mass beyond R_{HO}.

Before continuing, it is of interest to inquire why Page with essentially similar data obtained a much lower value of M/L_B for spiral-spiral pairs. According to Peterson, the difference is due to Page's weighting scheme, which set weights inversely proportional to the square of the variance in $(\Delta v)^2$ due to observational error. Small observed values of Δv therefore are given very high weight, which acts to reduce the calculated M_T and hence M/L_B. Page's method also gives highest weight to pairs with small separations (large $1/r_p$). If M/L_B does increase with radius, this effect would further shift M/L_B to systematically smaller values. Both Turner and Peterson have subjected their data to Page's scheme of analysis and confirm the fact that the method yields spuriously low values.

The high M/L_B values determined from binary galaxies are subject to several potential sources of uncertainty. The first arises from observational errors in the velocity differences. Let us suppose that there were no actual increase in M/L_B beyond R_{HO}. Then velocities would decline approximately as $r_p^{-1/2}$, and at a distance of 100 kpc, we would predict orbital velocity differences of roughly 140 km s^{-1}. The observed velocity difference Δv is further reduced by the projection factor cos ϕ cos ψ, the mean value of which is roughly 0.46 (for circular orbits). The expected Δv on this hypothesis is therefore only ~ 65 km s^{-1}. Since this is comparable to the precision of typical optical velocities, it has therefore been argued that existing data are biased against low mass-to-light ratios.

Although simple, this argument approaches the problem in backwards fashion. The proper question is whether the full width of the observed histogram of Δv's is essentially all due to observational error, for only if this is the case can we substantially reduce the measured values of M/L_B.

Peterson's 21-cm velocities are most useful in answering this question because of their high accuracy, typically better than ± 20 km s^{-1}. M/L_B for this subsample (30 pairs) is actually slightly larger than for his sample as a whole. Furthermore, the Δv distribution for pairs with 21-cm velocities is very similar to that for pairs having at least one optical velocity. Both these tests indicate that the optical velocities are substantially correct. L. Schweizer (in preparation) and Karachentsev (1978) are currently collecting new, highly accurate velocities for binary galaxies which should fully resolve this question. For the moment, however, we are inclined to believe that the velocities are not at fault.

The second problem which might affect the results is contamination by spurious pairs. Turner was able to show conclusively that his sample is not appreciably contaminated by objects in the distant foreground or background. However, a great many of Turner's and Peterson's binaries are members of small groups of galaxies, in which the problem of contamination by foreground and background group members could be serious. Statistical estimates of the frequency of spurious pairs made to date are unsatisfactory because they do not include a probable spatial correlation between the target galaxy and contaminating galaxies. Furthermore, the relative velocities of group members, typically a few hundred km s^{-1}, are just in the range where much of the information in the Δv distribution resides.

Yahil (1977) has pointed out a disturbing fact which may be related to contamination problems. He has searched for a positive correlation between $F(M_T)$ and the combined luminosity of the pair. Even though variations in $\langle \eta \rangle$ produce a large spread in $F(M_T)$, Yahil predicts that there should be a marked correlation between $F(M_T)$ and luminosity, provided binaries have uniform M/L_B. For Turner's sample, no correlation is found, indicating that M/L_B must vary over at least one order of magnitude. The correlations between $F(M_T)$ and luminosity for the Peterson and Karachentsev samples appear similar, supporting this conclusion. To Yahil, this result suggests that the large-scale distribution of matter in the universe is not strongly coupled to the distribution of luminous matter, and that the concept of mass-to-light ratio is not useful on scales much larger than 10 kpc. Alternatively, one might conclude that the lack of correlation is caused by errors in $F(M_T)$ introduced by the inclusion of spurious pairs. Yet the observed distribution of radial separations, $D(R)$, suggests that the majority of pairs must be real. If they were chance alignments, $D(R)$ would increase roughly as R for small separations, whereas the observed distribution is peaked near zero, suggesting real physical association.

To investigate the contamination problem further, it would be extremely useful to compile a binary sample having significantly more stringent isolation criteria than those used heretofore. At the very least, one might test whether M/L_B is noticeably smaller for those binaries in existing samples having only distant neighbors but sizeable spatial separations. Note that an analysis confined to just those binaries that show obvious signs of interaction will not help to test the existence of dark material, since such pairs have separations not much larger than the Holmberg radius. They therefore should have rather small M/L_B.

For the moment we continue to assume that the masses for spiral-spiral pairs as measured by binary galaxies are real, but the exact value of the mass-to-light ratio remains somewhat doubtful until the problem of contamination is conclusively cleared up.

Turning now to binary mass determinations for early-type galaxies, we recall Page's finding that E and S0 galaxies have much larger M/L_B than spirals. Both Turner and Peterson obtained a similar result, although their measured differences are smaller: Turner finds the ratio to be 2.0 ± 0.5, while Peterson obtains 1.7 with larger errors.

Actually it seems more probable that most of the mixed E-spiral pairs identified to date are not in fact physically bound to one another. The evidence for this assertion can be found in Figure 4, which presents distributions of Δv for Turner and Peterson's data. For simplicity let us

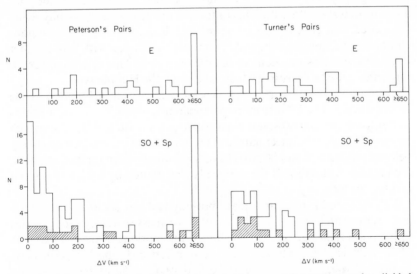

Figure 4 Histograms of Δv for binary galaxies in Turner and Peterson's samples, divided according to morphological type. Lower histograms represent pure spiral or S0 pairs; upper histograms represent pairs in which at least one member is an E. Shaded area refers to spiral pairs in which at least one member is an S0.

assume circular orbits, although our conclusion does not rest on this assumption. For a sample of binary galaxies that are real physical pairs, we expect that the distribution of Δv's will always be peaked at zero owing to the effect of the projection factor $\cos \phi \cos \psi$. This prediction is verified for spiral-spiral pairs (lower histograms). But the upper distributions, in which at least one member is an E galaxy, are nearly flat, with little or no peak at zero. The Kolmogorov-Smirnov test (Hollander & Wolfe 1973) shows quantitatively that the E and spiral distributions are unlikely to be drawn from the same population. The probability for the Turner sample is only 8% and that for the Peterson sample is only 3%. This test therefore confirms the fact that the samples really are quite different.

This disparity between E's and spirals was first noticed by K. C. Freeman and T. S. Van Albada (in preparation) for Turner's pairs. The existence of the same trend in Peterson's data, which is an essentially independent sample, is strong confirmation that the effect is real. We conclude that few if any of the pairs containing elliptical galaxies are physical associations. Virtually all the E pairs are members of groups or clusters, and it seems likely that they are due to chance superpositions of cluster members.

The great majority of these E pairs are mixed, that is, only one member is an E galaxy. Very few are EE pairs, and their small number does not allow us to test whether they, in contrast to the mixed pairs, are physically associated. One conclusion seems probable, however. Although luminous ellipticals and spirals are quite commonly associated with one another in groups, close associations in binaries are rare. This fact might be an important clue to processes which determine the Hubble type.

Histograms for S0 pairs are shown for comparison in Figure 4 as the hatched areas. The S0 distributions apparently resemble those of spirals more closely, so that S0-spiral binaries probably exist. It is this fact which is responsible for the queer nature of the E pairs having escaped discovery before now: since E and S0 pairs have traditionally been lumped together, the peculiar histogram of the ellipticals was diluted by the more normal one for the S0's.

If these mixed E pairs are indeed not physical associations, there exists at present virtually no reliable information on masses of early-type galaxies in binary systems. Jenner (1974) studied the motions of the companions of cD galaxies, but only ten pairs were included. Using his mean mass to obtain M/L_B on our system, we find $M/L_B \sim 50$ at 60 kpc spatial separation. However, if one system with extremely large Δv is omitted, M/L_B drops to only ~ 15. Smart (1973) obtained results consistent with Jenner's.

We found in Section 3 that data on M/L for S0 galaxies within R_{HO} are scanty, while those for ellipticals are nonexistent. The binary data are likewise fragmentary for these early morphological types. The results of Jenner and Smart, however, suggest that the mass-to-light ratio of E and S0 galaxies are broadly similar to those of later-type spirals.

Although the binary data seem to imply the existence of dark matter, we encounter a possible problem when trying to estimate the extent of the dark envelopes from these data. Such an estimate can be made in two ways. First, we have the results of Turner's model simulations, which ruled out envelopes larger than 100 kpc in extent. Second, we have the estimates of global M/L_B from Turner's halo model and also from Peterson's widely-spaced pairs. These both yield $M/L_B \gtrsim 35$ at large separations (the value is a lower limit because both estimates assume circular orbits). According to the usual version of the massive halo hypothesis, $M/L_B(R) \propto R$. Since the average value of M/L_B within R_{HO} is ~ 5 for spirals, the extent of the envelopes must be greater than or equal to $\sim 7\ R_{HO}$, or $\gtrsim 150$ kpc.

These estimates are in fair agreement with one another, but are marginally at variance with the conclusions of White & Sharp (1977), who pointed out that spherical halos around close binaries must interpenetrate strongly and that the effects of dynamical friction will be severe. In fact, two binaries ought to merge completely within an orbital period if their distance of closest approach is less than three times the half-mass radii ($r_{1/2}$) of the halos. For Turner's sample, this implies that the mean value of $r_{1/2}$ is less than 58 kpc, and hence that r, the outer boundary of the halo, is less than 116 kpc for the usual halo model. This limit must be reduced even further if the orbits are appreciably eccentric.

Even though this limit is barely consistent with Turner's estimate, many individual binaries must have true spatial separations much smaller than 100 kpc, and their envelopes should interpenetrate strongly. How are they then able to persist? Perhaps the outer envelope radius varies widely from galaxy to galaxy. Close pairs might then simply be those objects with initially small envelopes, the others having already merged long ago. This reasoning would suggest that the observed radial distribution function for binaries, $D(R)$, is strongly determined by the initial distribution function for the envelope radii themselves and that the binaries we see today are just those which were able to survive over a long period of time. In this regard, we recall the suggestion that merged galaxies become ellipticals (Toomre 1977, White 1978); this effect might then explain the rarity of E binaries.

As White and Sharp point out, the dynamics of binary galaxies, if analyzed from this more general point of view, might well place severe constraints on the distribution of unseen matter. N. Krumm (private communication) has emphasized the advantages of studying interacting pairs because of the information they afford in disentangling projection effects, which make the study of ordinary binaries so difficult. Tidal tails in interacting galaxies might have significantly different shapes if the gravitational effect of dark envelopes were included. In short, dynamical modelling of binary galaxies including the effects of halos seems a fruitful area for observer and theorist alike in the near future.

In summary, for spatial separations greater than 100 kpc, the binary data indicate $M/L_B \approx 35\text{--}50$, where the higher value applies if the orbits have moderate eccentricity ($\varepsilon \approx 0.7$).

6   DYNAMICS OF SMALL GROUPS OF GALAXIES

Over the past decade, our knowledge of the statistics of galaxy clustering has increased enormously, due in large part to the pioneering analysis of galaxy positions by Peebles and co-workers [see *Proceedings of I.A.U. Symposium No. 79, The Large Scale Structure of the Universe* (Longair & Einasto 1978) for discussions and references to earlier work]. The great majority of galaxies are apparently located in groups (Gott & Turner 1977, Soneira & Peebles 1977, Gregory & Thompson 1978). The definitive determination of group masses would therefore yield a representative estimate of the mass associated with galaxies in the universe as a whole.

Studies of galaxy clustering have demonstrated quite clearly that clusters exist on all scale sizes from the traditional great clusters like Coma down to binaries. Dynamically speaking, then, there exists a continuum, and there is no obvious dividing line between clusters and smaller groups. In practice, however, analyses of great clusters and small groups are unlike in a number of ways. Because the membership sample in groups is usually very small, the procedures for statistical mass determination and error estimate differ substantially from those in great clusters, where techniques based on large samples are more appropriate. Furthermore, the contrast of the cores of rich clusters against the background is much larger than for small groups, and the treatment of foreground-background contamination is therefore different. Finally, there are a number of dynamical processes more likely to occur in large clusters, where the densities are large and the crossing times short. Foremost among these are dynamical friction and tidal stripping, both of which can significantly rearrange the distribution of matter inside the cluster. For these reasons, we reserve the discussion of large clusters to Section 7, and treat here only the analysis of small groups.

Earlier work on groups of galaxies culminated in de Vaucouleurs' (1975) monumental listing of nearby associations, a systematic survey of volume density enhancements within 35 Mpc. Membership determination was based on absolute magnitude indicators as well as on redshifts. Sandage & Tammann (1975) presented a catalog selected in similar fashion. Despite the use of all available criteria in addition to redshift, the assignment of galaxies to groups remained a very difficult question and one without any obvious statistical solution. A major new development has been the identification of groups via a surface-density criterion only (Turner & Gott 1976). In this technique, one accepts the inevitable contamination by foreground and background galaxies as the price one must pay for a statistically well-defined and unbiased sample. However, one then must find a satisfactory method for dealing with the contamination problem.

6.1 Crossing Times

It has been obvious for a long while that the motions of galaxies in groups and clusters, if simply interpreted, imply the existence of a substantial amount of mass in addition to that traditionally associated with individual galaxies (Burbidge & Burbidge 1975). However, this conclusion depends entirely on the assumption that the groups are at least bound, if not in virial equilibrium. It is therefore necessary at the outset to decide on the nature of small groups: bound or unbound density enhancements? A useful concept here is the crossing time (Field & Saslaw 1971). If crossing times are short compared to the Hubble time, the groups must be bound; otherwise they would have dispersed long ago. It has sometimes been argued that groups may be unbound and "exploding" (e.g. Ambartsumian 1961), but the fact that the great majority of galaxies are group members makes this hypothesis unattractive as a general explanation.

One may choose various definitions of the velocity and radius in defining the crossing time, and these choices make significant and systematic differences in the final values. For example, one may use as R the mean harmonic radius, R_{VT}, of the group (Limber 1959). R_{VT} is the characteristic radius used in the virial theorem. Taking V_{VT} equal to the square root of the mass-weighted rms space velocity with respect to the center of mass, one has $t_{VT} = R_{VT}/V_{VT}$ for the virial crossing time. Using this definition of t, Turner & Sargent (1974) concluded that a large number of de Vaucouleurs' groups have long crossing times and are therefore not bound. However, Jackson (1975) demonstrated that t_{VT} is a poor estimator of the crossing time, yielding values that are systematically too large. He suggested instead a moment-of-inertia crossing time based on the moment-of-inertia radius. Rood & Dickel (1978a) have used the linear crossing time

$$t_L = \frac{2}{\pi} \frac{\langle r \rangle}{\langle V \rangle}, \tag{9}$$

where $\langle r \rangle$ is the average projected radial distance of group members from the center of mass and $\langle V \rangle$ is the average of the absolute value of the radial velocities with respect to the center of mass. Gott & Turner (1977) have adopted a similar definition of t_L. With these new definitions of crossing time, all three studies conclude that virtually all the groups identified by Sandage, Tammann, and de Vaucouleurs (STV) and Turner and Gott (TG) have crossing times significantly less than H_0^{-1}.

There are a number of effects which combine to bias these crossing times to systematically small values, notably the inclusion of nonmembers and the existence of binaries and subclusters, both of which increase the mean projected velocity. Even after all reasonable nonmembers are removed, however, the crossing times are still short, and it seems safe to assume that the great majority of these groups are bound density enhancements after all.

However, it is not clear that these groups have had time to reach virial equilibrium. Gott & Turner (1977) conclude that the median TG group is just now entering the virialized regime. Their calculation rests on necessarily rough dynamical estimates, however, which might not apply accurately to groups with only a few members. Given all the uncertainties, it seems possible that many of the looser groups have not virialized. We must therefore keep in mind that in some cases masses may be as much as a factor of two smaller than those obtained from the virial theorem.

6.2   Mass-to-Light Ratios

Limber (1959) derived the virial theorem as applied to small groups while Materne (1974) examined the treatment of observational errors. A number of uncertainties plague the analysis, as discussed by Aarseth & Saslaw (1972). Potential problems include observational errors, extrapolation to include the luminosity of faint members, and incomplete data. Furthermore, the virial theorem is based on the time-averaged energies, whereas we observe the group at just one moment in its history, when it may be out of equilibrium. Finally, we must employ projection corrections to convert the radial velocity dispersion and angular separations to their three-dimensional values. These mean projection corrections may be valid averages over a long period of time but are incorrect at any given moment. Derivation of the mean correction factors also involves assumptions as to the character of the motions, in particular that the projection corrections in velocity and radius are uncorrelated. This assumption is untrue for binaries, for example, and the standard virial formulation can substantially over- or underestimate binary masses, depending on the eccentricity of the orbits.

Taken together, these are serious difficulties. However, they are of minor importance compared to the twin problems of group membership and group definition. The group catalogs referred to above differ substantially in their membership assignments; these differences introduce uncertainties of a factor of 3–4 in the resultant mass-to-light ratios (see below).

As a result of a heightened awareness of these sources of error, a new consensus is emerging that the earlier piecemeal approach which attacked groups one at a time is unlikely to succeed. Since the analysis of any one group is subject to large statistical uncertainty, discussions of group properties must be based on a representative sample. On the observational side, we require accurate redshifts for a large, magnitude-limited sample of galaxies. Although not complete as yet, the available redshift sample is growing. Recent lengthy lists include redshifts in the RC2, velocities of galaxies in Gott-Turner groups measured by Kirshner (1977), accurate 21-cm redshifts for spiral galaxies (Dickel & Rood 1978, Shostak 1978, S. Peterson 1978, Thuan & Seitzer 1979), and many new optical redshifts by Sandage (1978). Several lengthy unpublished lists also exist (Tully & Fisher 1978, G. Knapp and W. L. W. Sargent, private communication, M. S. Roberts and co-workers, unpublished). A complete program to obtain a magnitude-limited sample of redshifts has been initiated by Davis and Huchra (Huchra 1978), but the final results are still a few years away. Taking an alternative approach, Gregory & Thompson (1978) have collected a redshift sample complete to a very deep limiting magnitude over a small region of sky.

On the theoretical side, we strongly believe that it will never be possible to assign individual galaxies to groups or fields in a definitive way. Any approach which relies solely on such group assignments is inevitably subject to insurmountable bias and cannot possibly yield reliable results. New theoretical methods are required which either avoid this bias altogether or correct for it in a statistical way.

Before discussing recent work along these lines, we summarize the results of traditional virial analyses of small groups. As was noted in previous sections, it is of paramount importance to employ a consistent system of masses and luminosities when comparing mass-to-light ratios derived by various workers. For example, the magnitude correction factor for Gott-Turner M/L_B's to our system is 0.50, and for Rood-Dickel values is 1.25. The local luminosity density on our system is $\sim 1.0 \times 10^8$

$L_\odot$ Mpc^{-3} (Gott & Turner 1976, Davis, Geller & Huchra 1978). With this value, M/L_{crit}, the mass-to-light ratio for a universe having critical density, is ~ 700. In such a universe, $\rho/\rho_{crit}(\equiv \Omega)$ is unity.

We confine ourselves to those investigations in which a sizeable number of groups have been treated simultaneously in homogeneous fashion. Results are collected in Table 5. Consider first the TG sample. The median M/L_B for the original TG groups including bogus members is 70. Gott & Turner (1977) later presented a list of revised groups, culled of obvious nonmembers. The median M/L_B for this sample is 30, a significantly lower value. Thus although the median M/L_B for the unculled groups is of theoretical interest because of its value as an unbiased estimator of *some* statistical property of the ensemble, by itself it has no obvious connection with the true M/L_B of the sample. In order to obtain M/L_B from the TG catalog alone, one must resort to the usual strategy of choosing members, accepting the inevitable bias therein. The great value of the unculled TG catalog is that it is an unbiased sample which can properly be compared with numerical simulations of galaxy clustering. In this role it appears to be quite powerful (see below).

Rood & Dickel (1978a) have determined M/L's for the culled TG groups and for STV groups, all required to have at least three members with measured velocities. Their median value for the TG groups is 40, not significantly different from the Gott-Turner value despite the inclusion of 40% more radial velocities. The median value for the STV groups, however, is 140. According to Rood and Dickel, this difference is due to different definitions of what constitutes a group: many of the STV groups break up into subgroups using the TG prescription. This is one

Table 5   Mass-to-light ratios of small groups

Source	Median M/L_B	No. groups
Turner & Gott 1976:		
All groups, unculled	70	39
Culled groups	30	48
Rood & Dickel 1978a:		
Turner-Gott groups, culled	40	29
Sandage-Tammann-de Vaucouleurs groups	140	63
Materne & Tammann 1974	~ 260	14
Tammann & Kraan 1978	~ 40[a]	7
Tully & Fisher 1978	~ 40	9

[a] Mean value.

more illustration of the extent to which the final value of M/L depends on how the groups are defined.

Table 5 also includes M/L_B values from Materne & Tammann (1974), Tammann & Kraan (1978), and Tully & Fisher (1978), all based on a selection of nearby, rather poor groups. The values in the table are very approximate because these authors did not give M/L_B directly. Thus, M/L_B had to be inferred from M_{VT}/M_L, the ratio of virial mass to luminous mass.

Gott & Turner and Rood & Dickel have emphasized the existence of missing mass outside of galaxies, while the remaining three authors have minimized its importance. Yet as Table 5 indicates, the observational data seem to be similar in all cases. The disagreements among authors stem in part from differences in adopted magnitude conventions, and have been removed by placing all values of M/L_B on a common system, as in Table 5. To a large extent, however, the disagreement is philosophical: Gott & Turner and Rood & Dickel have placed greatest weight on the median values, which appear to support the existence of unseen matter. The remaining authors instead have emphasized those groups with small M/L_B and have criticized the rest as being contaminated or not in equilibrium.

Derived M/L's have also been questioned because of observational errors in the radial velocities (Karachentsev 1978, Materne & Tammann 1975, Tully & Fisher 1978). Group velocity dispersions are in many cases less than 100 km s^{-1} (Tammann & Kraan 1978), conceivably too small to be accurately measured using conventional optical velocities, which often have errors of the same magnitude. However, we believe that velocity uncertainties are not likely to grossly inflate the median M/L_B for the following reasons: first, the velocities used by Rood & Dickel are substantially more accurate than those used by Gott & Turner, yet their median M/L_B is slightly higher: second, an increasingly large number of velocities are very accurate 21-cm redshifts and 21-cm checks of optical velocities in the mean show good agreement (Rood & Dickel 1976); and third, the really high values of M/L_B are virtually all associated with groups having large velocity dispersions of several hundred km s^{-1}, much greater than the measuring errors. The radial velocities are therefore not a likely source of error. Nevertheless, we are still left with the membership question, which cannot be resolved in any convincing way on a group-by-group basis.

Attempting to break this deadlock, Aarseth and co-workers (Aarseth, Gott & Turner 1979, Turner et al. 1979) have recently introduced a new method based on N-body simulations of galaxy clustering. The resultant models are based on a variety of initial conditions, obtained by varying the mass-to-light ratio for galaxies, initial density fluctuation spectrum, starting redshift, and peculiar velocities of galaxies. Group catalogs are constructed for the model universes in a manner identical to that of the original TG catalog, and the two sets of group properties are compared. Since no culling is performed in either case, contamination enters equally in both analyses. By comparing models and data in exactly parallel fashion, one ought to obtain a useful measurement of M/L_B for groups free of bias. Turner et al. point out that an N-body model with Ω equal to 0.1 ($M/L_B = 70$) exhibits group membership characteristics very similar to those of the real universe. In particular, the distribution of galaxies in redshift space is quite different from the true spatial distribution, and group assignments based on radial velocities would often be in error.

For the same model, Turner and collaborators have emphasized that the median M/L_B for the simulated unculled groups agrees well with the true (i.e. model) M/L_B, both being equal to 70. However, this good agreement is largely a coincidence. The binary galaxies in this model have highly eccentric orbits, and their mass is badly underestimated as a result. The median M/L_B for binaries is only 23, considerably less than the model value of 70. On the other hand the groups with three or more members are strongly affected by contamination. Their median M/L_B is 200, much larger than the model value. The two errors combine fortuitously so as to make the median M/L_B of all groups together equal to the true value of the model. Moreover, nearly all TG groups in the real universe have three or more members. Therefore, if the model can be taken as a guide, it corroborates our previous conclusion that contamination in the real unculled TG groups is serious and significantly biases the median M/L_B to higher values. Furthermore, if we restrict our attention to groups with three or more members, the spread in the model mass-to-light ratios is significantly smaller than is observed in the real universe. This result supports the suggestion of Rood & Dickel (1978b) that there exists an intrinsic spread in M/L_B among groups which is larger than can be attributed to the various sources of error.

These model experiments are in their infancy and can surely be improved. Future calculations should include a realistic mass spectrum for galaxies, the effect of massive envelopes on the interactions of galaxies, and a more complete examination of initial conditions. The method of comparison between the real and simulated data might also be refined. Nevertheless, as an unbiased statistical approach to the problem, the present computations represent an original and promising line of attack which should be vigorously pursued.

6.3   Compact Groups

Because of possible interpenetrations of nonluminous massive envelopes in compact groups of galaxies, these groups provide a severe challenge to any theory of gravitational encounters between galaxies. Rose (1979) and Rose & Graham (1979) have recently reviewed the various hypotheses for the origin of such groups. A possible problem with these groups is their short lifetimes; if the member galaxies really do have massive envelopes, the galaxies should coalesce rapidly due to dynamical friction. Alternatively, perhaps these compact groups survive just because their galaxies have abnormally small envelopes.

Reviews of compact groups have been given by Karachentsev (1966) and by Burbidge & Sargent (1971); Rose (1977) has conducted a new survey for such groups. The problem of discrepant redshifts has long plagued the dynamical study of compact groups. Rose (1977) suggests that they can be explained simply on the basis of chance projections, but Nottale & Moles (1978) feel that discrepant cases are too frequent to be explained in this way.

For many of the groups, however, the velocity dispersion appears normal, and the virial theorem should apply. Regardless of the origin of these compact configurations, one would predict small mass-to-light ratios for those groups because the spatial separations between the member galaxies are too small to sample the gravitational potential of extended, nonluminous envelopes. This prediction appears to be consistent with the available information. The median M/L_B for the compact groups listed by Burbidge & Sargent is ≤ 30 (an exact value cannot be specified because the magnitude system was not fully described), while Rose & Graham find values of only 2.3 and 12.2 for two southern compact groups.

6.4   The Local Group

The seminal paper on Local Group dynamics was written by Kahn & Woltjer (1959). Lynden-Bell & Lin (1977) and Yahil, Tammann & Sandage (1977) have recently given thorough rediscussions of the problem, while de Vaucouleurs, Peters & Corwin (1977) have given a new solution for the solar motion relative to Local Group galaxies.

The dynamical analysis of the Local Group is deeply entwined with the determination of the velocity of the sun's motion about the galactic center. The Milky Way and M31 together dominate the kinetic and potential energies of the Local Group to the extent that the problem becomes in essence an ordinary two-body interaction. The velocity of M31 relative to the galaxy is therefore essential to a knowledge of their mutual orbit. By sheer bad luck, the apparent radial velocity of M31 with respect to the sun is in large measure simply a reflection of the sun's motion about the galactic center. We may take comfort in the fact that the coincidence is lessening with time: in 40 million years or so the two problems will be geometrically independent. For the present, however, we must struggle to disentangle the two motions.

The magnitude of the sun's orbital velocity is a matter of dispute, but the values widely discussed range between 220 km s^{-1} (Section 2.2) and 300 km s^{-1} (Lynden-Bell & Lin 1977, Yahil et al. 1977). Using these values to correct the radial velocity of M31 to the galactocentric value, we obtain -125 km s^{-1} and -60 km s^{-1}, both negative. This is the key point: no matter how large a rotational velocity for the sun is assumed, within reasonable limits, we cannot convert the apparent approaching motion of M31 into one of recession. Barring a theoretically implausible "slingshot" effect in which M31 or the Milky Way caromed off some third galaxy in the past, the velocity of approach of the two galaxies must arise from their mutual gravitational interaction. Hence, there must have been time for the orbital motion to "turn around" during the lifetime of the universe, and this requirement in turn makes the galaxies considerably more massive than mere boundedness of the orbit would imply. We review this argument in some detail because, even though it was outlined quite clearly in Kahn & Woltjer's original paper (see also Peebles 1971), one still sees it ignored today in favor of estimates based on energy considerations alone. This omission seems hardly reasonable given the lack of any other convincing theory for the motion of approach of the two galaxies.

We assume that the orbital motion is radial and that the orbital time is 2×10^{10} yr. Using convenient formulae given by Gunn (1974), we calculate the total mass of M31 plus the Milky Way to be $\geq 2.9 \times 10^{12} \, M_{\odot}$ and $\geq 1.1 \times 10^{12} \, M_{\odot}$ for local circular velocities of 220 and 300 km s^{-1}, respectively. The luminosity of M31 is $2.7 \times 10^{10} \, L_{\odot}$ on our system, and that of the Milky Way is $2.0 \times 10^{10} \, L_{\odot}$ (Section 2.2). We then obtain mass-to-light ratios of ≥ 60 and ≥ 25 respectively, values quite typical of small groups and binary galaxies. These values are lower limits because the assumption of radial motion yields the minimum possible mass. These results are quite consistent with the value of $M/L_B \lesssim 70$ for the Milky Way alone found in Section 2.2.

We have so far neglected a second consideration. The orbital solution described above must also be consistent with the observed motion of the sun with respect to the center of the mass of the Local Group, which is assumed to be the center of mass of M31 and the Milky Way. Using this additional constraint, one obtains best-fit solutions of the solar orbital velocity close to 300 km s^{-1} (Lynden-Bell & Lin 1977, Yahil et al. 1977), although 220 km s^{-1} is still within the 90% probability contour. A more accurate value for the solar motion relative to Local Group galaxies would help greatly to narrow these possibilities, but it probably will never be forthcoming—there are simply too few Local Group members to serve as referents.

6.5   Conclusion

It is still impossible to give a definitive value for the mass-to-light ratios of small groups of galaxies because the remaining uncertainties are large. As a temporary expedient, one can use the numerical simulation of Aarseth et al. to estimate the effect of contamination on the median M/L_B for unculled TG groups. This is the only bias-free method available at present. We employ the model with $\Omega = 0.1$, including only those simulated groups with three or more members. The median M/L_B for this subsample is · 200, compared to the true value of 70 for the model as a whole. The median M/L_B for observed unculled TG groups having three or more members is 80. Assuming the simulated universe is a fair match to the real one, we apply the correction factor 70/200 to this value, obtaining 30 as the best unbiased estimate for the median M/L_B of small groups. This value is to be compared with values of 30 and 40 for the culled TG groups found by Gott & Turner and Rood & Dickel, and 90 for both the TG and STV groups having the best membership assignments (Rood & Dickel 1978a). These values probably encompass the allowable range. Though still very uncertain, these estimates seem distinctly higher than M/L_B within R_{HO} for individual spirals (~ 5) yet very compatible with the range of 35–50 estimated for binaries in the previous section. The mass-to-light ratios in groups therefore do not imply the existence of additional dark matter beyond that sampled by the orbits of wide binary galaxies.

7   GALAXY MASSES FROM CLUSTER MEMBERSHIP

With the recognition of galaxy clusters as dynamical entities, it was realized that the internal kinematics could be used to measure average masses of galaxies. Zwicky (1933) and Smith (1936) applied the virial theorem to the Virgo cluster using the small sample of radial velocities then available and found that the virial mass exceeded by a factor of several hundred that expected from the sum of the masses of individual galaxies. In a classic paper, Zwicky (1937) derived an M/L_B of 500 for the

Coma cluster which he compared with the $M/L_B \simeq 3$ for the solar neighborhood. Thus the "missing mass," or more properly "invisible mass," problem was born. Since then, our understanding of the structure and properties of clusters of galaxies has grown enormously (see reviews by Abell 1975, van den Bergh 1977, Bahcall 1977). However, the problem of invisible mass in the great regular clusters has not abated.

7.1 The Virial Theorem

In applying the virial theorem to large clusters, it is assumed that the cluster is in a stationary state. The kinetic energy is obtained from an appropriately weighted determination of the space velocity dispersion V^2 (Rood, Rothman & Turnrose 1970; see also Rood & Dickel 1978a). Choosing a proper weighting scheme is not a trivial point; for example, Chincarini & Rood (1977) show that the calculated mass of Abell 194 (hereafter, A194) varies by almost a factor of three among plausible weighting schemes. The issue is whether to weight by observed luminosities or by masses inferred from mean mass-to-light ratios. The latter method introduces great uncertainty since mean M/L's are poorly known for early-type galaxies, which are a significant fraction of the membership in large clusters.

As with small groups, a correction must also be applied to convert the observed projected velocity dispersion σ^2 to V^2, which will in general depend on the types of orbits in the cluster. If the orbits are primarily radial and we determine σ^2 from the cluster core, then V^2 may only slightly exceed σ^2, while if the orbits are nearly circular with an isotropic distribution, $V^2 \simeq 3\,\sigma^2$ (Abell 1977). Thus the kinetic energy term may be uncertain by as much as a factor of 3, although Rood (1970) has suggested that usually $2 < V^2/\sigma^2 < 3$ and that a ratio of 2.1 is most appropriate for the Coma cluster.

In calculating the potential energy, the mass distribution of the cluster is usually taken to be the same as that of the galaxies. Schwarzschild (1954) showed that the number of galaxies per unit area, S, along a strip passing a minimum distance q from the cluster core could be used to find the effective virial radius,

$$R_{\mathrm{VT}} = 2\,\frac{(\int S\,dq)^2}{\int S^2\,dq}. \tag{10}$$

Unfortunately, S is difficult to measure precisely, as it depends on the statistical correction for background galaxies, which is most important near the outer parts of the cluster and can introduce significant errors in R_{VT} (Rood et al. 1972). As an alternative to Equation (10), the observed surface density profile may be inverted to give the space density distribution $\rho(r)$, and the potential integral then can be explicity evaluated (Oemler 1974, Abell 1977). Once R_{VT} has been found, the virial mass of the cluster can be calculated in the usual way as $M_{VT} = V^2 R_{VT}/G$.

In this section we use luminosities on the V system since virtually all cluster work has employed this convention. Furthermore, most cluster photometry is necessarily somewhat less precise than that for nearby, bright galaxies, and thus the derived M/L are more uncertain than those found by other methods. We convert to M/L_B at the end of the section.

We have recomputed virial masses and M/L's in a homogeneous way for seven large clusters. We have taken σ^2 from Yahil & Vidal (1977) and cluster luminosities and virial radii from Oemler's study of galaxy populations in 15 clusters. In order to illustrate a standard virial mass, we have assumed $V^2/\sigma^2 = 3$. The dispersions have been multiplied by $(1+z)^{-1}$ to correct for redshift (Harrison 1974, Faber & Dressler 1977), and cluster distances are based on mean radial velocities. We find the median $M/L_V \approx$ 290, with a range of 165–800. This result is typical of values quoted in the literature (Bahcall 1975) and is substantially larger than M/L's of individual galaxies found in Table 2.

However, there is considerable room to maneuver within the framework of the virial theorem. The ratio M/L could be reduced by a factor of 2 or 3 if our guess concerning velocity isotropy is incorrect. An additional decrease might be obtained by mass-weighting the velocity dispersion and by correcting the luminosities for halos of galaxies and faint cluster members. Chincarini & Rood (1977) show that these considerations can reduce M/L_{pg} for A194 to 36 and that of Coma to 170. It is also possible that irregular clusters such as Hercules are not yet in virial equilibrium. The simple requirement for boundedness would reduce the mass by a factor of two. Hercules would then have a visual mass-to-light ratio of 270, rather typical of other large clusters.

7.2   Other Methods of Mass Determination

Unfortunately, the use of the virial theorem to measure cluster masses is sensitive to the structure of the outer parts of the cluster (Rood et al. 1972). An alternative is to consider only the well observed core regions. Zwicky (1937, 1957) found that the number-density profile of the Coma cluster could be fit by a bounded isothermal sphere. Using such a model, Bahcall (1975) derived a mean core radius, r_c, of 0.25 ± 0.04 Mpc for 15 clusters. Dressler (1978a) also found a small dispersion in core radius but his mean value was 0.5 Mpc. Avni & Bahcall (1976) have tested the measuring techniques for r_c on simulated clusters. They found that the existence of cluster cores is real, not an artifact of the data analysis, but that the close fit of Bahcall's clusters to a common isothermal model is due to the fitting procedure used.

If r_c and the core velocity dispersion, σ_c, are known for a cluster, a dynamical model can be used to calculate the central mass density. The King (1966, 1972) models were shown by Rood et al. (1972) to provide a reasonable representation of both the run of surface density and of velocity dispersion with radius in the Coma cluster. With this method, M/L_V values of 280, 350, and 250 have been derived for the cores of Coma (Rood et al. 1972), Perseus (Bahcall 1974), and Virgo (van den Bergh 1977). The value for Coma has been rescaled using a revised core velocity dispersion of 1260 km s^{-1} (Gregory & Tifft 1976).

Core-fitting procedures are subject to operational difficulties which Rood et al. clearly describe: first, the core radius is difficult to determine (e.g. the factor of two difference in mean r_c between Bahcall and Dressler); second, due to small-number statistics, the core velocity dispersion is uncertain and the core luminosity distribution is grainy. The situation is often further complicated by the presence of centrally located luminous galaxies, which make it difficult to measure the core luminosity accurately. To avoid the problem of graininess in the number density, Dressler (1978b) used the smoothed central number density from the model fit, assigned a mean luminosity per galaxy using a Schechter (1976) luminosity function, and applied a 20% correction to convert from isophotal to total luminosities. From a sample of nine rich clusters with velocity dispersions primarily taken from Faber & Dressler (1977), he found a median M/L_V equal to 270. For comparison, he computed a global M/L_V for each cluster using a fit to a de Vaucouleurs' (1948) law (see Young 1976). This approach yielded a median M/L_V equal to 280. The global and core-fitting methods are therefore in good agreement for these clusters. Rood et al. also analyzed the Coma cluster with a global de Vaucouleurs law and found $M/L_V = 200$.

With the increasing sophistication of computer N-body models, it has become possible to custom build a model to represent a specific cluster. White (1976b) extended the N-body program of Aarseth (1969) to produce a realistic dynamical model of the Coma cluster using 700 particles. The model was scaled to Oemler's gravitational radius and the radial velocities of Rood et al. and Gregory (1975). From the model, White (1976b) found $M/L_V = 258 \pm 36$. Melnick, White & Hoessel (1977) have shown that the value of M/L_V is reduced slightly to 207 ± 47 if L_V is corrected to the total light in the central regions of Coma.

All of these analyses of the Coma cluster yield values of M/L_V close to 250. The exception is Abell's (1977) study, which concludes that $M/L_V =$

120. This result is due to two factors. First, the derived mass is small because Abell assumed $V^2/\sigma^2 = 2$ rather than 3. More important, however, the form of the luminosity function chosen by Abell results in a luminosity twice as large as Oemler's. Relative to Oemler, Abell counts more galaxies with $13.5 < m_v < 14.5$. Furthermore, Oemler's faint-end extrapolation adds fewer galaxies than actually counted by Abell. Both effects contribute roughly equally to the net difference, with the excess at the bright end being somewhat more significant. Godwin & Peach (1977) also measured the luminosity function in Coma and found more galaxies with $15 \leq m_v \leq 17.5$ than either Abell or Oemler.

It suffices to note that luminosity functions of clusters are still not well known. This problem is especially severe in poorly studied clusters, where one must assume a universal luminosity function to estimate the contribution of faint galaxies. Furthermore, there is good evidence for significant cluster-to-cluster variations in the luminosity function, and luminosity functions may not be amenable to simple analytic forms (Dressler 1978a).

Ignoring uncertainties in the total luminosity for the moment, we emphasize the good agreement between the mass-to-light ratios for Coma found by many different authors using various techniques. Furthermore, these various approaches weight subsets of the data in very different fashions, some emphasizing the core, others the outer regions. It is indeed remarkable that these techniques yield such consistent values.

7.3 Uncertainties

Adopting the mass-to-light ratio of Coma as typical of great clusters, we have $M/L_V \approx 250$ and $M/L_B \approx 325$, much larger than those of individual galaxies and significantly larger than those of binaries and small groups. Without doubt, the single aspect of the data that contributes most to the large M/L in clusters is their large velocity dispersions. We must therefore consider the possible effects of contamination and subclustering on this parameter.

To assess contamination, the cluster must be viewed in the context of its environment. The N-body calculations of Aarseth et al. (1979) are appropriate here since they model a representative volume of space rather than an isolated cluster. For their model with $\Omega = 0.1$, the largest cluster has $M/L_B = 320$, much larger than the model value of 70 (Turner et al. 1979). This discrepancy is entirely a result of contamination. Turner et al. point out that substantial contamination is to be expected in rich clusters because of their large angular extent, and on this basis question the belief that the great clusters provide a reliable estimate of masses of galaxies. They also show that because of the high intrinsic velocity dispersions in clusters, noncluster members in a large volume of the Hubble flow can cause confusion. Because of this effect, Turner et al. suggest that velocity sampling alone cannot eliminate contamination. L. Thompson (private communication) has likewise speculated that the traditional Coma sample is contaminated by unbound members of the Coma supercluster.

To assess the severity of this effect, we need better dynamical models of clusters and their environs, plus a very deep and complete survey of radial velocities in several cluster complexes. Unfortunately this material is not yet available. For the present, however, we are inclined to think that contamination is not solely responsible for large cluster dispersions. Our belief rests principally on the core properties of such clusters, where the density contrast is large and the contamination therefore smaller. As we have seen, the core fitting procedures yield mass-to-light ratios just as high as those from global methods. Furthermore, in Coma the velocity dispersion declines markedly outside the core region, whereas one would in general expect the reverse if contamination were significant. Finally, virtually all the galaxies in the Coma core are of early morphological type, a much larger fraction than those either in small groups or in the extended supercluster (Gregory & Thompson 1978). These galaxies therefore appear to be a distinct population physically located in the core itself.

Holmberg (1961) suggested that cluster velocity dispersions might be spuriously inflated through the inclusion of binaries and subclusters. However, since binaries and small groups generally have dispersions ≤ 300 km s^{-1}, this effect cannot produce the large dispersions of several hundred km s^{-1} typical of great clusters.

X-ray observations of clusters may soon provide an independent check on cluster potential and kinetic energies. With the discovery of iron-line emission in clusters of galaxies (Mitchell et al. 1976, Serlemitsos et al. 1977), the probability that cluster X rays originate from thermal bremsstrahlung in a hot intracluster medium (ICM) has greatly increased. If the emission is thermal, the X-ray temperature is determined by the cluster potential. For gas in equilibrium, the X-ray temperature should follow $T \propto V^2$ (or σ^2) (Mushotzky et al. 1978, Jones & Forman 1978). Although the current data are limited, there is a good correlation between the observed X-ray temperatures and σ^2 in X-ray clusters (Mitchell, Ives & Culhane 1977, Jones & Forman 1978, Mushotzky et al. 1978). A strong statement would be premature, but the currently available data are consistent with the relation $V^2 = 3 \sigma^2$, where σ is the usual line-of-sight velocity dispersion. With more observations, we expect that modelling of the ICM will provide an accurate measurement of cluster potentials and thus of cluster masses.

Finally, the considerable range in M/L_V among large clusters themselves suggests that the amount of unseen matter might vary significantly from cluster to cluster. Rood (1974) and Rood & Dickel (1978a) have emphasized this possibility and noted that the virial mass discrepancy is strongly correlated with the velocity dispersion for both small groups and large clusters.

7.4 Conclusion

For the present, we conclude that the available data continue to support the existence of high cluster mass-to-light ratios, with $M/L_B \approx 325$. This value at first sight looks significantly larger than the values of 30–90 which we determined for binaries and small groups. However, part of the discrepancy is due to stellar population differences between the early-type galaxies in clusters and the spirals in small groups. The theoretical and observational data in Table 2 suggest that M/L_B for the stellar population in spirals is only half that in ellipticals and S0's. So, to properly compare small groups with great clusters, we must increase M/L_B for small groups from 30–90 to 60–180. Thus roughly 20% to 50% of the total mass can plausibly be associated with galaxies.

A further small correction must be made for ionized gas. Conventional X-ray measurements of cluster cores suggest that the mass of ionized gas is probably $\sim 10\%$ of the virial mass (Lea et al. 1973, Field 1974, Gull & Northover 1975, Malina et al. 1978), in agreement with the requirements imposed by the detection of microwave diminution by hot gas in clusters (Birkinshaw, Gull & Northover 1978). Adding 10% to our previous total, we can account for roughly 30% to 60% of the virial mass, leaving a net discrepancy of approximately a factor of two. Although this difference is uncomfortably large, real difficulties still remain in the determination of cluster M/L ratios (e.g. the luminosity function). The reality of excess unseen mass in great clusters relative to small groups must therefore still be considered uncertain at the present time.

We note in passing that a controversial detection of large X-ray halos around clusters has just been reported by Forman et al. (1979). While the gas in these halos potentially could bind the cluster as a whole, the halo has little dynamical influence on the core and thus cannot ease the M/L problem in the central regions.

White (1976a, 1977) has shown through N-body models that the dark material in clusters cannot all be attached to individual galaxies, or else a marked degree of radial mass segregation should be observed owing to dynamical friction. Since little segregation is apparent in real clusters, the hidden matter must be distributed rather uniformly throughout the cluster as a whole. Hence, galaxies in clusters cannot have large massive envelopes still attached. Gallagher & Ostriker (1972) and Richstone (1975, 1976) have suggested that high-velocity encounters between galaxies might liberate these envelopes through tidal shocks, thus spreading the dark matter throughout the cluster. The exact interrelationship between the competing processes of tidal disruption and dynamical friction remains to be worked out.

8   CODA

After reviewing all the evidence, it is our opinion that the case for invisible mass in the Universe is very strong and getting stronger. Particularly encouraging is the fact that the mass-to-light ratio for binaries agrees so well with that for small groups. Furthermore, our detailed knowledge of the mass distribution of the Milky Way and Local Group is reassuringly consistent with the mean properties of galaxies and groups elsewhere. In sum, although such questions as observational errors and membership probabilities are not yet completely resolved, we think it likely that the discovery of invisible matter will endure as one of the major conclusions of modern astronomy.

In addition to the dynamical evidence, there are other indirect indications of dark material in galaxies. The most important of these are the stability analyses of cold, self-gravitating axisymmetric disks (e.g. Ostriker & Peebles 1973, Hohl 1976, Miller 1978), which show them to be susceptible to bar-formation if not stabilized by a hot dynamical component. This hot component may or may not be related to massive envelopes.

Although present data give us little information on the shape of massive envelopes, further study of the outermost hydrogen in spirals may tell us more about this question. For example, the apparent lifetime of warps in many spirals poses severe theoretical difficulties as long as it is assumed that disks are self-gravitating (e.g. Binney 1978, Bosma 1978). This problem would not arise if the warps existed within the potential of a nearly spherical massive envelope. The precession of the warp due to the torque of the disk would then be much smaller, and the warp would be very long-lived. Alternatively, Binney (1978) has suggested that a warp might actually be driven by a triaxial dark halo. Finally, z-motions of H I far from the nucleus can be used to measure the space density of matter in the plane and thus to set limits on the flattening of the envelope.

Despite the general lack of observational evidence on the shapes of massive envelopes, there exists a strong consensus among theorists that they cannot be very flat. It is widely suggested that the large radial extent of the dark material relative to the luminous matter is due to the dissipationless collapse of the invisible matter. If so, a thin disk is unlikely, and we would more plausibly expect a thickened mass distribution spheroidal or triaxial in shape.

Suggestions as to the identity of the unseen matter include massive neutrinos (Cowsik & McClelland 1972, Gunn et al. 1978), faint stars (Ostriker, Peebles & Yahil 1974), black holes (Truran & Cameron 1971), and comets (Tinsley & Cameron 1974). Many attempts have been made to detect luminous matter in the halos of edge-on galaxies (e.g. Freeman, Carrick & Craft 1975, Gallagher & Hudson 1976, Hegyi & Gerber 1977, Kormendy & Bruzual 1978, Spinrad et al. 1978). Although faint luminosity has been found in some cases, it can plausibly be identified with the normal spheroidal stellar component.

Further progress in the study of unseen matter will continue to be made by mapping the gravitational potential using all observable test particles. Massive envelopes may well have a significant effect on the shapes and velocities of bridges and tails created in tidal encounters. The embarrassingly short theoretical lifetimes of binary galaxies and compact groups require careful consideration, as does the hypothesized stripping of extended halos and subsequent redistribution of the dark matter during the collapse of dense clusters. Most important, we need to know whether luminosity is a good indicator of mass density over scales greater than a few kiloparsecs. If a sizeable fraction of the mass in the universe is uncorrelated with the visible light, our dynamical analyses might be greatly in error. For example, our basic model for the formation of a group or cluster as a dissipationless collapse of noninteracting mass points might need serious revision. It is to be hoped that the systematic redshift surveys now in progress, coupled with more realistic theoretical simulations of galaxy interactions, will eventually yield definitive answers to these and related questions.

ACKNOWLEDGMENTS

We would like to thank C. Cox and G. McLellan for their aid in preparing the manuscript. We also thank R. Webbink for useful discussions and permission to reference his unpublished work. M. Roberts suggested the term "massive envelope." We are also grateful to S. Wyatt, N. Krumm, I. Iben, A. Faber, and A. Whitford for their critical comments on the original draft.

184 FABER & GALLAGHER

Literature Cited

Aaronson, M. 1978. *Ap. J. Lett.* 221: L103
Aaronson, M., Huchra, J., Mould, J. R. 1979. *Ap. J.* 229: 1
Aarseth, S. J. 1969. *MNRAS* 144: 537
Aarseth, S. J., Gott, J. R., Turner, E. L. 1979. *Ap. J.* 228: 664
Aarseth, S. J., Saslaw, W. C. 1972. *Ap. J.* 172: 17
Abell, G. O. 1975. In *Galaxies and the Universe*, ed. A. Sandage, M. Sandage, J. Kristian, p. 601. Chicago: Univ. Chicago Press
Abell, G. O. 1977. *Ap. J.* 213: 327
Allen, C. W. 1973. *Astrophysical Quantities*, p. 162. London: Athlone
Allen, R. J. 1975. In *La Dynamique des Galaxies Spirales: Proc. C.N.R.S. Colloq.* 241: 157
Ambartsumian, V. A. 1961. *Astron. J.* 66: 536
Arp, H. C. 1966. *Atlas of Peculiar Galaxies.* Pasadena: California Institute of Technology
Arp, H. C. 1974. *Proc. IAU Symp.* 63: 61
Avni, Y., Bahcall, J. N. 1976. *Ap. J.* 209: 16
Bahcall, J. N., Sarazin, C. L. 1977. *Ap. J. Lett.* 213: L99
Bahcall, N. A. 1974. *Ap. J.* 187: 439
Bahcall, N. A. 1975. *Ap. J.* 198: 249
Bahcall, N. A. 1977. *Ann. Rev. Astron. Astrophys.* 15: 505
Balkowski, C. 1973. *Astron. Astrophys.* 29: 43
Bertola, F., Capaccioli, M. 1975. *Ap. J.* 200: 439
Bertola, F., Capaccioli, M. 1977. *Ap. J.* 211: 697
Bertola, F., Capaccioli, M. 1978. *Ap. J.* 219: 404
Binney, J. 1976. *MNRAS* 177: 19
Binney, J. 1978. *MNRAS* 183: 779
Birkinshaw, M., Gull, S. F., Northover, F. J. E. 1978. *MNRAS* 185: 245
Bosma, A. 1978. *The distribution and kinematics of neutral hydrogen in spiral galaxies of various morphological types.* PhD thesis. Groningen Univ., Groningen. 186 pp.
Bosma, A., Ekers, R. D., Lequeux, J. 1977. *Astron. Astrophys.* 57: 97
Bosma, A., van der Hulst, J. M., Sullivan, W. T. 1977. *Astron. Astrophys.* 57: 373
Brandt, J. C. 1960. *Ap. J.* 131: 293
Brosche, P., Einasto, J., Rümmel, U. 1974. *Veröff des Astronomischen Rechen—Instituts Heidelberg*, No. 26
Burbidge, E. M., Burbidge, G. 1975. In *Galaxies and the Universe*, ed. A. Sandage, M. Sandage, J. Kristian, p. 81. Chicago: Univ. Chicago Press
Burbidge, E. M., Burbidge, G. R., Prendergast, K. H. 1960. *Ap. J.* 131: 282

Burbidge, E. M., Burbidge, G. R., Prendergast, K. H. 1961. *Ap. J.* 134: 874
Burbidge, E. M., Burbidge, G. R., Prendergast, K. H. 1965. *Ap. J.* 142: 154
Burbidge, E. M., Sargent, W. L. W. 1971. In *Nuclei of Galaxies*, ed. D. J. K. O'Connell, p. 351. Amsterdam: North-Holland
Burbidge, G. R. 1975. *Ap. J. Lett.* 196: L7
Burstein, D. 1978. *The Structure and Metallicity of S0 Galaxies.* PhD thesis. Univ. Calif., Santa Cruz
Burstein, D., Heiles, C. 1978. *Ap. J.* 225: 40
Burton, B. 1974. In *Galactic and Extragalactic Radio Astronomy*, ed. G. L. Verschuur, K. I. Kellermann, p. 82. New York: Springer
Carter, D. 1977. *MNRAS* 178: 137
Chincarini, G., Rood, R. T. 1977. *Ap. J.* 214: 351
Cowsik, R., McClelland, J. 1972. *Phys. Rev. Lett.* 29: 669
Davis, M., Geller, M. J., Huchra, J. 1978. *Ap. J.* 221: 1
de Harveng, J. M., Pellet, A. 1975. *Astron. Astrophys.* 38: 15
Dessureau, R. L., Upgren, A. R. 1975. *Publ. Astron. Soc. Pac.* 87: 737
de Vaucouleurs, G. 1948. *Ann. Astrophys.* 11: 247
de Vaucouleurs, G. 1974. *Proc. IAU Symp.* 58: 1
de Vaucouleurs, G. 1975. In *Galaxies and the Universe*, ed. A. Sandage, M. Sandage, J. Kristian, p. 557. Chicago: Univ. Chicago Press
de Vaucouleurs, G., de Vaucouleurs, A. 1972. *Mem. R. Astron. Soc.* 77: 1
de Vaucouleurs, G., de Vaucouleurs, A., Corwin, H. G., Jr. 1976. *Second Reference Catalogue of Bright Galaxies.* Austin: Univ. Texas Press
de Vaucouleurs, G., Freeman, K. C. 1973. *Vistas Astron.* 14: 163
de Vaucouleurs, G., Peters, W. L., Corwin, H. G., Jr. 1977. *Ap. J.* 211: 319
Dickel, J., Rood, H. 1978. *Ap. J.* 223: 391
Dressler, A. 1978a. *Ap. J.* 223: 765
Dressler, A. 1978b. *Ap. J.* 226: 55
Dressler, A., Rose, J. 1979. Preprint
Einasto, J. 1972. *Proc. Eur. Astron. Mtg., 1st*, ed. L. N. Mavridis, p. 291. Berlin: Springer
Einasto, J., Kaasik, A., Saar, E. 1974. *Nature* 250: 309
Emerson, D. T. 1976. *MNRAS* 176: 321
Faber, S. M., Balick, B., Gallagher, J. S., Knapp, G. R. 1977. *Ap. J.* 214: 383
Faber, S. M., Burstein, D., Dressler, A. 1977. *Astron. J.* 82: 941

Faber, S. M., Dressler, A. 1977. *Astron. J.* 82:187
Faber, S. M., Jackson, R. E. 1976. *Ap. J.* 204:668
Field, G. B. 1974. *Proc. IAU Symp.* 63:13
Field, G. B., Saslaw, W. C. 1971. *Ap. J.* 170:199
Fisher, J. R., Tully, R. B. 1975. *Astron. Astrophys.* 44:151
Fitzgerald, M. P., Jackson, P. D., Moffat, A. F. J. 1978. *Proc. IAU Symp.* 77:31
Ford, H. C., Jacoby, G., Jenner, D. C. 1977. *Ap. J.* 213:18
Forman, W., Jones, C., Murray, S., Giacconi, R. 1979. *Ap. J. Lett.* 225:L1
Freeman, K. C. 1970. *Ap. J.* 160:811
Freeman, K. C. 1975. In *Galaxies and the Universe*, ed. A. Sandage, M. Sandage, J. Kristian, p. 409. Chicago: Univ. Chicago Press
Freeman, K. C., Carrick, D. W., Craft, J. L. 1975. *Ap. J. Lett.* 198:L93
Gallagher, J. S., Hudson, H. S. 1976. *Ap. J.* 209:389
Gallagher, J. S., Ostriker, J. P. 1972. *Astron. J.* 77:288
Gliese, W. 1969. *Veröff. Astron. Rechen-Inst. Heidelberg,* No. 22
Goad, J. W. 1976. *Ap. J. Suppl.* 32:89
Godwin, J. G., Peach, J. V. 1977. *MNRAS* 181:323
Gott, J. R. III, Turner, E. L. 1976. *Ap. J.* 209:1
Gott, J. R. III, Turner, E. L. 1977. *Ap. J.* 213:309
Graboske, H. C., Grossman, A. S. 1971. *Ap. J.* 170:363
Greenstein, J. L., Boksenberg, A., Carswell, R., Shortridge, K. 1977. *Ap. J.* 212:186
Greenstein, J. L., Neugebauer, G., Becklin, E. E. 1970. *Ap. J.* 161:519
Gregory, S. A. 1975. *Ap. J.* 199:2
Gregory, S. A., Thompson, L. A. 1978. *Ap. J.* 222:784
Gregory, S. A., Tifft, W. G. 1976. *Ap. J.* 206:934
Gull, S. F., Northover, F. J. E. 1975. *MNRAS* 173:585
Gunn, J. E. 1974. *Comments Astron. Astrophys.* 6:7
Gunn, J. E., Knapp, G. R., Tremaine, S. 1979. Preprint
Gunn, J. E., Lee, B. W., Lerche, I., Schramm, D. N., Steigman, G. 1978. *Ap. J.* 223:1015
Harrison, E. R. 1974. *Ap. J. Lett.* 191:L51
Hartwick, F. D. A., Sargent, W. L. W. 1978. *Ap. J.* 221:512
Hegyi, D. J., Gerber, G. L. 1977. *Ap. J. Lett.* 218:L7
Hills, J. G. 1978. *Ap. J.* 219:550
Hintzen, P., Strittmatter, P. A. 1974. *Ap. J. Lett.* 193:L111
Hohl, F. 1976. *Astron. J.* 81:30

Hollander, M., Wolfe, D. A. 1973. *Nonparametric Statistical Methods.* New York: Wiley
Holmberg, E. 1937. *Lunds Ann.,* No. 6
Holmberg, E. 1954. *Medd. Lunds Astron. Obs., Ser. I* 86:1
Holmberg, E. 1958. *Medd. Lunds Astron. Obs., Ser. II,* 136
Holmberg, E. 1961. *Astron. J.* 66:620
Hoxie, D. T. 1970. *Ap. J.* 161:1083
Huchra, J. 1978. *Proc. IAU Symp.* 79:271
Huchtmeier, W. K. 1975. *Astron. Astrophys.* 45:259
Illingworth, G. 1977. *Ap. J. Lett.* 218:L43
Jackson, J. C. 1975. *MNRAS* 173:41P
Jenner, D. C. 1974. *Ap. J.* 191:55
Joeveer, M., Einasto, J. 1976. *Tartu Astron. Obs. Teated* 54:77
Jones, C., Forman, W. 1978. *Ap. J.* 224:1
Jones, D. H. P. 1976. In *The Galaxy and the Local Group. Royal Greenwich Obs. Bull.* No. 182, ed. R. J. Dickens, J. E. Perry, p. 1
Kahn, F. D., Woltjer, L. 1959. *Ap. J.* 130:705
Karachentsev, I. D. 1966. *Astrofizika* 2:81
Karachentsev, I. D. 1970. *Astron. Zh.* 47:509
Karachentsev, I. D. 1972. *Catalog of Isolated Pairs of Galaxies in the Northern Hemisphere, Soobshch. Spec. Astrophys. Obs.* 7:3
Karachentsev, I. D. 1977. *Proc. IAU Colloq.* 37:321
Karachentsev, I. D. 1978. *Proc. IAU Symp.* 79:11
Karachentsev, I. D., Pronik, V. I., Chuvaev, K. K. 1976. *Astron. Astrophys.* 51:185
Kerr, F. J., Westerhout, G. 1965. In *Galactic Structure,* ed. A. Blaauw, M. Schmidt, p. 167. Chicago: Univ. Chicago Press
King, I. R. 1966. *Astron. J.* 71:64
King, I. R. 1972. *Ap. J. Lett.* 174:L123
King, I. R. 1977. In *Highlights of Astronomy,* ed. E. A. Müller, 4 (Part II):41. Dordrecht: Reidel
King, I. R. 1978. *Ap. J.* 222:1
King, I. R., Minkowski, R. 1972. *Proc. IAU Symp.* 44:87
Kirshner, R. 1977. *Ap. J.* 212:319
Knapp, G. R., Kerr, F. J., Williams, B. A. 1978. *Ap. J.* 222:800
Kormendy, J. 1977a. *Ap. J.* 217:406
Kormendy, J. 1977b. *Ap. J.* 218:333
Kormendy, J., Bruzual, G. 1978. *Ap. J. Lett.* 223:L63
Kumar, S. S. 1969. In *Low Luminosity Stars,* ed. S. S. Kumar, p. 255. New York: Gordon & Breach
Larson, R. B., Tinsley, B. M. 1978. *Ap. J.* 219:46
Lea, S. M., Silk, J., Kellogg, E., Murray, S. 1973. *Ap. J. Lett.* 184:L105

Light, E. S., Danielson, R. E., Schwarzschild, M. 1974. *Ap. J.* 194:257

Limber, D. N. 1959. *Ap. J.* 130:414

Longair, M., Einasto, J., eds. 1978. *Proc. IAU Sump. No. 79.* Dordrecht: Reidel

Luyten, W. J. 1968. *MNRAS* 139:221

Luyten, W. J. 1974. In *Highlights of Astronomy,* ed. G. Contopoulos. 3:389. Dordrecht: Reidel

Luyten, W. J. 1975. *Proper Motion Survey with the Forty-Eight Inch Schmidt Telescope X L.* Minneapolis: Univ. Minnesota Press

Lynden-Bell, D., Lin, D. N. 1977. *MNRAS* 181:37

Malina, R. F., Lea, S. M., Lampton, M., Bowyer, S. 1978. *Ap. J.* 219:795

Materne, J. 1974. *Astron. Astrophys.* 33:451

Materne, J., Tammann, G. A. 1974. *Astron. Astrophys.* 37:383

Materne, J., Tammann, G. A. 1975. *Proc. Eur. Astron. Mtg., 3rd,* ed. E. K. Kharadze, p. 455. Tbilisi: Abastumani Ap. Obs.

Mathews, W. G. 1978. *Ap. J.* 219:413

McCuskey, S. W. 1966. *Vistas Astron.* 7:141

Melnick, J., White, S. D. M., Hoessel, J. 1977. *MNRAS* 180:207

Miller, R. H. 1978. *Ap. J.* 223:122

Mitchell, R. J., Culhane, J. L., Davison, P. J. N., Ives, J. C. 1976. *MNRAS* 175:29P

Mitchell, R. J., Ives, J. C., Culhane, J. L. 1977. *MNRAS* 181:25P

Morton, D. C., Andereck, C. D., Bernard, D. A. 1977. *Ap. J.* 212:13

Morton, D. C., Chevalier, R. 1972. *Ap. J.* 174:489

Morton, D. C., Chevalier, R. 1973. *Ap. J.* 179:55

Mushotzky, R. F., Serlemitsos, P. J., Smith, B. W., Boldt, E. A., Holt, S. S. 1978. *Ap. J.* 225:21

Newton, K., Emerson, D. T. 1977. *MNRAS* 181:573

Neyman, J., Page, T. L., Scott, E. 1961. *Astron. J.* 66:633

Noerdlinger, P. D. 1975. *Ap. J.* 197:545

Nordsieck, K. H. 1973. *Ap. J.* 184:735

Nottale, L., Moles, M. 1978. *Astron. Astrophys.* 66:355

Oemler, A., Jr. 1974. *Ap. J.* 194:1

Oemler, A., Jr. 1976. *Ap. J.* 209:693

Oort, J. H. 1965. In *Galactic Structure,* ed. A. Blaauw, M. Schmidt, p. 455. Chicago: Univ. Chicago Press

Ostriker, J. P., Caldwell, J. A. R. 1979. *Proc. IAU Symp.* 84: In press

Ostriker, J. P., Peebles, P. J. E. 1973. *Ap. J.* 186:467

Ostriker, J. P., Peebles, P. J. E., Yahil, A. 1974. *Ap. J. Lett.* 193:L1

Page, T. 1952. *Ap. J.* 116:63

Page, T. 1961. *Proc. Berkeley Symp. Math.*

Stat. Prob., 4th 111:277

Page, T. 1962. *Ap. J.* 136:685

Pease, F. G. 1916. *Proc. Natl. Acad. Sci. US* 2:517

Pease, F. G. 1918. *Proc. Natl. Acad. Sci. US* 4:21

Peebles, P. J. E. 1971. *Physical Cosmology,* p. 81. Princeton: Princeton Univ. Press

Peterson, C. J. 1978. *Ap. J.* 222:84

Peterson, S. 1978. *A Study of Binary Galaxies: Total Mass and the Ratio of Mass to Luminosity.* PhD thesis. Cornell Univ., Ithaca, NY 150 pp.

Poveda, A. 1958. *Bol. Obs. Tonantzintla y Tacubaya* 17:3

Richstone, D. 1975. *Ap. J.* 200:535

Richstone, D. 1976. *Ap. J.* 204:642

Roberts, M. S. 1969. *Astron. J.* 74:859

Roberts, M. S. 1975a. *Proc. IAU Symp.* 69:331

Roberts, M. S. 1975b. In *Galaxies and the Universe,* ed. A. Sandage, M. Sandage, J. Kristian, p. 309. Chicago: Univ. Chicago Press

Roberts, M. S., Whitehurst, R. N. 1975. *Ap. J.* 201:327

Rogstad, D. H., Lockhart, I. A., Wright, M. C. H. 1974. *Ap. J.* 193:309

Rogstad, D. H., Shostak, G. S. 1971. *Astron. Astrophys.* 13:99

Rogstad, D. H., Shostak, G. S., Rots, A. H. 1973. *Astron. Astrophys.* 22:111

Rogstad, D. H., Wright, M. C. H., Lockhart, I. A. 1976. *Ap. J.* 204:703

Rood, H. J. 1970. *Ap. J.* 162:333

Rood, H. J. 1974. *Ap. J.* 194:27

Rood, H. J., Dickel, J. 1976. *Ap. J.* 205:346

Rood, H. J., Dickel, J. 1978a. *Ap. J.* 224:724

Rood, H. J., Dickel, J. 1978b. Preprint

Rood, H. J., Page, T. L., Kintner, E. C., King, I. R. 1972. *Ap. J.* 175:627

Rood, H. J., Rothman, V. C., Turnrose, B. E. 1970. *Ap. J.* 162:411

Rose, J. A. 1977. *Ap. J.* 211:311

Rose, J. A. 1979. *Ap. J.* 231. In press

Rose, J. A., Graham, J. A. 1979. *Ap. J.* 231. In press

Rubin, V. C. Ford, W. K., Jr. 1971. *Ap. J.* 170:25

Rubin, V. C., Ford, W. K., Jr., Thonnard, N. 1978. *Ap. J. Lett.* 225:107

Rubin, V. C., Peterson, C. J., Ford, W. K., Jr. 1976. *Bull. Am. Astron. Soc.* 8:297

Salpeter, E. E. 1978. *Proc. IAU Symp.* 77:23

Sancisi, R. 1976. *Astron. Astrophys.* 53:159

Sancisi, R. 1978. *Proc. IAU Symp.* 77:27

Sandage, A. R. 1973. *Ap. J.* 183:711

Sandage, A. R. 1978. *Astron. J.* 83:904

Sandage, A. R., Tammann, G. A. 1975. *Ap. J.* 196:313

Sandage, A. R., Tammann, G. A. 1976. *Ap. J.* 210:7

Sargent, W. L. W., Schechter, P. L., Boksenberg, A., Shortridge, K. 1977. *Ap. J.* 212: 326

Sargent, W. L. W., Young, P. J., Boksenberg, A., Shortridge, K., Lynds, C. R., Hartwick, F. D. A. 1978. *Ap. J.* 221:731

Savage, B. D., Bohlin, R. C., Drake, J. F., Budich, W. 1977. *Ap. J.* 216:291

Schechter, P. 1976. *Ap. J.* 203:297

Schechter, P., Gunn, J. E. 1979. *Ap. J.* 229. In press

Schmidt, M. 1965. In *Galactic Structure*, ed. A. Blauuw, M. Schmidt, p. 513. Chicago: Univ. Chicago Press

Schwarzschild, M. 1954. *Astron. J.* 59:273

Schweizer, F. 1976. *Ap. J. Suppl.* 31:313

Schweizer, F. 1978. *Ap. J.* 220:98

Segalowitz, A. 1976. PhD thesis. Leiden Univ., Leiden

Seielstad, G. A., Wright, M. C. H. 1973. *Ap. J.* 184:343

Serlemitsos, P. J., Smith, B. W., Boldt, E. A., Holt, S. S., Swank, J. H. 1977. *Ap. J. Lett.* 211:L63

Shane, W. W. 1975. In *La Dynamique des Galaxies Spirals: Proc. C.N.R.S. Colloq.* 241:217

Shostak, G. S. 1973. *Astron. Astrophys.* 24: 411

Shostak, G. S. 1978. *Astron. Astrophys.* 68: 321

Shostak, G. S., Rogstad, D. H. 1973. *Astron. Astrophys.* 24:405

Sion, E. M., Liebert, J. 1977. *Ap. J.* 213:468

Slipher, V. M. 1914. *Lowell Obs. Bull.* II, No. 12

Smart, N. C. 1973. PhD thesis. Cambridge Univ.

Smith, S. 1936. *Ap. J.* 83:23

Soneira, R., Peebles, P. J. E. 1977. *Ap. J.* 211:1

Spinrad, H., Ostriker, J. P., Stone, R. P. S., Chiu, L.-T. G., Bruzual, G. 1978. *Ap. J.* 225:56

Starikova, G. A. 1960. *Astron. Zh.* 37:476

Stebbins, J., Kron, G. E. 1957. *Ap. J.* 126:266

Straka, W. C. 1971. *Ap. J.* 165:109

Strom, S. E., Strom, K. M. 1978. *Astron. J.* 83:73

Tammann, E. A., Kraan, R. 1978. *Proc. IAU Symp.* 79:71

Thuan, T. X., Seitzer, P. 1979. Preprint

Tinsley, B. M., Cameron, A. G. W. 1974. *Astrophys. Space Sci.* 31:31

Toomre, A. 1963. *Ap. J.* 138:385

Toomre, A. 1977. In *The Evolution of Galaxies and Stellar Populations*, ed. B. M.

Tinsley, R. B. Larson, p. 401. New Haven: Yale Univ. Press

Truran, J. W., Cameron, A. G. W. 1971. *Astrophys. Space Sci.* 14:179

Tully, R. B., Bottinelli, L., Fisher, J. R., Gougenheim, L., Sancisi, R., van Woerden, H. 1978. *Astron. Astrophys.* 63:37

Tully, R. B., Fisher, J. R. 1978. *Proc. IAU Symp.* 79:31

Turner, E. L. 1976a. *Ap. J.* 208:20

Turner, E. L. 1976b. *Ap. J.* 208:304

Turner, E. L., Aarseth, S. J., Gott, J. R., Mathieu, R. D. 1979. *Ap. J.* 228:684

Turner, E. L., Gott, J. R. III. 1976. *Ap. J. Suppl.* 32:409

Turner, E. L., Sargent, W. L. W. 1974. *Ap. J.* 194:587

van Albada, G. D., Shane, W. W. 1976. *Astron. Astrophys.* 42:433

van de Kamp, P. 1971. *Ann. Rev. Astron. Astrophys.* 9:103

van den Bergh, S. 1977. *Vistas Astron.* 21:71

van der Kruit, P. C., Allen, R. J. 1978. *Ann. Rev. Astron. Astrophys.* 16:103

Veeder, G. J. 1974. *Ap. J. Lett.* 191:L57

Vorontsov-Velyaminov, B. 1959. *Atlas of Interacting Galaxies.* Moscow: Sternberg Astronomical Institute

Warner, P. J., Wright, M. C. H., Baldwin, J. E. 1973. *MNRAS* 163:163

Wegner, G. 1974. *MNRAS* 166:271

Weliachew, L., Sancisi, R., Guélin, M. 1978. *Astron. Astrophys.* 65:37

White, S. D. M. 1976a. *MNRAS* 174:19

White, S. D. M. 1976b. *MNRAS* 177:717

White, S. D. M. 1977. *MNRAS* 179:33

White, S. D. M. 1978. *MNRAS* 184:185

White, S. D. M., Sharp, N. A. 1977. *Nature* 269:395

Wielen, R. 1974. In *Highlights of Astronomy*, ed. G. Contopoulos, 3:395. Dordrecht: Reidel

Williams, T. B. 1977. *Ap. J.* 214:685

Wolf, M. 1914. *Vierteljares schr. Astron. Ges.* 49:162

Yahil, A. 1977. *Ap. J.* 217:27

Yahil, A., Tammann, G. A., Sandage, A. R. 1977. *Ap. J.* 217:903

Yahil, A., Vidal, N. V. 1977. *Ap. J.* 214:347

Young, P. J. 1976. *Astron. J.* 81:807

Young, P. J., Sargent, W. L. W., Boksenberg, A., Lynds, C. R., Hartwick, F. D. A. 1978. *Ap. J.* 222:450

Zwicky, J. 1933. *Helv. Phys. Acta* 6:110

Zwicky, F. 1937. *Ap. J.* 86:217

Zwicky, F. 1957. *Morphological Astronomy.* Berlin: Springer

Ann. Rev. Astron. Astrophys. 1979. 17:189–212

DIGITAL IMAGING TECHNIQUES

✷2149

W. Kent Ford, Jr.

Department of Terrestrial Magnetism, Carnegie Institution of Washington,
5241 Broad Branch Road, N.W., Washington, D.C. 20015

INTRODUCTION

Astronomers attempting to obtain more efficient detecting systems often acquire instead a better understanding and appreciation of the information storage capacity of photographic emulsions. The information content of plates from wide-field Schmidt telescopes and from high-dispersion coude spectrographs is awesome to those working with 256×256 formats or 1024 linear elements. Recent advances in astronomical photography at low light levels are reviewed elsewhere in this volume by Smith & Hoag (1979). However, a different sort of information from that which is readily obtained from photographs is often required by the astronomer who seeks data in the form of redshifts, surface brightness distributions, spectral line ratios, etc. As one astronomer put it: "The observer wants answers, not information!"

Digital imaging techniques have been developed for several reasons including an impatience to have the results of observations, "the answers," quickly available in numerical form suitable for quantitative interpretation. A major goal in the design of these systems is to increase the accuracy of the observational data. This objective requires a highly stable, reproducible transfer characteristic relating the digital output to the flux of the optical input. A linear transfer characteristic greatly simplifies the operating system but is not absolutely essential. It is essential that the system as a whole, i.e., detector plus digital memory, be capable of very high storage capacity per picture element. A reproducible, linear transfer characteristic along with storage of images in digital representation in memory permits point-by-point calibration of the response of the system to standard sources and the subtraction of background due to the night sky and to thermal processes in the detector. Very modest computational capability is required for on-line estimation of the statistical precision of

189

the data. Hence, the observer can more closely control the use of the available telescope time.

In this paper we review the various techniques that are currently employed by astronomers in detecting and recording images in digital form. Both one and two dimensional systems are included because of the general importance of spectroscopic data. Our emphasis is on observations at low light levels, and hence this review is largely limited to stellar ground-based systems rather than solar or planetary observations or spaceborn systems.

This review is a continuation of the survey of image tube systems made in *Annual Review of Astronomy and Astrophysics* by Livingston (1973), but is limited to those imaging systems whose output is digital rather than analog in form. These systems have evolved greatly in the six years since Livingston's review, and many are no longer primarily experimental in nature.

Boyce (1977) presented a general review of low-light-level detectors in ground-based astronomy from photographic plates to CCD's. Carruthers (1977) reviewed electronic imaging for space science with emphasis on the basic principles of various types of detectors. The proceedings of a symposium held at the University of British Columbia (edited by Glaspey & Walker 1973) contain good reviews of television-type sensors. Many aspects of detector technology are described in the published preprints of papers presented at the Seventh Symposium on Photo-Electronic Image Devices (edited by McMullan & Morgan 1978). Also, in another area outside the scope of this review, the proceedings of IAU Colloquium No. 40 on astronomical detectors with linear response, edited by Duchesne & Lelievre (1976), contain many papers on electronographic cameras in which photographic emulsions are exposed directly to photoelectrons and after development are digitized.

AN OVERVIEW OF TYPES OF DIGITAL SYSTEMS

For the purpose of this review, digital imaging techniques are divided according to the type of detector that is employed into four categories: television camera tubes, silicon arrays, hybrid systems, and digital image tubes. The choice of these particular divisions is somewhat arbitrary, but they serve as a convenient framework for discussing the characteristics of various digital systems.

The first category is composed of television camera tubes [such as the Silicon Target Vidicon, Silicon Intensifier Target Vidicons (SIT tubes), and secondary-electron-conduction target tubes (SEC tubes)] that have been adapted to the low-light-level requirements of astronomers.

The silicon arrays are solid-state devices used without intensification.

These include self-scanned diode arrays, charge injection devices (CID's), and charge-coupled devices (CCD's).

The hybrid systems are characterized by having the output of an image intensifier read and converted into an electrical signal by a TV-type tube, an image dissector, or a solid-state array.

The digital image tubes are custom-made devices having a photo-cathode and an electron-sensitive detector sealed in a single envelope. They are electron bombarded silicon (EBS) devices.

A distinction is often made between those devices whose electrical output is an analog signal that is digitized and event-counting systems whose output is a series of discreet electrical pulses that are counted. In general, the digital TV systems and silicon arrays are used in the analog output mode. The hybrid systems with intensifier front ends and the digital image tubes are usually (but not always) operated in a mode that permits the actual counting of photoelectron events. In principle pulse-counting photometric systems have a small advantage at the low-light-level limit (on the order of $\sqrt{2}$) over analog integration systems due to the equal weighting of each photoelectron in the counting process. Actually a practical advantage of greater importance has often been the relative stability of calibration of pulse counting over systems relying largely on amplifier stability. However, modern FET amplifiers and hybrid-circuit analog-to-digital converters can be sufficiently stable that other performance characteristics are likely to determine the choice between analog and pulse-counting systems.

The relative performance of a detector is measured by its Detective Quantum Efficiency (DQE). The significance of the DQE is that it is a measure of efficiency of an actual system relative to that of an idealized, perfect photon-counting system. This figure of merit is defined as being the ratio of the square of the signal-to-noise ratio obtained in the output of a detector to the square of the signal-to-noise ratio of the input flux of photons from a calibrated source.

$$DQE = (S/N)^2_{out}/(S/N)^2_{in}.$$

Unfortunately, the general lack of availability of well-calibrated, low-light-level sources makes a careful determination of DQE a nontrivial task.

In cooled, pulse-counting systems the noise can be small relative to signals corresponding to very low levels of input light. The emphasis in the design of these systems is on maintaining high counting efficiency and therefore a DQE near that expected from the responsivity of the detector. On the other hand, the analog systems have readout noise that degrades the DQE for very small signals.

TELEVISION CAMERA TUBES

In the camera tubes used in commercial television, electrical charge representing the image at the input of the device accumulates and is stored on the surface of a target. The vidicon family of camera tubes, for example, has photoconductive targets. The target is externally biased with respect to the cathode of an electron gun. The target is scanned by an electron beam from the gun and is thereby charged to the potential of the cathode. Photogenerated charge locally discharges the photoconductive target. An electrical output signal is generated in the readout when the electron beam recharges the target in a raster scan. Even though these highly developed devices generate video pictures of excellent visual quality, they are poorly suited for producing quantitative data at low light levels. [See, for example, Livingston (1973).]

Photometric difficulties arise basically because the output signal representing a picture element is not completely independent of the charge stored in adjacent areas of the target. Much of the problem is inherent in the electron beam readout and the raster scanning process. The lack of sharply defined edges in the profile of charge distribution in the read beam produces an asymmetrical output from a raster scan of a symmetrical image on the target. The geometrical integrity of the output can also be compromised by an actual "bending" or displacement of the beam by the charge accumulated on adjacent areas of the target and, in some types of targets, lateral charge leakage. These subtle effects which creep into the photometric data are unrelated to the stability of the sweep circuits driving the scanning beam.

The output from a camera tube is the charge measured in the target bias circuit as the electron beam recharges the potential distribution on the target representing the image. The electrons from the hot electron gun have a distribution in energy characteristic of their thermal origin and an additional spread in axial velocity due to aberrations in the electron optics. For small residual images the spread in energy in the beam is comparable to the local surface potential on the target. Consequently only those electrons in the higher energy portion of the distribution are able to land on the target. This beam-landing characteristic is responsible for incomplete recharging of low level images during readout and is called capacitive lag. Thus, in order to maintain a linear transfer characteristic and efficient readout of the signal, an increment of target bias, or "pedestal," is usually introduced prior to readout to insure adequate beam landing. Recharging of the target occurs as long as there are electrons available in the beam with sufficient energy to land. Hence there is always a statistical uncertainty in the amount of charge deposited on the target. This beam-landing noise contributes, along with preamplifier noise, to the readout noise.

"Vidicon" is a general term for a camera tube with a photosensitive target. The target both senses the image and stores the photogenerated electrical image. Standard vidicons have photoconductive targets and are not used in low-light-level imaging. Vidicons with lead oxide targets are widely used in commercial television and are usually known by the manufacturer's trade name. ("Plumbicon" is Philip's registered trade name and "Visticon" is RCA's.) This type of tube is used in the hybrid intensifier systems described in a later section. The silicon diode vidicon has a photosensing target consisting of a mosaic array of silicon diodes. It is of particular interest because of its high quantum efficiency peaking at 0.8 μ and spectral response extending beyond 1.0 μ.

Other camera tubes separate the image conversion and charge storage functions. In an SEC camera tube photoelectrons from a semitransparent photocathode are accelerated and focussed onto a target with a high secondary electron emission yield. A target prepared by charging to gun potential by scanning with the electron beam is discharged by a process of secondary electron conduction. The video signal is generated as the scanning electron beam replenishes the charge on the target. The conversion of the energy of the accelerated primary photoelectrons through secondary emission provides an effective charge gain for small signals of less than 100. The gain, however, depends on the potential across the SEC layer and is therefore signal dependent, decreasing for larger surface potentials. This leads to a nonlinear response function that saturates for large signals. The detailed operation of the SEC targets is described by Goetze (1966) and Boerio et al. (1966).

An EBS or SIT camera tube also uses a photocathode and a separate target for charge storage. In this case the target is a silicon diode array. The diodes are pn junctions formed by photolithographic techniques at a density of roughly half a million diodes per cm^2. In operation the diodes are reverse biased by the electron beam, and they effectively become, because of the depletion region formed in the bulk silicon, an array of charged storage capacitors. Photoelectrons impinging on the bulk target create electron-hole pairs. This is an effective gain mechanism as one electron-hole pair is created for every 3.6 eV of absorbed energy. Typically about 1000 charge carriers are collected for each photoelectron. This charge diffuses through the silicon and contributes to the discharge of the capacitance associated with the diodes. As with other vidicons the video output is generated as the scanning electron beam replenishes the charge.

With this introduction to a few members of the camera tube family we turn now to their astronomical application in digital imaging.

SEC Camera Tubes

At Princeton University Lowrance et al. (1976) have had an active program to develop large format camera systems for space astronomy. As a result of this program Westinghouse SEC tubes have been used in many ground-based applications (Morton 1972, 1973). The SEC tube is operated in an integration mode. After the target has been through an erase and prepare cycle the scanning beam is turned off and the exposure accumulates on the target. Cooling is not required for long exposures except to remove the heat generated by the focus solenoid for the image section and to chill the bialkali photocathode. The capability to make long exposures was exploited, for example, by Lowrance et al. (1972) and by Morton & Andereck (1976) in high-dispersion coude spectroscopy with the Hale telescope. The spectral input was sampled by 25-μ elements stored in 1024 bins. The system had a 60% (square wave test pattern) response at 10 line pairs per millimeter. The target gain varied from 78 for weak exposures to 40 for "full target" exposures.

The photometric properties of the Princeton SEC system have been analyzed by Crane (1973) and Zucchino (1976). The observed S/N ratio in the SEC exposures approach noise-limited performance for optimum exposures on the order of 10^3 photoelectrons per pixel or about 3% photometric accuracy. The performance deteriorates for higher than optimum exposures. The Princeton SEC camera was used, for example, by Crane (1975) for detailed surface photometry of a SBO galaxy, NGC 2950, and by Hoffman & Crane (1977) for determining isophotal magnitudes for several thousand galaxies in six clusters. In these particular programs the SEC camera provided calibrated, large format arrays of digital photometric data rather than great gains in speed or accuracy.

Chiu (1977) describes the operation and performance of an experimental SEC camera system on the McMath telescope at Kitt Peak. The efficiency and resolution of the system have been applied to a program of high-dispersion stellar spectrophotometry by Chiu et al. (1977).

Silicon Target Vidicon

The target of the silicon target vidicon is the photosensing element in this tube type. It has high quantum efficiency in the near-infrared, but there is no mechanism to provide electron gain. In order to make use of the near-infrared response McCord & Westphal (1972) developed a cooled silicon vidicon camera that permitted integrations of several hours. McCord & Frankston (1975) describe the calibration and operational procedures required to achieve high photometric accuracy with their system. Hunten et al. (1976) developed a similar camera at Kitt Peak. Interference fringes in the silicon target become a major problem when the devices are used in the infrared (Title 1974).

The performance characteristics of the silicon target vidicon have been analyzed by Crane & Davis (1975), by Hunten & Stump (1976), and by Diner & Westphal (1977). With suitable calibration it is possible to do surface photometry with better than 1% accuracy with this tube, but the lack of internal gain tends to limit its use to relatively bright objects. (See for example the review of uses by Westphal & Kristian 1976.) At CTIO Aikens & Lasker (1973) and Ingerson et al. (1976) have adapted silicon target vidicons with enhanced red sensitivity to spectroscopic work.

EBS/SIT Camera Tubes

Camera tubes with electron-bombarded silicon targets (EBS or SIT tubes) have photoemissive cathodes and gain provided by the loss of energy of photoelectrons impinging on the target. Westphal (1973) first demonstrated the sensitivity of the SIT in an integrating camera system used at the prime focus of the Hale Telescope. Thermal leakage in the target was suppressed by operating the tube in a dry ice cooled camera head, so that the charge storage capacity of the target could be used for long integrations. The linearity of the system was demonstrated by laboratory measurements and by comparison of SIT photometry with stellar photoelectric sequences (Westphal & Kristian 1976). The system was made sufficiently stable that reproducible transfer functions could be derived from flat field calibrations made at the beginning and end of the night's observations.

Westphal, Kristian & Sandage (1975) exploited these properties of the SIT in an efficient prism spectrograph used for obtaining absorption-line redshifts of faint galaxies. The linearity and stability of the transfer function permit subtraction of the night sky spectrum. A more general purpose spectrograph utilizing a SIT camera has been developed by Gunn (1977).

A SIT camera system is used extensively at CTIO for spectrophotometry and redshift measurements (Osmer 1977, Ingerson et al. 1976). The camera system is described by Aikens & Lasker (1973). The precision of the spectrophotometric capability of a commercially available SIT camera is reported by Weller et al. (1977) and by Jeffers & Weller (1978).

SILICON ARRAYS

Silicon arrays for digital imaging are solid-state detectors that convert an optical image directly to an electrical output. The silicon array photo-sensors are attractive devices for astronomical observations because of several of their inherent characteristics. Silicon detectors have a high responsive quantum efficiency especially in the red and near-infrared. The arrays, which are fabricated on a single wafer of silicon, have a well-defined, stable geometry. These advantages are achieved at the expense of problems with readout noise, with pixel-to-pixel variations in sensitivity and dark current that must be carefully calibrated, and with interference fringes within the array from the input image.

At present, three different types of silicon arrays are available for digital imaging. These are the self-scanned photodiode arrays, the charge-coupled devices, and the charge injection devices. These differ in the way that the photogenerated signal is accumulated and transferred out. Two-dimensional arrays have been developed for television applications. These generally have storage elements on the chip in addition to image-sensing, or charge-accumulating, elements. Other devices made especially for image digitizing may require a shutter (or a pulsed input image) to prevent smearing of the image in the output.

These are relatively new and unfamiliar devices that are evolving rapidly. They have just begun to produce useful astronomical observations. Many of the principles of operation of the various sorts of silicon arrays are illustrated in a review by Livingston (1976). Here we summarize some of the distinctive characteristics of each type of device.

Self-Scanned Photodiode Arrays

A self-scanned photodiode array consists of a series of pn junction photo-diodes, each of which is connected to an output line via a low-leakage transitor switch. These switches are activated sequentially by means of an on-chip shift register. For operation as a sensor each diode is reverse biased creating a region depleted of charge and is then left floating. The depletion region can be filled, in time, by thermally generated charge (dark current) and by photogenerated charge. A video output is produced by clocking a switching signal through the shift register. Thus each diode is in turn connected to the output line and thereby recharged. The output signal is the charge required to recharge the diode. The capacitance of the video line is high (~ 100 pf) since it is the sum of the capacitance of all the FET switches in parallel. Since the preamp noise is directly proportional to the line capacitance, significant noise reduction is possible by arranging to read out the array on multiple video lines to reduce the capacitance associated with each line.

The development of self-scanned photodiode arrays has been pioneered by the Reticon Corporation, Sunnyvale, California. Many groups have used "Reticons" for astronomical observations. Of these, some representative systems illustrate their characteristics.

Tull & Nather (1973) and Vogt, Tull & Kelton (1978) have adapted Reticons for use in high-precision spectrophotometry. Their 1024-element array installed at the coude spectrograph of the 2.7-m telescope at McDonald Observatory is cooled to $-100°C$ to $-150°C$ to suppress thermal leakage and thus permit long exposures. A mini-computer is used to control the detector system and to store data. This is a highly successful operating system that is productive in a wide variety of high-dispersion spectroscopic programs. These include measure of interstellar isotopic abundance ratios, Zeeman analysis of late-type stars and measures of stellar abundance ratios. The system is suitable for high exposure, high signal-to-noise ratio work. Vogt et al. (1978) show that, for exposures corresponding to 10^7 photons per diode, the total noise is essentially just the shot noise of the input photons.

Livingston et al. (1976) utilized the photometric properties and red spectral response of a pair of 512-element Reticons in a solar magneto-graph at Kitt Peak National Observatory. The instrument, which has been in operation since 1974, is used to map the magnetic field across the disk of the sun. The operation of the instrument as a magnetograph requires accurately measuring the intensity differences in right and left circularly polarized components that are of the order of 3×10^{-4}. Exposures of 1/60 sec yield signals of 4×10^6 electrons or about 20% of a fully saturated exposure. Scanning the entire solar disk requires 21 contiguous sweeps and a total of 45 minutes.

Another example of the use of the self-scanning diode array is an attempt to incorporate a Reticon into a camera for low-dispersion spectroscopy. Buchholz et al. (1976) and Walker et al. (1976) at the University of British Columbia built a Schmidt camera of the conventional optical design that has provision for the operation of a cooled 1024-element Reticon at the camera focus. The sensor was cooled with a liquid nitrogen cold probe. The entire cold assembly is sufficiently small that it can be placed interior to the camera at the conventional Schmidt focus. This elegant optical scheme presents problems, however, with long signal lines leading from the chip to the electronics located outside of the camera. Also it is difficult to avoid some frosting in the system in spite of care in sealing and drying the interior of the camera.

The small physical size of the Reticon was also exploited by Campbell

(1977) in the construction of a system attached to a small spectrograph. This system had, for various reasons, high readout noise and was therefore useful only at moderately high light levels. However, a spectrophotometric capability was achieved in a compact system.

The lesson to be learned from these examples is that the presently available self-scanned photodiode arrays can be used for precision spectrophotometry at high signal-to-noise ratios in applications where exposures of 10^5 photons or greater are feasible. Their successful use requires well-executed electronics to minimize readout noise, reliable and stable cooling to minimize dark current, and, of course, adequate data handling capability for calibration. Their use is probably limited to linear arrays. Area arrays of up to 50 × 50 elements are available but have the inherent disadvantage that surface electrodes block 50% of the sensor area.

Charge-Coupled Devices (CCD's)

CCD's are silicon arrays that are based on the principle that charge accumulated in a spatially defined potential well at the surface of a semiconductor can be moved about by moving the local potential minimum. The charge accumulated in each element is moved unidirectionally to the adjacent element and then to the next and so on until it arrives at the input of a sensitive preamp. The operation of the device depends on the ability to transfer charge from one element to another with very high efficiency. Thus, a large number of transfers can be made without degrading the amplitude of the signal and consequently losing resolution within the image. Two-dimensional imaging is achieved by one of several geometrical layouts. The commercial interest in CCD's stems from their compactness and sensitivity. The dark current at normal ambient temperatures is sufficiently low for many applications at standard television frame rates. At temperatures below $-100°C$, the dark current is reduced to the point that integrations of several hours are possible.

Movement of charge along a semiconductor, which is the essential part of the charge-coupled device, is achieved with electrodes deposited on the surface of a silicon chip. A local potential well is created by applying the appropriate voltages to a pattern of electrodes deposited on an insulating surface on a wafer of silicon. This is a metal, oxide, and silicon or MOS structure. Three electrodes typically define one element of the array, and every third electrode is connected. A voltage on the center electrode creates a potential well under that electrode; for p-type silicon a positive voltage creates a region depleted of charge. This depletion region is filled in time by thermal diffusion of electron (i.e. dark current) and by photogenerated electrons. After a period of integration, the collected charge is moved along the device by applying a three-phase voltage in sequence to the three lines. The potential well is defined in the directions perpendicular to the transfer by "channel stops," i.e. potential barriers implanted into the array. Two-phase devices, which require somewhat simpler external electronics, are built by making the wells asymmetric.

The efficiency of transfer of charge in the CCD is limited largely by losses of electrons to traps associated with the semiconductor-oxide interface. These traps, or fast interface states, can retain a small portion of charge from one pixel and release it later to the signal from adjacent pixels. Two methods are used to circumvent the problem. The device can be operated with a uniform bias illumination so that there is always some charge filling the interface states. Alternatively a buried channel can be fabricated by implanting an n-type layer between the oxide insulator and the p-type silicon. Both methods reduce the available charge capacity of the wells.

High transfer efficiency in CCD's makes feasible two-dimensional arrays with a large number of picture elements. The product of the fraction of charge loss at each transfer times the number of transfers is a measure of the degradation of the spatial response of the device. To achieve compatibility with standard television frame times, storage elements are included on the chip in addition to the sensor elements. In one arrangement the storage registers are interlined line by line with the sensing elements, thus cutting the sensitive area of the array by one-half. Another method relies on transferring the contents of an entire frame to an adjacent frame storage register. No storage is required for applications where the exposure can be externally terminated so as not to smear the readout.

CCD's can be illuminated from the front by imaging through transparent electrodes and oxide layer. Alternatively, the bulk silicon is made sufficiently thin that charges generated from backside illumination are collected in the depletion region. The difficulty in this is that an inherently rugged wafer must be uniformly thinned to a rather thin 10 μ, making mounting and cooling more difficult. A decided benefit is that the blue spectral response is considerably enhanced by backside illumination. Moreover, CCD's that are used as charge sensitive image detectors are more resistant to radiation damage to the charge-coupling structure when back-illuminated by an electron image. (See section on digital image tubes below.)

The first generally available CCD's were the 100 × 100 arrays made by Fairchild Camera Instrument Corporation in 1973. These are buried-channel devices illuminated through semitransparent electrode structures. The device is designed with interline transfer organization so only half of the illuminated area produces signal charge. These devices were evaluated for various astronomical purposes by the Panoramic Detector Group at KPNO (Lynds 1975), at Princeton by Renda & Lowrance (1975), also at Princeton by Loh & Wilkinson (1976), and at the University of Maryland by Currie (1975). These early tests generally were directed toward measuring and understanding various sources of noise in the system. Loh (1976) and Beck & Wilkinson (1976) made use of the photometric properties of the CCD in a digital camera used as an area photometer. The output noise of their system for 20-minute integrations is dominated by photon shot noise from the sky at a level which is 0.5% of the sky flux. A similar system is under development at the Center for Astrophysics (R. Leach, personal communication).

An extensive program of development of CCD's for scientific imaging from space has been undertaken at Texas Instruments for the Jet Propulsion Lab with NASA support. Vescelus & Antcliffe (1976) describe some of the tradeoffs and choices in sensor technology that were considered in designing first a 100×160 and then a 400×400 imaging array. These were thinned, backside-illuminated devices with 3-phase, buried-channel transfer. Cooling the device lowered the dark current at the expected rate and a simple on-chip preamp allowed low noise readout.

A breadboard camera from JPL was used by Smith (1976) at Mt. Lemmon Observatory for planetary imagery in the narrow spectral passbands in the near IR. Particularly interesting were exposures of Uranus in the 890-nm CH_4 absorption band showing considerable structure.

In collaboration with F. Landauer the JPL CCD's have been used at the Hale Observatories by Oke (1977) and by Westphal & Kristian (1977). The photometric imaging properties were investigated by Kristian and Westphal who found no departure from linearity over a 12-mag range and agreement with a photoelectric sequence by Sandage to within 2%. The dynamic range and spatial resolution were exploited by Young et al. (1978) in area photometry of the nucleus of M87.

In 1977 JPL with NSF support made a CCD camera available for evaluation by astronomers at Lick, CTIO, and KPNO. Many problems with fixed pattern noise were encountered but useful data were obtained before the device was accidentally destroyed.

In summary, the present status of the CCD as an astronomical sensor is that a number of evaluations have demonstrated its good photometric properties. The successful use of devices that are generally available today requires working around characteristics inherent to the television-compatible design. The use of CCD's in the space program may lead to more general availability of these devices for digital imaging.

Charge-Injection Devices (CID's)

The charge-injection device (CID) (Michon & Burke 1973, Michon et al. 1975) is an interesting silicon array that superficially resembles a CCD. Each pixel consists of two capacitors formed by a pair of transparent metal oxide electrodes. One of the pair is connected to the capacitors of other pixels in that row of the array; the second of the pair is connected to the capacitors of the other pixels in that column. This allows each pixel to be addressed individually by row and column. The row and column lines are biased with a larger voltage applied to the row electrode. To read out the accumulated charge the voltage on the selected column and row electrodes is simultaneously set to zero and the net charge injected is measured by integrating the displacement current. Alternatively, a nondestructive readout can be made by transferring all stored charge from the row well to the column well and measuring a potential that is proportional to the accumulated charge. In contrast with the CCD, charge is held where it is generated so that charge transfer loss is minimized. The nondestructive read capability of the CID can be used to great advantage by summing repeated reads to the readout noise contribution. Ultra-low-noise readout is difficult because the on-chip preamp sees the capacitance associated with a whole column (typically 25 pf).

Planetary imaging with a tunable filter by Wattson et al. (1975) demonstrated the usefulness of CID for digital imaging. They illustrate, for example, the apparent absence of Saturn's ball in the wavelengths of the deep methane absorption.

At KPNO the panoramic detector group has explored a number of uses of the CID. Aikens et al. (1976a,b) report on the operation of the GE CID for photometric imaging of high precision. They cool the chip to $-196°C$ to make dark current negligible in exposures up to one hour. A full exposure is represented by 6×10^6 charges and the RMS noise is 1000 charges.

HYBRID SYSTEMS

In the systems considered above, an analog video signal is digitized. In an ideal low-light-level system, photoelectron events are counted in a geometrically well-defined array of pixels. If it is established that the counting of events is a noise-free process, then in a given pixel, the accumulated count, be it large or small, is an accurate measure of the signal in that pixel. In the late 1950s and early 1960s, image intensifiers were developed that had sufficient gain that individual photoelectron events were evident at the output screen. Since that time, many systems have been devised to count individual photoelectron scintillations displayed at the phosphor screen of an image intensifier. These systems, consisting of a high-gain image intensifier and some device for detecting discrete photoelectron events, are called hybrid systems. The scintillation detecting device may be a sensitive television sensor tube (e.g. Plumbicon or SIT), an image dissector (which may be thought of for this purpose as a scanning photomultiplier), or a silicon array.

The implementation of a hybrid system involves the practical consideration of count-rate limitations, scintillation spot size relative to the size of the sensor elements, and memory access times. Hybrid systems have therefore generally been most successful for observations in a limited format of very low photon fluxes.

Intensifier-Camera Tube Systems

The feasibility of using television type sensors to detect photoelectron events from an image intensifier has been explored by several groups. One early study that was particularly relevant was Mende's (1971) analysis of the single photoelectron recording efficiency of his 4-stage EMI magnetically-focussed intensifier, lens-coupled to a Plumbicon. Mende used uniform bias illumination on the Plumbicon to improve the signal from the scintillations occurring randomly in a dark field. He also analyzed the effect of the scintillation spot size and persistence on the observed pulse height distribution. He concluded quite correctly that the practical deficiency of the system was that the television system made equal samples of lines and frames in space and time. Single scintillations, occurring randomly in time and location, are likely to be sampled partially by adjacent lines and succeeding frames. Much of the work since 1971 in developing event-counting systems has been directed towards circumventing these factors which degrade the system performance.

In the University College London "Image Photon Counting System" (Boksenberg 1972, Boksenberg & Burgess 1973), many innovative features directly attack these problems. Boksenberg's system consists of a 4-stage EMI cascade intensifier with a gain of 10^7 lens-coupled to a Philips Plumbicon. To locate accurately the position of each photoelectron, a peak detector and a fast logic unit, operating on the video signal from successive scan lines, generate an address that represents the position of the center of the scintillation. The space on the detector associated uniquely with an address can be made small compared to the width of the average scintillation thereby achieving higher spatial resolution than could be obtained by analog integration. This technique makes it feasible to use a very high gain intensifier in the front end of the system so that even the scintillations in the weaker portion of the brightness distribution are well above the noise in the detector system. At the same time sufficient resolution is achieved with the aid of the centroiding system so that data can be usefully acquired in 2048 channels, or twice the number of television scan lines.

Fort, Boksenberg & Coleman (1976) have analyzed the effect of centroiding and sampling time (i.e. frame time) on the DQE of a event-counting system. They derive an expression for DQE as a function of the count rate per pixel per frame time. This is just an evaluation of the effect of losing counts due to the probability of having more than one event per pixel per frame time in a system that counts only zeros or ones. According to this analysis there is essentially no loss in DQE for count rates of 0.01 counts per pixel frame time or less and a 10% loss in DQE at 0.2 counts per pixel per frame.

These considerations lead to an understanding of the success of Boksenberg's well-travelled system in faint object spectroscopy. For example, in their 1973 observations of the absorption-line redshifts in the spectrum of the QSO PKS 0237-23, Boksenberg & Sargent (1975) obtained useful data at the Hale telescope with a signal on only 200 to 500 counts per channel after sky subtraction. Greenstein et al. (1977) studied the rotation and gravitational redshifts of white dwarfs by utilizing the improved resolution of the centroiding process. With the Isaac Newton telescope at Herstmonceux Boksenberg et al. (1975) made spectrophoto-metric observations of 3C 273 consisting of many thousand counts per channel. These examples are only a sample of the types of observations undertaken with the University College Image Photon Counting System.

Hybrid systems for photoevent counting have been built using other combinations of intensifiers and television tubes. For example, Cenalmor (1976) and Cenalmor et al. (1978) describe a system consisting of a micro-channel plate intensifier coupled by fiber optics to a SIT tube (Thomson CSF "Nocticon"). Boulesteix (1978) has explored various astronomical applications for this equipment. He reports a linear response to levels equivalent to an average rate of 1/60 event per pixel per frame. A 256-square format is used for photometric imaging. Narrow-band (10 Å) inter-ference filter exposures of galaxies e.g. M51) and large HII regions (e.g. NGC604 in M33) have been made at the 1.9-m telescope at Haute Provence with a focal reducing camera to match the scale of the telescope to the detector.

Gilbert, Angel & Grandi (1976) have a three-stage intensifier tube lens coupled to an RCA SIT tube that can be operated in a photoelectron counting mode. In order to extend the range of the instrument beyond the limits imposed by coincidence losses an alternative mode of detection is available. In the alternative mode a high speed sample-and-hold circuit and analog-to-digital converter generate a digital word that is added to the contents of memory. At very low counting rates this digital word just represents the pulse height or brightness associated with single photo-electron events, and a histogram of a number of these words is a measure of their brightness distribution. With increasing input light multiple events per pixel per readout gives, in principle, multiple peaks in the distribution. The addition of these words from successive scans gives a measure statistically of the number of events sampled. This method of digitization is implemented in a carefully engineered video camera in operation at Kitt Peak National Observatory (Lynds 1975, Strom 1976). This camera consists of an electrostatically-focussed image tube fiber-optically coupled to a 16-mm SIT tube. This ISIT combination is available as RCA type 4849.

The electron gain of the KPNO system is such that single photoelectron events are five or more times the rms system noise of approximately 800 electrons per pixel per read. The camera is operated in what Lynds describes as an equilibrium mode in which the tube is read every few seconds. The ambient scene illumination provides the required target bias for efficient readout. The digitized output is summed in a fast external memory until the desired signal-to-noise ratio is achieved. The rationale behind this mode of operation is that once the charge distribution on the target has reached equilibrium the output signal should depend linearly on the charge input to the target since the previous reading of the target.

Butcher & Oemler (1978) describe the photometry of galaxies with the KPNO video camera. At the 2.1-m, $f/7.5$ telescope, five to ten photo-events are accumulated per pixel per read period. Data is collected in a 256-square array in a fast external memory. Significant variations in sensitivity as a function of radial position caused by the pincushion distortion of the first stage are not corrected for by division by a flat field frame and are corrected for separately. Reduction of photometric data is relatively straightforward because of the stability of the system parameters.

Colgate, Moore & Colburn (1975) coupled an ITT F4089 magnetically-focussed image tube to a SIT tube (RCA 4826) to permit fast, on-line computer processing of the digital images. Sandford et al. (1976) and Gow et al. (1976) describe a camera system consisting of an S1 electrostatically-focussed tube fiber-optically coupled to a RCA 40-mm-input SIT tube (C21145).

In summary, then, the scintillation-counting imaging systems based on intensifier-camera tube combinations are particularly effective at low count rates. The detector package in these systems tends to be bulky. The major technical problems to overcome have involved image instabilities in the camera system and inefficient readout of the scintillation images stored on the camera-tube target (lag and splitting of the images by the scan lines). In some systems higher resolution is achieved by centroiding; in others, an increase in dynamic range at high count rates is achieved by a pseudo analog digitizing method.

Intensifier-Silicon Array Systems

There are several advantages to using a silicon array to count bright scintillations from a high-gain intensifier system. The pixels in which scintillations are detected are stable and have well-defined geometries. Single line formats for spectroscopic applications are readily implemented with the available arrays. However, with the presently available arrays, a very high gain intensifier system is required to assure that single photoelectron events are efficiently detected above system noise.

Shectman & Hiltner (1976) at the University of Michigan and Shectman (1976) at the Hale Observatories have built faint-object multichannel spectrometers based on high-gain intensifier chains lens-coupled to Reticon arrays. In these innovative systems a pair of three-stage, electro-statically-focussed Varo tubes are fiber-optically coupled together. Single photoelectron events appear as bright scintillations that are more than 100 microns in diameter and which, because of the compounding of the persistence of six phosphor screens, are detectable for tens of milliseconds. Each event, then, may appear in several successive readouts of the array. To avoid the excessive coincidence losses that normally would result from these long decay times a digital storage system is used to subtract the signal from the previous read of a pixel from the current read of the pixel. The resulting signal depends on the rise time of the scintillation brightness and the repetition rate of the readout, and the fixed pattern noise of the array is cancelled out. A centroiding scheme is used to determine the location of events thus detected.

Several versions of Shectman's system have been built. For example, Davis and Latham have recently constructed a spectrometer specifically for measuring redshifts of faint galaxies with the 1.5-m telescope at Mt. Hopkins. In this device a three-stage intensifier stack is coupled by fiber optics to a dual 936-element Reticon array. Chaffee (1978) finds that the efficiency of the detector system for spectrophotometry as determined from measured noise is within a factor of 1.6 of that expected from the recorded counts. Similar systems are being evaluated at Steward Observatory (Hege et al. 1978) and at Mt. Stromlo Observatory (Stapinski et al. 1978).

At present the astronomical use of these event-counting systems is limited to spectroscopic applications by coincidence loss considerations. This may change with the development of high-gain, fast-decay intensifiers and low-noise CCD's. For example, Airey, Morgan & Ring (1978) have described a system that is being developed at Imperial College that will consist of a microchannel plate intensifier lens coupled to a CCD.

Intensifier-Image Dissector Systems

The Lick image dissector scanner developed by Robinson & Wampler (1972) has been a tremendously successful instrument. It efficiently produces two lines of digitized spectra at moderate resolution. The system is heavily used by many astronomers in a great variety of programs. An image dissector, unlike a camera tube or solid state array, has no inherent storage capability, but gives a measure of the present brightness of an input image point by point. Unlike other hybrid systems the efficiency of an intensifier-image dissector system depends on the scintillations displayed by the intensifier having appreciable lifetimes. In most high-gain systems this is assured by the convolution of the persistence characteristics of the phosphors. Repeated, short-duration samples of the brightness of a pixel give a statistically defined measure of the average rate at which scintillations are occurring in that pixel. Sampling with a frequency at least comparable with the time associated with the decay of the average scintillation to half brightness improves the probability that each event makes some statistical contribution to the measure of the average rate.

The present configuration of the Lick system is described by Miller, Robinson & Wampler (1976). Three electrostatically-focussed fiber-optic image tubes are coupled to an ITT image dissector having a fiber-optic input window. The image dissector scans both of two spectra in a period of above five milliseconds. Photoelectrons from the cathode of the image dissector are counted in a fast four-bit scaler and these counts are summed in a 4096-channel external memory. Thus each of the two spectra are represented by 2048 channels. The response of the system is linear to relatively bright limits which makes it feasible to use a set of relatively bright standards for spectrophotometry.

An intensifier-image dissector scanner, which uses a three-stage magnetically-focussed intensifier, a transfer lens, and a conventional image dissector, has been built at Kitt Peak National Observatory. The operation of the system is well documented in a KPNO observer's manual. Observations with an early version of this system are reported, for example, by Ford & Jenner (1976). Other intensifier-image dissector scanners based on the Lick design have been built by Robinson and Wampler for the Anglo-Australian Telescope; Rybski, Mitchell, and

Montemayor at McDonald, and by Byard at Ohio State University. Rybski, Van Citters & Benedict (1976) at McDonald Observatory have constructed a 64 × 64 element area photometer that utilizes the intensifier-dissector combination. Gaskell & Robinson (1978) report the experimental use of the Lick system as an imaging area photometer.

DIGITAL IMAGE TUBES

A more direct technique for generating digital images is to detect photoelectrons with an imaging charge-sensitive detector incorporated in an image tube. Photoelectrons are accelerated through a few tens of kilovolts and focussed on a charge-sensitive silicon device. In silicon one electron-hole pair is created for each 3.6 eV of energy absorbed, or about 2800 charge carriers for each 10 kV of absorbed energy. This charge can be routed directly to a low-noise preamplifier rather than being read with an electron beam as in the SIT tube. The digital image tubes described here all operate in an electron bombarded silicon (EBS) mode. They differ in the methods that are used to route the charge and to convert it to a digital representation of the image.

The Digicon

Beaver & McIlwain (1971) at the University of California at San Diego first demonstrated the astronomical use of a digital image, the Digicon, in a 40-channel spectrum analyzer. The construction of the Digicon tube is described by Beaver (1973) and Choisser (1976). The present Digicon is a 212-element device (Beaver et al. 1976b) that has been used extensively at the Steward Observatory. In these tubes photoelectrons are accelerated through 20 kV and focussed onto an monolithic array of reverse-biased silicon diodes. The charge pulse from single photoelectrons in an element is detected and counted by external electronic systems connected to that element. Thus an array of N elements requires N connecting pins from the tube leading to N preamplifiers, amplifiers, discriminators, and counters. This parallel output approach is practical with present hybrid and integrated circuit technology for arrays of hundreds of elements. (Digicons with 512 channels are being developed for the Faint Object Spectrograph and High-Resolution Spectrograph on the Space Telescope. See Beaver 1978 and Harms 1978.)

In spite of the complexity of the large number of parallel circuits, this approach has many advantages. Single photoelectrons are counted with high efficiency in spatially well-defined pixels. The upper limit to the counting rate in any channel is set by the width of the amplified pulse and is independent of the counting rate in other channels. External deflection coils are provided for periodically displacing the electronic image along the diode array in steps that are a fraction of a diode width. By storing the acquired counts for each displacement in a separate array of bins, the input image can be oversampled by the diode array. Deflection perpendicular to the linear array can be used to switch alternately between two input apertures for sky background subtraction. These features, along with the high DQE of the system, have made it possible to use the Digicon for spectropolarimetry, area polarimetry, and area photometry of faint objects, as well as for spectroscopy. A good example of the low-light-level capability of the Digicon is given by Beaver et al. (1976a) in the determination of redshift of the QSO 0938 + 119. Observations were made with a 200-element Digicon and a low-dispersion prism system. Entrance apertures were arranged so that the sky spectrum was observed with half of the array simultaneously with the QSO spectrum on the other half. The counting rate from the QSO was about 1/10 of that of the sky and sufficient counts were obtained in an hour of integration to adequately define the spectrum.

Self-Scanned Digicon

Tull, Choisser & Snow (1975) collaborated in the development of a self-scanned Digicon in which the photoelectron image is accumulated and read out with a 1024-element linear Reticon. The external circuitry required to operate such a device is less than that for the parallel output Digicon. The construction and operation of the self-scanned Digicon is described by Tull et al. (1975) and Choisser (1976). The array is read out repetitively in short intervals. The video signal from each diode is digitized and added to the sum of the previous reads of that pixel. Tull et al. showed that in this method of operation the total noise is dominated by photo-electron shot noise (with little contribution from readout noise) for rates as low as one photoelectron per diode per read period. This performance is more than adequate for many types of astonomical spectroscopy.

Self-scanned Digicons have been used at the coude spectrograph of the 2.7-m telescope at McDonald for many high-dispersion spectrophotometry programs. (See, for example, Tull 1976 and Tull & Vogt 1977). These clearly demonstrate that spectra with a high signal-to-noise ratio can be routinely obtained by this technique. Unfortunately internal shorts and problems with cathode stability have limited the effectiveness and life of the first of these devices.

The possibility of using a Reticon as an electron counting device has been investigated in laboratory experiments by Mende & Shelley (1975), Tull (1976), and by Mende & Chaffee (1977). The distribution of pulse amplitude shows the single photoelectron peak separated from a peak due to readout noise but not fully resolved. This indicates that it may be feasible to do true photoelectron counting with this approach.

EBS-CCD's

There is considerable interest in the use of CCD's in the EBS (electron bombarded silicon) mode of operation. In this type of device the output signal is generated by a CCD rather than by a diode array as in the Digicon or by a Reticon-type detector as in the self-scanned Digicon. This approach offers the possibility of reading out with low readout noise the charge from a large number of pixels in a two-dimensional format. The gain provided by EBS operation is sufficient that for many applications the device gives adequate performance at standard television frame rates at room temperature (Caldwell & Boyle 1976).

A photon-counting array photometer is being developed by Currie (1976). The photometer is based on a Digicon with a 100 × 100 Fairchild CCD array (Currie & Choisser 1976). The system operates in a fast-scan, photoelectron-counting mode. A counting efficiency of 84% and a dark current equivalent to one electron per sec per pixel has been achieved with the preliminary versions of the tube.

Other experimental data on EBS-CCD's have been reported by Sobieski (1976) and Williams (1976) who evaluated engineering samples of tubes made by ITT and Varo that had three-phase, buried-channel CCD's from Texas Instruments incorporated in them. Chips made by Texas Instruments for JPL have also been tested in a demountable system by Hier et al. (1978). Pulse height distributions of single photoelectron events were analyzed in these tests. Also Zucchino & Lowrance (1978) reported tests on a Varian electrostatic tube with 30:1 demagnification that has a 160 × 100 element Texas Instruments CCD providing the readout.

CONCLUSION

This review is not an all-inclusive tabulation of existing digital systems. Many systems are not mentioned, and in some cases, many years of innovative development work are inadequately summarized in a sentence or two. An attempt has been made here, however, to lead the nonspecialist through a broad range of techniques used for digital imaging. That more detailed measurements of performance are not available for some systems is explained by the understandable interest in using the devices that work well for scientific observations rather than for laboratory evaluation. Nevertheless, as diverse digital imaging techniques are developed it is becoming increasingly important to characterize accurately the efficiency and limitations of the various systems.

Literature Cited

Aikens, R. S., Harvey, J. W., Lynds, C. R. 1976a. See Duchesne & Lelievre 1976. pp. 25.1–14

Aikens, R. S., Lasker, B. M. 1973. See Glaspey & Walker 1973, pp. 65–68

Aikens, R. S., Lynds, C. R., Nelson, R. E. 1976b. *Proc. Soc. Photo-Optical Instrum. Eng.* 78:65–72

Airey, R. W., Morgan, B. L., Ring, J. 1978. See McMullan & Morgan 1978, pp. 329–34

Beaver, E. A. 1973. See Glaspey & Walker 1973, pp. 55–63

Beaver, E. A. 1978. See McMullan & Morgan 1978, p. 393

Beaver, E. A., Harms, R., Hazard, C., Murdoch, H. S., Carswell, R. F., Strittmatter, P. A. 1976a. *Ap. J. Lett.* 203: L5–7

Beaver, E. A., Harms, R. J., Schmidt, G. W. 1976b. *Adv. Electron. Electron Phys.* 40B: 745–63

Beaver, E. A., McIlwain, C. E. 1971. *Rev. Sci. Instrum.* 42:1321–24

Beck, S. C., Wilkinson, D. T. 1976. *Bull. Am. Astron. Soc.* 8:350

Boerio, A. H., Beyer, R. R., Goetze, G. W. 1966. *Adv. Electron. Electron Phys.* 22A: 229–39

Boksenberg, A. 1972. *Conf. Auxil. Instrum. for Large Telescopes*, ed. S. Lanstsen, A. Reiz, pp. 295–319. Geneva: ESO/CERN

Boksenberg, A., Burgess, D. E. 1973. See Glaspey & Walker 1973, pp. 21–43

Boksenberg, A., Sargent, W. L. W. 1975. *Ap. J.* 198:31–34

Boksenberg, A., Shortridge, K., Fosbury, R. A. E., Penston, M. V., Savage, A. 1975. *MNRAS* 172:289–303

Boulesteix, J. 1978. See McMullan & Morgan 1978, pp. 313–19

Boyce, P. B. 1977. *Science* 198:145–48

Buchholz, V. L., Walker, G. A. H., Glaspey, J. W., Isherwood, B. C., Lane-Wright, D. 1976. *Adv. Electron. Electron Phys.* 40B: 879–85

Butcher, H., Oemler, A. Jr. 1978. *Ap. J.* 219: 18–30

Caldwell, L., Boyle, J. 1976. *Proc. Soc. Photo-Optical Instrum. Eng.* 78:10–13'

Campbell, B. 1977. *Publ. Astron. Soc. Pac.* 89:728–32

Carruthers, G. R. 1977. *Astronaut. Aeronaut.* Oct 1977:56–68

Cenalmor, V. 1976. See Duchesne & Lelievre 1976, pp. 16.1–14

Cenalmor, V., Lamy, P. H. L., Perrin, J. M., Nguyen-Trong, T. 1978. *Astron. Astrophys.* 69:411–19

Chaffee, F. H. Jr. 1978. See McMullan & Morgan 1978, pp. 341–46

Chiu, H. Y. 1977. *Appl. Opt.* 16:237–43

Chiu, H. Y., Adams, P. J., Linsky, J. L., Basri, G. S., Moran, S. P., Hobbs, R. W. 1977. *Ap. J.* 211:453–62

Choisser, J. P. 1976. *Adv. Electron. Electron Phys.* 40B:735–43

Colgate, S. A., Moore, E. P., Colburn, J. 1975. *Appl. Opt.* 14:1429–36

Crane, P. 1973. See Glaspey & Walker 1973, pp. 391–414

Crane, P. 1975. *Ap. J.* 197:317–28

Crane, P., Davis, M. 1975. *Publ. Astron. Soc. Pac.* 87:207–16

Currie, D. G. 1975. *Symp. Charge-Coupled Device Technol. Sci. Imaging Appl.* JPL SP 43–21, pp. 80–90

Currie, D. G. 1976. See Duchesne & Lelievre 1976, pp. 30.1–21

Currie, D. G., Choisser, J. P. 1976. *Proc. Soc. Photo-Optical Instrum. Eng.* 78:83–94

Diner, D. J., Westphal, J. A. 1977. *Icarus* 32:299–313

Duchesne, M., Lelievre, G., eds. 1976. *Proc. IAU Colloq. No. 40. Applications Astronomiques des Récepteurs d'Images à Réponse Linéaire.* Paris: Meudon

Ford, H. C., Jenner, D. C. 1976. *Ap. J.* 208:683–87

Fort, B., Boksenberg, A., Coleman, C. 1976. See Duchesne & Lelievre 1976, pp. 15.1–14

Gaskell, C. M., Robinson, L. B. 1978. See McMullan & Morgan 1978, pp. 301–5

Gilbert, G. R., Angel, J. R. P., Grandi, S. 1976. *Adv. Electron. Electron Phys.* 40B: 699–710. See also Gilbert, G. R., Angel, J. R. P., Grandi, S. A., Coleman, G. D., Strittmatter, P. A., Cromwell, R. H., Jensen, E. B. 1976. *Ap. J. Lett.* 206: L129–31

Glaspey, J. W., Walker, G. A. H., eds. 1973. *Proc. Symp., Univ. British Columbia, 15–17 May 1973: Astronomical Observations With Television-Type Sensors.* Vancouver

Goetze, G. W. 1966. *Adv. Electron. Electron Phys.* 22A:219–27

Gow, C. E., Sandford, M. T., Honeycutt, R. K., Jekowski, J. P. 1976. See Duchesne & Lelievre 1976, pp. 21.1–16

Greenstein, J. L., Boksenberg, A., Carswell, R., Shortridge, K. 1977. *Ap. J.* 212:186–197

Gunn, J. E. 1977. *Carnegie Inst. Washington Yearb.* 76:168

Harms, R. 1978. See McMullan & Morgan 1978, pp. 395–99

Hege, E. K., Cromwell, R. H., Woolf, N. J.

1978. See McMullan & Morgan 1978, p. 347
Hier, R. G., Beaver, E. A., Schmidt, G. W., Schmidt, G. D. 1978. See McMullan & Morgan 1978, pp. 401–6
Hoffman, A. A., Crane, P. 1977. *Ap. J.* 215: 379–400
Hunten, D. M., Nelson, B. E., Stump, C. J. Jr. 1976. *Appl. Opt.* 15:2264–67
Hunten, D. M., Stump, C. J. Jr. 1976. *Appl. Opt.* 15:3105–10
Ingerson, T. E., Lasker, B. M. Osmer, P. S. 1976. See Duchesne & Lelievre 1976, pp. 20.1–10
Jeffers, S., Weller, W. G. 1978. See McMullan & Morgan 1978, pp. 287–91
Livingston, W. C. 1973. *Ann. Rev. Astron.. Astrophys.* 11:95–114
Livingston, W. C. 1976. See Duchesne & Lelievre 1976, pp. 22.1–13
Livingston, W. C., Harvey, J., Slaughter, C., Trumbo, D. 1976. *Appl. Opt.* 15:40–52
Loh, E. D. 1976. *Bull. Am. Astron. Soc.* 8:350
Loh, E. D., Wilkinson, D. T. 1976. *Bull. Am. Astron. Soc.* 8:350
Lowrance, J. L., Morton, D. C., Zucchino, P., Oke, J. B., Schmidt, M. 1972. *Ap. J.* 171:233–51
Lowrance, J. L., Zucchino, P., Williams, T. B. 1976. See Duchesne & Lelievre 1976. pp. 18.1–22
Lynds, C. R. 1975. *Q. Rep. Kitt Peak Natl. Obs.* Apr–Jun, pp. 3–6
McCord, T. B., Frankston, M. J. 1975. *Appl. Opt.* 14 : 1437–46
McCord, T. B., Westphal, J. A. 1972. *Appl. Opt.* 11:522–26
McMullan, D., Morgan, B. L., eds. 1978. *Preprints of papers presented at Symp. Photo-Electronic Image Devices, 7th.* Imperial College, London
Mende, S. B. 1971. *Appl. Opt.* 10:829–37
Mende, S. B., Chaffee, F. H. 1977. *Appl. Opt.* 16:2698–2702
Mende, S. B., Shelley, E. G. 1975. *Appl. Opt.* 14:691–97
Michon, G. J., Burke, H. K. 1973. *IEEE Int. Solid-State Circuits Conf. Digest*, pp. 138–39
Michon, G. J., Burke, H. K., Brown, D. M. 1975. *Proc. Symp. Charge-Coupled Device Technol. Sci. Imaging Appl.* JPL SP 43-21, pp. 106–15
Miller, J. S., Robinson, L. B., Wampler, E. J. 1976. *Adv. Electron. Electron Phys.* 40B: 693–98
Morton, D. C. 1972. *Conf. Auxiliary Instrum. Large Telescopes*, ed. S. Lanstsen, A. Reiz, pp. 317–31. Geneva: ESO/CERN
Morton, D. C. 1973. See Glaspey & Walker 1973, pp. 193–97
Morton, D. C., Andereck, C. D. 1976. *Ap. J.* 205:356–9
Oke, J. B. 1977. *Carnegie Inst. Washington Yearb.* 76:167
Osmer, P. S. 1977. *Ap. J.* 214:1–9
Renda, G., Lowrance, J. L. 1975. *Proc. Symp. Charge-Coupled Device Technol. Sci. Imaging Appl.* JPL SP 43-21, pp. 91–105
Robinson, L. B., Wampler, E. J. 1972. *Publ. Astron. Soc. Pac.* 84:161–6
Rybski, P. M., Van Citters, G. W., Benedict, G. F. 1976. See Duchesne & Lelievre 1976, pp. 54.1–20
Sandford, M. T., Gow, C. E., Jekowski, J. P. 1976. *Rev. Sci. Instrum.* 47:486–92
Shectman, S. 1976. *Carnegie Inst. Washington Yearb.* 75:319–20
Shectman, S., Hiltner, W. A. 1976. *Publ. Astron. Soc. Pac.* 88:960–65
Smith, A. G., Hoag, A. A. 1979. *Ann. Rev. Astron. Astrophys.* 17:43–71
Smith, B. A. 1976. *Proc. Conf. Charge-Coupled Device Technol. Appl.* JPL SP 43-40, pp. 135–38
Sobieski, S. 1976. *Proc. Soc. Photo-Optical Instrum. Eng.* 78:73–77
Stapinski, T. E., Rodgers, A. W., Ellis, M. J. 1978. See McMullan & Morgan 1978, pp. 335–39
Strom, S. 1976. *Q. Rep. Kitt Peak Nat. Obs.* Oct–Mar, pp. 5–8
Title, A. M. 1974. *Sol. Phys.* 35:233–37
Tull, R. G. 1976. See Duchesne & Lelievre 1976, pp. 23.1–19
Tull, R. G., Choisser, J. P., Snow, E. H. 1975. *Appl. Opt.* 14:1182–89
Tull, R. G., Nather, R. E. 1973. See Glaspey & Walker 1973, pp. 171–91
Tull, R. G., Vogt, S. S. 1977. *Ap. J. Suppl. Ser.* 34:505–64
Vescelus, F. E., Antcliffe, G. A. 1976. *Proc. Soc. Photo-Optical Instrum. Eng.* 78:60–64
Vogt, S. S., Tull, R. G., Kelton, P. 1978. *Appl. Opt.* 17:574–92
Walker, G. A. H., Buchholz, V., Fahlman, G. G., Glaspey, J., Lane-Wright, D., Mochnaki, S., Condal, A. 1976. See Duchesne & Lelievre 1976, pp. 24.1–24
Wattson, R. B., Harvey, P., Swift, R. 1975. *Proc. Symp. Charge-Coupled Device Technol. Sci. Imaging Appl.* JPL SP 43-21, pp. 70–79. See also Wattson, R. B., Rappaport, S. A., Frederick, E. E. 1976. *Icarus* 27:417–23
Weller, W., Herbst, W., Jeffers, S. 1977. *Publ. Astron. Soc. Pac.* 89:935–38
Westphal, J. A. 1973. See Glaspey & Walker 1973, pp. 127–36

Westphal, J. A., Kristian, J. 1976. See
Duchesne & Lelievre 1976, pp. 19.1–24
Westphal, J. A., Kristian, J. 1977. *Carnegie
Inst. Washington Yearb.* 76:167
Westphal, J. A., Kristian, J., Sandage, A.
1975. *Ap. J. Lett.* 197:L95–98
Williams, J. T. 1976. *Proc. Soc. Photo-
Optical Instrum. Eng.* 78:78–82

Young, P. J., Westphal, J. A., Kristian, J.,
Wilson, C. P., Landaner, F. P. 1978.
Ap. J. 221:721–30
Zucchino, P. 1976. *Adv. Electron. Electron
Phys.* 40A:239–52
Zucchino, P., Lowrance, J. L. 1978. See
McMullan & Morgan 1978, pp. 415–18

Ann. Rev. Astron. Astrophys. 1979. 17:213–40
Copyright © 1979 by Annual Reviews Inc. All rights reserved

THE VIOLENT INTERSTELLAR MEDIUM[1]

✻2150

Richard McCray and Theodore P. Snow, Jr.

Department of Physics and Astrophysics, Joint Institute for Laboratory
Astrophysics and Laboratory for Atmospheric and Space Physics,
University of Colorado, Boulder, Colorado 80309

I. INTRODUCTION

Our conception of the structure of the interstellar medium (ISM) is undergoing a revolution. A few years ago, it was thought that the ISM consisted of cool ($T \lesssim 100$ K) "clouds" embedded in a substrate of a warm ($T \approx 10^4$ K) intercloud medium of partially ionized HI and of HII, and that the main source of heating was ionization by low energy cosmic rays, soft X-ray photons, or extreme ultraviolet photons. An adequate source of ionizing particles or radiation has not been found, however. Instead, a radically different picture has developed as a result of observations of interstellar O VI with the ultraviolet spectrometer on the *Copernicus* spacecraft and measurements of the soft X-ray background radiation, both of which indicate the presence of a hot ($T \approx 10^6$ K) "coronal" phase of the ISM. If observations of relatively local (within a few hundred parsecs) gas reliably indicate the global structure, this coronal gas may occupy a large fraction of the interstellar volume. Indeed, it may be the fundamental substrate in which the warm intercloud medium and the cool clouds are embedded.

A growing body of evidence from radio, optical, and UV astronomy indicates that shocks of high velocity, i.e. $V_s \gtrsim 20$ km s^{-1}, occur commonly in the ISM. Since the thermal expansion of HII regions cannot yield such velocities, these shocks must have a more dramatic origin, such as supernova explosions or the powerful winds from early-type stars. In some cases, the shocks appear to have structure organized over large regions of the sky, resulting perhaps from the cumulative effects of many supernovae. The high velocity shocks are probably responsible for producing and maintaining the coronal interstellar gas. In addition,

[1] Of the many colleagues who have helped us with this work, we wish to thank Dr. J. Michael Shull especially for his careful reading of the manuscript and many valuable comments. This work was supported by grants from the NSF and NASA.

0066-4146/79/0915-0213$01.00

they may play a role in accelerating cosmic rays and in destroying interstellar grains.

The idea that coronal temperatures may exist in the ISM began with Spitzer's (1956) classic paper on the galactic corona. The essential point is that coronal gas radiates very inefficiently by bremsstrahlung; this represents a sharp contrast with cooler gas, which contains many partially ionized atoms that can radiate efficiently following collisional excitation. As a result, coronal gas, once created in the ISM, may persist for millions of years, even without a heat source.

Cox & Smith (1974) pointed out that galactic supernovae may occur at a rate sufficient to produce an interconnecting "tunnel" system of coronal gas, so that the ISM has the morphology of swiss cheese. As with most good ideas, the point is obvious in hindsight. It was well-known that the interiors of supernova remnants consisted of coronal gas, and that such gas should persist; a simple counting argument then shows that the interiors are likely to overlap. McKee & Ostriker (1977) have developed this idea further, suggesting that the volume fraction of the hot gas may be so great that the ISM consists of disconnected regions of HI and HII embedded in a substrate of low density coronal gas. Whether or not the cooler gas is connected makes little qualitative difference, since this gas plays no structural role. The important qualitative property shared by the two models is that the *coronal* gas is connected, so that much of the energy of shock waves resulting from supernova explosions remains in the interconnected coronal region, where it is not easily lost to radiation.

As in the solar corona, thermal conduction plays a major role in the thermodynamics of the hot ISM. Because of the strong temperature dependence of the conductivity ($K(T) \propto T^{5/2}$), thermal conduction is unimportant in the cooler ($T \lesssim 10^4$ K) ISM. But conduction at the interfaces with the cooler regions may be the dominant cooling mechanism of the hot component, and may cause the "evaporation" of interstellar clouds. The O VI observed in the interstellar absorption spectra of hot stars probably originates in these conductive interfaces. Some fraction of the absorption may come from the conductive interface caused by the action of the stellar wind of the background star on its interstellar environment, which produces an "interstellar bubble" (Castor, McCray & Weaver, 1975) with a structure similar to a supernova remnant. The rest comes from other conductive interfaces in the intervening ISM.

This "violent interstellar medium," comprising high-velocity and high-temperature gas, has major implications for galactic ecology: the cycle of interstellar grains, the propagation of cosmic rays, and the flow of gas through the galactic spiral structure and between the galactic disk and corona. In Section II we review the observational evidence for high-velocity and high-temperature gas, and in Section III we describe the important physical processes in and theoretical models for the medium.

II. THE OBSERVATIONAL EVIDENCE

Although the full scope of high-energy phenomena in the interstellar medium has only recently begun to be appreciated, various observational manifestations have been noticed over the past four decades. In the 1930s, Beals (1936) observed velocity component structure in interstellar CaII lines, and a variety of other studies in all parts of the spectrum have since provided additional evidence of violent activity in space.

The observable phenomena associated with this activity include high-velocity gas, extreme physical conditions, abundance anomalies, and the distribution morphology and energetics of aggregates of material. The data that provide insight into each of these aspects of the ISM are described in this section.

A. High-Velocity Gas

As mentioned in the foregoing paragraph, the existence of interstellar motions at substantial velocities has been recognized since the first observation of velocity component structure by Beals (1936). Later, Sanford (1939), Adams (1949), Münch (1957), and Münch & Zirin (1961) catalogued large numbers of interstellar lines of NaI and CaII with velocity structure, showing that while the strongest components had velocities due primarily to galactic rotation, weaker components with peculiar velocities with respect to the Local Standard of Rest (LSR) were common. The survey of Adams (1949), for example, showed that about one half of some 300 OB stars have complex structure in the interstellar CaII H and K lines. Velocities of order 20 km s^{-1} (LSR) were typical, and values of 50 km s^{-1} or greater were occasionally found.

Thanks primarily to the work of Hobbs and co-workers (Hobbs, 1974, 1978, and references cited therein), very high-resolution (~ 1 km s^{-1}) data have become available over the past decade, revealing velocity structure in much more detail than was possible with the previous photographic spectroscopy.

Meanwhile, 21-cm observations of neutral hydrogen also showed evidence for velocity components in a patchy interstellar medium, beginning with the work of Muller, Oort & Raimond (1963). Later Blaauw & Tolbert (1966) and Oort (1966) found a preponderance of negative-velocity components at high galactic latitudes, suggesting a large-scale disturbance of the interstellar gas. Heiles & Jenkins (1976) have mapped the high-velocity HI gas for galactic latitudes $b^{II} \geq 10°$ and Verschuur (1975) has summarized observations of 21-cm components seen with high velocity out of the plane. The 21-cm data clearly show regions of the galaxy which contain numerous fast-moving clouds. Recently, Dickey, Salpeter & Terzian (1977, 1979), and Payne et al. (1978), using high velocity-resolution data, have found for several directions in space a multiplicity of 21-cm velocity components, with LSR velocities ranging up to ± 30 km s^{-1}.

From all of these velocity data obtained with classical techniques, one gets a distinct impression of an energetic, dynamic interstellar medium. This has been borne out by space observations made in ultraviolet wavelengths with *Copernicus*. Because of the presence in the UV of strong resonance transitions of common elements in a wide range of ionization states, these observations are considerably more sensitive tracers of high-velocity gas than either the optical absorption measurements or the 21-cm emission data. Shull & York (1977) and Shull (1977a) found high-velocity SiIII absorption components in the lines of sight of several high galactic latitude objects, and more recently Cohn & York (1977), Cowie & York (1978a), and Cowie, Songaila & York (1979) have mapped a variety of high-velocity interstellar clouds which are seen most strongly in SiIII, CII, CIII, and NII absorption. Predominantly negative velocities as high as 120 km s^{-1} (LSR) have been found. Very recently, Cowie, Laurent, & Vidal-Madjar (1979) have found that these clouds can be observed in HI absorption, via high Lyman-series lines of HI which are weak enough so that the lower-velocity components do not obscure the high-velocity absorption.

The moderate- and high-velocity clouds seen in radio, optical, and ultraviolet generally have different characteristics. The 21-cm observations and optical absorption data usually indicate the presence of cool gas, while the ultraviolet measurements of SiIII lines refer to warmer gas. Some attempts have been made to identify specific velocity components in different regimes. For example, Habing (1969) and Hobbs (1971) found a few velocity and spatial coincidences between HI clouds and known high-velocity CaII components. More recently, Giovanelli et al. (1978) have carried out 21-cm observations in directions known from optical and ultraviolet data to intercept high-velocity clouds, finding a number of coincidences. The combination of radio and optical data allows detailed mapping of the high-velocity gas.

B. *Physical Characteristics of the Observed Gas*

The mere presence of high-velocity gas suggests the energetic processes which produce it, but more important is the physical state of the gas that has been subjected to these processes. A great deal of information on temperatures, densities, and ionization conditions has been derived from observations in all wavelengths.

The most obvious interstellar gas is that which is contained in cold clouds. The data permit extensive analysis of their interior conditions, particularly since the advent of ultraviolet absorption-line observations. In the past few years, similar information has become available for the low-density, hot material. It has become apparent that several different regimes are present in the interstellar gas, although their distributions and filling factors are not yet well known. In the following, two of these regimes will be identified with their inferred physical conditions, while a discussion of the interrelationship of these and the cooler regimes will be reserved for Section III.

A striking new observational development in the past few years has been the discovery of the coronal gas whose existence was first inferred by Spitzer (1956). Both ultraviolet spectroscopy and broad-band X-ray surveys have contributed to our growing awareness of this basic component of our galaxy.

The first ultraviolet indications of the existence of the coronal gas were reported by Rogerson et al. (1973), Jenkins & Meloy (1974), and by York (1974), who found O VI resonance doublet absorption at 1032 and 1037 Å in the spectra of a number of stars. The initial hypothesis that this absorption had a nonstellar origin was confirmed in later work by York (1977) and by Jenkins (1978a,b), primarily on the basis of a non-correlation between the O VI and stellar velocities. Column densities of order 10^{13-14} cm^{-2} prevail for the stars surveyed by Jenkins (1978a), with evidence being found for a patchy distribution of this gas (Jenkins 1978b). Distances up to 500 pc have been well-sampled in the O VI survey, with some coverage of distances as great as 1–2 kpc. Principal features of the O VI absorption are that it is seen in most directions, its strength correlates only weakly with distance, and the lines are relatively broad for interstellar absorption. Both the degree of ionization and the line widths argue for gas temperatures of 2.5–7 × 10^5 K (e.g. York 1977).

Although a stellar origin for the O VI absorption was quickly ruled out, there remains some uncertainty concerning possible circumstellar origins in material swept up by stellar winds (Castor, McCray & Weaver 1975, Weaver et al. 1977). The detailed statistical study of Jenkins (1978b), along with the existence of O VI absorption in the spectra of stars with weak or nonexistent winds, appear to favor an interstellar origin, however. It is evident that peculiar conditions can produce enhancements of O VI gas, as noted for the Vela supernova remnant by Jenkins, Silk & Wallerstein (1976a,b), where either X-ray photoionization or shocks from the supernova explosion have produced this gas.

Evidence for coronal gas at even higher temperature is found in the diffuse soft X-ray background (Williamson et al. 1974). Observations of this background have been reviewed recently by Kraushaar (1977) and by Tanaka & Bleeker (1977). The thermal nature of this emission has been confirmed by Inoue et al. (1979), who have observed an emission line at 0.57 keV due to O VII. A variety of experiments operating in different energy bands have derived temperatures ranging from several times 10^5 K to a few times 10^6 K (e.g. Levine et al. 1977, de Korte et al. 1976, Yentis, Novick & Vanden Bout 1972, Cash, Malina & Stern 1976, Inoue et al. 1979). These derived temperatures are sensitive to the abundances of heavy elements in the gas, which strongly affect the emissivity, and to the assumed theoretical models for X-ray emission from coronal gas (Section III.A). However, it is clear that there are real spatial variations in temperature of at least a factor of two (Burstein et al. 1977).

As was the case with the ultraviolet absorption measures of O VI, the soft X-ray background data refer only to the local portion of the galaxy. In this case the range is limited by the X-ray opacity of the interstellar gas (predominantly due to HeI and HeII). Data on this opacity compiled by Brown & Gould (1970), Fireman (1974), Cruddace et al. (1974), and Ride & Walker (1977), combined with interstellar HI column density data given by Bohlin, Savage & Drake (1978), show that typically the X-ray background at 200 eV comes from coronal gas within 100 pc of the sun. The soft X-ray background is rather patchy as well. The distance probed varies with direction, and can be less than 100 pc in directions containing moderately dense clouds. In terms of measurable parameters, an optical depth $\tau = 1$ at 200 eV occurs for $N(H) = 1.3 \times 10^{20}$ cm^{-2}, or $E(B-V) = 0.02$. The soft X-ray opacity inferred from observations of HI absorption is a lower limit, since substantial additional X-ray opacity may be contributed by helium, carbon, and oxygen in HII regions and in H_2 clouds along the line of sight.

The moderate- to high-velocity gas discussed in Section II.A is distinct from the coronal gas. This 20,000–100,000 K material, which was found by Shull & York (1977) and Shull (1977a) and later surveyed by Cohn & York (1977), shows up most strongly in absorption lines of the ions SiIII, CII, CIII, and NII. The SiIII line at 1206 Å is a particularly sensitive tracer because of its large oscillator strength and because SiIII is the dominant ionization stage for the appropriate temperature range. The electron densities inferred for this gas from the population of fine-structure levels in CII is $n_e \lesssim 1$ cm^{-3}.

C.   Grains Under Duress

The equilibrium between the gas and the dust is apparently strongly affected by the energetic events which produce the extreme physical conditions we have been describing. This was first hinted at by the ground-based absorption measurements of Routly & Spitzer (1952) and later by Siluk & Silk (1974), who found an enhanced ratio CaII/NaI in high-velocity clouds. Despite uncertainties in interpretation due to possible ionization effects, it was suggested that the enhancement of CaII relative to NaI could result from disruption of grains, releasing calcium into the gas phase.

Since then, the *Copernicus* data have confirmed that depletion of many elements is prevalent, and that the degree of depletion varies strongly with the element. Calcium (seen optically) is one of the most highly-depleted species. In addition, evidence of the dependence of depletion on physical conditions has been appearing in the ultraviolet data and in high-resolution ground-based observations. For example, de Boer et al. (1974), Stokes & Hobbs (1976), and Stokes (1978) found variable ratios of Fe/H and Ti/H; and York (1976), Shull & York (1977), and Shull (1977a) have found that the refractory species, usually highly depleted, are enhanced in high-velocity clouds relative to normally undepleted elements. Jenkins, Silk & Wallerstein (1976b) found no depletion in the high-velocity gas in the Vela supernova remnant, an example where the cause of the acceleration is known. Shull, York & Hobbs (1977) have summarized results for several elements, showing that each is enhanced in high-velocity clouds with respect to sulfur, which is usually undepleted. Cowie & York (1978b) have found that the abundance of silicon is essentially cosmic in clouds with $V \gtrsim 50$ km s^{-1}. Since the dominant ionization stage was observed for each of these elements, the remaining ambiguities concerning possible velocity dependence of the ionization equilibrium could be removed, and it was safely concluded that whatever has caused the high velocities in these clouds has also disrupted the grains, returning material to the gas phase.

There remain some uncertainties in interpreting the observations, however. One is that the true velocities of individual clouds could be rather different from the velocities inferred from the line displacements. Another is that the present velocity of a cloud may not accurately reflect its past history; the cloud could have been accelerated to a higher velocity in the past than it now has. In addition, questions persist concerning the degree to which the grains are destroyed. These will best be answered when hydrogen column densities are known in more cases for the same velocity components for which the heavy element abundances are measured, allowing determinations to be made of the absolute degree of depletion in the separate clouds. Such data are available for a few well-separated components (Shull & York 1977, Giovanelli et al. 1978), but it would be useful to have this information for moderate-velocity clouds as well, where the separation of components is small. Possibly high spatial- and velocity-resolution HI data, obtained toward stars whose interstellar absorption profiles can be analyzed, could provide the needed data. An attempt to determine absolute depletions for individual velocity components has been made by Snow & Meyers (1979), who analyzed ultraviolet absorption profiles of the ζ Ophiuchi line of sight and compared the results with published HI data. Within the observational uncertainties, it appeared that most elements in the high-velocity components are present in their solar ratios with hydrogen, except possibly for species such as calcium and titanium, which are normally among the most strongly depleted. Thus, some residual grain cores apparently can survive the acceleration process that produces cloud velocities of order 20–30 km s^{-1}. It will be interesting to extend such analyses to higher-velocity clouds, for which prospective grain destruction mechanisms (cf Section III.B) are likely to be more effective.

D. Distribution, Morphology, and Kinetics

To place in proper perspective the assorted interstellar regimes is a difficult task. We have seen that a wide variety of physical conditions and velocities exist in a confusing assortment of regions of varying sizes and shapes. In this section we attempt to summarize the observational data on the locations, interactions, and motions of the various kinds of material. It is this overall picture which best emphasizes the tremendous departure from earlier perceptions of the ISM.

In terms of volume, the hot coronal gas responsible for the diffuse soft X-ray background appears dominant, with a filling factor $f \approx 50\%$ inferred from the X-ray data (Kraushaar 1977); there is considerable uncertainty in this figure, due to the unknown degree to which this coronal gas may be nonuniformly distributed. The O VI gas, which is typically much cooler, has a filling factor $f \approx 20\%$ if it is uniformly distributed within the discrete domains where it is found (Jenkins 1978b), but much less if it exists on the surfaces of clouds (Cowie, Jenkins et al. 1979).

The warm ($T \approx 10,000$ K) and cool ($T \lesssim 100$ K) components both exist in clouds, often the same ones (Cowie & York 1978a), and in many cases the O VI velocities correlate well with those of the warm gas absorption lines (Cowie, Jenkins et al. 1979), implying that O VI is present at the boundaries of clouds. The filling factor for these clouds is evidently quite small ($f \approx 1$–10%), although they contain most of the total mass.

Very recently, Savage & de Boer (1979) have obtained spectra of two hot stars in the Large Magellanic Cloud with the *International Ultraviolet Explorer*. The spectra show absorption lines of C IV and Si IV that are much stronger than those typically seen in stars in the Milky Way, and imply column densities $N(C\ IV) \approx 10^{14}$ cm^{-2} and $N(Si\ IV) \approx 2.5 \times 10^{13}$ cm^{-2} toward the LMC. The features indicate gas with temperature $\approx 10^5$ K in a galactic corona with a scale height ~ 2 kpc below the galactic plane.

To summarize, from an observational point of view, a picture has emerged in which a substantial fraction of the volume of the ISM is million-degree gas. Immersed within this gas are clouds (or aggregates of clouds) which contain the cold gas whose presence has been known for decades. Warm regions, such as those producing the observed SiIII absorption, are often associated with clouds. The O VI gas, with a temperature of a few times 10^5 K, most likely exists in the conductive interfaces between the clouds and the ambient coronal gas. In this picture, the observed column density of O VI is determined by the number of interfaces intercepted by the line of sight, whereas the column densities of the cold and warm components are determined by the sizes of the clouds.

Apparently it is not realistic to view the clouds as isolated, independent islands scattered randomly through a sea of coronal gas. Over 20 years ago Münch (1957) noted a tendency for expanding negative-velocity sheets of material to exist in front of OB associations. Barnard's Loop has been recognized as a large-scale structure for a much longer time, and in the past two decades, 21-cm observations have revealed a number of coherent structures in the interstellar medium (Heiles 1976). Recently, the ultraviolet absorption-line data have revealed sheet-like conglomerates of clouds, such as "Orion's Cloak" (Cowie, Songaila & York 1979) and the clouds towards Perseus OB2 (Snow 1976, 1977), both of which apparently coincide with known 21-cm emitting shells (Sancisi 1974, Sancisi et al. 1974). Reynolds & Ogden (1978) have found Hα, [NII], and [OIII] emission from a similar structure associated with two young clusters of stars. Analyses of velocity correlations have revealed often that the same structures are seen in various wavelengths (e.g. Giovanelli et al. 1978, Cowie, Songaila & York 1979), even including soft X-rays in some cases (de Korte et al. 1976, Hayakawa et al. 1977), indicating that they contain a wide variety of temperature and density regimes. Typically these sheets of material have total masses of order $10 M_\odot$ and kinetic energies (due to bulk motion) of 10^{47} ergs. Velocities (LSR) of 20–40 km s^{-1} are typical.

We must emphasize that the soft X-ray and presently-available ultraviolet data probe only a rather local portion of the galaxy. Because of this, it is difficult to assess the statistics of these large-scale aggregates of clouds. The 21-cm data give one the impression that our neighborhood is not unusual, however. The *International Ultraviolet Explorer* will play an important role in removing this uncertainty.

A few of the observed expanding sheets are outstanding in terms of extent and richness of observational manifestation. Barnard's Loop, for example, is a large ringlike structure covering some 13° on the sky which shows up in Hα emission (Pickering 1890, Barnard 1895, Reynolds & Ogden 1979), ultraviolet scattering (O'Dell et al. 1967, Carruthers & Opal 1977), and optical and ultraviolet absorption lines and 21-cm emission at a variety of velocities (Adams 1949, Hobbs 1969, Cohn & York 1977, Cowie, Songaila & York 1979, Menon 1958). Velocity components ranging from ≈ 20 km s^{-1} to 100–120 km s^{-1} have been found. The highest-velocity component is identified by Cowie, Songaila & York (1979) as a radiative shock, while the assorted lower-velocity components are probably embedded clouds which have been ionized by the shock. The highest-velocity component has a total mass of ≈ 100 $M_\odot$ and a kinetic energy of about 10^{49} ergs, while a lower-velocity shell identified by Reynolds & Ogden (1979) has a kinetic energy of order 10^{51} ergs.

Another prominent region of moderate- and high-velocity clouds and ionized gas is the Gum Nebula, and the Vela supernova remnant embedded within it. The nebula as a whole is some 30° across, and shows up clearly in Hα emission (e.g. Sivan 1974, Reynolds 1976a,b). The Vela remnant is characterized by complex filamentary structures, and a multitude of velocity components which must arise in thin sheets (Wallerstein & Silk 1971, Jenkins, Silk & Wallerstein 1976a,b), with velocities of expansion as great as 180 km s^{-1}. Highly-ionized species such as O VI and N V are abundant in the Vela remnant. Again the interpretation of the ultraviolet data involves a shock with a variety of clouds lying behind which have been ionized and compressed by it. The total kinetic energy in this case is about 5×10^{50} ergs (Jenkins, Silk & Wallerstein 1976b). Energies of order 10^{50} to 10^{51} ergs may be typical of these expanding conglomerates of gas. Heiles (1976) found a total kinetic energy of about 4×10^{50} ergs for a particularly striking closed loop structure seen in 21-cm data. (See note added in proof, p. 238.)

III. THEORETICAL INTERPRETATION

In this section we describe the physical processes characterizing high-temperature and high-velocity gas. First, we consider the ionization and emissivity of coronal gas, determined by the atomic processes of the abundant elements. These results are used to characterize the behavior and appearance of high-velocity shocks. Then, we describe the interfaces between coronal gas and cooler gas, whose structure is controlled by thermal conduction. We describe how these processes manifest themselves in hydrodynamical models for the action of supernova explosions and stellar winds on the ISM. Finally, we discuss the recent attempts to synthesize all these processes into a global model for the ISM.

A. *Ionization and Emissivity of Coronal Gas*

Recently, several calculations of the ionization balance, radiative cooling, and emergent spectrum from a low-density, high-temperature gas of cosmic abundances have appeared. These calculations incorporate extensive atomic data; cross sections for dozens of ions and several hundred emission lines are involved. The results are in substantial agreement, although some detailed differences do appear as a result of uncertainties in cosmic abundances and atomic data.

The ionization equilibrium of a hot optically thin plasma is established through a balance between collisional ionization and radiative plus dielectronic recombination; for low densities it is a function of temperature alone. Results of such calculations have been given by Jordan (1969, 1970) for the most abundant trace elements and by Landini & Fossi (1972) for several other elements. Results of more recent calculations have been reported by Summers (1974a,b) for several trace elements and by Jacobs et al. (1977a,b) for Fe and Si. At high electron densities, say, $n_e > 10^8$ cm^{-3}, dielectronic recombination is suppressed and the ionization equilibrium becomes a function of density as well as temperature (Summers 1974a,b). In the results reported by Summers each element tends systematically to be less ionized at a given temperature than in the results reported by Jordan and by Jacobs et al., leading to the suspicion that the collisional ionization rates adopted by Summers have been underestimated.

The total radiative cooling rate has been calculated by Raymond, Cox & Smith (1976). At temperatures $T_6 = T/(10^6 \text{ K}) \geq 30$, all abundant elements except iron are almost fully ionized, and the dominant atomic cooling mechanism is bremsstrahlung, which has a rate coefficient (such that the radiated power per unit volume is given by $n_e n \Lambda$):

$$\Lambda_{ff} \approx 2.3 \times 10^{-24} \, T_6^{1/2} \text{ erg cm}^3 \text{ s}^{-1}. \tag{1}$$

At lower temperatures, the radiative cooling is dominated by collisional excitation of lines; a power-law fit to the detailed results of Raymond et al. gives the line cooling rate

$$\Lambda_L \approx 1.6 \times 10^{-22} \, T_6^{-0.6} \text{ erg cm}^3 \text{ s}^{-1}, \tag{2}$$

which is accurate to within roughly a factor of 2 over the temperature range $0.1 \leq T_6 \leq 30$. The spectral emissivity of such a coronal plasma has been calculated by Mewe (1972, 1975), Kato (1976), Raymond & Smith (1977), and Sarazin & Bahcall (1977) for $\lambda < 250$ Å, and by Stern, Wang & Bowyer (1978) for $\lambda > 250$ Å. For temperatures $3 \lesssim T_6 \lesssim 30$, the emergent radiation is dominated by K (~ 2 Å) and L (12–16 Å) lines of iron and by K lines of S, Si, and Mg in the range 3.5–7.5 Å. At lower temperatures the spectral range in which the dominant power emerges shifts dramatically; for example, for $0.3 \lesssim T_6 \lesssim 1$ the dominant emission occurs in the range 150–250 Å, and for $T_6 \lesssim 0.3$, in the range > 500 Å. The power radiated in the observationally favorable band 44–70 Å is less than 20% of the total at the optimum temperature $T_6 \approx 1$, and negligible at $T_6 \lesssim 0.3$.

The temperature of a coronal gas can be determined by fitting data from two or more channels of a broad-band X-ray detector to a theoretical emission spectrum convolved with detector response. In doing so, it is necessary to use the detailed theoretical emission spectra rather than a thermal bremsstrahlung spectrum, because emission lines dominate for $T_6 \lesssim 30$. Temperatures inferred from such a procedure are limited in accuracy to about $\pm 10\%$ as a result of uncertainties in atomic rates; a greater uncertainty may result from the sensitivity of the inferred temperature to the assumed cosmic abundances. More precise model-independent temperature determinations may be obtained in the future by observations of X-ray line ratios at high spectral resolution (Bahcall & Sarazin 1978).

Time-dependent effects resulting from changes in the heating rate of a coronal gas may cause the radiative cooling and line emission of a coronal gas to differ significantly from the results quoted above, which were calculated on the assumption of a stationary equilibrium between ionization and recombination. The establishment of ionization balance tends to lag behind temperature changes due to sudden heating or radiative cooling. As a result, gas that is subject to sudden heating tends to remain under-ionized at a given temperature and can radiate as much as ten times the power in emission lines during the transient phase as a stationary gas at the same temperature (Shapiro & Moore 1977). Conversely, a gas that is cooling by radiative relaxation will cool more slowly than the equilibrium cooling rates would imply (Kafatos 1973, Shapiro & Moore 1976).

Ostriker & Silk (1973) and Burke & Silk (1974) have suggested that inelastic collisions of thermal ions with interstellar dust grains can be an important cooling mechanism for coronal gas. If graphite grains are present with normal abundance relative to gas, the effective dust cooling rate coefficient is $\Lambda_d \approx 1.0 \times 10^{-23} T_6^{3/2}$ erg cm^3 s^{-1}, which may dominate for $T_6 \gtrsim 4$. In this case, the thermal energy escapes as far-infrared

($\lambda \sim 100\mu$) continuum radiation rather than as X-ray or UV lines. However, the importance of this mechanism is limited because the grains are not likely to survive in the coronal gas. If they are not destroyed in the shocks that create the coronal gas (see below), they are destroyed by sputtering as they cool the gas. A simple estimate of the maximum fractional heat loss of coronal gas due to graphite grains is given by $\Delta Q/Q \approx X'_C/S$, where X'_C is the fractional abundance of carbon atoms in grains relative to hydrogen atoms in gas, and S is the sputtering yield per hydrogen atom collision with grains. Taking $X'_C = 3 \times 10^{-4}$ and a low estimate $S = 10^{-2}$ (Draine & Salpeter 1979a), gives a fractional heat loss $\Delta Q/Q = 0.03$, for which the grains are completely destroyed. According to Draine & Salpeter, a refractory grain of radius r_g should be destroyed by sputtering in coronal gas of atomic density n_0 in a time $t \approx 2 \times 10^4 \, n_0^{-1} \, (r_g/0.01 \, \mu m)$ yr.

B. High-Velocity Shocks

A high-velocity shock in the ISM has four important zones. The upstream flow may be heated and ionized by precursor radiation emitted by hot shocked gas. If the gas is partially ionized, electrostatic or magnetic plasma instabilities make the electron and ion distribution functions isotropic in a very thin zone (McKee 1970, Tidman & Krall 1971). Following, there is another thin zone in which the thermal energy of the shocked ions is shared with the electrons by Coulomb collisions.

If the Mach number is large, the gas density is raised across these zones to a value $n_1 = 4n_0$, where n_0 is the upstream atomic density. The temperature is raised to a value

$$T_1 = 3\mu V_s^2/16k \tag{3a}$$

$$\approx 1.4 \times 10^5 \, V_{100}^2 \text{ K,} \tag{3b}$$

where V_s is the shock velocity, $V_{100} = V_s/(100 \text{ km s}^{-1})$, k is Boltzmann's constant, and the mean molecular weight μ has the value $\mu \approx 0.6 \, m_H$ appropriate for HII (for HI, take $\mu \approx 1.3 \, m_H$).

The most interesting region from the observational point of view is the zone following the shock, where the temperature decreases from T_1 to some terminal temperature, T_f, by radiative losses. This relaxation occurs in a layer of thickness

$$L_1 \approx 3kT_1V_s/[4n_1\Lambda(T_1)] \tag{4a}$$

$$\approx 6.6 \times 10^{16} \, V_{100}^{4.2} \, n_0^{-1} \text{ cm} \tag{4b}$$

for $0.8 < V_{100} < 14$, using the cooling rate $\Lambda_L(T)$ given by Equation (2). This rough estimate may differ significantly from results of detailed numerical calculations. In particular, departures from ionization equilibrium behind the shock may cause the true value of $\Lambda(T)$ to exceed the assumed value by as much as an order of magnitude, and the true value of L_1 would be correspondingly less.

The gas in this cooling zone usually is thermally unstable, and will collapse to a high density at which the sum of the gas pressure and the magnetic pressure balance the upstream ram pressure. Without a magnetic field, the final density contrast will reach a value $n_f/n_0 = \mu V_s^2/kT_f$; but a magnetic field whose component parallel to the shock in the upstream flow is B_0 will cushion the collapse and limit the compression factor to a value $n_f/n_0 = (8\pi n_0\mu)^{1/2}V_s/B_0$. The thermal instability can amplify upstream density fluctuations that enter the shock, so that the dense downstream flow may break up into sheets or filaments (McCray, Stein & Kafatos 1975, Chevalier & Theys 1975).

The detailed structure of the radiative cooling zone following a high velocity shock was first calculated by Cox (1972) for $V_s = 70$ to 140 km s^{-1}. Many of the resulting structural details and observational consequences, such as cooling length L_1, column densities of ions, and line emissivities, are sensitive to the assumed ionization structure immediately behind the shock. Cox assumed that this was given by the results of a stationary ionization equilibrium calculation for coronal gas at a temperature T_1, then followed the time-dependent recombination and cooling of the gas downstream. Raymond (1979) improved the theory by assuming that the post-shock ionization structure was that of a typical HII region (i.e. HII, HeI or HeII, SiII, etc.) and by including more extensive and accurate atomic data. The theory has been further advanced by Shull & McKee (1979), who have calculated the structure of high-velocity ($V_s \leq 130$ km s^{-1}) shocks entering interstellar HI by solving the transfer of ionizing radiation both upstream and downstream as well as the time-dependent ionization rate equations and gas dynamical equations. Their calculations are the first in which the ionization structure of the gas entering the shock is determined self-consistently by the precursor radiation from the shock itself. The results show that, for $V_s \leq 90$ km s^{-1}, precursor radiation is unimportant, but the structure of the radiative layer is sensitive to the ionization state of the upstream gas. For $V_s > 90$ km s^{-1}, precursor ionization begins to become important, and for $V_s \gtrsim 110$ km s^{-1}, the upstream gas has the ionization structure of a typical HII region (since the precursor radiation is mostly in the form of lines and continuum with $hv \lesssim 50$ eV). If, as may frequently occur, the upstream gas is already photoionized by a hot star, the precursor radiation is redundant.

High-velocity shocks can be observed spectroscopically in UV absorption lines or in infrared, optical, or UV emission lines. Because cooling and recombination times are inversely proportional to density (cf. Equations 4a,b), the column densities of multiply-ionized trace elements in the downstream gas are almost independent of n_0, and the surface brightness of permitted emission lines increases linearly with n_0. Forbidden line strengths have a more complicated density dependence. The shock velocity V_s at which an appreciable column density or line emission of a given ion appears can be estimated roughly by finding the temperature at which the abundance of that ion reaches a maximum in the coronal equilibrium results (cf Section III.A), then finding V_s from Equation (4b). For example, SiIII becomes prominent for $V_s \gtrsim 50$ km s^{-1} (Shull 1977b), O VI, for $V_s \gtrsim 150$ km s^{-1}, and FeXIV, for $V_s \gtrsim 400$ km s^{-1}. Typically, line emissivities and column densities reach a maximum near the characteristic velocity and decrease for greater velocities; but the velocity dependences are complicated. Accurate quantitative results have been obtained by Raymond (1979) and by Shull & McKee (1979) for V_s up to 200 km s^{-1}, but are not yet available for greater V_s. The ions produced in the downstream cooling region have velocities ranging from $3V_s/4$ to V_s, which should be easily separable from the velocities of ions produced ahead of the shock by precursor ionization. The latter may be obscured by the contribution from normal HII regions.

The observed correlation between depletions of refractory elements and velocity (Section II.C) has stimulated several investigations of mechanisms which might return these elements from grains to the gas phase via interstellar shocks. Since the motion of grains is not tightly coupled to that of the gas, large relative velocities among grains and between grains and gas may be generated by the passage of a shock. The grains may be destroyed subsequently by grain-grain collisions or by sputtering (Aannestad 1973, Jura 1976, Barlow & Silk 1977). However, these mechanisms alone do not appear capable of returning a sufficient fraction of refractory materials to the gas to account for the observed pattern of depletions at intermediate (20–100 km s^{-1}) velocities. Efficient destruction of refractory grains (e.g. silicates, graphite, or iron) requires relative grain-gas velocities $\gtrsim 200$ km s^{-1}.

The solution to this dilemma was provided by Spitzer (1976), who suggested that the grains are accelerated by the betatron mechanism as a result of compression of the interstellar magnetic field in the shock. The grains, which are expected to carry an electrical charge (Spitzer 1978, Draine & Salpeter 1979a), gyrate in Larmor orbits with initial velocity $V_g \approx 3V_s/4$ following the passage of the shock. Then, provided that the Larmor radius $R_L < L_1$, the grains can be accelerated as the magnetic field is compressed further in the radiative cooling zone following the shock. In this way grain velocities reaching $V_g \approx 2V_s$ can be generated in typical shocks. Shull (1977c) showed that the resulting grain-grain collisions could return typically a few percent of the grain mass to the gas in shocks with velocities $V_s \approx$ 20–50 km s^{-1}. However, if the shocks are propagating into a medium ionized by starlight, the enhanced Coulomb drag of the grains can suppress the betatron acceleration and prevent their destruction. Cowie (1978) showed that at somewhat higher shock velocities magnetically accelerated grains are destroyed more efficiently by sputtering, and that destruction yields $\approx$ 10–60% were possible of refractory grains in shocks with $V_s \gtrsim$ 50 km s^{-1}. The calculated destruction yields and velocity thresholds depend somewhat on grain composition and the assumed sputtering law. These results have been confirmed in more detailed calculations by Shull (1978) and by Draine & Salpeter (1979b). Shull pointed out that the process becomes ineffective for low-density grains or grains with radius $\lesssim 10^{-6}$ cm, which are rapidly stopped by Coulomb drag.

C.   Physics of Conductive Interfaces

The thermal conductivity of a fully ionized gas is $K = CT^{5/2}$, where $C \approx 0.6 \times 10^{-6}$ ergs cm^{-1} s^{-1} (K)$^{-7/2}$ (Spitzer 1962), and the corresponding mean free path of the conducting electrons is $\lambda \approx 10^4 \ T^2 \ n^{-1}$ cm. Because of the strong temperature dependence of this coefficient, thermal conduction plays a negligible role in the ISM for $T \lesssim 10^4$ K, but may play a dominant role in the coronal gas.

Cowie & McKee (1977) have investigated how the classical conductivity formula above breaks down when the temperature gradient is sufficiently great that the heat flux, $q_{cl} = K \nabla T$, becomes comparable to the "saturated" value $q_{sat} \approx 0.4(2kT/\pi m_e)^{1/2} n_e kT$, where the heat flux is limited by the density n_e of electrons available to transport thermal energy. Thermal conduction may also be suppressed by a magnetic field, which greatly inhibits heat transport across field lines, but the suppression should not be significant unless the temperature gradient is nearly perpendicular to the field lines.

Let us first consider the idealized model of a stationary, plane-parallel interface of thickness R at constant pressure p, whose hot boundary is maintained at a temperature T_h. If there are no radiative losses, the temperature profile is given by $T(z) = T_h(z/R)^{2/5}$, where z is the distance from the cold boundary. In order to maintain this structure, a heat flux

$$q = 2CT_h^{7/2}/5R \ \text{erg cm}^{-2} \ \text{s}^{-1} \tag{5}$$

must enter the hot boundary. The thermal energy that flows into the interface causes "evaporation," whereby the cool gas diffuses into and merges with the hot gas. Therefore, this stationary model requires an evaporative mass flux

$$\dot{m} = 4\mu C T_h^{5/2}/(25\ kR)\ \mathrm{g\ cm^{-2}\ s^{-1}} \qquad (6)$$

to flow through the system from the cold boundary to the hot boundary. Multiplying Equations (5) and (6) by $4\pi R^2$ gives, respectively, the correct total heat flow Q into a spherical cloud of radius R in an infinite medium of asymptotic temperature T_h, and the total evaporative mass loss rate $\dot{M}$ of the cloud (Cowie & McKee 1977). Such a mass loss rate could cause a cloud of atomic density n_c (cm^{-3}) and radius R_c (pc) to evaporate in a time $t \approx 3 \times 10^5\ n_c R_c^2 T_6^{-5/2}$ yr, where $T_6 = T_h/10^6$ K.

In a realistic astrophysical context, these results are modified by the effects of saturation of thermal conduction, radiative losses, and finite cloud radius. Saturation sets in when $\lambda(T_h) > R_c$, or $R_c < 3 \times 10^{-3}\ T_6^2\ n_h^{-1}$ pc (Cowie & McKee 1977) where n_h is the atomic density of the hot gas. In that case the velocity of the evaporative flow approaches the isothermal sound speed $c_h = (kT_h/\mu)^{1/2}$ and the pressure of the cool gas becomes significantly greater than that of the hot gas. The mass flux is limited to a few times $\mu n_h c_h$.

Radiative cooling can become an important heat sink in the interface, where $T_6 < 1$, and the strong line cooling (Equation 2) may dominate the divergence of the conductive heat flux. This happens for $R_c > R_{cr}$, where $R_{cr} \approx 0.1\ T_6^2\ n_h^{-1}$ pc (McKee & Cowie 1977). The effect is to increase the temperature gradient in the interface and to suppress the evaporation; indeed, for $R_c > R_{cr}$ the evaporative flow is actually reversed, and the coronal gas condenses onto the cloud surface as a result of radiative losses in the interface. In this case the dominant energy source of the radiation from the interface is the condensation flow rather than thermal conduction from the coronal gas, which is rather insensitive to the radiative losses.

If the gas were subject to some kind of local heating, for example by low energy cosmic rays, stationary equilibrium temperatures for which radiative cooling is balanced by the local heating might be possible (cf. Section III.E). In that case, the equilibrium temperatures T_h and T_{cl} (for the hot gas and the cool clouds, respectively) and the structure of the conductive interface depend only on the local heating rate Γ and the pressure p of the medium, provided that $R > R_{cr}$. Evaporation or condensation will then occur at the conductive interface, depending on whether the pressure p is less than or greater than a "vapor pressure" $p_v(\Gamma)$ (Zel'dovich & Pikel'ner 1969, Penston & Brown 1970, Graham & Langer 1973).

It is likely that highly ionized trace elements such as O VI are present in detectable quantities in conductive interfaces, where the temperature varies over the range $10^4 < T < 10^6$ K. For example, if collisional ionization equilibrium holds and radiative losses in the interface are negligible, the column density of O VI in the interface is roughly (Castor, McCray & Weaver 1975, McKee & Cowie 1977)

$$N_{OVI} \, (cm^{-2}) \approx 1.6 \times 10^{17} \, X_O \, n_h \, T_6^{-3/2} \, R_{pc}. \tag{7}$$

Given a fractional abundance $X_O = 4.4 \times 10^{-4}$, one finds that the O VI column density should be in the range $N_{OVI} \gtrsim 10^{12}$ cm^{-2} observable by UV absorption spectroscopy if $T_6 \approx 1$ and $n_h R_{pc} > 0.1$.

Radiative cooling in the interface and departures from ionization equilibrium can modify the column densities of O VI and other ions appreciably. For example, Weaver et al. (1977) found that radiative cooling narrows the interface and reduces N_{OVI} by a factor 0.5; but they also found that departures from ionization equilibrium raised N_{OVI} by a factor 3.4, resulting in a net increase by a factor 1.7. The latter effect was a consequence of the fact that the conductive interface in their model was evaporating, so that oxygen atoms flowing toward higher temperature remained in the ionization stage O VI for a greater distance than the ionization equilibrium assumption would imply. In other circumstances, for example if condensation rather than evaporation occurs at the interface, the quantitative effect of time-dependent ionization would be different.

The column densities of other ions, such as CIV, NV, SiIV, and SIV, which might also be observable, were calculated by Weaver et al. to be one to two orders of magnitude less than N_{OVI}. By observing ratios of column densities, one might in principle obtain further information about a conduction front—for example, whether it is evaporating or condensing. However, the necessary coverage of parameter space in the theoretical calculations is lacking, and any inference from observations would be clouded by uncertainties in the relative cosmic abundances and by the possibility that the column densities of highly ionized trace elements are contaminated as a result of photoionization by XUV or X-ray sources (e.g. McCray, Wright & Hatchett 1977).

D. Energy Sources

1. SUPERNOVAE   Of the possible sources of high-temperature and high-velocity gas in the ISM, supernovae are almost certainly dominant (e.g. Salpeter 1976, 1979, Chevalier 1977). A supernova explosion that releases an energy E_0 at time $t = 0$ into a medium of uniform density ρ_0 will form a spherical blast wave of radius (Sedov 1959):

$$R_s = 1.15(E_0/\rho_0)^{1/5} t^{2/5} \tag{8a}$$
$$\approx 13(E_{51}/n_0)^{1/5} t_4^{2/5} \, pc, \tag{8b}$$

where $E_{51} = E_0/(10^{51}$ erg s$^{-1})$ and $t_4 = t/(10^4$ yr). The numerical co-efficient of Equation (8a) is increased to 1.27 if thermal conduction is sufficiently rapid to maintain an isothermal interior (Solinger, Rappaport & Buff 1975). The blast wave has a very low-density, high-temperature interior, and most of the swept-up mass is confined to a shell of thickness $\Delta R \approx R_s/12$. During this adiabatic phase, the shell has a constant kinetic energy $E_k = 0.28 E_0$.

Radiative losses become important when $\Delta R \approx L_1$. From Equations (4b) and (8b) we find that this occurs at a radius

$$R_c \approx 20 \, n_0^{-0.42} \, E_{51}^{0.29} \text{ pc.} \qquad (9)$$

Thereafter the shell expands as a pressure-driven snowplow according to the law (McKee & Ostriker 1977, Shull & Silk 1979)

$$R_s \approx 8.4 \, (R_c^2 E_{51}/n_0)^{1/7} \, t_4^{2/7} \text{ pc,} \qquad (10)$$

and the kinetic energy of the shell decreases as $(R_c/R)^2$. Equations (9) and (10) agree very well with results from numerical hydrodynamical calculations (Chevalier 1974).

A supernova shell is most evident optically when $R > R_c$, and the post-shock region has collapsed to high density and low temperature. The kinetic energy in the shell at that time is typically a few percent of E_0. Fitting the observed velocity, radius, and density of optical filaments to Equation (10) [not Equation (8)] leads to estimates of $E_0 \approx 10^{51}$ ergs for supernova outbursts (Chevalier 1977).

These results are modified when a supernova blast wave propagates into an inhomogeneous medium (McKee & Ostriker 1977). Embedded small clouds may be overtaken by the blast wave as it propagates through a low-density substrate. McKee, Cowie & Ostriker (1978) have shown how these clouds are accelerated by the passage of the shock, and have suggested that this mechanism accounts for the high-velocity filaments in the Cygnus Loop and the Vela SNR. If these clouds remain in the hot interior, they may evaporate or condense, and enhance the radiation losses of the system. The consequences are uncertain, as they depend on the spectrum of cloud sizes that remain in the interior of the shell.

2. STELLAR WINDS   Another cause of high-velocity and high-temperature gas is the action of winds from OB stars on the ISM. It is now established (Conti 1978) that an O star, with an initial main sequence mass greater than, say, 20 $M_\odot$, will lose a substantial fraction of its mass in a powerful stellar wind. Infrared observations (Barlow & Cohen 1977, Barlow 1979) indicate stellar wind mass loss rates $\dot{M} \approx 1.4 \times 10^{-12} \, L_*^{1.1} \, (M_\odot \text{ yr}^{-1})$, where L_* is the stellar bolometric luminosity in solar units. Ultraviolet observations (Abbott 1978) indicate that the terminal velocity, V_w, of the

wind is roughly three times the stellar escape velocity. These results, together with the mass-luminosity relation for massive stars, imply that the mechanical power of the stellar wind, $L_w = \dot{M}V_w^2/2$, is given approximately by $L_w \approx 3 \times 10^{35} (M_*/20M_\odot)^{2.3}$ ergs s^{-1}, or roughly 1.3×10^{-3} L_* for $M_* = 20 M_\odot$. Calculations of the evolution of massive stars, including the effects of stellar wind mass loss (de Loore et al. 1977, Chiosi et al. 1978), indicate that the integrated luminosity of such stars is roughly $L_* t \approx 3 \times 10^{-3} M_* c^2$, which implies that the net mechanical energy imparted to the stellar wind is given by $E_w \approx 10^{50} (M_*/20M_\odot)^{1.2}$ ergs. Therefore, the wind of a massive star can impart a mechanical energy to the ISM that is comparable to the energy $E_0 \approx 10^{51}$ ergs, typical of a supernova outburst.

Castor, McCray & Weaver (1975) and Weaver et al. (1977) have studied the structure and evolution of the "interstellar bubble" that is created by the action of such a stellar wind on the ISM. A relatively small fraction of the interior volume of the bubble is occupied by the stellar wind itself. The rest is a region of hot ($T \approx 10^6$ K), low-density ($n \approx 10^{-3}$ cm^{-3}) gas composed of shocked stellar mixed with interstellar gas. Most of the interstellar gas that is swept up by the shock resides in a thin shell of HII and possibly HI or H_2, depending on the ionizing luminosity of the central star. Except for the presence of the central OB star, the optical and soft X-ray appearance of the bubble is very similar to that of a supernova shell in the radiative stage.

Given a constant wind luminosity $L_{36} = L_w/(10^{36}$ ergs s$^{-1})$, the evolution of the bubble is given by the law

$$R_s \approx 27 \, L_{36}^{1/5} \, n_0^{-1/5} \, t_6^{3/5} \text{ pc}, \tag{11}$$

where $t_6 = t/(10^6$ yr). The fundamental qualitative difference between such a bubble and a supernova shell is that the energy of the stellar wind is slowly fed into the hot interior of the bubble, where it is not readily lost to radiation. Consequently, the kinetic energy of the expanding shell, E_k, remains a constant fraction $E_k = 0.20 \, L_w t$ of the mechanical energy supplied by the wind. Therefore, the typical radius, velocity, and kinetic energy of the expanding shell created by the stellar wind are comparable to those of a supernova shell in the radiative phase, even though the mechanical energy E_0 of the supernova outburst may be several times greater than the wind energy $E_w = L_w t$.

As with supernova shells, the greatest uncertainty in the structure and evolution of a bubble arises from uncertainty in the structure of the ambient gas. If the bubble expands into an inhomogeneous medium, dense clouds may be overtaken by the expanding shell and remain in the hot interior, where they can rob the system of thermal and kinetic energy by enhancing radiative losses. If such radiative losses are very effective, the shell will be driven directly by the momentum of the stellar wind and will expand according to the law (Steigman, Strittmatter & Williams 1975),

$$R_s \approx 16 \, (L_{36}/n_0)^{1/4} \, (V_w/1000 \text{ km s}^{-1})^{-1/4} \, t_6^{1/2} \text{ pc.} \tag{12}$$

In this case the kinetic energy of the expanding shell is a small fraction of the wind energy, $E_k = (\dot{R}_s/V_w)L_w t$. If the exciting star is moving through the ambient ISM, the resulting bubble is distorted and the star is not centrally located (Weaver et al. 1977).

Interstellar bubbles may be observed in a variety of ways. Castor et al. (1975) pointed out that there should be a column density $N_{OVI} \approx 2 \times 10^{13}$ cm^{-2} (cf. Equation 7) of O VI in the conductive interface between the shell and the hot interior of the bubble. Since the theoretical value is very insensitive to the parameters of the system and is in the range inferred from the UV absorption lines observed by the *Copernicus* spectrometer, Castor et al. suggested that the lines were formed in the bubbles. If so, the velocities of the absorption lines should be blueshifted relative to the stellar velocity, typically by 10–20 km s^{-1} (stellar motion can modify this conclusion). At present, the O VI observations do not seem to support this hypothesis (see Section II.B and Jenkins 1978b,c). Since the observed O VI column densities can be several times greater than the value that can be produced in a bubble, it seems reasonable that the light from an OB star may intersect several conductive interfaces in the ISM, not due to bubbles (McKee & Ostriker 1977). However, the requirements of absorption spectroscopy of O VI with the *Copernicus* spectrometer bias the observations toward hot OB stars, which are likely to have strong stellar winds. What fractions of the observed O VI absorption lines are caused by these winds remains an outstanding observational question.

Other UV absorption lines may be used as tracers of the expanding shell surrounding the bubble. For example, Jura (1975a,b) has inferred from the populations of rotationally excited H_2 molecules observed in the spectra of OB stars that the molecules are in a thin sheet of high-pressure gas near the exciting stars. The tendency of the H_2 lines to be blueshifted relative to the stars indicates expansion. Hollenbach et al. (1976) have interpreted these observations in the context of the bubble theory. This interpretation might be tested by searching for velocity correlations between rotationally excited H_2 and CH$^+$ if, as Elitzur & Watson (1978) suggest, the CH$^+$ is formed in shocks by the endothermic reaction $C^+ + H_2 \rightarrow CH^+ + H$ (but cf. de Jong 1979). It might also be possible to distinguish the HII regions in the expanding shells from other intervening HII by observing the lines of excited fine structure levels of ions such as NII*, which are indicators of high-pressure HII regions.

Optical evidence for shell structure in HII regions abounds. Good examples in our own galaxy are found in the Orion Nebula (Gull & Sofia 1979), the Bubble Nebula, NGC 7635 (Icke 1973), and the Rosette Nebula (Mathews 1966). Hα photographs of the Magellenic Clouds (Lasker 1976) and other nearby galaxies (Courtes 1977) show many ring-shaped emission nebulae. We think it unlikely that they are supernova shells because HII regions, caused by stars lasting more than 10^6 yr, should be much more common than supernova shells, lasting a few times 10^4 yr.

Certainly the wind from a massive star will create a large cavity of low density in the ISM before the star dies. This fact helps to solve a puzzle regarding the Cygnus Loop (Tucker 1971), namely, that the velocity of the blast wave inferred from the temperature of the observed soft X rays exceeds the velocities of the optical filaments by roughly a factor of 3. McKee & Cowie (1975) suggested that the optical filaments are dense clouds in an inhomogeneous medium, which have suddenly decelerated the blast wave. Their explanation, while reasonable, is not completely satisfying because it begs the question of why the clouds appear to have arranged themselves in such a nicely spherical shell. The answer we suggest is that the Cygnus Loop was created by the progenitor of the supernova, and not by the supernova itself.

3. REPEATED SUPERNOVAE   The giant expanding systems mentioned in Section II, such as the Gum Nebula and the Orion's Cloak complex, are difficult to account for with a single supernova explosion. In each case the estimated kinetic energy of the system, $E_k \gtrsim 3 \times 10^{50}$ ergs, implies an outburst energy $E_0 \gtrsim 10^{52}$ ergs, uncomfortably large for a supernova explosion. We favor the suggestion (Cowie, Songaila & York 1979, Reynolds & Ogden 1979) that these structures are the result of *repeated* supernova explosions. The ages of these expanding systems, estimated roughly from $t \sim R/\dot{R}$, are a few million years. The Gum Nebula contains the O5 supergiant, ζ-Puppis, the massive Wolf-Rayet binary, γ-Velorum, a B-association, and two supernova remnants, the Vela SNR and Puppis A, whose ages are $\lesssim 10^4$ yr. These young SNR's cannot be responsible for the Gum Nebula, but the fact that two events have occurred so recently makes it easy to believe that several more events could have occurred within the last few million years. Similarly, the suggestion that several supernova events might be responsible for the Orion complex is strongly supported by the existence of three runaway OB stars which have apparently originated in the I-Ori OB association within the past few million years. The most spectacular manifestation of the cumulative effect of many supernova events may be the region of Constellation III in the Large Magellenic Cloud (Westerlund & Mathewson

1966). If the great loop (diameter ≈ 1 kpc) of OB stars, HI emission, and supernova remnants is a real association, the inferred energy is $E_0 \approx 10^{54}$ ergs! (See note added in proof, p. 238.)

The structure and evolution of these giant expanding systems are very similar to those of an interstellar bubble. Each time a supernova outburst occurs within the system, the blast wave propagates through the hot interior without radiative losses, depositing $0.72\,E_0$ as thermal energy in the interior. The remaining $0.28\,E_0$ of kinetic energy of the blast wave is lost to radiation when it encounters the dense outer shell. As with a bubble, the dissipation of interior thermal energy is limited by the rate of thermal conduction into the shell. Therefore, the system resulting from the cumulative effect of many supernova explosions, repeating at a mean rate t_{SN}^{-1}, evolves according to Equation (11), with the replacement $L_w = E_0/t_{SN}$.

E. Global Models

Ultimately, we would like to have a global model that accounts quantitatively for the observed distribution of physical parameters and for the cycle of mass and energy in the ISM. This goal has proved elusive. A decade ago, Field, Goldsmith & Habing (1969) proposed a model composed of cool ($T \lesssim 100$ K) clouds in pressure equilibrium with a substrate of warm ($T \approx 10^4$ K) "intercloud medium" consisting of partially ionized HI. The energetics of the system were controlled by a balance between radiative cooling and heating by a hypothetical flux of low-energy (~ 2 MeV) cosmic rays (cf. Dalgarno & McCray 1972). However, this model has not been supported by ultraviolet absorption observations of HD and H_2, whose ratio implies a cosmic ray ionization rate in dense clouds that fails to meet the requirements of the model by a factor $\gtrsim 10$ (Watson 1978).

The fact that the observed temperature distribution of the ISM favors the discrete ranges $T \lesssim 100$ K, $T \approx 10^4$ K, and $T \gtrsim 10^6$ K follows generally from the atomic properties of the gas, which control its radiative cooling, and is not specific to the model for heating of the gas. For example, a bimodal distribution of temperature similar to that obtained by Field et al. was also obtained by Gerola et al. (1974) for a radically different time-dependent model in which the medium was heated impulsively by hypothetical sudden bursts of X rays from supernovae. Spitzer (1956) was the first to appreciate that the radiative cooling properties of interstellar gas also favored the existence of a coronal phase of the ISM with $T \gtrsim 10^6$ K. McCray & Buff (1972) noticed that such a coronal phase fits naturally into the cosmic ray heating model of Field et al. Coronal gas in the galaxy has a scale height $\delta \approx 5T_6$ kpc; therefore, its existence in the disk implies a hot galactic corona (Spitzer 1956, Weisheit & Collins 1976, Chevalier & Oegerle 1979).

Soon after observations of soft X-ray emission in the galactic plane began to demand serious consideration of the properties of coronal gas in the ISM, Cox & Smith (1974) proposed a model of an interstellar "tunnel" system of interconnecting supernova remnants (SNR's). They began with the recognition that the hot low-density gas in the interior of a SNR radiates so inefficiently that the system will persist for a time $\tau \gtrsim 4 \times 10^6$ yr. They assume that supernova outbursts occur at a rate r per unit volume in the galactic disk with a random distribution in space and time. The fraction f of interstellar space occupied by the SNR's, whose average final volume is V_{SNR}, is related to the porosity parameter $q = r\tau V_{SNR}$ by $f \approx q$ for $q \ll 1$; Cox & Smith estimate $q \approx 0.1$. However, once $q \gtrsim 0.1$ there is a substantial chance for a given SNR to encounter another before it grows to its final volume; if that happens, the old SNR can be rejuvenated by the blast wave from the new one, which propagates preferentially into the coronal gas. In this way, a substantial fraction of the blast energy of supernovae can be diverted into an interconnecting tunnel system that threads the ISM like the holes in Swiss cheese. The development and maintenance of such a tunnel system has been demonstrated in Monte Carlo simulations by Smith (1977).

Cox & Smith also suggested that the coronal gas within the tunnels was the source of the observed O VI absorption lines. However, Shapiro & Field (1976) showed that this interpretation required an unrealistically high interstellar gas pressure. More likely, the O VI is located in the conductive interfaces at the tunnel walls.

The most ambitious attempt so far to construct a global model for the ISM is that of McKee & Ostriker (1977). These authors have combined the theory of SNR evolution (Section III.D) and the theory of conductive interfaces (Section III.C) within the framework of the Cox & Smith model. The resulting model consists of cool clouds embedded in a coronal substrate which occupies a large fraction of the interstellar volume. A stationary cycle of mass flow is maintained in a balance between evaporation of clouds and the addition of matter to the clouds from the coronal gas by the sweeping action of supernova blast waves. McKee & Ostriker note that comparable fractions of the supernova power source may be lost to heating the galactic corona, to dissipation of the blast waves by clouds (which act as shock-absorbers), and to radiative losses at the conductive interfaces between the clouds and the coronal gas. By neglecting the former two effects, they are able to obtain analytical formulae for the mean pressure, density, and temperature of clouds and coronal gas and for the fractional volume occupied by each.

According to this model, the clouds consist of a cold ($T \lesssim 100$ K) core surrounded successively by a layer of warm ($T \approx 10^4$ K) HI, partially ionized and heated by soft X rays and cosmic rays as in the model of Field et al. (1969), and then by a layer of warm HII. [De Jong (1979) has pointed out that absorption lines of CH^+ may provide an important diagnostic of the warm HI layer.] The outer layers comprise a large fraction of the cloud volume but not of the mass. By assuming a reasonable spectrum of cloud sizes, a mean interstellar density, a mean supernova rate and blast energy, McKee & Ostriker are able to construct a model for the ISM that conforms with many observations, such as O VI absorption, soft X-ray emission, pulsar dispersion measures, and 21-cm emission and absorption. The model even accounts roughly for the random motions of interstellar clouds, established by a balance between acceleration by shocks (McKee, Cowie & Ostriker 1978) and cloud-cloud collisions. Also, Cox (1979) has suggested that the dissipation of shock waves by the cool clouds may be a significant heating mechanism for interstellar HI.

While there is undoubtedly much truth in the theory of McKee & Ostriker, it would be premature to take the model literally. For example, we can point to two effects that may introduce substantial uncertainties into the results. The first is the morphology of the clouds. McKee & Ostriker assumed spherical clouds, but modern observations as well as the theory itself imply that the "clouds" are actually more like sheets (*stratus* rather than *cumulus*), which must have vastly different surface-to-volume ratios from spheres. How sensitive is the theory to that? McKee & Ostriker mention that the assumed spectrum of cloud sizes (which affects the same ratio) is an uncertainty of their theory, but they do not clearly state the quantitative importance of the effect. A second uncertainty stems from the assumed statistical distribution of supernova events. Certainly, the progenitors of supernovae are highly correlated with each other and with concentrations of gas, since Population I stars are born in clusters from dense gas clouds (Scott, Jensen & Roberts 1977). How are the porosity and morphology of the model affected by these correlations? We have already mentioned in Sections II.D and III.D the observational evidence for large coherent structures in the ISM resulting from spatially correlated supernovae. These are but two of the several aspects of the McKee-Ostriker model that merit further investigation.

This newly emerging picture of the ISM has many implications that deserve further investigation. Shu (1978) has pointed out that if the coronal gas fills a substantial fraction of the ISM, the interstellar clouds are not highly compressed by galactic shocks as they enter spiral arms. This makes it difficult to understand the pattern of star formation in spiral galaxies. The presence of the coronal gas also has major implications for the origin and propagation of cosmic rays. Resonant trapping by hydromagnetic waves in the ISM appears to play a major role in cosmic ray confinement (cf. Wentzel 1974, Skilling 1975a,b,c). The mechanism is most effective in the coronal gas, where the waves are not easily damped. Furthermore, Bell (1978a,b) and Blandford & Ostriker (1978) have shown that cosmic rays can be accelerated by shocks in the ISM; the theory predicts a cosmic ray energy spectrum that agrees with observations. Finally, Silk (1975) has suggested that the damping of hydrodynamic and hydromagnetic waves can be an important nonionizing heat source of the ISM. It appears that the violent interstellar medium will play a major role in the future development of theories for the structure and evolution of galaxies.

NOTE ADDED IN PROOF   C. Heiles (1979, *Ap. J.* 229:533) presents new evidence of expanding HI shells seen in 21-cm emission with radii ranging from 10 pc to 1.2 kpc. The interpretation of these structures in terms of blast wave models (cf Section III.D) implies explosion energies ranging from 10^{49} to 3×10^{53} ergs. Heiles discusses the relation of shell structures seen in 21-cm emission to similar structures seen in radio continuum emission and in H_α emission, and the correlation of the shell structures with stellar associations.

Literature Cited

Aannestad, P. 1973. *Ap. J. Suppl.* 25:223
Abbott, D. 1978. *Ap. J.* 225:893
Adams, W. S. 1949. *Ap. J.* 109:354
Bahcall, J. N., Sarazin, C. L. 1978. *Ap. J.* 219:781
Barlow, M. J. 1979. In *Mass Loss and Evolution of Early-Type Stars, I.A.U. Symp. No. 83*, ed. P. Conti. Dordrecht: Reidel
Barlow, M. J., Cohen, M. 1977. *Ap. J.* 213:737
Barlow, M. J., Silk, J. 1977. *Ap. J. Lett.* 211:L83
Barnard, E. E. 1895. *Pop. Astron.* 2:151
Beals, C. S. 1936. *MNRAS* 96:661
Bell, A. R. 1978a. *MNRAS* 182:147
Bell, A. R. 1978b. *MNRAS* 182:443
Blaauw, A., Tolbert, C. R. 1966. *Bull. Astron. Inst. Neth.* 18:405
Blandford, R. D., Ostriker, J. P. 1978. *Ap. J. Lett.* 221:L29
Bohlin, R. C., Savage, B. D., Drake, J. F. 1978. *Ap. J.* 224:132
Brown, R. L., Gould, R. J. 1970. *Phys. Rev. D* 1:2252
Burke, J. R., Silk, J. 1974. *Ap. J.* 190:1
Burstein, P., Borken, R. J., Kraushaar, W. L., Sanders, W. T. 1977. *Ap. J.* 213:405
Carruthers, G. R., Opal, C. B. 1977. *Ap. J. Lett.* 212:L27

Cash, W., Malina, R., Stern, R. 1976. *Ap. J. Lett.* 204:L7
Castor, J., McCray, R., Weaver, R. 1975. *Ap. J. Lett.* 200:L107
Chevalier, R. A. 1974. *Ap. J.* 188:501
Chevalier, R. A. 1977. *Ann. Rev. Astron. Astrophys.* 15:175
Chevalier, R. A., Oegerle, W. R. 1979. *Ap. J.* 227:398
Chevalier, R. A., Theys, J. C. 1975. *Ap. J.* 195:53
Chiosi, C., Nasi, E., Sreenivasan, S. R. 1978. *Astron. Astrophys.* 63:103
Cohn, H., York, D. G. 1977. *Ap. J.* 216:408
Conti, P. S. 1978. *Ann. Rev. Astron. Astrophys.* 16:371
Courtes, G. 1977. In *Topics in Interstellar Matter*, ed. H. van Woerden, p. 290. Dordrecht: Reidel
Cowie, L. L. 1978. *Ap. J.* 225:887
Cowie, L. L., Jenkins, E. B., Songaila, A., York, D. G. 1979. *Ap. J.* 232: In press
Cowie, L. L., Laurent, C., Vidal-Madjar, A., York, D. G. 1979. *Ap. J. Lett.* 229:L81
Cowie, L. L., McKee, C. F. 1977. *Ap. J.* 211:135
Cowie, L. L., Songaila, A., York, D. G. 1979. *Ap. J.* 230:469
Cowie, L. L., York, D. G. 1978a. *Ap. J.* 220:129

Cowie, L. L., York, D. G. 1978b. *Ap. J.* 223: 876
Cox, D. P. 1972. *Ap. J.* 178:143
Cox, D. P. 1979. *Ap. J.* In press
Cox, D. P., Smith, B. W. 1974. *Ap. J. Lett.* 189:L105
Cruddace, R., Paresce, F., Bowyer, S., Lampton, M. 1974. *Ap. J.* 187:497
Dalgarno, A., McCray, R. A. 1972. *Ann. Rev. Astron. Astrophys.* 10:375
de Boer, K. S., Morton, D. C., Pottasch, S. R., York, D. G. 1974. *Astron. Astrophys.* 31:405
de Jong, T. 1979. *Astron. Astrophys.* In press
de Korte, P. A. J., Bleeker, J. A. M., Deerenberg, A. J. M., Hayakawa, S., Yamashita, K., Tanaka, Y. 1976. *Astron. Astrophys.* 48:235
de Loore, C., De Greve, J. P., Lamers, H. J. G. L. M. 1977. *Astron. Astrophys.* 61:251
Dickey, J. M., Salpeter, E. E., Terzian, Y. 1977. *Ap. J. Lett.* 211:L77
Dickey, J. M., Salpeter, E. E., Terzian, Y. 1979. *Ap. J.* 228:465
Draine, B., Salpeter, E. E. 1979a. *Ap. J.* 231:77
Draine, B., Salpeter, E. E. 1979b. *Ap. J.* 231:438
Elitzur, M., Watson, W. D. 1978. *Ap. J. Lett.* 222:L141
Field, G. B., Goldsmith, D. W., Habing, H. J. 1969. *Ap. J. Lett.* 155:L49
Fireman, E. L. 1974. *Ap. J.* 187:57
Gerola, H., Kafatos, M., McCray, R. 1974. *Ap. J.* 189:55
Giovanelli, R., Haynes, M. P., York, D. G., Shull, J. M. 1978. *Ap. J.* 219:60
Graham, R., Langer, W. D. 1973. *Ap. J.* 179:469
Gull, T. R., Sofia, S. 1979. *Ap. J. Lett.* 230:782
Habing, H. J. 1969. *Bull. Astron. Inst. Neth.* 20:177
Hayakawa, S., Kato, T., Nagase, F., Yamashita, K., Murakami, T., Tanaka, Y. 1977. *Ap. J. Lett.* 213:L109
Heiles, C. 1976. *Ap. J. Lett.* 208:L137
Heiles, C., Jenkins, E. B. 1976. *Astron. Astrophys.* 46:333
Hobbs, L. M. 1969. *Ap. J.* 157:135
Hobbs, L. M. 1971. *Ap. J.* 166:333
Hobbs, L. M. 1974. *Ap. J.* 191:381
Hobbs, L. M. 1978. *Ap. J. Suppl.* 38:129
Hollenbach, D., Chu, S. I., McCray, R. 1976. *Ap. J.* 208:458
Icke, V. 1973. *Astron. Astrophys.* 26:45
Inoue, H., Koyama, K., Matsuoka, M., Ohashi, T., Tanaka, Y., Tsunemi, H. 1979. *Ap. J. Lett.* 227:L85
Jacobs, V. L., Davis, J., Kepple, P. C., Blaha, M. 1977a. *Ap. J.* 211:605
Jacobs, V. L., Davis, J., Kepple, P. C.,

Blaha, M. 1977b. *Ap. J.* 215:690
Jenkins, E. B. 1978a. *Ap. J.* 219:845
Jenkins, E. B. 1978b. *Ap. J.* 220:107
Jenkins, E. B. 1978c. *Comments Astrophys.* 4:121
Jenkins, E. B., Meloy, D. A. 1974. *Ap. J. Lett.* 193:L121
Jenkins, E. B., Silk, J., Wallerstein, G. 1976a. *Ap. J. Lett.* 209:L87
Jenkins, E. B., Silk, J., Wallerstein, G. 1976b. *Ap. J. Suppl.* 32:681
Jordan, C. 1969. *MNRAS* 142:501
Jordan, C. 1970. *MNRAS* 148:17
Jura, M. 1975a. *Ap. J.* 197:575
Jura, M. 1975b. *Ap. J.* 197:581
Jura, M. 1976. *Ap. J.* 206:691
Kafatos, M. 1973. *Ap. J.* 182:433
Kato, T. 1976. *Ap. J. Suppl.* 30:397
Kraushaar, W. L. 1977. Review lecture, 149th Meeting, AAS
Landini, M., Fossi, B. C. M. 1972. *Astron. Astrophys. Suppl.* 7:291
Lasker, B. M. 1976. In *The Galaxy and the Local Group, R.G.O. Bull. No. 182*, eds. R. J. Dickens, J. E. Perry, p. 186. Herstmonceaux
Levine, A., Rappaport, S., Halpern, J., Walter, F. 1977. *Ap. J.* 211:215
Mathews, W. G. 1966. *Ap. J.* 144:206
McCray, R., Buff, J. 1972. *Ap. J. Lett.* 175:L65
McCray, R., Stein, R. F., Kafatos, M. 1975. *Ap. J.* 196:565
McCray, R., Wright, C., Hatchett, S. 1977. *Ap. J. Lett.* 211:L29
McKee, C. F. 1970. *Phys. Rev. Lett.* 24:290
McKee, C. F., Cowie, L. L. 1975. *Ap. J.* 195:715
McKee, C. F., Cowie, L. L. 1977. *Ap. J.* 215:213
McKee, C. F., Cowie, L. L., Ostriker, J. P. 1978. *Ap. J. Lett.* 219:L23
McKee, C. F., Ostriker, J. P. 1977. *Ap. J.* 218:148
Menon, T. K. 1958. *Ap. J.* 127:28
Mewe, R. 1972. *Solar Phys.* 22:459
Mewe, R. 1975. *Solar Phys.* 44:376
Muller, C. A., Oort, J. H., Raimond, E. 1963. *C. R. Acad. Sci. Paris* 257:1661
Münch, G. 1957. *Ap. J.* 125:42
Münch, G., Zirin, H. 1961. *Ap. J.* 133:11
O'Dell, C. R., York, D. G., Henize, K. G. 1967. *Ap. J.* 150:835
Oort, J. H. 1966. *Bull. Astron. Inst. Neth. Suppl.* 18:421
Ostriker, J. P., Silk, J. I. 1973. *Ap. J. Lett.* 184:L113
Payne, H. E., Dickey, J. M., Salpeter, E. E., Terzian, Y. 1978. *Ap. J. Lett.* 221:L95
Penston, M. V., Brown, F. E. 1970. *MNRAS* 150:373
Pickering, W. H. 1890. *Sidereal Messenger* 9:2

240   MᶜCRAY & SNOW

Raymond, J. 1979. *Ap. J. Suppl.* 39:1
Raymond, J. C., Cox, D. P., Smith, B. W. 1976. *Ap. J.* 204:290
Raymond, J. C., Smith, B. W. 1977. *Ap. J. Suppl.* 35:419
Reynolds, R. J. 1976a. *Ap. J.* 203:151
Reynolds, R. J. 1976b. *Ap. J.* 206:679
Reynolds, R. J., Ogden, P. M. 1978. *Ap. J.* 224:94
Reynolds, R. J., Ogden, P. M. 1979. *Ap. J.* In press
Ride, S. K., Walker, A. B. C. 1977. *Astron. Astrophys.* 61:339
Rogerson, J. B., York, D. G., Drake, J. F., Jenkins, E. B., Morton, D. C., Spitzer, L. 1973. *Ap. J. Lett.* 181:L110
Routly, P. M., Spitzer, L. 1952. *Ap. J.* 115:227
Salpeter, E. E. 1976. *Ap. J.* 206:673
Salpeter, E. E. 1979. *I.A.U. Symp. No. 84, Large Scale Characteristics of the Galaxy,* ed. W. B. Burton, p. 245. Dordrecht: Reidel
Sancisi, R. 1974. *I.A.U. Symp. No. 60, Galactic Radio Astronomy,* ed. F. J. Kerr, S. C. Simonson, p. 115. Dordrecht: Reidel
Sancisi, R., Goss, W. M., Anderson, C., Johansson, L. E. B., Winnberg, A. 1974. *Astron. Astrophys.* 35:445
Sanford, R. F. 1939. *Publ. Astron. Soc. Pac.* 51:238
Sarazin, C. L., Bahcall, J. N. 1977. *Ap. J. Suppl.* 34:451
Savage, B. D., de Boer, K. S. 1979. *Ap. J. Lett.* 230:L77
Scott, J. S., Jensen, E. B., Roberts, W. W. 1977. *Nature* 265:123
Sedov, L. 1959. *Similarity and Dimensional Methods in Mechanics.* New York: Academic
Shapiro, P. R., Field, G. B. 1976. *Ap. J.* 205:762
Shapiro, P. R., Moore, R. T. 1976. *Ap. J.* 207:460
Shapiro, P. R., Moore, R. T. 1977. *Ap. J.* 217:621
Shu, F. H. 1978. *I.A.U. Symp. No. 77, Structure and Dynamics of Nearby Galaxies,* ed. E. M. Berkhuijsen, R. Wielebinski, p. 139. Dordrecht: Reidel
Shull, J. M. 1977a. *Ap. J.* 212:102
Shull, J. M. 1977b. *Ap. J.* 216:414
Shull, J. M. 1977c. *Ap. J.* 215:805
Shull, J. M. 1978. *Ap. J.* 226:858
Shull, J. M., McKee, C. F. 1979. *Ap. J.* 227:131
Shull, J. M., Silk, J. 1979. *Ap. J.* In press
Shull, J. M., York, D. G. 1977. *Ap. J.* 211:803
Shull, J. M., York, D. G., Hobbs, L. M. 1977. *Ap. J. Lett.* 211: L139

Silk, J. 1975. *Ap. J. Lett.* 198:L77
Siluk, R. S., Silk, J. 1974. *Ap. J.* 192:51
Sivan, J. P. 1974. *Astron. Astrophys. Suppl.* 16:163
Skilling, J. 1975a. *MNRAS* 172:557
Skilling, J. 1975b. *MNRAS* 173:245
Skilling, J. 1975c. *MNRAS* 173:255
Smith, B. W. 1977. *Ap. J.* 211:404
Snow, T. P. 1976. *Ap. J.* 204:759
Snow, T. P. 1977. *Ap. J.* 216:724
Snow, T. P., Meyers, K. A. 1979. *Ap. J.* 229:545
Solinger, A., Rappaport, S., Buff, J. 1975. *Ap. J.* 201:381
Spitzer, L. 1956. *Ap. J.* 124:20
Spitzer, L. 1962. *Physics of Fully Ionized Gases.* New York: Wiley
Spitzer, L. 1976. *Comments Astrophys.* 6:177
Spitzer, L. 1978. *Physical Processes in the Interstellar Medium.* New York: Wiley
Steigman, G., Strittmatter, P. A., Williams, R. E. 1975. *Ap. J.* 198:575
Stern, R., Wang, E., Bowyer, S. 1978. *Ap. J. Suppl.* 37:195
Stokes, G. M. 1978. *Ap. J. Suppl.* 36:115
Stokes, G. M., Hobbs, L. M. 1976. *Ap. J. Lett.* 208:L95
Summers, H. P. 1974a. *MNRAS* 169:663
Summers, H. P. 1974b. *Appleton Lab. Rep.* CR74.102. Culham, England
Tanaka, Y., Bleeker, J. A. M. 1977. *Space Sci. Rev.* 20:815
Tidman, D. A., Krall, N. A. 1971. *Shock Waves in Collisionless Plasmas.* New York: Wiley
Tucker, W. H. 1971. *Science* 172:372
Verschuur, G. L. 1975. *Ann. Rev. Astron. Astrophys.* 13:257
Wallerstein, G., Silk, J. I. 1971. *Ap. J.* 170:289
Watson, W. D. 1978. *Ann. Rev. Astron. Astrophys.* 16:585
Weaver, R., McCray, R., Castor, J., Shapiro, P., Moore, R. 1977. *Ap. J.* 218:377, errata, 220:742
Weisheit, J. C., Collins, L. A. 1976. *Ap. J.* 210:299
Wentzel, D. G. 1974. *Ann. Rev. Astron. Astrophys.* 12:71
Westerlund, B. E., Mathewson, D. S. 1966. *MNRAS* 131:371
Williamson, F. O., Sanders, W. T., Kraushaar, W. L., McCammon, D., Borken, R., Bunner, A. N. 1974. *Ap. J. Lett.* 193:L133
Yentis, D. J., Novick, R., Vanden Bout, P. 1972. *Ap. J.* 177:375
York, D. G. 1974. *Ap. J. Lett.* 193:L127
York, D. G. 1976. *Ap. J.* 204:750
York, D. G. 1977. *Ap. J.* 213:43
Zel'dovich, Ya. B., Pikel'ner, S. B. 1969. *Sov. Phys.-JETP* 29:170

Ann. Rev. Astron. Astrophys. 1979. 17:241–74
Copyright © 1979 by Annual Reviews Inc. All rights reserved

GLOBULAR CLUSTERS IN GALAXIES

×2151

William E. Harris
Department of Physics, McMaster University, Hamilton, Canada

René Racine
Département de physique, Université de Montréal, Montréal, Canada

INTRODUCTION

Globular clusters are among the most versatile of astronomical objects. Within our own Galaxy, they play central roles in galactic structure, the evolution of low-mass and metal-poor stars, the dynamics of stellar systems, the early chemical and dynamical history of the Galaxy, stellar pulsation, and most recently the nature of certain X-ray sources. But as remnants of the earliest (?) star formation epoch, they are also believed to occur almost universally in galaxies of sufficient size. The purpose of this review is to assess current knowledge about globular-cluster systems viewed as subsystems of their parent galaxies.

Modern understanding of the structure of the Galactic globular-cluster system begins with the classic studies of Shapley (1918), and knowledge of clusters in distant galaxies (M31 and beyond) dates from the equally monumental work of Hubble (1932). Despite this honorable history, detailed photometric and spectroscopic studies of clusters in any galaxies more distant than the Magellanic Clouds have taken place only in the last decade, largely because of the imposing observational problem of measuring extremely faint and small clusters superimposed on the backgrounds of large galaxies. This subject, then, is a young one and is seen here in a phase of rapid development. We shall be concerned primarily with what cluster systems have revealed about the structure, chemical composition, and extent of galaxy haloes, their intrinsic luminosity and color distribution, the kinematics of cluster systems, and the sizes of total cluster populations as functions of galaxy size and type. Inevitably, we must bypass any discussion of the wealth of data specifically on individual clusters.

241

0066-4146/79/0915-0241$01.00

Figure 1   The largest system of globular clusters known, in the giant elliptical galaxy M87. The reproduction is from a 45-minute exposure on IIIaJ + GG385, taken by M. G. Smith with the 4-meter telescope at Cerro Tololo.

In the subsequent sections, we first discuss the compilation of data for each known globular-cluster system in turn, beginning with our own Galaxy and proceeding outward. In the final section we synthesize several integrated properties of the various known cluster systems. We shall find that, despite some significant differences, systems of globular clusters are much more similar to one another than are their parent galaxies.

1 THE GALAXY

1.1 *Fundamental Data*

The cluster system in our own Galaxy remains the one for which most complete knowledge exists, and provides the basis of comparison for all other systems. In Table 1 we summarize in catalog form the raw data relevant to this review for the 131 known Galactic globular clusters. The tabular data—which form the most important material of this section—are described below.

Columns 1–4 Cluster identification (NGC or other number, and commonly used name if any), and galactic coordinates (l,b) in degrees. The identification list follows Kukarkin (1974) with the addition of clusters 0423-21 (Schuster & West 1977) and 1730-33 (Liller 1977).

Column 5 Apparent visual distance modulus $(m-M)_V$ taken from the compilation of Harris (1976), with several significant more recent changes and additions; sources for these are given below the table. Distances for 78 clusters are now based exclusively on color-magnitude diagrams, and the others on secondary methods (Harris 1976). All are reduced to the same homogeneous system of Harris, assuming $M_V(HB) = 0.6$ for the horizontal branch in all clusters.

Column 6 Integrated absolute magnitude $M_V = V_t - (m-M)_V$ where V_t is the total apparent magnitude of the cluster in the sense defined by Peterson & King (1975). Here, we collected anew all available photoelectric concentric-aperture magnitude measurements $V(r)$ for each cluster (sources given in the table), and fitted them to the King (1966a) structural curves to obtain V_t. This could be done for a total of 92 clusters. For six clusters entirely without useful photoelectric magnitudes, we adopted the photographic m_{pg} data of Christie (1940) and Sawyer & Shapley (1927) transformed into the photoelectric scale. Finally, for the sparse objects Pal 3, 4, 5, 12, and 13 the integrated magnitudes and colors were obtained by direct addition of stars from the color-magnitude diagrams.

Columns 7–8 Integrated apparent colors of the clusters in the *UBV* system. The intrinsic colors are known to be approximate indicators

Table 1 Fundamental data for Galactic globular clusters

NGC	Name	l	b	$(m-M)_V^a$	M_V^b	$B-V^b$	$U-B^b$	E_{B-V}	R	[Fe/H][c]	$\log r_t$	c	v_{LSR}^d
104	47 Tuc	305.9	-44.9	13.46	-9.43	0.89	0.37	0.04	8.2	-0.44	1.64	2.03	-21
288		149.7	-89.4	14.70	-6.60	0.66	0.09	0.03	12.3	-1.41	1.19	0.99	-47
362		301.5	-46.3	14.90	-8.32	0.76	0.14	0.04	10.2	-1.20	1.01	1.70	218
1261		270.6	-52.1	15.70	-7.32	0.70	0.14	0.02	16.1	-1.24	0.90	1.30	41
Pal 1		130.0	19.1	18.7:				0.12	52.0:	-1.9:	0.7:	1.50	10
0423-21	Eridanus	251.1	-41.3					0.03					
Pal 2		170.5	-9.0			1.9:	1.8:	1.2:			0.67		-140
1851		244.5	-35.0	15.40	-8.10	0.78	0.22	0.07	16.4	-1.29	0.91	1.83	299
1904	M79	227.2	-29.3	15.65	-7.65	0.63	0.03	0.01	20.0	-1.58	1.03	1.60	181
2298		245.6	-16.0	15.80	-6.40	0.73	0.19	0.11	17.9	-1.41	0.86:	1.2:	48
2419		180.4	25.3	19.94	-9.57	0.66	0.07	0.03	101.3	-2.00	1.03	1.41	-23
2808		282.2	-11.3	15.52	-9.22	0.93	0.29	0.22	11.5	-1.09	1.15	1.75	89
Pal 3		240.3	41.9	20.0:	-5.3:			0.03	99.4:	-2.2:	0.63	0.97	3
3201		277.2	8.6	14.15	-7.40	0.98	0.37	0.21	9.7	-1.26	1.56	1.52	478
Pal 4		202.3	71.8	19.85	-5.65	0.80		0.00	96.3	-2.4:	0.49	0.76	157
4147		252.9	77.2	16.28	-6.02	0.60	0.07	0.02	20.2	-1.77	0.86	1.51	174
4372		301.0	-9.9	14.90	-7.10	0.97:	0.32	0.45	7.8	-1.7:	1.5:		59
4590	M68	299.6	36.0	15.01	-6.81	0.63	0.04	0.03	10.2	-2.04	1.47:	1.62:	-125
4833		303.6	-8.0	14.90	-7.55	0.96	0.28	0.38	7.6	-2.15	1.08:	0.95:	197
5024	M53	333.0	79.8	16.34	-8.62	0.64	0.10	0.05	18.1	-1.85	1.34	1.67	-88
5053		335.6	79.0	16.00	-6.20	0.64	0.06	0.03	16.2	-2.09	1.14	0.75	
5139	ω Cen	309.1	15.0	13.92	-10.27	0.79	0.19	0.11	7.1	-1.6:p	1.74	1.36	224
5272	M3	42.2	78.7	15.00	-8.65	0.69	0.10	0.01	12.4	-1.57	1.59	1.90	-143
5286		311.6	10.6	15.61	-7.99	0.87	0.29	0.27	7.5	-1.38	1.08	1.75	40
5466		42.1	73.6	15.96	-6.86	0.71	0.04	0.05	15.3	-1.91	1.32	1.15	127
5634		342.2	49.3	16.90	-7.33	0.67	0.12	0.07	17.5	-1.70	0.92:	1.6:	-58
5694		331.1	30.4	17.80	-7.60	0.69	0.07	0.08	26.1	-1.91	1.18	1.78	-179
IC 4499		307.4	-20.5	17.12	-6.52	0.88	0.36:	0.24	15.4	-1.0:	1.20	1.10	
5824		332.6	22.1	17.32	-8.32	0.75	0.15	0.14	17.1	-1.67	1.30	2.51	-56
Pal 5		0.9	45.9	16.75	-5.00	0.70		0.03	16.5	-1.24	1.26:		
5897		342.9	30.3	15.60	-7.05	0.75	0.08:	0.06	6.9	-1.45	1.06	0.82	
5904	M5	3.9	46.8	14.51	-8.76	0.71	0.12	0.03	6.7	-1.25	1.46	1.78	59
5927		326.6	4.9	16.10	-7.77	1.31	0.84	0.55	5.0	-0.67	1.06:	1.45:	-85
5946		327.3	4.2	16.7:	-7.05:	1.24	0.54	0.56	5.3:	-1.5:	0.8:	1.6:	
5986		337.0	13.3	15.90	-8.78	0.90	0.30	0.27	4.5	-1.26	1.10	1.42	5
1608+15	Pal 14	28.8	42.2	19.2:				0.03	60:	-1.6:			98
6093	M80	352.7	19.5	15.28	-8.08	0.85	0.24:	0.21	3.2	-1.54	0.97	1.88	28
6101		317.7	-15.8	15.70	-6.40	0.68:	0.10	0.08	8.6	-1.8:			
6121	M4	351.0	16.0	12.73	-6.80	1.03	0.44	0.35	7.0	-1.30	1.64	1.45	73
6139		342.4	6.9	16.93:	-7.75:	1.39	0.73	0.68	3.0:	-1.27:	1.10	1.75	25
6144		351.9	15.7	15.69	-6.57	1.01	0.45	0.36	2.8	-0.9:	0.8:	0.97:	
6171	M107	3.4	23.0	15.03	-6.90	1.14	0.55	0.37	4.3	-0.79	1.36	1.50	-135
6205	M13	59.0	40.9	14.35	-8.49	0.69	0.03	0.02	9.1	-1.42	1.43	1.55	-225
6218	M12	15.7	26.3	14.30	-7.70	0.82	0.21	0.19	5.1	-1.64	1.26	1.31	21
6229		73.6	40.3	17.50	-8.07	0.71	0.05:	0.01	30.6	-1.44	0.75	1.41	-137
6235		358.9	13.5	16.31:	-6.16:	1.04	0.40	0.38	2.8:	-1.2:			
6254	M10	15.1	23.1	14.05	-7.48	0.92	0.23	0.26	5.5	-1.43	1.38	1.49	82
6256	Trz 12	347.8	3.4	16.6:		1.68	1.04						
Pal 15		18.9	24.3					0.09:					
6266	M62	353.6	7.3	15.38	-8.78	1.17	0.52	0.46	3.2	-1.14	1.03	1.63	-66
6273	M19	356.9	9.4	16.35	-9.20	1.00	0.37	0.38	2.4	-1.61	1.24	1.50	135
6284		358.4	9.9	15.89:	-6.94:	0.95	0.37	0.27	2.0:	-1.01:	1.0:	1.9:	32
6287		0.1	11.0	15.92:	-6.67:	1.20:	0.65	0.36	1.7:	-0.39:	1.15:	1.65:	
6293		357.6	7.8	15.41:	-7.21:	0.98	0.29	0.34	2.0:	-1.86:	1.2:	1.8:	-63
6304		355.8	5.4	15.50	-7.08	1.33	0.85	0.58	3.7	-0.37	1.2:	1.6:	-89
6316		357.2	5.8	16.99:	-7.99:	1.27	0.66	0.48	3.5:	-0.44:	1.25:	1.6:	
6325		1.0	8.0	16.70	-6.00	1.54:	0.88:	0.80	2.5	-0.7:	1.1:	1.6:	
6333	M9	5.5	10.7	15.34:	-7.41:	0.94	0.30	0.36	2.7:	-1.81:	1.19	1.60	236
6341	M92	68.4	34.9	14.50	-7.98	0.62	0.00	0.01	10.0	-2.12	1.22	1.78	-98
6342		4.9	9.7	17.5:	-7.60:	1.29	0.73	0.49	6.7:	-0.41	1.05:	1.6:	
6352		341.4	-7.2	14.47	-6.32	1.06	0.63	0.25	4.3	-0.06	1.05:	1.1:	
6355		359.6	5.4	16.6:	-7.00:	1.46	0.76	0.76	2.3:	-1.05:	1.15:	1.6:	
6356		6.7	10.2	17.07	-8.67	1.11	0.60	0.28	8.6	-0.37	0.92:	1.3:	45
Trz 2		356.3	2.3								0.4:		
6362		325.5	-17.6	14.65	-6.35	0.85	0.28	0.12	5.6	-0.9:	1.22	1.02	-18
6366		18.4	16.0	15.1:	-5.10:	1.47	0.98	0.65	5.6:	-0.1:	1.35:	1.45:	
Trz 4		356.0	1.3										
HP 1		357.4	2.1								0.7:		
1730-33	Liller 1	354.8	-0.2										
6380		350.3	-3.6								0.8:		
6388		345.5	-6.7	16.83	-9.98	1.17	0.66	0.32	6.3	-0.48	0.92	1.75	86
Trz 1		357.6	1.0								0.7:		
Ton 2		350.8	-3.4								0.8:		
6397		338.2	-12.0	12.30	-6.65	0.75	0.15	0.18	7.1	-1.83	1.64	1.86	20
6401		3.5	4.0	16.7:	-7.20:	1.58	0.89	0.79	2.3:	-0.7:	0.95:	1.45:	
6402	M14	21.3	14.8	16.90	-9.34	1.28	0.64	0.58	4.4	-1.28	1.00	1.10	-88
Pal 6		2.1	1.8	18.1:				1.8:	6.1:		1.1:		
6426		28.1	16.2	17.30	-6.10	1.03	0.34	0.40	9.6	-1.35:	1.0:	1.45:	
Trz 5		3.8	1.7			2.77	2.1:	1.8:			0.6:		
6440		7.7	3.8	16.40	-6.75	1.98	1.51	1.11	4.8	-0.28	0.9:	1.6:	-84

Table 1   (*continued*)

NGC	Name	l	b	$(m-M)_V^a$	M_V^b	$B-V^b$	$U-B^b$	E_{B-V}	R	$[Fe/H]^c$	$\log r_t$	c	v_{LSR}^d
6441		353.5	-5.0	16.50	-9.08	1.28	0.83	0.45	1.9	-0.24	0.88	1.70	20
Trz 6		358.6	-2.2								0.4:		
6453		355.7	-4.0	16.4:	-6.50:	1.28:	0.65:	0.67:	2.1:	-1.1:	0.8:		
6496		348.1	-10.0	15.0:	-5.80:	0.93	0.42:	0.07	2.4:	-0.1:	1.1:		
Trz 9		3.6	-2.0										
6517		19.2	6.8	18.10:	-7.80:	1.79	0.94	1.14	3.2:	-1.33	1.2:	2.0:	
6522		1.0	-3.9	15.64	-7.04	1.22	0.67	0.50	2.6	-1.04	0.85:	1.45:	-17
6528		1.1	-4.2	16.40	-6.90	1.45	1.10	0.65	1.8	-0.43	0.75:	1.45:	133
6535		27.2	10.4	16.35:	-5.75:	0.97	0.31	0.36	5.3:	-1.9:	0.75:	0.7:	
6539		20.8	6.8	15.7:	-6.10:	1.89	1.10:	1.22	6.9:	-1.2:	1.35:	1.6:	
6541		349.3	-11.2	14.60	-7.96	0.77	0.14	0.13	3.0	-1.59	1.50	2.00	-142
6544		5.8	-2.2	15.35	-7.10	1.36	0.67:	0.63	4.4	-1.02	1.1:	1.45:	-1
6553		5.3	-3.1	16.40	-8.15	1.62	1.24	0.79	3.3	-0.4:	1.0:	1.2:	-18
6558		0.2	-6.0	16.1:		1.09	0.48:	0.40	1.0:	-0.9:	0.8:		
IC 1276	Pal 7	21.8	5.7	18.50		1.74	1.0:	0.92	5.8	-0.8:	1.1:		
Trz 11		8.4	-2.2										
6569		0.5	-6.7	16.47:	-7.77:	1.34	0.63	0.63	1.6:	-0.54:	1.0:	1.45:	
6584		342.1	-16.4	16.17:	-6.99:	0.79	0.17	0.11	7.3:	-1.40	0.75:	1.2:	164
6624		2.8	-7.9	15.45	-7.13	1.10	0.57	0.25	1.4	-0.34	0.8:	1.2:	79
6626	M26	7.8	-5.6	14.99:	-8.06:	1.10	0.45	0.33	3.1:	-1.08	1.2:	1.6:	12
6637	M69	1.7	-10.3	15.60	-7.90	0.99	0.50	0.17	2.2	-0.47	1.0:	1.45:	68
6638		7.9	-7.2	15.66:	-6.51:	1.16	0.58	0.36	1.9:	-0.6:	0.77	1.30	-3
6642		9.8	-6.4	15.11		1.08	0.50:	0.36	3.2	-0.88:	0.9:		-75
6652		1.5	-11.4	16.23:	-7.32:	0.92	0.39	0.11	6.4:	-0.5:	0.8:	1.45:	-115
6656	M22	9.9	-7.6	13.55	-8.45	0.99	0.30	0.35	6.1	-1.69	1.52	1.24	-132
Pal 8		14.1	-6.8	18.4:		1.19	0.69:	0.30	22:		0.9:		
6681	M70	2.9	-12.5	15.40	-7.32	0.71	0.14	0.07	2.9	-1.17	1.05:	1.1:	208
6712		25.3	-4.3	15.51	-7.30	1.16	0.56	0.35	4.0	-0.43	1.1:	1.2:	-111
6715	M54	5.6	-14.1	17.11	-9.41	0.85	0.24	0.14	13.0	-1.55	0.87	1.85	132
6717	Pal 9	12.9	-10.9	16.55:		0.93	0.37	0.18	7.5:		0.8:		
6723		0.1	-17.3	14.80	-7.48	0.75	0.24	0.03	2.7	-0.85	1.10	1.20	5
6749		36.1	-2.2			1.76:	0.74:	0.96:			0.7:	1.2:	
6752		336.5	-25.6	13.20	-7.80	0.65	0.07	0.03	6.0	-1.62	1.54	1.84	-37
6760		36.1	-3.9	15.90	-6.80	1.66	1.00:	0.91	6.3	-1.06	1.03:	1.45:	
Trz 7		3.4	-20.0					0.12:					
6779	M56	62.7	8.3	15.60	-7.35	0.86	0.18	0.22	9.7	-1.79	1.0:	1.4:	-126
Pal 10		52.4	2.7	18.6:				1.2:			7.9:	0.8:	
1925-30		8.6	-20.8					0.11:				0.8:	
6809	M55	8.8	-23.3	13.80	-6.85	0.69	0.10	0.07	4.8	-1.78	1.27	1.03	178
Pal 11		31.8	-15.6	16.40				0.35:	6.7	-0.63	0.87	0.75	-54
6838	M71	56.7	-4.5	13.90	-5.60	1.13	0.54	0.28	7.6	-0.28	1.1:	1.2:	-1
6864	M75	20.3	-25.8	16.85	-8.30	0.87	0.28	0.17	11.7	-1.30	0.82	1.78	-187
6934		52.1	-18.9	16.22	-7.34	0.74	0.19	0.12	12.0	-1.38	0.95	1.35	-345
6981	M72	35.2	-32.7	16.29	-6.94	0.72	0.13	0.03	12.9	-1.27	0.94	1.19	-244
7006		63.8	-19.4	18.12	-7.52	0.74	0.17	0.13	32.1	-1.66	0.80	1.38	-371
7078	M15	65.0	-27.3	15.26	-8.91	0.68	0.06	0.12	10.3	-2.01	1.32	1.96	-97
7089	M2	53.4	-35.8	15.45	-8.95	0.67	0.08	0.06	10.5	-1.53	1.21	1.61	0
7099	M30	27.2	-46.8	14.60	-7.10	0.58	0.03	0.01	7.6	-2.03	1.20	2.11	-172
Pal 12		30.5	-47.6	16.46	-4.30	0.90	0.29:	0.02	15.6	-1.55	0.90	0.75	15
Pal 13		87.1	-42.7	17.10	-2.60	0.69:	-0.14:	0.05	25.7	-2.03	0.5:		-22
7492		53.3	-63.5	16.70	-5.20	0.48	0.17:	0.00	21.3	-2.0:	0.88	0.98	
Error				0.3	0.1	0.02	0.03	0.04	15%	0.2	0.1	0.2	20

Additional sources of compiled data:

[a] Distance moduli: Alcaino 1977a, 1978a; Canterna & Schommer 1978; Diamond 1976; Harris 1977, 1978; Harris & Canterna 1979a; Harris & Hesser 1976; Hartwick & Sargent 1978; Hesser et al. 1977; Lee 1977a,b,c; Liller & Carney 1978; Rutily & Terzan 1977; Sandage & Hartwick 1977; Sandage et al. 1977; Searle & Zinn 1978.

[b] Aperture photometry: Bernard 1976; Corwin 1977; Gascoigne & Burr 1956; Harris & van den Bergh 1974; Illingworth 1976; Johnson 1959; King 1966b; Kron 1966, 1975; Kron & Mayall 1960; Racine 1975; Rousseau 1964; van den Bergh 1967, 1971, 1977a; van den Bergh & Hagen 1968; Zaitseva et al. 1974.

[c] Heavy-element abundances: Bell & Gustafsson 1976; Butler 1975; Canterna & Schommer 1978; Cohen 1978; Cowley et al. 1978; Harris 1978; Harris & Canterna 1979b; Helfer et al. 1959; Hesser et al. 1977 and references cited; Mallia 1977; Searle & Zinn 1978.

[d] Radial velocities: Da Costa et al. 1977; Gratton & Nesci 1978; Hartwick & Sargent 1978; Jenner & Kwitter 1977; Kinman 1959a; Mayall 1946; Smith et al. 1976; van den Bergh 1969; Zinn 1974.

of cluster metallicity, and for most of the extragalactic cluster systems, only the broadband colors are available to provide heavy-element abundance estimates. Here, all available photoelectric UBV measurements have again been correlated and averaged into a completely new list (sources as in Column 6 above).

Column 9   Foreground reddening $E(B–V)$, from Harris (1976) or the individual color-magnitude studies. For several remaining clusters a cosecant-law estimate $E(B–V) = 0.06$ (csc $|b| - 1$) was used.

Column 10   Galactocentric distance R, in kpc. We assume the Sun to be 9.0 kpc from the Galactic center (Harris 1976), and the ratio $A_V/E(B–V) = 3.2$.

Column 11   Estimate of the heavy-element abundance of each cluster, $[Fe/H] \equiv \log (Fe/H)_{cl} - \log (Fe/H)_{\odot}$. Where possible, these are based on averages of abundance determinations from individual cluster giants; primary references are given in the table. The relations between $[Fe/H]$, integrated spectral type, and integrated broadband colors (Harris & Canterna 1977, 1979b, Kron & Guetter 1976) were used to derive less precise $[Fe/H]$ estimates on the same scale for the other clusters.

Columns 12–13   The King (1966a) structural parameters: tidal radius r_t in arc min and central concentration $c = \log (r_t/r_c)$. These are taken for 105 clusters either from the extensive star-count data of Peterson & King (1975) and Peterson (1976), or from our own application of the King (1966a) curves to the concentric-aperture photometry (Column 6 above) to derive r_t and r_c. For 18 remaining objects we have taken r_t as transformed from Kukarkin (1974), although these are not used in subsequent calculations as they were obtained originally on different scales.

Column 14   The radial velocity v_{LSR}, reduced to the Local Standard of Rest (peculiar solar motion adopted as 20 km s^{-1} toward $l = 57°$, $b = 22°$; cf Delhaye 1965). Sources are as given in Table 1; we have not included the important but as yet incomplete new southern survey of J. E. Hesser & S. J. Shawl, which should yield major improvements in the velocity data for many clusters.

Because of the wealth of new data in almost every category, Table 1 is intended, where applicable, to replace earlier catalogues (e.g. Sawyer Hogg 1959, Arp 1965, Kukarkin 1974, Peterson & King 1975, Woltjer 1975, Harris 1976, Alcaino 1977b). But despite the dramatic current observational gains, the many blank or uncertain entries in Table 1 testify to the amount of fundamental work still needed. At the end of Table 1 we have listed internal errors for each column. These are not meant to apply rigorously to each cluster, but are simply our estimates of the typical reliability of each quantity. Values marked with colons are regarded as more uncertain than the quoted error by a factor of two or more.

1.2   Population and Space Distribution

The 131 objects in Table 1 can hardly make up all globular clusters now present in the Galaxy, both because many objects in the Galactic center region must be obscured from (optical) view and because the sky has not yet been completely searched for distant Palomar-type clusters. Recent discussions (Harris 1976, Oort 1977) conclude that the central regions especially may be hiding ~ 30–70 more clusters, mostly on the far side of the center, making the total population N_t (Galaxy) $\simeq$ 160–200. These estimates will remain uncertain until appropriate infrared searches in the Galactic center are completed.

The distribution of clusters through the halo, i.e. the number of clusters per unit volume, $\rho(R)$, has been extensively reviewed by Harris (1976) (see also Woltjer 1975, Sharov 1976, Oort 1977, de Vaucouleurs 1977a). Clusters of all metallicity types are strongly concentrated to the Galactic center, in roughly spherical symmetry. Harris (1976) shows that $\rho(R)$ can be described by a simple power law $\rho \sim R^{-n}$, where n gradually increases with R, from $n \simeq 3$ at $R \lesssim 10$ kpc up to $n \simeq 4$ for $R \gtrsim 40$ kpc. De Vaucouleurs (1977a) demonstrates in addition that $\rho(R)$ can also be reproduced accurately with the same type of $R^{1/4}$ law that applies to the haloes of many elliptical galaxies.

1.3   Distribution of Physical Properties with Distance

Early indications that certain intrinsic properties of the globular clusters (most noticeably, their integrated spectral types) changed with location in the Galaxy (Mayall 1946, Morgan 1956, Kinman 1959b) have grown into a more fully developed picture of cluster heavy-element abundance as a function of distance R. Recent discussions of this topic, with conflicting conclusions, are found in Harris (1976), Canterna & Schommer (1978), Cowley et al. (1978), Searle & Zinn (1978), Searle (1978), and Harris & Canterna (1979b). Although it is clear that the most metal-rich clusters ([Fe/H] > -1) can be found only in the inner halo ($R < 8$ kpc), a point of current dispute is whether the average cluster metallicity or range in [Fe/H] at a given R continues to decrease outward to the limits of the system. The abundance structure of the halo provides a basic test of protogalactic enrichment models (Larson 1976, Hartwick 1976, Searle & Zinn 1978).

The relevant data ([Fe/H] vs R from Table 1) are displayed in Figure 2. Though better observations for more outlying clusters are urgently needed to fill in the region $R > 20$ kpc, it is plain that the maximum [Fe/H] decreases roughly linearly with log R, even out to ~ 100 kpc. But even at the largest distances, a noticeable range of abundances and line strengths still appears to be present (Cowley et al. 1978, Canterna & Schommer 1978). Little can be said of the true minimum [Fe/H] vs R, since few abundance indicators used to date are sensitive to variations in [Fe/H] at such low metallicity; thus previous suggestions that [Fe/H]$_{min}$ remains roughly constant with R (Harris 1976) may prove invalid. Nevertheless, the lack of any known super-metal-poor objects with [Fe/H] $\lesssim -3$, —the fabled "Population III"—remains an intriguing problem.

The fairest summary of the current abundance-gradient situation may be that (a) both the total range in cluster abundances and the maximum abundance found at any R decrease steadily outward to at least 20 kpc, and likely further, and (b) measurable abundance differences between clusters persist at all distances, with [Fe/H]$_{min}$ not being well defined. Table 2 summarizes these points numerically.

The internal structural properties of the clusters—their luminosities M_V (or masses), radii r_t, and central concentrations c—are tied both to the cluster formation process and to subsequent dynamical evolution.

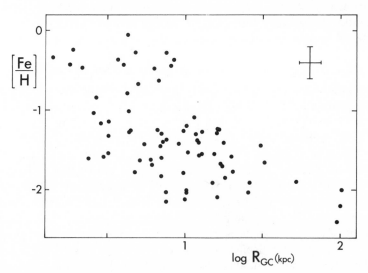

Figure 2   The relation between metallicity index [Fe/H] and galactocentric distance R_{GC} for globular clusters in the Galaxy. The typical internal errors for each point (± 0.2 in [Fe/H], $\pm 0^m3$ in distance modulus) are indicated at upper right.

Table 2 Mean physical properties of Galactic globular clusters[a]

R (kpc)	[Fe/H]	$\bar{M}_V$	$\bar{c}$	$\bar{r}_t$ (pc)	$\bar{e}$
0–3	-0.91 ± 0.10 (24)	-7.20 ± 0.18 (23)	1.51 ± 0.06 (20)	29 ± 3 (20)	0.32 ± 0.07 (9)
3–6	-0.99 ± 0.11 (25)	-7.43 ± 0.20 (23)	1.42 ± 0.06 (23)	27 ± 2 (23)	0.58 ± 0.04 (16)
6–10	-1.27 ± 0.11 (26)	-7.60 ± 0.24 (25)	1.47 ± 0.07 (24)	37 ± 3 (24)	0.65 ± 0.03 (18)
10–20	-1.52 ± 0.06 (23)	-7.48 ± 0.28 (23)	1.51 ± 0.09 (22)	61 ± 7 (23)	0.58 ± 0.06 (20)
>20	-1.90 ± 0.08 (11)	-6.39 ± 0.68 (9)	1.30 ± 0.11 (9)	100 ± 26 (9)	0.55 ± 0.09 (7)
All	-1.24 ± 0.05 (109)	-7.34 ± 0.12 (103)	1.46 ± 0.03 (98)	44 ± 4 (99)	0.57 ± 0.03 (70)

[a] Quoted errors represent the formal internal standard error of the *mean*. Numbers in parentheses are the total number of clusters in the bin.

Thus one might expect *a priori* that even if clusters were formed similarly at all distances, some trend of $M_V(R)$, $r_t(R)$, etc. might develop with time as a result of tidal shocks, dynamical friction, or various other processes (Spitzer 1958, King 1962, Ostriker et al. 1972, Tremaine et al. 1975, Tremaine 1976, Oort 1977, Surdin & Charikov 1977, Keenan 1978), though the actual efficacy of dynamical friction has been questioned (van den Bergh 1978). All clusters are driven by internal dynamical relaxation toward stellar evaporation, higher central concentration, and internal stratification of stars by mass (for comprehensive reviews with references, see Lightman & Shapiro 1978 and Saslaw 1973).

Clusters now near the Galactic center also have shorter core-relaxation times (van den Bergh 1978, King 1978b). The overall result of all these effects, over sufficient time, would be to flatten out the central $\rho(R)$ distribution as well as to change the mean cluster luminosity at different distances.

Interestingly, in reality few trends with R are evident within our own Galaxy aside from the obvious ones that the inner clusters tend to have systematically smaller core and tidal radii, and that the extremely sparse Palomar-type objects are presently found only at large distances. No strong trends with R are found in the central concentration c, or in the mean cluster luminosity $\bar{M}_V$ or detailed luminosity function $\phi(M_V)$. This latter point bears on the practical question of the universality of the globular-cluster luminosity distribution depending on what region of the halo or what type of galaxy is studied (Section 6.2 below).

1.4 *Kinematics and Orbital Characteristics*

A long-standing problem has been to estimate the true sizes and eccentricities of globular-cluster *orbits* (Kurth 1960, Schmidt 1956, von Hoerner 1955, Kinman 1959c, Matsunami 1963, Innanen 1967, Peterson 1974, Racine & Harris 1975, House & Wiegandt 1977). Since cluster-cluster encounters are too rare to produce any internal relaxation of the system over 10^{10} yr (Woolley 1961), the present distribution of cluster orbits should be close to that prevailing at their formation, except possibly in the very central regions and the disk, where dynamical friction may regularize the orbits (Tremaine et al. 1975, Keenan 1978). In addition, close orbital encounters with the Magellanic Clouds may be frequent enough to have altered the orbits of one or two outer clusters drastically (Innanen & Valtonen 1977).

In general, no complete solution for the orbit of a given cluster is possible since the tangential velocity is usually unknown. A statistical approach (Peterson 1974) can be used to calculate the most probable orbital eccentricity e_p for a given cluster, knowing its present distance R, radial velocity V_r, and tidal radius r_t (and hence perigalacticon distance R_p). Peterson (cf his Figure 4), using the Schmidt (1965) mass model and data for 30 clusters, found an average $\bar{e}_p = 0.64$ with dispersion $\sigma_e \simeq 0.2$. Thus, few clusters have extremely large or small eccentricities, with 2/3 in the middle range $0.4 < e_p < 0.8$. Matsunami (1963) and House & Wiegandt (1977) reached similar conclusions from somewhat different approaches. For the 49 clusters in Table 1 with accurate (R, r_t) we find results for $\bar{e}_p$ and σ_e virtually identical to Peterson's, by using a simple point-mass model for the Galaxy following Hartwick & Sargent (1978).

Although the foregoing method is preferred for estimating the orbit of an individual object, the mean results can be checked by a second, velocity-independent approach. For any elliptical orbit of semimajor axis a and eccentricity e, the time-averaged mean distance $\bar{R}$ from the galactic center is $\bar{R} = R_p(1 + \frac{1}{2}e^2)/(1 - e)$ (van de Kamp 1964). We may then blindly assume for every cluster that $\bar{R} \simeq R$ (now) and solve for e, knowing R_p. Over a sufficient sample of clusters the average $\bar{e}$ thus obtained will be a valid estimate for the group (after adding a second-order correction to account for the nonuniform spread of time intervals spent on each part of an elliptical orbit). We can apply this method to seventy clusters. Our results for $\bar{e}$, broken into distance groups, are summarized in Table 2. No significant trends with metallicity are found ($\bar{e} = 0.52 \pm 0.06$ for 17 clusters with $[Fe/H] > -1$, and 0.57 ± 0.03 for 53 more metal-poor clusters). We further find that $\bar{e}$ does not vary with distance R except for the (uncertain) innermost group where it is about half as large. If real, this latter difference may indicate the stronger "smoothing" effect of dynamical friction on orbits near the center, or a slightly later dynamical epoch of formation where the collapsing gas had settled into less eccentric orbits.

Both approaches suggest that a typical globular-cluster orbit has $\bar{e} \sim 0.6$ almost independent of distance. Although this eccentricity is appreciable, it should not be thought of as representing a "nearly rectilinear," plunging orbit as in the viewpoint of Eggen et al. (1962): with $e = 0.6$, the major/ minor axial ratio is still only $a/b \simeq 4/3$. The statement $\bar{e}(R) \sim$ constant also has the interesting consequence that the angular momentum per unit mass h carried by individual clusters increases outward, since $h^2 = G\mathcal{M}_{gal}R_p(1 + e)$. If $\mathcal{M}_{gal}$ increases roughly as $R^{0.5 - 1.0}$ (see below), then $h \propto R^{1/2}$. This qualitative effect is predicted in the protogalaxy models of Larson (1975, 1976), in which gas cloud collisions within the collapsing galaxy create a net outward transport of angular momentum until star formation is largely complete.

The *kinematics* of the cluster systems have also been frequently discussed (Schmidt 1956, Arp 1965, Woltjer 1975, and especially Kinman 1959c), the most complete up-to-date treatment being by Hartwick & Sargent (1978). Their results confirm earlier conclusions that the system possesses differential rotation: the inner, more metal-rich subsystem has a significantly higher overall rotation speed. But virtually all parts of the system appear to have at least some rotation relative to an external rest frame, if the orbital speed of the Sun in the Galaxy is 200–250 km s^{-1}. A possible reason for why we do not see any obvious rotational flattening of the inner subsystem (Harris 1976) has been discussed by Woolley (1961).

Hartwick & Sargent (1978) have also used their new radial-velocity measures for the outer Palomar and dwarf spheroidal systems to investigate the velocity dispersion of the cluster system and hence the mass of the Galaxy at different R. Depending on assumptions about the velocity ellipsoid (orbit eccentricity distribution), they find $\mathcal{M}_{gal}$ increases roughly as $R^{0.5}$ or $R^{1.0}$, up to $\sim(3–10) \times 10^{11}\ \mathcal{M}_{\odot}$ outside 60 kpc. More simply, if $\mathcal{M}_{gal}$ within $R \simeq 30$ kpc is any less than about $4 \times 10^{11}\ \mathcal{M}_{\odot}$ we would face the improbable situation that at least four clusters (Pal 3, 4, NGC 1851, 5694) would have unbound orbits from their radial velocities alone. The cluster velocity data therefore provide one of the most important direct indications that the outer halo contains substantial mass. The result (several $10^{11}\ \mathcal{M}_{\odot}$) is an order of magnitude higher than that obtained by simply scaling up the total mass in the cluster system by the solar-neighborhood ratio of Population II field-star mass density to cluster density (Harris 1976). In turn, it casts some doubts on the frequent assumption that the density distribution of halo mass mimics the cluster distribution $\rho(R)$.

2 THE MAGELLANIC CLOUDS

The cluster populations in the Large and Small Magellanic Clouds (LMC, SMC) present unique challenges: here, we must face squarely the problem of defining a globular cluster. The rather clear population separation between open and globular clusters within the Galaxy fails completely for the Clouds, which contain many young and intermediate-age clusters structurally similar to conventional globular clusters (Bok 1966, Freeman 1974, Gascoigne 1971, Gascoigne & Kron 1952). In general, cluster

Table 3 Globular clusters in the Magellanic Clouds

Name	V_t^a	$(B-V)$	$(U-B)$	CMD[b]	Remarks
Large Magellanic Cloud					
NGC 1466	11.4	0.61:	0.13	$\sqrt{}$	RRLyr.; bright star superimposed
NGC 1751	11.4	0.83	0.38	—	
NGC 1754	11.0	0.72	0.17	—	
NGC 1786	10.1	0.76	0.14	—	
NGC 1835	9.8	0.72	0.12	—	
NGC 1841	12.6	0.72	0.04	$\sqrt{}$	
NGC 1916	9.4:	0.78	0.19	—	
NGC 1978	9.9	0.79	0.24	?	RRLyr.; carbon stars
NGC 2019	10.6	0.75	0.18	—	
NGC 2121	11.2	0.86	0.25	?	
NGC 2155	11.9	0.80	0.23	?	
NGC 2173	11.6	0.83	0.28	?	
NGC 2210	10.2	0.71	0.11	$\sqrt{}$	
NGC 2257	13.5	0.68	—	$\sqrt{}$	RRLyr.
SL-868	11.2	0.60	0.00	—	
Anon 14	12.7	0.72	0.20	?	
Hodge 11	12.1	0.61	−0.06	$\sqrt{}$	
Small Magellanic Cloud					
NGC 121	10.6	0.78	0.14	$\sqrt{}$	RRLyr.; carbon stars
NGC 152	12.3	0.70	0.20	—	
NGC 339	11.9	0.69	0.09	$\sqrt{}$	
NGC 361	11.8	0.76	0.14	$\sqrt{}$	
NGC 416	11.0	0.77	0.19	—	
L 1	12.0	0.72	0.21	$\sqrt{}$	
L8/K3	11.4	0.67	0.12	$\sqrt{}$	
L58/K37	14.2	0.63	—	$\sqrt{}$	
L 68	12.9	0.80	0.17	—	
L113	12.9	0.73	0.09	—	
Possible Small Magellanic Cloud candidates					
NGC 419	10.0	0.67	0.25	?	carbon stars
HC 276#81	11.2	0.77	—	—	Kron & Mayall 1960
Anon	11.5	0.69	—	—	Gascoigne & Lynga 1963

[a] Obtained by extrapolating concentric-aperture $V(r)$ photometry via the King (1966a) curves where possible, otherwise, by subtracting the mean difference $\langle V(r) - V_t \rangle$ for the given aperture sizes. The most frequently used aperture is 60" diameter, for which $\langle V(r) - V_t \rangle = 0.7 \pm 0.1$.
[b] Color-magnitude diagram characteristics. ($\sqrt{}$): globular-cluster-like; (?): available CMD inconclusive; (—) no data available.

formation appears to have proceeded more steadily in the Clouds (particularly the LMC), and no large age gap separates the "true" globulars from the oldest open clusters, as it does in the Galaxy (Demarque & McClure 1977, van den Bergh 1975). Color-magnitude diagrams (CMD) might eventually provide definitive classification criteria but are extremely difficult to carry out in the crowded star fields of the Clouds. To date, calibrated CMDs are available for only a few LMC and SMC clusters clearly showing characteristic sequences of globular clusters (Arp 1958, Freeman & Gascoigne 1977, Gascoigne 1966, Hodge 1960b, 1971a, Tifft 1962, Walker 1970, 1972). Hesser et al. (1976) have also obtained uncalibrated but useful diagrams for numerous LMC clusters.

We have adopted integrated UBV colors as our primary classification criterion here, partly because we prefer, conservatively, to define globular clusters as restricted to the oldest group of clusters formed in any galaxy, and partly because integrated broad-band photometry is the only available criterion in still more distant galaxies (see Hartwick & Cowley 1978 for similar views). Photoelectric UBV magnitudes and colors are available for 141 LMC and 52 SMC clusters (Kron & Mayall 1960, Gascoigne 1966, van den Bergh & Hagen 1968, Bigay & Bernard 1974, Bernard 1975, 1976, Alcaino 1978b). Of these, we retained as globular-cluster candidates those with $(B–V)_0 > 0.5$ (38 in LMC, 13 in SMC). We next eliminated those whose $(U–B)$ index is >0.10 mag redder than normal for a Galactic globular of the same $(B–V)$. This leaves a final adopted list of 17 LMC and 10 SMC clusters (Table 3) which we define as most likely to be "globular" in the Galactic sense. Three additional SMC objects with insufficient or conflicting data are also listed. In Table 3, non-NGC clusters are from the lists of Lindsay (1956, 1958), Kron (1956), or Hodge (1960a). Additional references or classification criteria are listed in the last column.

All the red clusters are plotted in the two-color plane of Figure 3. Our rejected candidates (crosses in Figure 3c) are seen to occupy a region below the intrinsic globular line which is normally populated by old open clusters such as M67 or NGC 188 (Gray 1965, Goodenough & Hartwick 1970), so we identify them as mostly somewhat younger objects than the "true" globular clusters. Virtually all of these intermediate-age clusters are in the LMC sample; the SMC appears to contain relatively fewer of them. Despite the crudeness of our color criterion, additional tests [CMD morphology, occurrence of RR Lyrae variables, spectrophotometric indices (Danziger 1973)] yield no conflicting classifications except for NGC 1852 (colors like those of an old globular, but CMD unlike one). As yet, virtually no radial velocity information is available for any of the clusters and so kinematical criteria cannot be applied to refine the separation of the samples. We also note that our separation of the red LMC clusters into "globular" and "intermediate-age" objects has, interestingly, generated two groups whose luminosity distributions $\phi(V_t)$ are quite different. For the globulars, ϕ peaks symmetrically near $V_t \sim$ 11.3 (cf Figure 6c), while for the other objects ϕ increases monotonically faintward. This is additional evidence—since V_t was not a classification parameter—that the red clusters contain more than one distinct population.

The populous blue $[(B-V) < 0.5]$ clusters in the LMC are often

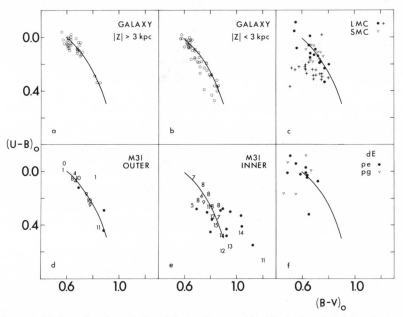

Figure 3   Color-color diagrams for globular cluster systems in the Local Group galaxies; all plotted points are photoelectric measurements except in panel (f). The intrinsic two-color line for the Galaxy (e.g. Racine 1973) is drawn in each panel. (a) Galactic globular clusters further than 3 kpc from the galactic plane. (b) Galactic globular clusters within 3 kpc of the plane; this sample contains most of the metal-rich clusters in the Galaxy. (c) Globular clusters in the Magellanic Clouds. Clusters in the LMC believed to be "nonglobular" (intermediate-age clusters or others; see text Section 2.1) are plotted as crosses. Clusters from Table 3 adopted as globular are plotted as open or closed symbols. (d) M31 globular clusters that lie outside the projected disk and are unreddened within M31 itself. Compare with panel (a). Those with measured line-strength index L (van den Bergh 1969) are plotted by their value of L. (e) M31 globular clusters seen projected on the M31 disk; compare with panel (b) for the analogous region in the Galaxy. (f) Globular clusters in the dwarf ellipticals NGC 147, 185, 205, and Fornax. Those with only photographic $U-B$ values are plotted as open symbols.

referred to as "young globular clusters" (Freeman 1974, Freeman & Munsuk 1973) in a stellar dynamical sense. Their ages are $\lesssim 10^8$ years, and some rival or exceed the brightest old globular clusters in luminosity. But their luminosities will fade with time as the upper main sequence evolves away. After 10^{10} yr, they will have dimmed by ~ 5 magnitudes (cf Larson & Tinsley 1978) and so will be much fainter and less conspicuous than the Cloud clusters which are *now* $\sim 10^{10}$ yr old. Thus by far the most massive Cloud clusters are just those that we have selected as globular, probably coeval with the Clouds themselves, and there appears no compelling evidence for believing that comparably massive clusters have formed since then.

Even for our restricted sample of old globulars in the Clouds, certain peculiarities in detail remain when compared with the Galactic globulars, which indicate residual differences in chemical composition or evolutionary characteristics. The same relation between CMD morphology and metallicity is not followed (cf the references cited); and three clusters (NGC 419, 1783, 1846)—sometimes considered globular (van den Bergh 1968) but rejected here on the basis of their integrated colors—possess extremely red stars at the tip of the giant branch (Feast & Lloyd-Evans 1973).

In Table 3, the 17 LMC clusters have a mean luminosity $\langle V_t \rangle = 11.21 \pm 0.27$ and the 13 SMC clusters have $\langle V_t \rangle = 11.82 \pm 0.30$. The 0.6 ± 0.4 mag difference between the two averages is consistent with the larger distance modulus of the SMC and with the assumption that the intrinsic luminosity distribution $\phi(M_V)$ is the same for both (Section 6.2 below). The LMC clusters display a significant range in intrinsic colors (Figure 3c) and hence presumably in metallicity. This is supported by the multiband photometry and line-strength indices of Danziger (1973), which also show substantial differences from cluster to cluster. By contrast, the SMC clusters have a much narrower color range and more uniform (lower) metallicity. However, no globular clusters in either Cloud are known to lie beyond the metallicity range seen in the Galactic globular clusters, and most fall in the range $-2 < [\text{Fe/H}] < -1$.

In summary, it appears entirely possible to isolate a sample of clusters in both the LMC and SMC that occupies the familiar ranges of integrated color, luminosity, and approximate chemical composition and that we have defined as "globular." This rather crude classification of their properties will, however, doubtless be replaced by a much firmer analysis resulting from their current intensive study with the new large telescopes in the Southern Hemisphere.

Finally, we may roughly estimate the total expected number of "true" globular clusters in the Clouds under our operating criteria. In the LMC, available photometric surveys of compact clusters reach $V \simeq 13.5$. Since most clusters of all kinds brighter than $V \sim 12.5$ have been observed, the present sample of 17 globulars should be 75% complete if their luminosity distribution resembles the one in the Galaxy. We adopt a total population $N_t = 23 \pm 5$. In the SMC, the larger distance is offset by its generally smoother background and the observers' efforts to push photometry somewhat deeper. A complete sample might therefore reach $N_t = 15 \pm 5$. These estimates are little more than educated guesses but will suffice for later discussion (Section 6.3).

3   THE ANDROMEDA GALAXY

3.1   Fundamental Data

Edwin Hubble (1932) first suggested that many of the small objects seen projected on the disk of M31 were its globular clusters. Walter Baade (1944a,b) later confirmed this identification through their incipient resolution on deep photographs. For entirely different reasons than in the Magellanic Clouds, the identification of a pure sample of globular clusters in M31 again presents an extremely difficult observational problem. Photographic images of the M31 clusters are less than 10 arc sec wide, and because their brightest stars are resolvable only in the best seeing conditions it is difficult to distinguish globular clusters from distant background spheroidal galaxies. Clusters projected against the irregular, patchy disk of M31 itself are additionally hard to distinguish, and can be confused with intermediate-age or compact young star clusters in the disk unless photometry or spectroscopy of their integrated light is available. Sargent et al. (1977) discuss in more detail the problems of image classification.

Hubble's original list of 140 objects was more than doubled with the later discoveries of many additional clusters by several authors (Seyfert & Nassau 1945, Mayall & Eggen 1953, Kron & Mayall 1960, Vetešnik 1962a,b, Baade & Arp 1964, Sharov 1973, 1976). The major recent survey of Sargent et al. (1977) yielded discoveries of many new clusters well out in the M31 halo. Sargent et al. also completely revised and compiled all previous lists of objects, eliminating known nonglobular clusters in the process (mainly by the selection criterion of diffuse and symmetric image structure, since no other useful data are available for most of these objects). Their final list contains 355 globular-cluster candidates on a uniform selection scale. Of these, eight are assigned to NGC 205 and we may reject 22 more whose UBV colors are too blue. Searle (1978) states that early results from spectrophotometry of a large subset of the Sargent et al. list indicate roughly 20% to be either background elliptical galaxies (for the halo regions well away from the M31 disk) or inter-mediate-age clusters within the disk itself. In summary, we may therefore estimate that the Sargent et al. list, which includes all surveys to date, probably contains 300 ± 30 "true" globular clusters in the sense defined previously.

Photoelectric UBV measurements for 77 clusters (not all globular) are available from the combined work of Kron & Mayall (1960), Hiltner (1960), Kinman (1963), and van den Bergh (1969), while photographic photometry for a total of 246 is given by Veteŝnik (1962a,b) and Racine (1965). We find the photometry of Sharov et al. (1976, 1977) and the photographic photometry of Kinman (1963) to contain sizeable systematic and random errors especially in the colors, and these are not used here. Comparison of the various photoelectric surveys shows satisfactory systematic agreement (to 0^m03 or better in all indices) and random errors in V, $(B–V)$ of $\sigma \simeq 0.07$, while the internal random errors of the averaged photographic data are roughly twice as large. After eliminating other types of objects from the photometric list, we are left with 188 out of the Sargent et al. survey that have UBV photometric indices.

The only other data yet available for a reasonably large number of M31 globular clusters are radial-velocity measurements for 43 objects and spectroscopic line-strength estimates for 37, both by van den Bergh (1969). Our current understanding of the kinematics and abundance structure of the M31 cluster system rests heavily on this single study. A major new velocity and abundance survey has been started by Searle (1978), which will eventually add considerably to both these areas.

3.2 Systematic Properties

Van den Bergh (1969) first indicated that the integrated colors of the M31 globular clusters—and by assumption their chemical composition—were distributed more heavily toward the red, metal-rich side and did not obey the same correlation in detail with galactocentric distance as in the Galaxy. Figure 3d,e displays the $(U–B, B–V)$ diagram for the M31 globular clusters with photoelectric color measurements, divided into two groups called "inner" (those projected on or near the M31 disk) and "outer" (those well beyond the disk and presumably unreddened within M31 itself). For comparison we show the same distribution for the Galactic globular clusters (data from Table 1), divided analogously into two "projected" spatial groups. Important differences between the two galaxies are immediately evident: the outer M31 clusters are more or less uniformly distributed over the full range of intrinsic colors; whereas few clusters redder than $(B–V)_0 \sim 0.75$ (or more metal-rich than $[Fe/H] \sim -1$) are known in the outer Galaxy sample. The inner M31 clusters display more scatter due to both larger photometric errors and effects of cluster reddening within M31, but their distribution suggests that this inner region may lack the blue, low-metallicity component $[(B-V) < 0.7]$ that is known to extend right into the center of our own Galaxy (cf Figure 2). The mean intrinsic color of the outer M31 sample including only photoelectric measures and corrected for the M31 foreground reddening is $\langle B-V \rangle_0 = 0.74 \pm 0.02$, $\langle U-B \rangle_0 = 0.14 \pm 0.04$, which is slightly redder than the mean colors of the Galactic globular clusters at all radii ($\langle B-V \rangle_0 = 0.70 \pm 0.01$, $\langle U-B \rangle_0 = 0.13 \pm 0.01$).

The conclusion that the M31 clusters are more strongly weighted toward metal-richer objects is additionally supported by the distribution of the spectroscopic line-strength index L (van den Bergh 1969, also plotted in Figure 3d,e) as well as by the Searle (1978) spectrophotometry, scanner abundance studies, and Washington-system photometry for a small number of clusters (Spinrad & Schweizer 1972, Harris & Canterna 1977). Searle (1978) concludes that the mean abundance for the outer M31 sample is near $[Fe/H] \simeq -1.1$ as opposed to $\simeq -1.5$ for the outer Galaxy halo. In addition, plots of L or intrinsic color with projected galactocentric distance reveal no strong trend with distance, unlike what is observed within our Galaxy. However, the *very* strongest-line M31 clusters ($L > 11$, possibly including some with $[Fe/H] > 0$) are still found predominantly in the inner region.

The M31 cluster system is the only one outside the Galaxy for which statements can currently be made about *kinematics*. The cluster velocities measured by van den Bergh (1969) have been analyzed by Hartwick & Sargent (1974) to determine the velocity dispersion within the system and the total mass of M31 out to the radial limit of the survey. By dividing the sample into metallicity groups, they showed that the more metal-poor samples have the highest velocity dispersion, and that the total mass of M31 out to $r \sim 20 \, \text{kpc}$ is $\sim 3 \times 10^{11} \, \mathcal{M}_\odot$. Thus to first approximation the kinematics appear to resemble the Galactic cluster system.

Determining the intrinsic luminosity distribution $\phi(M_V)$ for the M31 clusters is impeded not only by the incompleteness and photometric errors mentioned, but by the necessity of making individual absorption corrections for reddening within M31 itself. To minimize the latter problem we have selected a sample consisting of 46 "outer" clusters (unreddened within M31) plus 68 "inner" clusters with low reddening $[(B-V) < 0.9]$ for which photometry is available. Statistically corrected magnitudes $V_0 = V - 3[(B-V) - 0.82]$ were then used for the inner clusters since the photometric errors did not permit accurate individual dereddening. The resulting $\phi(M_V)$ histogram for 114 clusters is shown in Figure 6 and discussed further in Section 6.2. The faint-magnitude incompleteness of the photometry causes a steep drop for $M_V > -6$, but the bright half is consistent with a peaked distribution with mode at $M_V \sim -7.5$.

Estimates of the true total number of M31 globular clusters must be made by taking the observed list of candidates (Sargent et al.) and correcting it for (a) contamination by background galaxies, old open clusters, etc., (b) incompleteness at faint magnitudes, and (c) incompleteness for the areas of the sky around M31 not yet surveyed. From the radial area distribution of the clusters (Section 6.3 below), we estimate that the area not covered by the Sargent et al. survey contains 35 ± 6 more clusters above the survey limit. The limiting magnitude itself is not accurately known but must be near $V \simeq 18$ ($M_V \simeq -6.6$). If $\phi(M_V)$ is the same in M31 and in the Galaxy (Figure 6 and Section 6.2) with the peak frequency at $\langle M_V \rangle \simeq -7.3$, then the observed M31 sample is only $\sim 70\%$ of the total over all magnitudes. Combining these corrections with the 20% fraction of contamination by other types of objects (Searle 1978) we finally estimate $N_t = 450 \pm 50$ as the total population of M31 globular clusters. This number exceeds the Galactic population ($N_t = 180 \pm 20$) by a factor of 2.5 ± 0.5. Though this ratio may be surprisingly large considering the roughly similar sizes of the two major Local Group spirals, it may simply indicate that the spheroidal component (central bulge and halo) is relatively more dominant in M31.

The projected density profile for the M31 globular-cluster system (cf Figure 8) follows a $r^{1/4}$ law in its outer region (de Vaucouleurs & Buta 1978; Section 6.3 here) but becomes increasingly flatter for $r < 3$ kpc. If this were due to increasing incompleteness of the Sargent et al. (1977) survey, then some 140 objects (70% of the expected total above the survey limit), including 20 brighter than $V = 16$, would have been missed in the extensively searched inner 30 arc min diameter area of M31. This much incompleteness appears highly unlikely. We conclude that, at least in part, the flattening of the density profile at $r < 3$ kpc is real.

4   DWARF GALAXIES

Aside from the two dominant large spirals and the Magellanic Clouds, other Local Group members that contain globular clusters of their own include the Sc spiral M33 and the dwarf ellipticals NGC 147, 185, 205, and Fornax. The available and rather meager information (in most cases including only UBV magnitudes and colors) for these systems is summarized in Table 4 and Figure 3f. The cluster identifications in Table 4 follow the authors cited below, and in the last column we give the projected galactocentric distances for each cluster.

Table 4   Globular clusters in dwarf galaxies

Galaxy	Cluster ID	V	$B–V$	$U–B$	r (kpc)
M 33	e	16.15	0.77	0.16	
	m	17.48	0.80	0.22	
	n	16.76	1.16	0.52	
	q	17.16	0.74	0.13	
	s	18.26	1.26	0.60	
	Mc	15.92	0.58	—	
Fornax	1	15.2:	0.51	0.03	2.1
	2	13.6	0.66	0.06	1.2
	3	12.5	0.65	0.03	0.8
	4	13.5	0.74	0.09	0.2
	5	13.4	0.61	0.04	2.1
	6	14.0:	—	—	0.3
NGC 147	I (H1)	17.66	0.80	0.45	0.0
	II (H2)	18.38	—	—	0.4
	III (H3)	16.98	0.64	0.00	0.7
	IV (H4)	21.10	—	—	0.3
NGC 185	I (H1)	18.38	0.79	0.15	0.6
	II (3)	19.7	0.94:	—	0.5
	III (H4, Bc)	16.78	0.69	0.06	0.2
	IV	—	—	—	0.7
	V (H5, Ba)	16.72	0.71	0.05	0.8
	VI	—	—	—	0.6
NGC 205	41 (H VIII)	16.68	0.58	0.00:	1.8
	51 (H V)	16.73	0.57	—	0.2
	54 (H IV)	18.52	0.51	—	0.2
	55	—	—	—	0.3
	56 (H VII)	17.24	0.76	0.06:	0.3
	57 (H VI)	17.86	0.72	−0.05:	0.2
	61 (H I)	16.80	0.55	0.23:	1.0
	63 (H II)	16.69	0.64	0.24:	0.6

In M33 the only published data remain the photometric surveys of Hiltner (1960) and Kron & Mayall (1960) for 23 selected clusters down to $V \sim 18.5$. Of these, most are open clusters of various ages, but three (e, m, and q) have color indices similar to normal metal-poor globular clusters. A possible fourth has no $(U–B)$ index, and two others have colors resembling highly reddened globular clusters. Scaling up Hiltner's discovery frequency of $\sim 5/23$ by the estimate that M33 contains ~ 100 clusters of all types (Kron & Mayall 1960) then yields the very uncertain prediction that M33 contains $N_t = 20 \pm 10$ true globular clusters. Much more observation will be needed before the clusters in this system are at all well understood.

The Fornax system is the largest dwarf spheroidal galaxy in the Local Group, but the smallest galaxy known to have globular clusters of its own (Hodge 1961, 1971b). For its six clusters UBV magnitudes and colors are available from the work of Hodge (1961, 1965, 1969), Kron & Mayall (1960), Demers (1969), and de Vaucouleurs & Ables (1970), as summarized in Table 4. It is interesting that even in such a small system, the clusters exhibit a noticeable range in metallicity, demonstrated by spectra and photometry of their integrated light (van den Bergh 1969, Danziger 1973, Harris & Canterna 1977). Cluster No. 4 has a metallicity approximately as high as M3, while the others appear to have much lower abundances, in the M92–M15 range. The color-magnitude diagrams for several of the clusters are currently under study by Demers and collaborators.

Hubble (1932) and Baade (1944a) made the first discoveries of clusters in the small elliptical galaxies surrounding M31. Currently, eight clusters are believed to belong to NGC 205 (Sargent et al. 1977). An additional cluster (H III), seen in projection against NGC 205, is assigned to the M31 system because of its high metallicity and discrepant radial velocity (van den Bergh 1969). Photoelectric $(B–V)$ colors of the clusters from the observations of Kron & Mayall (1960), van den Bergh (1969), and Hodge (1973) are summarized in Table 4, while the $(U–B)$ values listed are from the photographic data of Racine (1965) and Vetešnik (1962a).

From the surveys of Hodge (1976) and Ford et al. (1977), four clusters have been found in NGC 147. For these, useful photoelectric UBV colors are available for only two (Table 4), but magnitudes for all four.

In NGC 185 (which forms an optical pair with NGC 147, though Ford et al. 1977 argue that they are not physically bound), recent searches for globular clusters are described by Hodge (1974) and by Ford et al. (1977). Comparison of their two discovery lists suggests that NGC 185 contains perhaps six clusters (rejecting Hodge No. 2, and Ford et al. Nos. VII and VIII as probable galaxies). The available photoelectric photometry (Hodge 1974, Kron & Mayall 1960) for only four of them is again summarized in Table 4.

Photometry for most of the clusters in the three M31 companions is made uncertain by the varying background of their underlying galaxies and by the faintness of some of the clusters themselves. Nevertheless, the color indices (Figure 3f) suggest that all of their clusters fall near the bluer end of the intrinsic two-color line, and that in this sense they can be regarded as normal, metal-poor clusters. The present uncertainties in the data make it impossible to draw any further conclusions about the total range of abundances that each of these three galaxies has been able to generate in its clusters.

None of the several remaining small galaxies in the Local Group is definitely known to contain true globular clusters. The bright, compact dwarf elliptical M32 is superimposed on the disk of M31, making it impossible to identify any of the handful of M31 clusters in the area as belonging to M32. In any case, much of the M32 outer envelope has apparently been tidally stripped (Faber 1973, van den Bergh 1975) and so much of its hypothesized original cluster population may have been torn away.

The two faint irregulars NGC 6822 and IC 1613, according to Hodge (1977, 1978), are believed to lack globular clusters since the few cluster-like objects discovered in them resemble open clusters or the intermediate-age Magellanic Cloud clusters. The slightly more distant WLM irregular system (Ables & Ables 1977) does contain a single candidate, a surprisingly bright cluster at $M_B \simeq -9.2$, $(B-V)_0 \simeq 0.61$. However, whether this is a true globular cluster must await further data.

5   BEYOND THE LOCAL GROUP

5.1   Globular Clusters in Virgo Galaxies

Baum (1955) and Sandage (1961) discussed the numerous faint, stellar objects seen around the giant Virgo elliptical M87 as being its globular

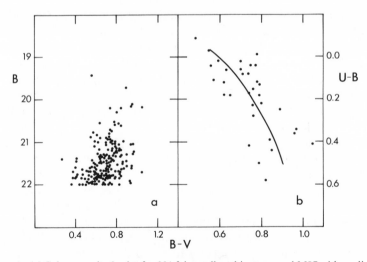

Figure 4   (a) Color-magnitude plot for 201 faint stellar objects around M87 with negligible proper motion ($\mu < 0.''005$ yr^{-1}). Almost all these objects are M87 globular clusters, since the proper-motion criterion eliminates most foreground stars. (b) Color-color diagram for the M87 clusters; the intrinsic color line is the same as in Figure 3. The large observed range in intrinsic color is comparable with the M31 clusters (cf Figure 3d,e).

clusters, and Racine (1968a,b) first showed that their colors were consistent with this interpretation. Subsequent UBV photometric surveys of the M87 clusters were carried out by Hanes (1971) and by Ables et al. (1974). Figure 4 shows representative color-magnitude and color-color diagrams for 201 of these objects, where the UBV data are from Hanes (1971) and membership in M87 is confirmed by their negligible proper motion, $\mu < 0\rlap{.}''005$ yr^{-1} (Prociuk 1976). Though the color calibrations are still uncertain (cf Ables et al. 1974 and Hanes 1977b) the UBV colors fall in the same range as do the Local Group clusters (Figure 3). Hanes (1977a) also discussed the effects of observational errors on arguments for their photometric similarity. Spectrophotometry for three of the brightest clusters ($B \sim 21$) by Racine et al. (1978) indicates that their mean metallicity is higher than in the Galactic halo clusters and more nearly resembles the M31 halo instead.

For years M87 remained the only Virgo member about which anything concerning globular clusters was known. However, the recent single-color photometric survey by Hanes (1977b,c) for several more large Virgo galaxies, both ellipticals and spirals, demonstrated that globular clusters could indeed be found frequently in large numbers. Hanes' data concerning the luminosity distributions $\phi(m)$ and the total cluster populations in each Virgo galaxy, compared with the Local Group systems, will be discussed in more detail in Section 6. Most recently, programs to obtain *multicolor* photometry for large samples of clusters (and hence crude information about metallicity distributions in the halo of each galaxy) have been initiated by Harris and collaborators for three of the largest Virgo systems, M87, M49 (NGC 4472), and M104 (NGC 4594). Preliminary results for M87 alone indicate that the mean $UBVR$ colors of the clusters decrease systematically from the center out to $r \sim 4$ arc min (25 kpc), and hence may represent an abundance gradient.

5.2   Outside the Virgo System

The Virgo galaxies quite obviously represent only a fraction of the cluster systems available for study beyond the Local Group. Several large E or S0 galaxies at roughly half the Virgo redshift including NGC 3115, 3377, 3379, 3384, and 4278 are noted to contain clusters (Sandage 1961, de Vaucouleurs & de Vaucouleurs 1965). These would be attractive to study particularly since they are ~ 1 mag closer than Virgo and provide correspondingly better chances to reach the turnover of the luminosity function. They may also yield further information on the relative effects of galaxy environment since none of them are members of large galaxy clusters. A preliminary single-color study of the clusters in NGC 3115 (Strom et al. 1977) indicates that their luminosity distribu-

Table 5 Globular-cluster systems in 27 galaxies

	Type	$(m-M)_V$	E_{B-V}	M_V	N_{obs}	N_t
Local Group						
M31	S	24.6	0.11	−21.1	300:	450
Galaxy	S	—	—	−20:	131	180
LMC	I	18.7	0.06	−18.5	17	23
SMC	I	19.1	0.04	−16.8	10	15
M33	S	24.5	0.03	−18.9	6	20
Fornax	dSph	21.0	0.02	−13.6	6	6
NGC 147	dE	24.6	0.15	−14.9	4	4
NGC 185	dE	24.6	0.15	−15.2	6:	7:
NGC 205	dE	24.6	0.11	−16.4	8	8
NGC 6822	I	24.2	0.30	−15.7	0	0
IC 1613	I	24.5	0.03	−14.8	0	0
WLM	I	26.2	0.06	−16.0	1:	1:

	Type	M_V	N_{obs}	N_t
Virgo System[a]				
NGC 4216	S	−21.0	21	520
NGC 4340	E	−19.9	26	650
NGC 4374 (M84)	E	−21.6	98	2500
NGC 4406 (M86)	E	−21.7	108	2600
NGC 4472 (M49)	E	−22.5	1700	4200
NGC 4486 (M87)	E	−22.3	6000	15000
NGC 4526	E	−21.3	87	2200
NGC 4564	E	−20.0	35	900
NGC 4569 (M90)	S	−21.4	32	800
NGC 4594 (M104)	S	−22.6	290	2800
NGC 4596	E	−20.4	82	2000
NGC 4621 (M59)	E	−21.1	63	1600
NGC 4636	E	−21.3	143	3600
NGC 4649 (M60)	E	−22.1	170	4200
NGC 4697	E	−21.6	72	1800

[a] $(m-M)_V = 30.9$, $E_{B-V} = 0.02$ assumed.

tion is at least roughly similar to that in M31 as it would be seen at 10 Mpc distance, which is consistent with the Hubble velocity of NGC 3115. At distances well beyond Virgo, one is restricted to studying only the bright tip of the $\phi(M_V)$ distribution, and hence clusters can be detected only in the dominant giant ellipticals of great clusters (cf Section 6.1 below). Dawe & Dickens (1976) have apparently detected cluster systems in three of the largest ellipticals of the Fornax I cluster (NGC 1374, 1379, 1399), at $V_r \sim 1500$ km s^{-1} or 40% further than Virgo. Interestingly, they do not find noticeable numbers of clusters (to $B \simeq 22$) around several other comparably bright Fornax ellipticals. The present distance record is held for the giant elliptical NGC 3311 (in Hydra I) by Smith & Weedman (1976). Here the brightest clusters are seen to appear at $B \simeq 23.5$ and increase rapidly in number to the plate limit ($B \simeq 24$). Smith & Weedman conclude that the number of clusters (i.e. excess stellar images surrounding the galaxy) is again as expected if, for example M87 were displaced outward by 2.4 mag, corresponding to the observed ratio of the Hydra/Virgo recession velocities.

It remains a problem that virtually all the reliably established very distant globular-cluster systems are in large ellipticals, where they are of course most easy to detect. But this is the one galaxy type not represented in the Local Group, and detailed information about clusters in even one distant spiral galaxy would provide a much more rigorous basis for comparison with the Local Group.

6 GENERAL FEATURES OF GLOBULAR-CLUSTER SYSTEMS

6.1 *Total Populations and Dimensions of Parent Galaxies*

Table 5 summarizes available data on the number of actually observed globular clusters in 27 galaxies along with our estimated total populations N_t for each. Similar but less extensive data were discussed by Jaschek (1957) and Wakamatsu (1977b). Individual members of the Local Group were analyzed above. For Virgo galaxies Hanes' (1977a,b) counts were used, updated by the deeper ($B_{lim} \sim 24$) counting surveys of Harris & Smith (1976) for M87, of Harris & Petrie (1978) for M49, and of Harris et al. (1979) for M104. Virgo statistics are strongly dependent on the adopted field corrections; a new analysis of this question results in our quoted numbers (Table 5) being somewhat larger than the published net counts for M87. The total populations for M49, M87, and M104 were estimated from the limiting magnitude of the deep surveys, assuming V (peak) = 23.6 and $\sigma = 1^m2$ for the luminosity distribution of Virgo globular clusters. Hanes's data for other galaxies were correspondingly scaled up by the factor (25 ×) by which N_t for these three galaxies exceeded Hanes's field corrected counts. No other corrections for the limited angular field of the surveys have been applied, but these are expected to be small except for M87 (see Section 6.3).

The correlation between N_t and the luminosities of the parent galaxies is illustrated by Figure 5. Hanes (1977a) has already presented a discussion of this relation for Virgo galaxies. Over more than a factor of 1000, N_t for elliptical (E) galaxies is approximately proportional to the galaxy's luminosity. Spirals and irregulars are too bright for their estimated N_t. M31 would conform to the E relation had we adopted for M_V the luminosity of its spheroidal component (de Vaucouleurs 1958). Wakamatsu (1977b) has also shown that statistical reductions of other spiral luminosities to their spheroidal component produces similar results. This is, of course, because the luminous young stellar populations in the disk of spiral galaxies or in irregular systems represent a small fraction of the total mass of these galaxies and are genetically unrelated to globular clusters.

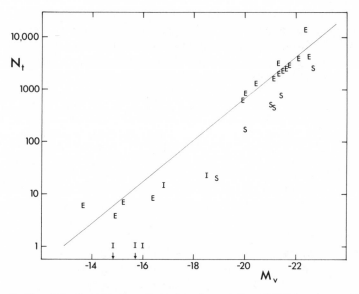

Figure 5   Correlation between the estimated total population of globular clusters N_t and the integrated absolute magnitude M_V of the parent galaxy. The different symbols are E for elliptical galaxies, S for spirals, and I for irregulars. The straight line represents direct proportionality between N_t and galaxy luminosity, as discussed in Section 6.1, and is drawn through the elliptical-galaxy sample. The spirals and irregulars all fall to the right of the line, presumably because their spheroidal (Population II) component represents only a fraction of their total luminosity (or mass).

Although Figure 5 might suggest that N_t is controlled by the total mass of the parent galaxy (cf Hanes 1977a), the data are not accurate enough to state that N_t is strictly proportional to L. The line $\log N_t = -0.4$ $(M_V - 12.9)$ is shown in the figure; this relation implies that globular clusters contribute approximately one half of one per cent of the total light of E galaxies. According to Faber & Jackson (1976) the mass-to-light ratio $\mathcal{M}/L$ for spheroidal galaxies tends to vary as $L^{1/2}$. If N_t were strictly proportional to $\mathcal{M}_{gal}$ this would require $\log N_t \propto 0.6\ M_V$. This is not consistent with the correlation shown in Figure 5. One could speculate that dwarf elliptical galaxies are more efficient progenitors of globular clusters—having similar $\mathcal{M}/L \sim 2$ (Hodge 1971b, Illingworth 1976). Alternatively they may have lost a large fraction of the initial mass they had when their globular clusters were formed.

M87 has many more globular clusters for its luminosity than normal E galaxies. Evidence that our estimate of N_t (M87) is still too low will be presented later. The extremely large population of globular clusters in M87 could be due to its privileged central position and low peculiar velocity in the Virgo cluster (van den Bergh 1977b). Alternatively, the M87 mass-to-light ratio may be abnormally large (Harris & Petrie 1978).

6.2 Luminosity Functions

The brightest globular clusters are more luminous than any other stellar "standard candles" except supernovae and have long been potentially attractive extragalactic distance indicators. Their fundamental advantages for this purpose are well described by Hanes (1977b), and current agreements or disagreements with other recent methods for calibrating the Hubble constant are well known (Tammann 1976, Fisher & Tully 1977, Peebles 1978). Only a brief discussion of the present state of the globular-cluster technique can be given here.

Early attempts to calibrate the Virgo distance by assuming a fixed luminosity for the brightest clusters in the Galaxy, M31 and M87 (Baum 1955, Sandage 1968, Racine 1968b), later failed when it was established that this maximum luminosity depended on galaxy size (de Vaucouleurs 1970, Harris 1974, Hodge 1974, Hanes 1977b,c). De Vaucouleurs (1977b, 1978) has demonstrated how a population-corrected brightest-cluster method can still be used. But ideally one should calibrate the intrinsic magnitudes of the clusters in a distant galaxy by fitting their *entire* observed luminosity distribution $\phi(m)$ to the familiar nearby galaxies.

The raw $\phi(M_V)$ distributions for Local Group galaxies are displayed in Figure 6. The four dE's are lumped together as are the two Magellanic Clouds, and the M31 sample is restricted to 114 clusters of low reddening as described in Section 3.2. For the Galaxy the distribution is seen to conveniently match a Gaussian curve (de Vaucouleurs 1970, 1977b, Hanes 1977b,c), although we emphasize that no compelling reason exists to adopt this model as *physically* real. The curve shown in Figure 6 has its peak at $\langle M_V \rangle = -7.3 \pm 0.1$ and dispersion $\sigma_V = 1.20 \pm 0.05$. A least-squares fit to the Galaxy data points for $M_V < -6.0$ (93 clusters) in the manner described by Hanes (1977c) yields a Gaussian with $\langle M_V \rangle = -7.34$, $\sigma_V = 1.17$, while adding the dE and the Magellanic Clouds samples to the Galaxy (139 clusters total) gives $(-7.28, 1.23)$. Applying the same Gaussian curve to the other Local Group members in Figure 6, we conclude that within the current uncertainties the hypothesis of a universal $\phi(M)$ for globular clusters is entirely reasonable for a remarkably wide range of parent galaxy types. Only the faint half of the M31 data, which is certainly incomplete, deviates significantly.

Figure 7 summarizes the totality of present $\phi(m)$ data by showing a fit of Virgo to the Local Group (cf Hanes 1979). To the Hanes data we have added two points on the integral $N(m)$ curve derived from deeper cluster counts in M87 (Harris & Smith 1976) and in M49 (Harris & Petrie

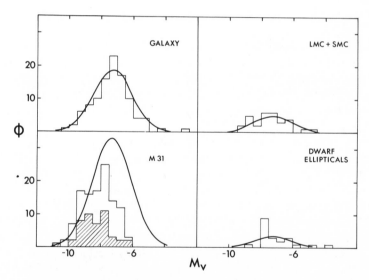

Figure 6 Luminosity distributions for globular-cluster systems in the Local Group galaxies. Here $\phi(M_V)$ is the number of clusters found in each half-magnitude interval for each galaxy sample. Gaussian model curves of peak $\langle M_V \rangle = -7.3$ and standard deviation $\sigma = 1.2$ mag are superimposed, suitably scaled to match the total cluster population N in each galaxy. For M31, the data are progressively more incomplete for $M_V \gtrsim -8$ (see text Section 6.2) and its curve is normalized to $N \simeq 200$. Here the shaded area of the histogram represents the 48 unreddened "outer" M31 clusters, and the unshaded part represents the 76 little-reddened clusters in the disk.

1978). These new counts reach one magnitude fainter than Hanes's limit and, being much closer to the peak of the $\phi(M_V)$ curve, provide a stronger constraint on the choice of the Virgo distance modulus. In Figure 7 we have more or less arbitrarily assumed $(m-M)_{V,\text{Virgo}} = 30.9 \pm 0.3$, corresponding to $H_0 = 75 \pm 10$ km s^{-1} Mpc^{-1}. The overall quality of the fit supports the assertion that at least the *shape* of the $\phi(M)$ curve is similar in all galaxies. The significant curvature of the mean line in this log-log plot shows also that $\phi(M)$ is not well fitted by a power law as has been claimed elsewhere (Tremaine et al. 1975, Tremaine 1976). As Hanes (1977c) emphasizes, it is important to realize that the distance fit can be carried out independently of the Gaussian model, although both the Local Group and the Virgo data fit this model satisfactorily.

We conclude that the observational evidence supports the concept of a universal luminosity function for globular clusters and that $\phi(m)$ is

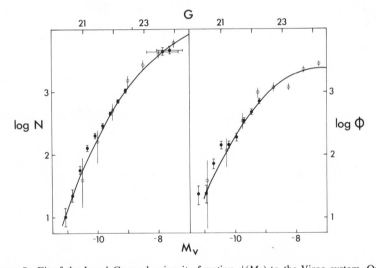

Figure 7   Fit of the Local Group luminosity function $\phi(M_V)$ to the Virgo system. Open circles denote the Local Group data combining ϕ for the Galaxy, dwarf ellipticals, Magellanic Clouds, and M31 (the latter corrected for incompleteness for $M_V > -8$). Closed circles represent the combined data for 6 giant Virgo ellipticals with the largest cluster populations (Hanes 1977c). The two populations have been fit assuming $(m-M)_V$ (Virgo) = 30.9, and the scale at top represents the corresponding apparent blue magnitudes G of Hanes's data. (*a*) The integrated luminosity function $N(M_V)$ (number of clusters brighter than M_V). Internal error bars are indicated for each data set, and the Gaussian model line described previously is drawn in. All the Virgo points for $G < 23$ are from Hanes (1977c), whereas the two extra points at upper right represent the Harris/Smith and Harris/Petrie cluster counts around M87 and M49, normalized to the same field size as the Hanes data. (*b*) The differential luminosity function $\phi(M_V)$ as defined in Figure 6, with the Gaussian curve again drawn in.

indeed a potentially important distance indicator in extragalactic studies. A crucial future test will be to observe the turnover of $\phi(m)$ in Virgo, which should occur at $V \sim 23.5$ if the present argument is correct.

6.3   Density Profiles

Figure 8 shows the projected density profile of four globular-cluster systems for which comprehensive data are available (Harris 1976, Sargent et al. 1977, Harris & Smith 1976, Harris & Petrie 1978). The outer regions of all these profiles are adequately represented by a de Vaucouleurs (1977a) $R^{1/4}$ law. They illustrate in a dramatic way the huge range in the characteristic sizes of globular-cluster systems. The $R^{1/4}$ lines drawn in Figure 8 have the following equations and effective radii R_e (Young 1976):

for the Galaxy, $\log \sigma = 3.4 - 2.57\,R^{1/4}$, $R_e = 3\,\text{kpc}$,
for M31, $\qquad \log \sigma = 3.1 - 2.03\,R^{1/4}$, $R_e = 7\,\text{kpc}$,
for M49, $\qquad \log \sigma = 2.7 - 1.33\,R^{1/4}$, $R_e = 40\,\text{kpc}$,
and for M87, $\quad \log \sigma = 2.5 - 1.05\,R^{1/4}$, $R_e = 100\,\text{kpc}$.

For each galaxy the zero points of $\log \sigma$ have been normalized to the total population N_t (Table 5) of the area surveyed. The quoted slope for the Galaxy is the same as de Vaucouleurs' (1977a). For M31, our corrections

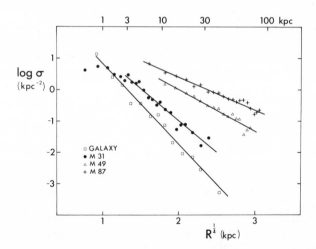

Figure 8   The radial density distribution of globular clusters in four major galaxies. Here σ is the number of clusters per kpc^2, plotted against projected galactocentric distance R in kpc. The data sources are Harris (1976, for the Galaxy), Sargent et al. (1977, and Section 3.2 for M31), Harris & Petrie (1978, for M49), and Harris & Smith (1976, for M87). The linear fits drawn through each distribution are the $R^{1/4}$ relations described in Section 6.3. Note the inner M31 points, which fall well below the $R^{1/4}$ curve, and also the vastly different scale sizes of each system.

for incompleteness of the survey (Section 3.2) at $R > 15$ kpc lead to a shallower profile, and hence a larger R_e, than obtained by de Vaucouleurs & Buta (1978).

The smaller size of the Galaxy's system would favor a type Sc rather than Sb for the Galaxy. The two gE systems have much larger R_e's than do the two spirals, and the M87 system is so large that our current estimate of $N_t = 15,000$ (Section 6.1) for $R < 90$ kpc may still fail to account for all its clusters by a factor of two or more. If no tidal radius limits the extent of this system, the M87 globular clusters would span much of the whole central Virgo cluster ($R \sim 1$ Mpc) before their projected density would drop to the level found in the outer halo of our Galaxy. Perhaps it will become preferable to think of these tens of thousands of globular clusters surrounding the central giant galaxy M87 as a system associated with the Virgo cluster itself!

A comparison of the density profiles in Figure 8 with the photometric profiles of the spheroidal components of M31 (de Vaucouleurs 1958), of M49 (King 1978a, Harris & Petrie 1978), and of M87 (de Vaucouleurs & Nieto 1978, King 1978a, Oemler 1976) reveals that the globular cluster systems have significantly larger characteristic sizes (shallower slopes) than do the parent galaxies. In earlier discussions (Harris & Smith 1976, Wakamatsu 1977a, Harris & Petrie 1978) it had been conventional to ascribe discrepant slopes between the cluster system and the halo inte-grated light to incompleteness of cluster counts (central regions) or to uncertain background corrections (outer regions). But the background counts are now better known, and, as has already been discussed for M31 (Section 3.2), the central regions would have many easily detectable bright clusters if the incompleteness fractions were as high as had been claimed. Our conclusion is therefore that the effect is at least partly real. This may indicate that globular clusters were formed at an earlier epoch than were the stars of the spheroidal component of galaxies, i.e. at a time when the protogalactic material was less centrally condensed.

As a final conclusion we may emphasize the impressive similarities displayed by systems of globular clusters in a vast diversity of galaxies. These systems appear to be inherently simple. Their total population is purely a function of the spheroidal mass of the parent galaxy; their density profiles are homologous, following the same law that applies to the simplest elliptical galaxies; they apparently share a unique luminosity (mass) function. Each individual system also presents peculiarities (notably in its metallicity characteristics) but here again some universal trends with galaxy mass or type are beginning to show. The overall picture of universality that emerges from this review suggests to us that as one learns more about globular clusters in galaxies, one uncovers important aspects of the earliest differentiated structures in the Universe—aspects of an epoch that initiated the birth of the galaxies themselves and that was not far removed from the primary instant of the Universe.

Literature Cited

Ables, H. D., Ables, P. G. 1977. *Ap. J. Suppl.* 34:245
Ables, H. D., Newell, E. B., O'Neil, E. J. 1974. *Publ. Astron. Soc. Pac.* 86:311
Alcaino, G. 1977a. *Astron. Astrophys. Suppl.* 27:255
Alcaino, G. 1977b. *Publ. Astron. Soc. Pac.* 89:491
Alcaino, G. 1978a. *Astron. Astrophys. Suppl.* 33:181
Alcaino, G. 1978b. *Astron. Astrophys. Suppl.* 34:431
Arp, H. C. 1958. *Astron. J.* 63:487
Arp, H. C. 1965. In *Galactic Structure*, ed. A. Blaauw, M. Schmidt, p. 401. Univ. Chicago Press, Ill. 606 pp.
Baade, W. 1944a. *Ap. J.* 100:137
Baade, W. 1944b. *Ap. J.* 100:147
Baade, W., Arp, H. C. 1964. *Ap. J.* 139:1027
Baum, W. A. 1955. *Publ. Astron. Soc. Pac.* 67:328
Bell, R. A., Gustafsson, B. 1976. In *The Galaxy and the Local Group*, ed. R. J. Dickens, J. E. Perry. *RGO Bull. No. 182*
Bernard, A. 1975. *Astron. Astrophys.* 40:199
Bernard, A. 1976. *Astron. Astrophys. Suppl.* 25:281
Bigay, J. H., Bernard, A. 1974. *Astron. Astrophys.* 33:123
Bok, B. J. 1966. *Ann. Rev. Astron. Astrophys.* 4:95
Butler, D. 1975. *Ap. J.* 200:68
Canterna, R., Schommer, R. 1978. *Ap. J. Lett.* 219:L119
Christie, W. H. 1940. *Ap. J.* 91:8
Cohen, J. 1978. *Ap. J.* 223:487
Corwin, H. G. 1977. *Astron. J.* 82:193
Cowley, A. P., Hartwick, F. D. A., Sargent, W. L. W. 1978. *Ap. J.* 220:453
Da Costa, G. S., Freeman, K. C., Kalnajs, A. J., Rodgers, A. W., Stapinski, T. E. 1977. *Astron. J.* 82:810
Danziger, I. J. 1973. *Ap. J.* 181:641
Dawe, J. A., Dickens, R. J. 1976. *Nature* 263:395
Delhaye, J. 1965. In *Galactic Structure*, ed. A. Blaauw, M. Schmidt, p. 61. Chicago: Univ. Chicago Press. 606 pp.
Demarque, P., McClure, R. D. 1977. In *The Evolution of Galaxies and Stellar Populations*, ed. B. M. Tinsley, R. B. Larson, p. 199. Yale Univ. Obs. 449 pp.
Demers, S. 1969. *Astrophys. Lett.* 3:175
de Vaucouleurs, G. 1958. *Ap. J.* 128:465

de Vaucouleurs, G. 1970. *Ap. J.* 159:435
de Vaucouleurs, G. 1977a. *Astron. J.* 82:456
de Vaucouleurs, G. 1977b. *Nature* 266:126
de Vaucouleurs, G. 1978. *Ap. J.* 224:14
de Vaucouleurs, G., Ables, H. D. 1970. *Ap. J.* 159:425
de Vaucouleurs, G., Buta, R. 1978. *Astron. J.* 83:1383
de Vaucouleurs, G., de Vaucouleurs, A. 1965. *Reference Catalog of Bright Galaxies.* Austin: Univ. Texas Press. 1st ed.
de Vaucouleurs, G., Nieto, J. L. 1978. *Ap. J.* 220:449
Diamond, G. E. 1976. MSc thesis. Univ. Toronto
Eggen, O. J., Lynden-Bell, D., Sandage, A. 1962. *Ap. J.* 136:748
Faber, S. M. 1973. *Ap. J.* 179:423
Faber, S. M., Jackson, R. E. 1976. *Ap. J.* 204:668
Feast, M. W., Lloyd-Evans, T. 1973. *MNRAS* 164:15P
Fisher, J. R., Tully, R. B. 1977. *Comments Astrophys. Space Sci.* 7:85
Ford, H. C., Jacoby, G., Jenner, D. C. 1977. *Ap. J.* 213:18
Freeman, K. C. 1974. In *Proc. ESO/SRC/CERN Conf. Research Programmes for Large Telescopes*, ed. A. Reiz, p. 177
Freeman, K. C., Gascoigne, S. C. B. 1977. *Proc. Astron. Soc. Aust.* 3:136
Freeman, K. C., Munsuk, C. 1973. *Proc. Astron. Soc. Aust.* 2:151
Gascoigne, S. C. B. 1966. *MNRAS* 134:59
Gascoigne, S. C. B. 1971. In *The Magellanic Clouds*, ed. J. Muller, p. 25. Dordrecht, Holland: Reidel
Gascoigne, S. C. B., Burr, E. J. 1956. *MNRAS* 116:570
Gascoigne, S. C. B., Kron, G. E. 1952. *Publ. Astron. Soc. Pac.* 64:196
Gascoigne, S. C. B., Lynga, G. 1963. *Observatory* 83:38
Goodenough, D. G., Hartwick, F. D. A. 1970. *Publ. Astron. Soc. Pac.* 82:921
Gratton, R. G., Nesci, R. 1978. *MNRAS* 182:61P
Gray, D. F. 1965. *Astron. J.* 70:362
Hanes, D. A. 1971. MSc thesis. Univ. Toronto
Hanes, D. A. 1977a. *MNRAS* 179:331
Hanes, D. A. 1977b. *Mem. R. Astron. Soc.* 84:45
Hanes, D. A. 1977c. *MNRAS* 180:309

Hanes, D. A. 1979. *MNRAS*. In press
Harris, H. C., Canterna, R. 1977. *Astron. J.* 82:798
Harris, W. E. 1974. PhD thesis. Univ. Toronto
Harris, W. E. 1976. *Astron. J.* 81:1095
Harris, W. E. 1977. *Publ. Astron. Soc. Pac.* 89:482
Harris, W. E. 1978. *Publ. Astron. Soc. Pac.* 90:45
Harris, W. E., Canterna, R. 1979a. Submitted for publication
Harris, W. E., Canterna, R. 1979b. *Ap. J. Lett.* 231. In press
Harris, W. E., Hesser, J. E. 1976. *Publ. Astron. Soc. Pac.* 88:377
Harris, W. E., Petrie, P. L. 1978. *Ap. J.* 223:88
Harris, W. E., Smith, M. G. 1976. *Ap. J.* 207:1036
Harris, W. E., Smith, M. G., van den Bergh, S. 1979. In preparation
Harris, W. E., van den Bergh, S. 1974. *Astron. J.* 79:31
Hartwick, F. D. A. 1976. *Ap. J.* 209:418
Hartwick, F. D. A., Cowley, A. P. 1978. Paper delivered at *NATO Adv. Study Inst. on Globular Clusters*. Cambridge Univ.
Hartwick, F. D. A., Sargent, W. L. W. 1974. *Ap. J.* 190:283
Hartwick, F. D. A., Sargent, W. L. W. 1978. *Ap. J.* 221:512
Helfer, H. L., Wallerstein, G., Greenstein, J. L. 1959. *Ap. J.* 129:700
Hesser, J. E., Hartwick, F. D. A., McClure, R. D. 1977. *Ap. J. Suppl.* 33:471
Hesser, J. E., Hartwick, F. D. A., Ugarte, P. 1976. *Ap. J. Suppl.* 32:283
Hiltner, W. A. 1960. *Ap. J.* 131:163
Hodge, P. W. 1960a. *Ap. J.* 131:351
Hodge, P. W. 1960b. *Ap. J.* 132:346
Hodge, P. W. 1961. *Astron. J.* 66:83
Hodge, P. W. 1965. *Ap. J.* 141:308
Hodge, P. W. 1969. *Publ. Astron. Soc. Pac.* 81:875
Hodge, P. W. 1971a. *Smithsonian Astrophys. Obs. Spec. Rep. No. 337*
Hodge, P. W. 1971b. *Ann. Rev. Astron. Astrophys.* 9:35
Hodge, P. W. 1973. *Ap. J.* 182:671
Hodge, P. W. 1974. *Publ. Astron. Soc. Pac.* 86:289
Hodge, P. W. 1976. *Astron. J.* 81:25
Hodge, P. W. 1977. *Ap. J. Suppl.* 33:69
Hodge, P. W. 1978. *Ap. J. Suppl.* 37:145
House, F., Wiegandt, R. 1977. *Astrophys. Space Sci.* 48:191
Hubble, E. 1932. *Ap. J.* 76:44
Illingworth, G. 1976. *Ap. J.* 204:73
Innanen, K. A. 1967. *Z. Astrophys.* 64:445
Innanen, K. A., Valtonen, M. J. 1977. *Ap. J.* 214:692

Jaschek, C. O. R. 1957. *Z. Astrophys.* 44:33
Jenner, D. C., Kwitter, K. B. 1977. *Bull. Am. Astron. Soc.* 9:287
Johnson, H. L. 1959. *Lowell Obs. Bull.* 4:117
Keenan, D. W. 1978. Paper delivered at *NATO Adv. Study Inst. on Globular Clusters*. Cambridge Univ.
King, I. R. 1962. *Astron. J.* 67:471
King, I. R. 1966a. *Astron. J.* 71:64
King, I. R. 1966b. *Astron. J.* 71:276
King, I. R. 1978a. *Ap. J.* 222:1
King, I. R. 1978b. *NATO Adv. Study Inst. on Globular Clusters*. Cambridge Univ.
Kinman, T. D. 1959a. *MNRAS* 119:157
Kinman, T. D. 1959b. *MNRAS* 119:538
Kinman, T. D. 1959c. *MNRAS* 119:559
Kinman, T. D. 1963. *Ap. J.* 137:213
Kron, G. E. 1956. *Publ. Astron. Soc. Pac.* 68:230
Kron, G. E. 1966. *Publ. Astron. Soc. Pac.* 78:143
Kron, G. E. 1975. Private communication
Kron, G. E., Guetter, H. H. 1976. *Astron. J.* 81:817
Kron, G. E., Mayall, N. U. 1960. *Astron. J.* 65:581
Kukarkin, B. V. 1974. *The Globular Star Clusters*. Sternberg State Astron. Inst. Moscow: Nauka. 135 pp.
Kurth, R. 1960. *Z. Astrophys.* 50:215
Larson, R. B. 1975. *MNRAS* 173:671
Larson, R. B. 1976. *MNRAS* 176:31
Larson, R. B., Tinsley, B. M. 1978. *Ap. J.* 219:46
Lee, S. W. 1977a. *Astron. Astrophys. Suppl.* 27:367
Lee, S. W. 1977b. *Astron. Astrophys. Suppl.* 28:409
Lee, S. W. 1977c. *Astron. Astrophys. Suppl.* 29:1
Lightman, A. P., Shapiro, S. L. 1978. *Rev. Mod. Phys.* 50:437
Liller, M. H., Carney, B. W. 1978. *Ap. J.* 224:383
Liller, W. 1977. *Ap. J. Lett.* 213:L21
Lindsay, E. M. 1956. *Irish Astron. J.* 4:65
Lindsay, E. M. 1958. *MNRAS* 118:172
Mallia, E. A. 1977. *Astron. Astrophys.* 60:195
Matsunami, N. 1963. *Publ. Astron. Soc. Jap.* 16:141
Mayall, N. U. 1946. *Ap. J.* 104:290
Mayall, N. U., Eggen, O. J. 1953. *Publ. Astron. Soc. Pac.* 65:24
Morgan, W. W. 1956. *Publ. Astron. Soc. Pac.* 68:509
Oemler, G. 1976. *Ap. J.* 209:693
Oort, J. H. 1977. *Ap. J. Lett.* 218:L97
Ostriker, J. P., Spitzer, L., Chevalier, R. A. 1972. *Ap. J. Lett.* 176:L51
Peebles, P. J. E. 1978. *Comments Astrophys. Space Sci.* 7:197

Peterson, C. J. 1974. *Ap. J. Lett.* 190:L17
Peterson, C. J. 1976. *Astron. J.* 81:617
Peterson, C. J., King, I. R. 1975. *Astron. J.* 80:427
Prociuk, I. 1976. MSc thesis. Univ. Toronto
Racine, R. 1965. MSc thesis. Univ. Toronto
Racine, R. 1968a. *Publ. Astron. Soc. Pac.* 80:326
Racine, R. 1968b. *J. R. Astron. Soc. Can.* 62:367
Racine, R. 1973. *Astron. J.* 78:180
Racine, R. 1975. *Astron. J.* 80:1031
Racine, R., Harris, W. E. 1975. *Ap. J.* 196:413
Racine, R., Oke, J. B., Searle, L. 1978. *Ap. J.* 223:82
Rousseau, J. 1964. *Ann. Astrophys.* 27:681
Rutily, B., Terzan, A. 1977. *Astron. Astrophys. Suppl.* 30:315
Sandage, A. 1961. *The Hubble Atlas of Galaxies. Carnegie Inst. Washington Publ. No. 618.* Washington DC
Sandage, A. R. 1968. *Ap. J. Lett.* 152:L149
Sandage, A. R., Hartwick, F. D. A. 1977. *Astron. J.* 82:459
Sandage, A., Katem, B., Johnson, H. L. 1977. *Astron. J.* 82:389
Sargent, W. L. W., Kowal, S. T., Hartwick, F. D. A., van den Bergh, S. 1977. *Astron. J.* 82:947
Saslaw, W. C. 1973. *Publ. Astron. Soc. Pac.* 85:5
Sawyer, H. B., Shapley, H. 1927. *Harvard Coll. Obs. Bull. No. 848*
Sawyer Hogg, H. B. 1959. *Handbuch der Physik*, ed. S. Flugge, 53:129. Berlin: Springer
Schmidt, M. 1956. *Bull. Astron. Inst. Neth.* 13:15
Schmidt, M. 1965. See Arp, 1965, p. 513
Schuster, H. E., West, R. M. 1977. *The Messenger*, No. 10, p. 13
Searle, L. 1978. Paper delivered at *NATO Adv. Study Inst. on Globular Clusters.* Cambridge Univ.
Searle, L., Zinn, R. 1978. *Ap. J.* 225:357
Seyfert, C. K., Nassau, J. J. 1945. *Ap. J.* 102:377
Shapley, H. 1918. *Ap. J.* 48:89
Sharov, A. S. 1973. *Sov. Astron.—AJ* 17:174
Sharov, A. S. 1976. *Sov. Astron.—AJ* 20:397
Sharov, A. S., Lyutyi, V. M., Esipov, V. F. 1976. *Sov. Astron. Lett.* 1:69
Sharov, A. S., Lyutyi, V. M., Esipov, V. F.

1977. *Sov. Astron. Lett.* 2:128
Smith, M. G., Hesser, J. E., Shawl, S. J. 1976. *Ap. J.* 206:66
Smith, M. G., Weedman, D. W. 1976. *Ap. J.* 205:709
Spinrad, H., Schweizer, F. 1972. *Ap. J.* 171:403
Spitzer, L. 1958. *Ap. J.* 127:17
Strom, K. M., Strom, S. E., Jensen, E. B., Moller, J., Thompson, L. A., Thuan, T. X. 1977. *Ap. J.* 212:335
Surdin, V. G., Charikov, A. V. 1977. *Sov. Astron.—AJ* 21:12
Tammann, G. A. 1976. See Bell & Gustaffson 1976. p. 136
Tifft, W. G. 1962. *MNRAS* 125:199
Tremaine, S. 1976. *Ap. J.* 203:345
Tremaine, S. D., Ostriker, J. P., Spitzer, L. 1975. *Ap. J.* 196:407
van de Kamp, P. 1964. *Elements of Astromechanics.* San Francisco: Freeman. 133 pp.
van den Bergh, S. 1967. *Astron. J.* 72:70
van den Bergh, S. 1968. *J. R. Astron. Soc. Can.* 62:145, 219
van den Bergh, S. 1969. *Ap. J. Suppl.* 19:145
van den Bergh, S. 1971. *Astron. J.* 76:1082
van den Bergh, S. 1975. *Ann. Rev. Astron. Astrophys.* 13:217
van den Bergh, S. 1977a. *Astron. J.* 82:796
van den Bergh, S. 1977b. *Vistas Astron.* 21:71
van den Bergh, S. 1978. Paper delivered at *NATO Adv. Study Inst. on Globular Clusters.* Cambridge Univ.
van den Bergh, S., Hagen, G. L. 1968. *Astron. J.* 72:569
Vetešnik, M. 1962a. *Bull. Astron. Inst. Czech.* 13:180
Vetešnik, M. 1962b. *Bull. Astron. Inst. Czech.* 13:218
von Hoerner, S. 1955. *Z. Astrophys.* 35:255
Wakamatsu, K. I. 1977a. *Publ. Astron. Soc. Pac.* 89:267
Wakamatsu, K. I. 1977b. *Publ. Astron. Soc. Pac.* 89:504
Walker, M. F. 1970. *Ap. J.* 161:835
Walker, M. F. 1972. *MNRAS* 156:459
Woltjer, L. 1975. *Astron. Astrophys.* 42:109
Woolley, R. v. d. R. 1961. *Observatory* 81:161
Young, P. J. 1976. *Astron. J.* 81:807
Zaitseva, G. V., Lyutyi, V. M., Kukarkin, B. V. 1974. *Sov. Astron.—AJ* 18:257
Zinn, R. 1974. *Ap. J.* 193:593

Ann. Rev. Astron. Astrophys. 1979. 17:275–308

STELLAR WINDS *2152

Joseph P. Cassinelli

Washburn Observatory, University of Wisconsin, Madison, Wisconsin 53706

1 INTRODUCTION

This review is concerned with theories explaining the continual expansion of the outer atmospheric layers of luminous early- and late-type stars. For these stars the winds are sufficiently massive to be optically thick in the opacity of strong resonance lines and in certain continua. Thus, the winds are detectable through the emergent stellar spectra. Analyses of the strengths of the spectral features indicate that the stars are losing mass at a rate as high as 10^{-5} $M_\odot$/yr. The shortward displacement of absorption lines yields information concerning the velocities of the winds. The terminal velocities are large for the early-type stars, 600–3500 km/sec, but are small for the K and M stars, 10–100 km/sec.

Much of the interest in stellar winds in the past decade was stimulated by the discovery of the high velocity outflows from O and B supergiants by Morton (1967) and his co-workers (Morton, Jenkins & Brooks 1969). Their rocket-ultraviolet observations showed broad P Cygni-shaped profiles of lines of moderate stages of ionization, such as C^{+3}. The ionization stages and the high flow velocities could not be explained by a simple extrapolation of solar wind theory, and this led to the development of radiatively driven wind theories. In contrast, the winds from evolved K and M stars are detected through narrow P Cygni lines in the cores of strong Fraunhofer lines, and the velocities are well below the photospheric escape speeds. Deutsch (1956) proved that the features represent an actual mass loss from the stars by showing that the expansion extends to several hundred stellar radii where the rate of expansion exceeds the local escape speed.

Although most of the recent work on stellar winds concerns giants and supergiants, there is indirect evidence for winds from stars on the main sequence. For example, stars with spectral type later than F5V have low rotation speeds (Kraft 1967), which can be explained as a consequence of loss of angular momentum through the coupling of the stellar magnetic field with the expanding coronal plasma. The mechanism for the mass and

275

0066-4146/79/0915-0275$01.00

angular momentum loss of these stars is probably the same as that for the sun, and is not considered in this review. The solar wind is treated in the books by Parker (1963), Brandt (1970), and Hundhausen (1972). Current solar wind research problems are reviewed by Holzer (1978) and Hollweg (1978).

Some observational information is presented in this review, but more comprehensive discussions of observational data are contained in the reviews by Conti (1978) and Reimers (1975, 1978) for early- and late-type stars, respectively. The effect of the mass loss by stellar winds on stellar evolution is studied by Chiosi, Nasi & Sreenivasan (1978), and the effects of a wind on the surrounding interstellar medium are studied by Weaver et al. (1977).

Stellar Wind Theories

Current stellar wind theories fall under three broad classes.

1. Radiative models. For the very luminous stars, radiative acceleration of the matter in the outer atmosphere can occur if there is sufficient opacity at wavelengths near flux maximum. In early-type stars, transfer of photon momentum to the gas occurs through the opacity of the many strong resonance lines in the ultraviolet. The progressive Doppler shifting of the line opacity into the unattenuated photospheric radiation field can result in a rapid acceleration to very high speeds. For late-type stars, the atmosphere is cool enough for grain formation. The radiative force on the infrared continuous opacity of the grains can drive the dust and surrounding gas outward. Unlike the line driven case the acceleration does not depend on velocity gradients, and, as a result, the rate of acceleration is slower and the terminal velocity is lower.

2. Coronal models. Stars that have convection zones, or some other source of acoustic or mechanical wave energy, are expected to have coronal zones as a result of wave dissipation. The outer atmosphere expands as a wind because of the gas pressure gradient. The process is assumed to be similar to that producing the solar wind. In practice, stellar wind theorists have used only the most elementary formulation of solar wind theory, i.e. they have assumed that the flow is that of a spherically symmetric single fluid, in steady state and with no magnetic field constraints. Processes such as those involving two-fluid flow and coronal hole geometry, which are known to be important in the sun, are not considered for two reasons: (a) the observational data are rather primitive compared with those available for the sun; (b) the winds have mass loss rates and densities that are orders of magnitude higher than in the sun. As a result, many of the solar phenomena associated with closed magnetic structures and large conductive losses may not be relevent. Since the electron densities

are higher, radiative cooling is enhanced and thus temperature structures quite different from that of the expanding solar corona may be produced.

3. Hybrid models. Because fully radiative models and fully coronal models cannot explain various observations, a third class of model, the hybrid corona-plus-radiatively-driven-wind models, are receiving some attention. For example, Hearn (1975a) has shown that the density in the flow from an OB supergiant is so large ($\sim 10^{10}$ cm^{-3} at $2R_*$) that rapid cooling of a coronal gas should occur. The temperature should drop rapidly in a "recombination region" at a height where mechanical deposition ceases to be effective. He suggested the flow is initiated by gas pressure gradients in the corona and is then accelerated to large velocities by the line acceleration mechanism, which is effective beyond the recombination regions. For the winds from the cool stars, Kwok (1975) appealed to turbulence in the chromospheric region to enhance the density at the dust condensation zone. Thus far, the hybrid models have been pursued primarily from a semiempirical standpoint. Cassinelli (1979) describes several studies to test the observational consequences of the hybrid model of Hearn (1975a). These studies are discussed in Section 3. Studies of the dynamical properties of the corona-plus-radiative-wind models will most likely be an active area of theoretical research in the next decade.

In addition to these three broad classes of theories, several other driving mechanisms have been investigated. The effects of rotation on radiatively driven winds have been studied by Marlborough & Zamir (1975). Belcher & MacGregor (1976) considered magnetically driven winds and the loss of angular momentum during the evolution of stars of spectral type F5 and later.

Stars Having Measured Mass Loss Rates

Many of the stars for which mass loss rates have been estimated are shown in Figure 1. The order of magnitude of the mass loss rate is indicated by the size of the symbols. Pre–main-sequence objects, which have very large mass loss rates for their luminosity, are indicated by the open circles. The mass loss rate estimates shown in the figure were taken from a number of sources. (a) For the early-type stars, most of the data for Of and OBA supergiants are available from the infrared survey of Barlow & Cohen (1977). Estimates from radio fluxes are available for P Cygni in Wendker et al. (1975) and for ζ Pup in Morton & Wright (1978). Line profiles were used to derive mass loss rates for τ Sco (Lamers & Rogerson 1978), the Be star ψ Per (Bruhweiler et al. 1978), the Wolf-Rayet star HD50896 (Rumpl & Cassinelli 1979), the central stars of planetary nebulae (Heap 1978), and pre–main-sequence stars (Garrison 1978, Kuhi 1964). (b) For the late-type stars, mass loss rates can be estimated from blueward

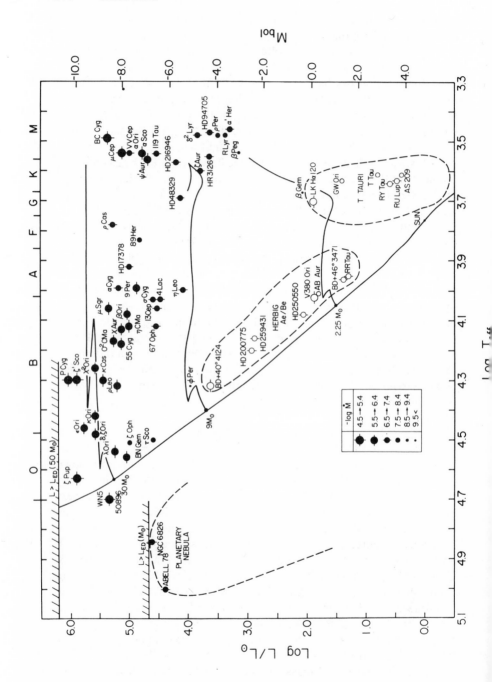

shifted cores in metallic resonance lines (Sanner 1976 and references in Reimers 1975), and from absorption seen in the spectra of a nearby companion (Bernat 1977, Reimers 1977, van der Hucht et al. 1979). The mass loss rate estimates for the late-type stars differ by one to two orders of magnitude, so averages of the logarithm of $\dot{M}$ were used for the estimate shown in Figure 1. The causes for the uncertainties are discussed in Section 4.

Note that most of the data concerning mass loss rates is for luminous stars earlier than about A5 and later than K4. Reimers (1977) completed a survey of stars in the intermediate spectral region and found that there are many stars with evidence for mass loss through shortward displaced absorption features in the cores of the Ca II H and K lines. However, it is difficult to derive reliable mass loss rates from that data alone. These problems are discussed in Section 3.

There are few supergiants of spectral class F and G, presumably because of the rapid evolution through this portion of the H-R diagram. Shown in Figure 1 are estimates for the mass loss rates from ρ Cas and 89 Her. The mass loss from these stars may not be steady but may occur in sporadic mass loss episodes (Sargent 1961, Sargent & Osmer 1969).

As a result of the gap in information concerning mass loss from stars of intermediate spectral type, the subject of stellar winds is currently divided into two parts: (a) the winds from hot stars for which radiation pressure on lines plays an important role in the dynamics, and (b) the winds from cool evolved stars for which coronae and dust are thought to be important. In Section 2 the basic stellar wind equations are presented, including terms to represent radiative acceleration and radiative and mechanical energy deposition. In Section 3 the winds of hot stars are considered. The theory of line driven winds is discussed, and a graphical solution of the nonlinear momentum equation is presented. Current difficulties with the theory and recent attempts to explain observed ionization anomalies are discussed.

Section 4 is concerned with the winds of cool evolved stars. Both coronal and dust driven models have been considered. Mullan's (1978) application of the Hearn (1975b) minimum energy coronal theory is discussed, as is Mullan's explanation of the locus in the H-R diagram that separates stars with large mass loss rates from those with undetectable

←*Figure 1* Location on the Hertzprung-Russell diagram of stars for which mass loss rates have been estimated. The size of the symbol indicates the magnitude of the mass loss rates in solar masses per year. Pre–main-sequence stars are shown with open circles. Eddington radiative instability limits for stars of 1 and 50 solar masses are indicated by the boundaries of the cross-hatched area. Evolutionary tracks for stars of 2.25, 9, and 30 solar masses are shown.

winds. A wide variety of observational techniques have been used to study the winds and the dust in the wind of the M stars. These techniques are reviewed and the dust driven wind theory is discussed briefly.

2   STELLAR WIND EQUATIONS

Most of the theoretical work thus far on stellar winds assumes a steady, radial, spherically symmetric flow. With these assumptions, the fluid flow equations expressing conservation of the mass, momentum, and energy of the gas may be written as

$$\dot{M} = 4\pi\rho \, vr^2 = \text{const}, \tag{1}$$

$$v\frac{dv}{dr} + \frac{1}{\rho}\frac{dp}{dr} + \frac{GM}{r^2} + g_R = 0, \tag{2}$$

$$v\frac{de}{dr} + p\,v\frac{d}{dr}\left(\frac{1}{\rho}\right) = -\left(\frac{1}{\rho}\right)(\nabla\cdot\mathbf{q}) = \frac{1}{\rho}\left(Q_A + Q_R - \nabla\cdot\mathbf{q_c}\right), \tag{3}$$

where ρ is the mass density, v is the radial speed, p is the gas pressure, e is the internal energy per unit mass, g_R is the acceleration produced by radiation pressure. The heat added to the gas is represented by the divergence of the energy flux $\mathbf{q}$, and Q_A is the energy deposited per volume by acoustic or mechanical energy, Q_R is the rate of deposition of radiative energy, and $\mathbf{q_c}$ is the conductive flux. The fluid equations, including a thorough discussion of the radiation terms, are derived in Chapters 14 and 15 of the stellar atmospheres text by Mihalas (1978). An elegant formulation of the equations, accounting for both mechanical energy deposition and mass deposition, is given in Holzer & Axford (1970).

The radiative acceleration is proportional to the radiative flux and thus tends to decrease as r^{-2}, as does gravity, g. Hence it is often convenient to use the ratio

$$\Gamma_R = g_R/g, \tag{4}$$

and to separate the radiative acceleration into its continuum and line contributions,

$$g_R = \frac{4\pi}{c}\int_{v=0}^{\infty} k_v H_v dv, \tag{5}$$

$$= \frac{k_F L}{4\pi cr^2} + \sum g_l, \tag{6}$$

where k_F is the continuum flux mean opacity (in cm^2/gm) and g_l is the acceleration due to the *l*th spectral line. The momentum equation may be rewritten as

$$v\frac{dv}{dr} + \frac{1}{\rho}\frac{dp}{dr} + \frac{GM}{r^2}\left(1-\Gamma\right) + \sum g_l = 0, \tag{7}$$

where

$$\Gamma = \frac{k_F L}{4\pi cGM}. \tag{8}$$

In the interior of the star, the net acceleration must be inward; the star is radiatively unstable if the ratio of the radiative acceleration on electron scattering to gravity, Γ_{es}, exceeds unity. This is the "Eddington limit" on the luminosity-to-mass ratio

$$\Gamma_{es} = \frac{\sigma L}{4\pi cGM} \lessgtr 1, \tag{9}$$

where σ is the electron scattering opacity per gram. The Eddington limits for stars of 1 $M_\odot$ and 50 $M_\odot$ are shown in Figure 1.

The momentum and gas energy equations (2, 3) can be combined in the form:

$$\frac{d}{dr}\left[\dot{M}\left(\frac{v^2}{2} + e + \frac{p}{\rho} - \frac{GM}{r}\right) + 4\pi r^2 |\mathbf{q}_c|\right] = \dot{M}g_R + \frac{\dot{M}}{\rho v}(Q_R + Q_A). \tag{10}$$

Letting E, be the energy per unit mass of the gas

$$E = \left(\frac{v^2}{2} + e + \frac{p}{\rho} - \frac{GM}{r}\right), \tag{11}$$

Equation (7) integrates to give

$$E = E_0 + \frac{1}{\dot{M}}\int_{r_0}^{\infty}(v\rho g_R + Q_R + Q_A)4\pi r^2\, dr + \frac{4\pi}{\dot{M}}\left[r^2|\mathbf{q}_c| - r_0^2|\mathbf{q}_c|^0\right]. \tag{12}$$

Thus, even though E is negative deep in the flow it may become positive farther out because of (*a*) deposition of radiative energy, (*b*) deposition of mechanical energy, or (*c*) momentum deposition by radiative acceleration of continuum and line opacity of the gas (Holzer 1977).

The various approximations employed in stellar wind theory can be summarized with reference to Equations (7) and (12).

1. In purely radiatively driven wind theories, it is assumed that there is no mechanical energy deposition, $Q_A = 0$, and that the temperature is

so low that conduction is unimportant, $|\mathbf{q}_c| = 0$. The time required by the material to gain and lose energy is short compared to flow time scales, hence the flow is nearly in radiative equilibrium. In this case the energy of the gas flow is much less than the radiative luminosity $\dot{M} E \ll L_*$ and

$$Q_R = \int_0^\infty 4\pi\kappa_v \, (J_v - B_v) \, dv \approx 0, \tag{13}$$

where κ_v, the absorptive opacity, and J_v, the mean intensity, are evaluated in the frame moving with the fluid (Castor 1972, Cassinelli & Castor 1973, Mihalas 1978). The radiative equilibrium condition, (13), can be used to derive the temperature distribution in the flow using temperature correction procedures if the wind is very thick, as in Wolf-Rayet stars (Cassinelli & Hartmann 1975) or in massive M supergiant flows (Menietti & Fix 1978). More commonly, it is assumed that the flow is isothermal with a temperature that is appropriate for a gas in radiative equilibrium (Klein & Castor 1978):

$$T_e \approx 0.8 \, T_{eff}. \tag{14}$$

For the isothermal case, Equation (7) has the form

$$\frac{r}{v}\frac{dv}{dr} = \left[2a^2 - \frac{GM}{r}\left(1 - \Gamma_R\right)\right]\Bigg/\left(v^2 - a^2\right), \tag{15}$$

where a is the isothermal sound speed $(a^2 = RT)$. Assuming for the moment that g_R does not depend on dv/dr, then this equation has the quotient form that is familiar from solar wind and de Laval nozzle theory (Brandt 1970). For transonic flow to occur it is necessary that the numerator vanish at the sonic point where $v = a$. Marlborough & Roy (1970) and Cassinelli & Castor (1973) discuss constraints on Γ_R to permit transonic flow: Γ_R must be less than unity at and below the sonic point in order for the velocity gradient to be positive. Beyond the sonic point, Γ_R may be larger than unity and large velocity gradients may be produced in the supersonic region. Equation (13), or a slight variant of it, was used by Lucy & Solomon (1970) for line driven winds, by Cassinelli & Hartmann (1975) for continuum driven winds of hot stars, and by Kwok (1975) and others for dust driven winds. In each of these cases, Γ_R was sufficiently close to unity at the sonic point that transonic flow could occur in a *cool* gas. That is, the effective escape speed $2GM(1 - \Gamma_R)/r$ is small and therefore the numerator of (13) can vanish for small a. For dust driven winds, Γ_R increases rapidly to near unity in the dust condensation region.

A quite different form of the momentum equation is considered in the line driven theory of Castor, Abbott & Klein (1975), who explicitly accounted for the dependence of the line force on dv/dr. In their theory $\sum g_l \propto (dv/dr)^\alpha$ where $\alpha \sim 0.7$ so the momentum equation is not linear and the topology of the solution is greatly different from the familiar X-type singularity associated with (15). This solution is discussed in Section 3.

2. In coronal wind models, it is common to assume that the energy deposition occurs in a thin shell at the base of the flow and to consider only the solution beyond that region, where $Q_A = 0$. For hot coronae, radiative heating by absorption is negligible, but significant radiative cooling can occur by recombination or through collisionally excited line radiation, so $Q_R = N_e^2 \Lambda$, where $\Lambda(T)$ is the cooling function (Cox & Tucker 1969). The spatial distribution of the mechanical deposition is unknown. Hearn proposed a "minimum flux coronal model" in which the pressure and temperature at the base of the corona are determined if the magnitude of the total mechanical deposition, i.e. $\int Q_A \, dV$, integrated over the volume of the corona, is known. This model is discussed in Section 4.

3. In the hybrid models such as the corona-plus-cool-wind model of Hearn (1975a) and Cassinelli & Olson (1979), constraints on the coronal emission measure $\int N_e^2 \, dV$ and hence on $\int Q_A \, dV$ can be derived from the requirement that the X-ray emission account for the anomalous ionization seen in the wind and that X rays satisfy observational limits. Details of the transition to the line driven flow have not been worked out, however it is probably adequate to assume that $Q_A = 0$ in the transition, that Q_R is the dominant energy loss term, and that the radiative acceleration increases rapidly in the recombination zone.

3   STELLAR WINDS OF EARLY-TYPE STARS

Observations

Winds from early-type stars are detected through the presence of broad P Cygni-shaped profiles, and are characterized by large mass loss rates, 10^{-8} to 10^{-5} $M_\odot$/yr, and by large flow velocities, 600 to 3500 km/sec. Although ground-based observations of emission lines in O and B supergiants indicated expanding atmospheres, not until the rocket-ultraviolet observations of Morton (1967) and his collaborators (Morton, Jenkins & Brooks 1969) was the extreme nature of the winds realized. Observations of the winds and the effects of mass loss on stellar evolution are described by Snow & Morton (1976), Conti (1978), and Hutchings (1978).

In the past few years a significant amount of data concerning mass loss rates, terminal velocity, and ionization conditions have been derived as a result of the completion of extensive surveys of the infrared continuum and ultraviolet line spectra.

1. Barlow & Cohen (1977) derived mass loss rates from their infrared continuum photometry of 44 early-type stars. They found that the mass loss rates are nearly proportional to the stellar luminosity ($\dot{M} \propto L^{1.1}$) and only slightly dependent on the effective temperatures of the supergiants. Mass loss rates can be derived from the long wavelength continua because the free-free opacity increases as λ^2, and at sufficiently long wavelength, optical depth unity occurs in the wind itself. Radio observations provide the most accurate method of determining mass loss rates, because the flux originates far enough out in the flow that the gas has reached terminal velocity, and the flux is nearly independent of the temperature in the flow. Radio fluxes have been used to derive $\dot{M}$ for only a few stars (Morton & Wright 1978) and much more information is expected in the future from VLA radio observations. The theory for deriving $\dot{M}$ from free-free continuum observations is given by Wright & Barlow (1975), Panagia & Felli (1975), and Cassinelli & Hartmann (1977).

2. The richest sources of information concerning the winds of hot stars are the accurate line profiles obtained from Copernicus. The P Cygni profile of OVI in ζ Pup is shown in Figure 2 (Morton 1976). The Copernicus spectra of 60 O and B stars are given in the catalogue of Snow & Jenkins (1977). The terminal velocities of the winds can be derived from the strongest P Cygni profiles if they have a vertical short wavelength edge to their absorption components. Abbott (1978a) derived an

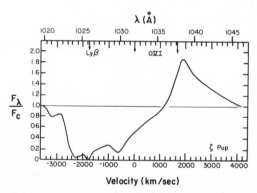

Figure 2   P Cygni profile of the OVI $\lambda\lambda 1031.945$, 1037.619 lines in the spectrum of ζ Pup (O4f) Morton (1976). The arrows indicate the rest wavelengths of the lines. The velocity scale at the bottom gives the Doppler displacement from the 1031.9-Å component. The core of the stellar Lyman β line is seen at -1900 km/sec. Interstellar lines have been eliminated.

empirical relation between terminal speed and photospheric escape speed, $v_\infty = 3\, v_{esc}$, for 34 O, B, A, and Wolf-Rayet stars.

3. Copernicus spectra can also be used to derive the relative abundance of various ionization stages of a given element and to provide some information on the velocity structure of the winds (Lamers & Morton 1976). Extensive families of theoretical P Cygni profiles were recently calculated (Olson 1978, Castor & Lamers 1979). Thus, good estimates of the ionization structure for the large number of hot stars may now be derived. The Copernicus spectra show strong lines of anomalously high stages of ionization, such as OVI in supergiants as late as B0.5Ia, and Si IV as late as B8I. Summaries of these anomalies along with two quite different explanations are presented by Lamers & Snow (1978) and Cassinelli & Olson (1979). The discovery of the high stage of ionization led to the realization that mechanical energy deposition occurs somewhere in the flow and revived interest in chromospheric and coronal models of the winds such as those proposed by Cannon & Thomas (1977) and Hearn (1975b).

Wind Models

Recent theoretical work has proceeded along two lines.

1. A purely radiatively driven wind theory has been developed by Castor, Abbott & Klein (hereafter CAK 1975, 1976). The high velocities and large mass loss rates are produced by the transfer of momentum from radiation to the gas, by way of line opacity. Abbott (1977, 1978b) and Castor (1978) recently refined the theory to include a large number of lines and a more accurate treatment of the radiation transfer in lines. The energy equation is not treated explicitly because there are a sufficient number of strong lines to drive the winds, even if the gas temperature is only roughly that which is expected in radiative equilibrium. The theory does not explain the ionization anomalies such as OVI.

2. In the development of empirical models, efforts have been made to derive the temperature and velocity structure of the winds from the observed line profiles and continuous energy distributions. Several models proposed to explain the observed ionization anomalies are presented in a review by Cassinelli, Castor & Lamers (1978). The models start with an assumed temperature and velocity structure, and then observational consequences are derived from theoretical calculations of optical and ultraviolet line profiles, and of calculations of continuum fluxes from X-ray through radio wavelengths. Lamers & Morton (1976) exposed several deficiencies in the line driven wind theory from such a process. Van Blerkom (1978) found from a study of Balmer line profiles that the velocity distribu-

tions in hot stars tend to be of two types, either "rapid accelerators," such as the winds of O stars, or "gradual accelerators," such as the massive wind of P Cygni.

Line Driven Wind Theory

A photon that is emitted from the photosphere at a frequency larger than that of a strong line may become resonant with that line opacity in a shell somewhat higher in the atmosphere, if the outer layers of the atmosphere are accelerating in the outward radial direction. If v is the original frequency of the photon and v_0 is the line central frequency, then the shell at which the photon may be scattered occurs where the flow velocity has reached the value

$$v(r) = c(v - v_0)/v_0.$$

The transfer of radiation through the shell is adequately described by the Sobolev escape probability theory (Sobolev 1960, Castor 1970, Mihalas 1978). Although a photon may scatter many times within the shell, the net momentum and the net energy imparted to the shell depend on the angle at which the photon leaves the shell after its last scattering.

The maximum mass loss rate that can be produced by one strong line, as for example the Lucy & Solomon (1970) case, can be derived by equating the final mass momentum flux $\dot{M}v_\infty$ to the photon momentum that is transferred by scattering all the stellar radiation between v_0 and $v_0 + \Delta v_{max}$, where $\Delta v_{max} = v_0 v_\infty/c$. Thus,

$$\dot{M}v_\infty = \frac{L_v \Delta v_{max}}{c} = \frac{L_v v_0}{c}\frac{v_\infty}{c},$$

and if v_0 is near the maximum of the stellar continuous energy distribution, this gives

$$\dot{M} \approx L/c^2. \tag{16}$$

Now if the entire spectrum is covered by nonoverlapping lines, i.e. adjacent lines separated by displacement v_∞, the "single scattering" maximum mass loss rate can be derived from the momentum flux of the entire stellar luminosity, $\dot{M}v_\infty = L/c$, giving

$$\dot{M}_{max} = L/v_\infty c. \tag{17}$$

This is a factor c/v_∞ (i.e. ~ 100) greater than that of Equation (16). For $L = 10^6 \, L_\odot$, $v_\infty/c = 0.01$, Equation (17) gives $\dot{M}_{max} = 7 \times 10^{-6} \, M_\odot/yr$, just about the mass loss rate observed from the most luminous Of and W-R stars and OB supergiants. The final kinetic energy luminosity of the star is, from Equation (17),

$$\frac{1}{2}\dot{M}\,v_\infty^2 = \frac{1}{2}\left(\frac{v_\infty}{c}\right)L, \tag{18}$$

or only about 0.5% of the radiative luminosity. So, although each photon has been scattered by one shell, it has not been destroyed. The photons have been redshifted by as much as $v_0 v_\infty/c$ and, except for the photons that are scattered back into the photosphere, they are seen as the "emission component" of the P Cygni profile, and as the residual radiation in the core of the P Cygni absorption component.

A photon that escapes from the resonant shell of one line may encounter another shell where it is resonant with a different line. Rybicki & Hummer (1978) developed a generalized form of the Sobolev theory to treat the problem of multiple resonance shells. Castor (1978) discusses the effects of multiple scattering on the flow dynamics and indicates that the extra momentum that can be imparted will not raise the maximum mass loss rate to more than about a factor of two or three above that given in Equation (17).

The acceleration imparted to a shell as a result of resonance line scattering of photospheric radiation is given by Lucy (1971) and Castor (1974). If πF_v is the radiative flux incident on the innerside of a shell that has a line opacity resonant to photons of frequencies $v \pm \Delta v_D/2$, then the radiative acceleration of the shell is

$$g_L = \kappa_L \frac{\pi F_v}{c} \Delta v_D \frac{[1-\exp(-\tau_L)]}{\tau_L}, \tag{19}$$

where κ_L is the opacity (cm^2/gm) at line center; $\Delta v_D = v_0\, v_{th}/c$, where v_{th} is the thermal velocity of the absorbing ion. The quantity τ_L is the effective optical thickness of the expanding shell and is given approximately by

$$\tau_L = \kappa_L\,\rho v_{th}\,\frac{dv}{dr}. \tag{20}$$

In Equation (19), τ_L/κ_L is the amount of mass in a column that can scatter the photons in Δv_D, and $1 - e^{-\tau_L}$ is the probability that scattering actually occurs. In two limiting cases we have for strong lines,

$$g_L = \frac{\pi F_v}{c}\frac{\Delta v_D}{\rho v_{th}}\frac{dv}{dr} \qquad \tau_L \gg 1, \tag{21}$$

and for weak lines,

$$g_L = \frac{\pi F_v}{c}\Delta v_D \kappa_L \qquad \tau_L \ll 1. \tag{22}$$

Note that the acceleration on strong lines is independent of the line strength, and thus the total acceleration due to strong lines is simply proportional to the *number* of strong lines. Abbott (1977) showed that the 18 most abundant elements have lines that can contribute to this number, and hence extensive line lists must be considered. The force on strong lines depends on dv/dr because a velocity gradient allows a line to move out from behind its own "shadow" and into frequencies where there is a strong continuum flux.

For realistic cases it is necessary to consider lines from many different elements and with a wide range in line strengths. It is, therefore, convenient to define a depth scale independent of line opacity;

$$t = \sigma \rho v_{\mathrm{T}} \left| \frac{dv}{dr} \right|, \tag{23}$$

where σ is the electron scattering opacity, v_{T} is the thermal speed of some reference ion, say oxygen, with mass m_0. Letting $\eta_i = \kappa_i/\sigma$ and $\mu_i = (m_0/m_i)^{1/2}$ where κ_i and m_i are the opacity and mass of the ion producing the ith line, then the total line acceleration is

$$G_{\mathrm{L}} = \sum_i (g_{\mathrm{L}})_i = \left(\frac{\pi F}{c} \right) \sum_i \frac{F_v^i}{F} \eta_i \mu_i \Delta v_{\mathrm{T}} \left(\frac{1 - \exp(-\eta_i \mu_i t)}{\eta_i \mu_i t} \right), \tag{24}$$

$$\approx \frac{\pi F}{c} \sigma \sum_i \frac{F_v^i}{F} \Delta v_{\mathrm{T}} \min \left(\frac{1}{t}, \eta_i \mu_i \right). \tag{25}$$

The coefficient of the summation is the radiative acceleration on electron scattering opacity, and the sum is referred to by CAK as "the radiation force multiplier."

$$M(t) = \sum_{\tau_i > 1} \frac{F_v^i \Delta v_{\mathrm{T}}}{F} \frac{1}{t} + \sum_{\tau_i < 1} \frac{F_v^i \Delta v_{\mathrm{T}}}{F} \frac{\tau_i}{t}. \tag{26}$$

This shows that if there were only strong lines the force multiplier would be proportional to t^{-1} and therefore depend linearly on dv/dr. The force due to strong lines is proportional to the number of strong lines $N(t)$. This number depends on the magnitude of t and on the distribution function giving the number of lines as a function of line opacity $f(\eta_i)$.

Using extensive lists of line strengths to give $f(\eta_i)$, Abbott (1977) found that the number of optically thick lines as a function of t can be fitted by $N(t) = N_0 t^\gamma$ where $N_0 \approx 10^3$, $\gamma \sim 0.2$ for flow from O stars with $T \sim 30{,}000$ K, and $N_e \simeq 10^{11}$ cm^{-3}. Thus, a realistic mixture of line opacity introduces a further dependence on t. CAK represented the radiation force multiplier by a fit of the form,

$$M(t) = k\,t^{-\alpha}, \tag{27}$$

and α is determined by the line mixture, ($\alpha \approx \gamma - 1$), and k is the "force constant." If all lines are thick $\alpha = 1$, if all lines are thin $\alpha = 0$. The value of α is an indicator of the fraction of the force that is produced by thick lines, and at very large distances α must drop to zero, where the last strong line becomes optically thin. Incorporating G_L into the momentum equation,

$$v\frac{dv}{dr} + \frac{1}{\rho}\frac{dp}{dr} + \frac{GM}{r^2}(1-\Gamma) - \frac{GM}{r^2}\,\Gamma\,M(t) = 0, \tag{28}$$

and assuming the flow is isothermal with the sound speed $a = (RT)^{1/2}$, and substituting

$$u = -\frac{2\,GM(1-\Gamma)}{a^2 r} = \frac{v_{esc}^2(r)}{a^2}, \tag{29}$$

$$w = v^2/a^2, \tag{30}$$

$$w' = \frac{dw}{du}, \tag{31}$$

the momentum equation becomes

$$\left(1 - \frac{1}{w}\right)w' + \left(1 + \frac{4}{u}\right) - C(w')^\alpha = 0$$

$$= F(u,w,w'). \tag{32}$$

This is the basic equation of the CAK theory. The constant, C, depends on k and the mass loss rate,

$$C = k\left(\frac{\Gamma}{1-\Gamma}\right)^{1-\alpha}\left(\frac{L/c}{\dot{M}v_T}\right)^\alpha. \tag{33}$$

Graphical Solution of the Castor, Abbott, and Klein Equation

Equation (32) reduces to the more familiar Parker solar wind equation for $C = 0$. In that simpler case, the equation is linear in w' and has a singularity at the sonic point ($w = 1$). The X-type singularity and the topology of the solutions, $v(r)$, are discussed in Holzer & Axford (1970).

As it stands, Equation (32) with $C \neq 0$ is nonlinear, so for a given radial "position," u, and velocity, w, there is not, in general, a unique solution for w'. Therefore, there is no analytic solution for $w(u)$ [or $v(r)$].

It is possible, nevertheless, to develop considerable insight into the line driven wind theory by considering a graphical solution of Equation (32). For this purpose it is sufficient to consider k and α to be constants, independent of radius. Following Abbott (1977), Equation (32) can be solved for some given (u,w) as follows. Let

$$F = F_1 - F_2 = 0, \tag{34}$$

$$F_1 = (1 - 1/w)w' + \left(1 + \frac{4}{u}\right), \tag{35}$$

$$F_2 = C(w')^\alpha. \tag{36}$$

F_1 is linear with respect to w'. Its slope is determined by w and its intercept is determined by the radial variable u. Several examples of F_1 are shown in Figure 3. The function F_2 is also plotted in Figure 3, where it is represented by the dashed line. The intersection(s) of the two functions F_1 and F_2 is the solution(s) for w'. There can be no solution, one solution, or two solutions, and in the latter case they will be called the "leftward" and "rightward" solutions.

Line A shows the function F_1 at small r where the flow is subsonic; note there is one solution for w'. As we integrate $w'(u)$ outwardly from deep in the subsonic region, there continues to be a unique solution for

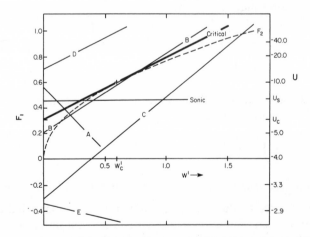

Figure 3 Graphical solution of the CAK equation of motion [Equation (32)]. The curved dashed line shows $F_2 (w')$. The solid straight lines indicate several forms of $F_1 (u, w, w')$. The slope of F_1 is determined by w, and the intercept of F_1 is determined by u. The scale on the right-hand side shows values of u corresponding to the intercepts on the F_1 scale. The graphical solution of the equation of motion for w' is the intersection of the line F_1 with the curve F_2.

w' until the sonic point is reached. As seen in the figure there may be two solutions for supersonic flow, and in order for w' to be continuous just beyond the sonic point it is necessary to choose the leftward root for w'. As the speed of the flow increases and the slope of F_1 increases it is possible for F_1 and F_2 to become tangential, i.e.

$$\frac{dF_1}{dw'} = \frac{dF_2}{dw'}. \tag{37}$$

For the solution to continue to higher speeds a regularity or critical point condition must also hold,

$$\frac{\partial F}{\partial u} + \frac{\partial F}{\partial w} w' = 0. \tag{38}$$

Beyond the critical point there are again two solutions (line B), but now the rightward root is the value of w' that will lead to large velocities at large r, where there is again only one solution for w' (line C). Also illustrated in Figure 3 are two cases for which there are no solutions, D and E, indicating that there are no solutions for supersonic flow at small r or for subsonic flow at very large r.

The graphical solution just described was carried out for the case $\alpha = 0.5$, $C = 0.76$, $v_{esc}(R_*) = 4.9$. These parameters are chosen for illustrative purposes, and not because they are realistic. The $\alpha = \frac{1}{2}$ case is especially simple because Equation (32) can be transformed into a quadratic equation for w'.

The plot of the numerical solution for v vs r is shown in Figure 4. The regions labeled A to E correspond to the lines with the same letters in Figure 3. Dashed lines in Figure 4 correspond to leftward intercept solutions for w', and solid lines correspond to rightward solutions. There are two solutions for every point in region B. The solutions have a cusp at the intersection with region D. Two cusps join only at the critical point and this allows a continuous transition from low speed to high speed flow.

Unlike the Parker solar wind equation there is no singularity at the sonic point, as can be seen directly from Equation (32). Abbott (1977) demonstrated that the critical point had several of the characteristics of the Parker sonic point. 1. The solution that passes the critical point is the only one that goes smoothly from low subsonic speeds near the surface of the star to high supercritical speeds far away from the star. This is analogous to Parker's unique solar "wind" solution that goes to supersonic speeds far from the star. 2. The presence of the radiation force modifies the speed of propagation of a disturbance, and the critical point occurs where the flow velocity equals the disturbance speed. Therefore in direct analogy with the Parker sonic point and with the "throat" of a de Laval nozzle, Abbott (1977) finds that "the critical point...(is) the point farthest downwind which is still able to communicate information to all parts of the flow."

Unlike the Parker case, there are no smooth transitions to decelerating (or breeze) flows. The radiation force is always in the outward direction and it therefore depends on the absolute value of w'. Thus, if the negative w' axis were shown in Figure 3, there would be a cusp in F_2 at $w' = 0$, and the only transitions to decelerating flow would have discontinuous velocity gradients. Only positive slopes are shown in Figure 4.

Now consider some prediction that can be made from the line driven theory. From the critical point conditions [Equations (34), (37), and (38)], several useful relations can be derived. Restricting attention to the CAK case for which $v_{esc} \gg a$ and for which the critical point, u_c, satisfies

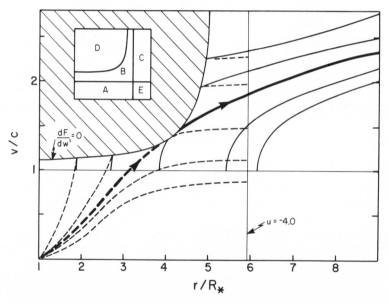

Figure 4 Topology of solutions to the equation of motion for line driven winds in a velocity versus radial distance plane with velocity given in units of the sound speed. The insert identifies regions in the plane, and the letters correspond to the graphical solution shown in Figure 3. In region *B* there are two solutions for the velocity gradient at any given v and r and these correspond to the case of two intersections of F_1 and F_2 in Figure 3. Dashed lines show results of integrations using the leftward solution for w' and solid lines show the results using the rightward solution. The only solution that goes smoothly from subsonic speeds close to the star to large velocities far from the star is shown by the heavy line.

$4/u_c \ll 1$, we get

$$C = \frac{1}{1-\alpha}\left(\frac{1-\alpha}{\alpha}\right)^{\alpha} \tag{39}$$

$$\dot{M} = \frac{Lk^{\frac{1}{\alpha}}}{v_T c}\left(\frac{\Gamma}{1-\Gamma}\right)\frac{1-\alpha}{\alpha}\,C^{-\frac{1}{\alpha}} \tag{40}$$

$$v^2 = v_0^2 + \frac{\alpha}{1-\alpha}\frac{2GM(1-\Gamma)}{R_*}\left(1 - \frac{R_*}{r}\right). \tag{41}$$

Using these equations, which were derived for the simple case with k and α constant, the theory can explain several of the empirical relations noted earlier. For $\alpha \approx 0.9$, Equations (33) and (39) indicate that $\dot{M} \propto L^{1.1}$, as derived empirically by Barlow & Cohen (1977) from their infrared survey. The terminal velocities are correctly predicted by Equation (40) to scale with $v_{\rm esc}$, in agreement with the empirical law, $v = 3v_{\rm esc}$, of Abbott (1978a).

More sophisticated versions of the theory recently developed by Abbott (1977) and Castor (1978) provide answers to several questions raised about the original CAK theory. Abbott (1977, 1978c) compiled extensive lists of lines and found that there are enough strong lines available to intercept sufficient flux to account for the observed mass loss rates. Furthermore, Abbott (1978b) derived the extent of a region in the H-R diagram in which the line acceleration in a static atmosphere exceeds gravity at large enough optical depth that the atmosphere must expand. This is a region in which the winds can be "initiated" solely by the radiation mechanism, and it corresponds well to the zone in the H-R diagram where outflows are actually observed. The effect of rotation would be to lower the boundary of the self-initiated wind region so that it includes the Be stars. This indicates that Be star winds should be emanating from a limited equatorial region. Castor (1978) comments on the effects of rotation on line driven winds of the more luminous stars and suggests that the mass loss rate may be the same at the polar and equatorial regions but the rate of increase of velocity is slower in the equatorial regions, leading to a density enhancement there.

While it is now, perhaps, clear that winds can be produced fully by radiative processes, whether or not they actually are is still under contention. Thomas (1978) criticizes the radiatively driven wind models for assuming laminar, steady state flow with a smooth transition to supersonic speeds. He contends that the mass loss is imposed by subphotospheric motions, and as a result, shocks will occur in the flow.

The CAK theory explains the observed mass loss rates and dependence on L and $T_{\rm eff}$, as well as the magnitude of the terminal velocity. These successes indicate that a major part of the mass loss mechanism has been identified. However, the presence of OVI in the spectra and the variability of the stellar winds (York et al. 1977, Stalio & Upson 1979) indicate that the picture is far from complete.

Semiempirical Models

The purely radiatively driven wind theory does not consider the possibility that mechanical energy deposition could affect the temperature and the dynamics of the flow. Several semiempirical models have been developed to derive the actual run of temperature in the wind of O stars and B supergiants. Cassinelli, Castor & Lamers (1978) review three quite different models that can explain the high stages of ionization observed in the UV spectra. The models can be used to make other predictions and they can then be refined or rejected on the basis of observational tests.

Lamers & Morton (1976) found that the ionization seen in the ultraviolet spectrum of ζ Pup, ranging from C^{+2} to O^{+5}, is the same as that of an optically thin plasma at $T = 2 \times 10^5$ K. They postulated that the wind has this elevated or "warm" temperature throughout the extended region in which OVI must exist. Lamers & Rogerson (1978) explained the OVI in τ Sco (BOV) by adopting the warm wind model. Lamers & Snow (1978) extended the elevated temperature idea to explain ionization anomalies in B supergiants. Castor, Abbott & Klein (1976) criticized the warm wind model because it ignored the diffuse radiation produced in the wind itself. Cassinelli, Olson & Stalio (1978) criticized it because the model could not explain the strong Hα emission from the winds of Of and OB supergiants. Furthermore, it is difficult to maintain an extended region at a temperature of 2×10^5 K because radiative cooling is especially effective at such temperatures (Cox & Tucker 1969). If the warm temperatures are to be maintained by the deposition of wave energy that emanates from the star, the luminosity of that mechanical energy would have to be about 5% the radiative luminosity of the star, and would have to be deposited over a spatially extended region of several stellar radii.

Castor (1978) found that the ionization could also be explained in a wind at a lower temperature of about 6×10^4 K, if due account was made for the diffuse radiation field in the optically thick winds. This lower temperature satisfies the Hα constraint and requires less energy deposition than a model with $T_e = 2 \times 10^5$ K. However, it does require *some* mechanical energy deposition, and this destroys the aesthetic appeal of the purely radiatively driven wind theory for which no source of energy or momentum deposition is required, other than that of the radiation field of a mechanically quiet star.

An alternative to the elevated wind temperature models is a hybrid model of the type first proposed by Hearn (1975a). Cassinelli, Olson &

Stalio (1978) and Cassinelli & Olson (1979) derived observational constraints on a hybrid corona-plus-cool-wind model from analyses of Hα
and ultraviolet profiles and from X-ray upper limits of ζ Pup from the
ANS satellite. In this model the winds are cool ($T_e \approx 0.8\ T_{eff}$) as expected
from radiative equilibrium, but there is a thin ($< 0.1\ R_*$) hot ($T_e \sim 5 \times 10^6$
K) corona at the base of the flow. The high ion stages are produced by
the Auger mechanism whereby two electrons are removed from C, N, and
O following the K shell absorption of X rays. The mechanical energy
deposition required to heat a corona that can produce enough X rays to
explain the ionization anomalies is small $\sim 10^{-4}\ L_*$. The model predicts
successfully that OVI should persist to B0.5Ia, NV to B2Ia, CIV and Si
IV to B8Ia, as is observed. The model also predicts that there will be an
emergent 2-keV X-ray flux that is large enough to be detected by the
HEAO-2 satellite just recently launched.

Thus, there are at least three very different temperature structures
compatible with the observed ionization anomalies, indicating that the
ultraviolet resonance lines alone are not accurate indicators of the temperature structure of the winds.

The infrared continuum spectrum, Hα line profiles, and He I lines
provide more direct information concerning the temperature of the winds.
As discussed by Castor (1978), these depend on both the temperature
and the velocity distributions in the wind. But fortunately, the emissivities
of the IR free-free continuum and the optical line strengths have different
functional dependences on T and v, so it should be possible to derive the
separated distributions for $T(r)$ and $v(r)$. Cassinelli & Hartmann (1977)
and Castor (1978) discuss the effects of temperature rises on the infrared
continua of hot stars for a given run of velocity. Cassinelli, Olson &
Stalio (1978) and Van Blerkom (1978) studied the dependence of Hα profiles
on the velocity structure in winds for a given temperature distribution.
However, thus far no complete study of both the IR continuum and Hα
profiles has been carried out. None of the three models that explain the
ionization anomalies has yet been ruled out. Only after this problem has
been settled can we focus attention on the actual source of the mechanical
heating.

To summarize, the massive winds of hot stars can be initiated and driven
out by radiative momentum transfer alone (Abbott 1978b). However, the
discovery of OVI and other high ionization states has led to the realization
that there is a source either of mechanical wave energy in the subphotospheric region, or of instability in the flow, that leads to high temperatures.
Observations by HEAO-2 and further simultaneous studies of infrared and
Hα emission are needed to derive the actual temperature structure in the
winds.

4 MASS LOSS FROM EVOLVED LATE-TYPE STARS

Observational Data on Mass Loss

Because of the development of new instrumentation, the number of observational studies of mass loss from cool stars has grown rapidly in the past few years. Surveys of the optical, infrared, and radio spectra are now sufficiently broad that general trends of mass loss rates and wind structures in the H-R diagram are becoming discernable. Some of the particularly useful results are summarized here; more complete reviews of optical, infrared, and radio studies are given by Reimers (1975, 1978), Merrill (1978), and Moran (1976) respectively.

OPTICAL SPECTRA   Very high resolution double pass echelle spectrographs were recently used to observe the cores of strong resonance lines and of lines from low excitation levels ($\lesssim 1$ eV) of metals (Bernat & Lambert 1976, Sanner 1976). Superimposed on the photospheric lines are often seen small P Cygni profiles, which, as usual, can be interpreted as having formed in the flow expanding from the star. An example is shown in Figure 5 for Mn I lines in α Ori (M2 Iab). The P Cygni profiles can be analyzed using moving atmosphere radiation transfer techniques to derive column densities, wind velocities, and mass loss rate estimates. Unfortunately,

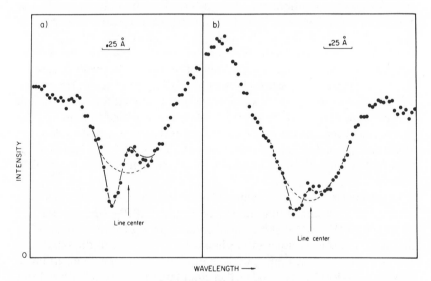

Figure 5 Circumstellar MnI λ4030.8-Å and λ4033.1-Å lines in α Ori, which have P Cygni profiles superimposed upon the background photospheric lines (Bernat & Lambert 1976).

the mass loss rates derived by different authors often disagree by one or two orders of magnitude (Sanner 1976, Bernat 1977, Hagen 1978). This disagreement arises because of the difficulty disentangling the P Cygni lines from the photospheric line, because of the uncertainties in the ionization fraction of the ion producing the line, and, perhaps primarily, because of disagreements as to the "inner boundary" of the line formation region and hence as to the total column mass density in the flow. Reimers (1978) expects the problem to remain unresolved until more UV observations become available and reveal many more resonance lines.

VISIBLE COMPANIONS   Another effective way of studying the structure of at least a few cool stars is through analysis of the lines seen superimposed on the spectrum of a near, visible companion. Notable examples of M giants with fainter, hotter companions are α Her (M5 II-III + GO III) and α Sco (M1.5 Iab + B2V). In fact, through a study of α Her Deutsch (1956) was first able to demonstrate that the low velocity flow from M giants actually represents a mass loss. He found that the matter that gave rise to the narrow displaced absorption lines in the spectrum of α^1 Her extended at least beyond the 180 stellar radii to the G companion, and since at that distance the expansion velocity of 10 km/sec was greater than the escape speed, the matter was lost to the interstellar medium. Recently, van der Hucht, Bernat & Kondo (1979) obtained the spectrum of α Sco via the balloon borne ultraviolet spectrometer (BUSS). As additional stages of ionization can be seen in the ultraviolet spectra, these authors were able to derive an improved model for the wind from the M star. They deduced that the entire H II region of the B2 V companion is contained within the cool wind of the M supergiant, 500 AU away, thus illustrating the very extended and massive nature of the outflow. The mass loss rate was derived to be 7×10^{-6} $M_\odot$/yr; more than an order of magnitude larger than the previous estimates by Sanner (1976) and Kudritzki & Reimers (1978).

INFRARED EXCESSES   Extensive surveys of the infrared spectra of luminous cool stars have been carried out since the discovery of the 10-μm silicate feature in μ Cep (M2Ia) and α Ori (M1Ib), and in several Mira variables by Woolf & Ney (1969). The current status is reviewed by Merrill (1978). The strength of the silicate bump can be fitted with theoretical models to derive the column density of the dust and to estimate mass loss rates (Gehrz & Woolf 1971, Hagen 1978). Because there are no sharp spectral features it is impossible to derive, through Doppler shifts, detailed information concerning the spatial distribution of the grains. It is usually assumed that the dust coexists with the gas that produces the P Cygni features discussed above. Of particular interest to wind theorists is the height above the photosphere at which the grains can form, the condensation radius. It is assumed to be the height at which the grain effective temperature is about 1000 K (Weymann 1978). This radius is very uncertain because it depends on unknown grain properties, such as the absorption coefficient in the near infrared (1–5 μ). Clean silicate grains with a low near-infrared opacity could form rather close to the star. However, observations of the near-IR excess and 10-μm bump indicate the presence of grains having a larger near-IR opacity, as do "dirty grains" (Jones & Merrill 1976) or silicate grains with near-IR band opacity (Hagen 1978). Figure 6 shows the regions in the H-R diagram where there is evidence for the presence of dust from infrared excesses. Dust is detected in all M supergiants and in M giants later than about M3. Dust is also detected around a few F, G, and K supergiants. However, it is unlikely that the temperatures of the winds around these stars fall below the grain condensation temperature. Stothers (1975) suggests that the infrared excesses in these stars arise in a fossil dust shell formed during a prior M supergiant phase of evolution.

MASER EMISSION   Strong maser lines of OH, H_2O, and SiO appear in the radio spectra of many Mira variables and in a few peculiar M supergiants such as VY CMa (Wilson & Barret 1972, Moran 1976, Winnberg 1978). Since the masers are pumped by infrared radiation from dust, the

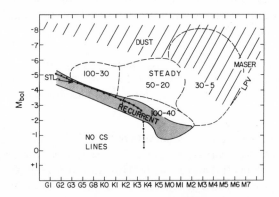

Figure 6   Mass loss domains for late type giants in the H-R diagram (Reimers 1977). At the bottom left-hand side there is no spectroscopic evidence for mass loss from Ca II K circumstellar lines. In the dotted strip, the ejection is recurrent: sometimes shortward shifted lines are seen and sometimes they are not. Above that region the flows are steady. The dashed lines mark various terminal velocity regimes (Reimers 1977, Weymann 1978). Regions in which dust is detected from M stars and from G and K supergiants are indicated. Maser emission is detected from Mira variables and from a few very luminous late M supergiants.

fraction of stars that are maser sources increases with decreasing effective temperature and increasing dust abundance. A good correlation is seen between variable infrared flux and OH maser emission (Moran 1976). Typically the emission due to one transition occurs in two velocity groups, which are displaced by equal amounts to both sides of the stellar velocity. This displacement can be explained by assuming that the maser emission is produced in the extended constant flow velocity regions in the stellar wind, for there is a large correlation path both in front of and behind the star as seen by the observer (Goldreich & Scoville 1976). The displacement of the lines gives the velocity flow. However, since the emission to each side of line center shows multiple velocity groups, the flow is probably clumpy. Furthermore, long baseline interferometer observations show a complicated surface spatial pattern (Reid, Muhlman & Moran 1977). These observations indicate that the simple spherically symmetric, uniform flow picture may be too naive. The major controversy about the winds of cool stars concerns the cause of the flow from the K and early M stars. Unfortunately, these stars do not show maser emission and thus the maser observations are of little help in resolving that problem.

CHROMOSPHERIC EMISSION   There is observational evidence for chromospheres in G, K, and early M giants and supergiants from the presence of emission in the cores of the Ca II, H, and K lines (Reimers 1975) and from He I, $\lambda 10830$ emission in the G and K stars (Zirin 1976). Reimers (1977) finds that stars above and to the right of a sloping line in the H-R diagram have expansion in the chromospheric layers, with speeds as high as one half the terminal speeds. This is seen as a blueward shift of the K_2 emission core and in the K_3 absorption component. Evidence for coronae in a few lower luminosity stars (giants) has been found through UV emission lines of high stages of ionization, such as OV and OVI (Gerola et al. 1974, Dupree 1975, Linsky & Haisch 1979).

Domains of Mass Loss in H-R Diagram

Through a careful analysis of the observational data Reimers (1977) identified several general properties of mass loss from cool stars that can serve as a guide to stellar wind theorists. The range in mass loss rates is roughly 10^{-8} to 10^{-5} $M_\odot/\text{yr}$ (Figure 1), and mass loss rates tend to increase toward the upper right-hand corner of the H-R diagram. Reimers (1978) gives empirical fit to that trend

$$\dot{M} \simeq 1 \times 10^{-13} \, L/gR, \qquad (42)$$

where L, g, and R are the stellar luminosity, surface gravity, and radius in solar units. This relationship can be rewritten as $\frac{1}{2}\dot{M}v_{\text{esc}}^2 = 10^{-13} \, L$, and implies that a constant fraction of the stellar luminosity is being used to provide the necessary energy for escape of the expanding material.

While essentially all M giants and supergiants show evidence for mass loss through the displaced narrow absorption in resonance line cores, the same is not true for the G and K stars. Reimers (1977) finds that these stars fall into two major mass loss domains, depending on whether the displaced Ca K absorption lines are either visible in the spectrum at all times or never visible. The separation between these two is surprisingly sharp, with a third minor group between them in which the lines are sometimes seen and sometimes not. The mass loss domains are shown in Figure 6. Also indicated in the figure is the not particularly uniform trend of expansion velocities derived from the displacement of the circumstellar lines. For the G and early K stars, it is not at all obvious whether the displacements should be interpreted as terminal velocities of the winds (Reimers 1977) or as the velocity of a shell at the wind interstellar medium interface (Mullan 1978). Assuming the velocities are, in fact, wind terminal speeds, Reimers finds a good correlation with the escape speeds of the stars. The velocities increase from the M supergiants (10 km/sec) through the M giants (25 km/sec) and late K supergiants and K giants (20 and 75 km/sec) to the sun with the relation

$$v_{wind} \propto v_{esc}^2, \tag{43}$$

and not linearly as might be expected. As both empirical relations [Equations (42) and (43)] extrapolate moderately well to fit the solar wind ($\dot{M}_\odot = 2 \times 10^{-14} M_\odot/yr$), Reimers suggests that the mass loss mechanism may be similar. However, Weymann (1978) notes several difficulties associated with the application of coronal theory to the M supergiants. In the next sections theoretical discussions are presented for a coronal model of G and K stars, and then for a model for dust driven mass loss of M stars.

Coronal Winds of G and K Stars

A coronal model, somewhat like that for the solar wind, is appropriate for stars not luminous enough for radiation pressure effects to be important or too hot for grain formation.

The nature of the coronal heating mechanism is not known and hence, as Mullan (1978) points out, the discussion of coronal properties must necessarily work within a framework in which neither the source nor the mode of coronal heating need be specified. Hearn (1975b) proposes such a framework in his "minimum flux coronal model."

Hearn considers three major sources of energy loss from a corona and expresses each as a flux away from the base of the corona, where the temperature and pressure are T_0 and P_0. First, there is a conductive flux, F_c, downward into the chromosphere-corona transition region. He assumes, on the basis of solar structure, that the transition region is at a constant pressure $P = P_0$, and he derives the conductive flux by equating the divergence of F_c to the radiative losses in the transition zone. Second, there is radiative energy lost from the coronal material itself, F_R. This depends on the temperature of the corona and on the density distribution by way of the coronal emission measure (N_e^2 Vol). To derive the emission measure, he assumes the coronae are isothermal and the relative velocity and density structure are known from the Parker solar wind theory (Parker 1958, Hundhausen 1972). Finally, there is the energy flux due to the wind itself, F_w, which again comes from the Parker theory. In this way all of the fluxes are expressed in terms of the two coronal parameters T_0 and P_0. In addition energy balance requires that the sum of the losses $F = F_c + F_R + F_w$, be equal to the energy input, which presumably is generated in some way in subphotospheric regions.

Hearn notes that while the total loss flux, F, is a monotonic function of P_0, it is not monotonic in T_0. For a given base pressure, there is a temperature, $T_{min}(P_0)$, for which the loss flux is a minimum. The minimum occurs because F_R decreases with increasing temperature, while F_w and F_c both increase with temperature. Hearn argues that the corona adjusts itself to the structure that minimizes the loss flux, so that $T_0 = T_{min}$. Hence the coronal structure is completely determined by the magnitude of the mechanical flux that enters from below. He identified three cases where a corona could exist as the input mechanical flux is increased. Class I coronal models are those in which the energy loss is dominated by conduction and radiative losses, with mass loss being relatively unimportant in the energy balance; he offers the solar wind as an example. Class II coronae are those in which most of the energy is lost by radiation and mass loss. Class III are coronae dominated by mass loss only; in such cases the sonic point tends to lie very close to the star.

Although Hearn originally developed the theory to explain the initiation of mass loss in OB supergiants, it has subsequently been used primarily for cooler less luminous objects. Haisch & Linsky (1976) modified the equations somewhat to better account for radiative losses in the supersonic wind and then applied the theory to a study of OVI emission seen in Capella. Mullan (1976) applied it to a wide range of giants and dwarf stars. Hearn & Mewe (1976) and Muchmore & Böhm (1978) used it to interpret X-ray data from the coronae of white dwarfs.

The theory has received criticism on several points, as summarized by Vaiana & Rosner (1978). It has not yet been proven that the minimal loss configuration is in fact required for stability as Hearn suggested. It is not entirely obvious that the structure is independent of the spatial distribution of the mechanical flux deposition (Endler, Hammer & Ulmschneider 1978). The assumption that the transition region is at a constant pressure is not valid in general (Antiarchos & Underwood 1978), and furthermore the structure of this lower region may be determined by closed magnetic structures, as in the sun (Vaiana & Rosner 1978). Mullan (1978) presents several arguments in defense of the theory; for example, stars that have large mass loss rates probably have a sufficiently large kinetic energy density to avoid significant closure into magnetic structures. It is in the treatment of the transition region that the theory is weakest. The energy deposition required to heat an element of matter to provide for the work done as it moves up through the transition region (enthalpy flux) is not accounted for explicitly. Mullan argues that this problem can be circumvented by considering in the theory only the mechanical flux that actually reaches the corona and by assuming that whatever the mechanical flux is, it must dissipate within the transition region an amount equal to the enthalpy flux generated there. Hearn (1979) is developing a more complete theory and is preparing a rebuttal to the criticism of the original paper.

In spite of the difficulties associated with the model, it is worth considering further, both because of an interesting application to cool stars by Mullan (1978) and because possible improvement of the theory can be seen.

As discussed before, a corona can have one of the three structures depending on stellar parameters such as escape speed and input mechanical flux. Mullan finds that an evolving star should shift from one of these classes to another as it moves toward the red giant stage, and eventually may enter the mass loss dominated category, Class III. Roberts & Soward (1972) pointed out that in stars with hot coronae and low surface gravity, the sonic point could move very close to the surface of the star and the wind would be supersonic essentially throughout the corona. Durney (1973) suggested that the transition to such a fully supersonic structure should be observable through a large increase in mass loss rate, and he derived from theoretical considerations a locus on the H-R diagram above which winds should have this structure and enhanced mass loss. Mullan (1978) calls the winds with negligibly thin subsonic regions and essentially completely supersonic flow "supersonic winds," in contrast with "transonic winds," which have a thicker subsonic region. He calls the locus in the H-R diagram the "Supersonic Transition Locus" (STL), above which the winds are "supersonic" and below which the winds are "transonic." In Durney's original derivation, the STL was approximately a horizontal line at $M_{\text{bol}} = -3$, and such a locus does not agree particularly well with any observational transition locus such as that of Reimers (1977) discussed before.

Mullan (1978) used a semiempirical approach in his derivation of the transition locus. The coronal base pressures, P_0, are determined from observational data, and the coronal temperatures, T_0, are determined from Hearn's minimum flux theory.

Kelch et al. (1978) derived the pressure at the top of the chromospheres, P_{TC}, in a large number of stars from chromospheric emission lines. They found that over a range of three orders of magnitude in surface gravity g, P_{TC} could be represented by a power law of the form $P_{TC} \propto g^{1/2}$. Mullan (1978) assumed that the pressure at the base of the corona is some constant fraction ($\sim 1/6$) of P_{TC}. With P_0 known, the Hearn theory allowed him to calculate T_0, the radius of the sonic point, R_S, and the mechanical flux that enters the corona. He then followed the evolutionary track of a star of mass M, in the H-R diagram, using data for T_{eff} and g from stellar evolution models (Iben 1974). As a star evolves, its radius grows faster than the radial distance to the sonic point. When the two radii, R_S and R_*, are equal, a point on the supersonic transition locus has been found. Other points on the STL are found by considering evolutionary tracks of stars with different masses.

While the formal mathematics of this approach is certainly obvious enough, the physical significance of the STL is not so clear. Mullan suggests the following scenario. Before the onset of the supersonic wind, the expansion is fed by purely coronal gas. When the sonic point overlaps the altitude to which chromospheric spicules can penetrate, there is an essentially discontinuous transition to a state in which the expansion is fed by chromospheric material. Typical coronal temperatures derived along the STL are about 4×10^5 K, and chromospheric temperatures are 8000 K. Thus there can be a jump in the density at the sonic point by a factor of 50 when the throat of the nozzle gains access to chromospheric material. Hence, there will also be a jump in mass loss rate by a factor of 50. This is quantitative restatement of Durney's speculation concerning the effect of the onset of supersonic winds. Mullan offers the following observational support for the idea. (a) The STL is found to match, reasonably well, Reimer's boundary of the large mass loss regime (Figure 6). (b) Stencel (1978) finds evidence for expansion of chromospheric layers beyond the STL, but negligible expansion prior to it. (c) Arcturus lies near the STL and sometimes shows phases of large mass loss rates, enhanced by an order of magnitude or more from the low mass loss rate phase.

Additional support for the presence of a sharp transition locus was recently discovered by Linsky & Haisch (1979) from IUE observations of transition region lines. High ionization stages, such as NV, CIV are seen in the spectra of stars below the STL, but these are absent above the STL. This perhaps indicates that the size of the transition region is much smaller in the mass loss dominated case.

Mullan derives an expression for the dependence of mass loss rate on stellar parameters (in solar units) for stars beyond the STL.

$$\dot{M} = 1.6 \times 10^{-9} M\sqrt{R} \qquad (44)$$

in solar masses per year. This expression shows qualitative agreement with what could be inferred from Figure 1, i.e. increasing mass loss to the upper right-hand corner, but it is functionally quite different from Reimer's expression [Equation (42)]. It also does not extrapolate back to fit the Sun, first because of the jump in density at the STL, and then because in the Hearn theory the mass loss rate decreases rapidly as the sonic point moves outward.

Since the Mullan explanation of the high mass loss regime depends so heavily on chromospheric spicules, it seems that the Hearn and Mullan theories could be greatly improved by incorporating the model of the solar transition region of Pneumann & Kopp (1978). These latter authors discuss the observational evidence for steady downflows in the solar transition region that have a mass flux comparable to the estimated upward mass flux in spicules. Thus the solar wind accepts only a small fraction of the upward spicule mass flux. This fraction presumably increases toward unity as the sonic point moves toward the transition region when the star evolves.

With the abundant observational data now available, the subject of coronal mass loss from cool evolved stars could profit from increased involvement in these studies by solar wind theorists.

Dust Driven Winds

After Woolf & Ney (1969) presented evidence for the presence of dust in the envelopes of red giant stars, it became plausible to attribute the expansion to radiation forces on the dust. Gilman (1972) showed that momentum could be transferred from the radiation field to the gas by grain-gas collisions and subsequent gas-gas collisions. The mechanism has now been investigated in several papers (Salpeter 1974, Kwok 1975, Jones & Merrill 1976, Goldreich & Scoville 1976, Lucy 1976, Menietti & Fix 1978). There are several new features to this type of radiatively driven wind theory: (a) grain condensation and growth, (b) radiation force on grains and the coupling of grain and gas momenta, and (c) the sputtering of grains due to a high grain drift velocity.

In order for mass loss rates of around 10^{-6} $M_{\odot}$/yr to be produced, grain condensation must commence at a height in the stellar atmosphere at which the density is still relatively large. For stars of spectral class M3 and later, dielectrics should be able to condense sufficiently close to the star, but for the early M and the K supergiants, the mechanism could be effective only if the atmospheric scale height is increased, say by turbulence (Kwok 1975, Weymann 1978).

After grains start to form and grow, they experience a very large radiative force and are assumed to be accelerated to the (local) terminal drift speed. Under these conditions, the momentum equation for dust simply equates the outward radiative force to the inward drag owing to collisions with gas particles. The momentum equation for the gas has the usual gas pressure gradient, gravity, an almost negligible radiation pressure term, and the outward drag by the dust grains. The last term can be replaced by the radiation force on grains, and thus the total radiation force, say Γ_R, of Equation (15), involves the sum of the grain and gas opacities. As before, the flow must obey the Marlborough & Roy (1970) constraint that Γ_R be less than unity at and below the sonic point, and be greater than unity beyond the sonic point.

In Kwok's theory the total opacity increases very rapidly near the dust condensation radius and the flow is accelerated to supersonic speeds. The drift velocity, v_D, of the grains relative to the gas increases as the matter flows out; $v_D \propto \rho^{-1/2}$. When v_D reaches speeds near 20 km/sec, the collisions with the gas particles are energetic enough for grain destruction to occur by way of sputtering. The acceleration to higher speeds ceases and the gas attains the relatively low terminal velocities of 5–50 km/sec detected in the core P Cygni lines and Maser line separations. If sputtering did not occur the flows would reach speeds of about 100 km/sec. These high speeds are, of course, not acceptable from an observational point of view.

Weymann (1978) and Hagen (1978) discuss several observational objections to the radiatively driven dust mechanism for explaining the flows from M stars. (a) If the dust is in the form of dirty silicates or more generally if the grains have significant near-IR absorptive opacity (Hagen 1978), then they cannot form and survive close to the star and push the mass flow off. Kwok (1978) counters this objection with the suggestion that clean dielectric grains form near the base of the flow, but they pick up impurities as they flow out. (b) Hagen (1978) argues that if the flows are driven by dust, there should be a correlation between the dust column density as measured by the 10-μ bump and mass loss rate, and she finds none. However, since the mass loss rate estimates continue to disagree by an order of magnitude or more and the nature of the near-IR opacity of grains is so uncertain, the lack of a good correlation is no surprise.

Furthermore, the degree of destruction of grains by sputtering should be considered in such an analysis. (*c*) Reimers (1978) finds evidence for expansion in chromospheric regions, with velocities as high as half the terminal speed, and since grains are not likely to exist in these regions, Hagen argues that dust cannot be responsible for the outflow. However, the dust does not have to "push" all the gas out. If a sufficiently large gas pressure gradient is produced as a result of the lifting of outer layers by way of the radiation process, expansion of the deeper layers will occur. Whether or not this is sufficient has yet to be demonstrated. (*d*) A particularly serious criticism of the model is that low velocity flows with apparently small mass loss rates have been detected in a few metal-poor population II stars, around which dust grains will not form (Hagen 1978). This indicates that the flows would occur even in absence of dust. If that is the case, then it would perhaps be best to consider the process of transferring momentum to the flow via dust opacity as a mass loss "enhancement" mechanism. The eventual solution of this problem is going to require a much better understanding of the transition from the thick chromospheric layers to the cool expanding envelope surrounding the stars.

ACKNOWLEDGMENTS

I am grateful to D. Abbott for many useful discussions and to W. Waldron for carrying through the graphical solution of the line driven wind equation of motion and for assistance in the preparation of the review. I am indebted to D. Abbott, J. Castor, and J. Fix for helpful comments on the manuscript, and to many authors for preprints of their research. This work was supported in part by the National Science Foundation under grant AST 76-15448 through the University of Wisconsin.

Literature Cited

Abbott, D. C. 1977. PhD thesis, Univ. Colorado
Abbott, D. C. 1978a. *Ap. J.* 225:893–901
Abbott, D. C. 1978b. In *Proc. IAU Symp. No. 83, Mass Loss and Evolution of O-Type Stars.* In press
Abbott, D. C. 1978c. *J. Phys. B.* 11:3479–97
Antiarchos, S. K., Underwood, J. H. 1978. *Astron. Astrophys.* 68:L19–22
Barlow, M. J., Cohen, M. 1977. *Ap. J.* 213: 737–55
Belcher, J. W., MacGregor, K. B. 1976. *Ap. J.* 210:498–507
Bernat, A. P. 1977. *Ap. J.* 213:756–66
Bernat, A. P., Lambert, D. L. 1976. *Ap. J.* 204:830–37
Brandt, J. C. 1970. *Introduction to the Solar*

Wind. San Francisco: Freeman
Bruhweiler, F. C., Morgan, T. H., van der Hucht, K. A. 1978. *Ap. J. Lett.* 225:L71–74
Cannon, C. J., Thomas, R. N. 1977. *Ap. J.* 211:910–25
Cassinelli, J. P. 1979. In *Proc. IAU Symp. No. 83, Mass Loss and Evolution of O-Type Stars.* In press
Cassinelli, J. P., Castor, J. I. 1973. *Ap. J.* 179:189–207
Cassinelli, J. P., Castor, J. I., Lamers, H. J. G. L. M. 1978. *Publ. Astron. Soc. Pac.* 90:496–505
Cassinelli, J. P., Hartmann, L. 1975. *Ap. J.* 202:718–32
Cassinelli, J. P., Hartmann, L. 1977. *Ap. J.* 212:488–93

Cassinelli, J. P., Olson, G. L. 1979. *Ap. J.* 229: 304–17

Cassinelli, J. P., Olson, G. L., Stalio, R. 1978. *Ap. J.* 220: 573–81

Castor, J. I. 1970. *MNRAS* 149: 111–27

Castor, J. I. 1972. *Ap. J.* 178: 779–92

Castor, J. I. 1974. *Ap. J.* 189: 273–83

Castor, J. I. 1978. In *Proc. IAU Symp. No. 83, Mass Loss and Evolution of O-Type Stars.* In press

Castor, J. I., Lamers, H. J. G. L. M. 1979. *Ap. J. Suppl.* 39: 481–512

Castor, J. I., Abbott, D. C., Klein, R. I. 1975. *Ap. J.* 195: 157–74

Castor, J. I., Abbott, D. C., Klein, R. I. 1976. In *Physique des Mouvements dans les Atmosphères Stellaires*, ed. R. Cayrel, M. Steinberg. Paris: Cent. Nat. Rech. Sci.

Chiosi, C., Nasi, E., Sreenivasan, S. R. 1978. *Astron. Astrophys.* 63: 103–24

Conti, P. S. 1978. *Ann. Rev. Astron. Astrophys.* 16: 371–92

Cox, D. P., Tucker, W. H. 1969. *Ap. J.* 157: 1157–67

Deutsch, A. J. 1956. *Ap. J.* 125: 210–27

Dupree, A. K. 1975. *Ap. J. Lett.* 200: L27–32

Durney, B. R. 1973. In *Stellar Chromospheres*, ed. S. D. Jordan, E. H. Avrett, pp. 282–85. NASA Sp-317

Endler, F., Hammer, R., Ulmschneider, P. 1978. *Astron. Astrophys.* Submitted for publication

Garrison, L. M. 1978. *Ap. J.* 224: 535–45

Gehrz, R. D., Woolf, N. J. 1971. *Ap. J.* 165: 285–94

Gerola, H., Linsky, J. L., Shine, R., McClintock, W., Henry, R. C., Moos, H. W. 1974. *Ap. J. Lett.* 193: L107–10

Gilman, R. C. 1972. *Ap. J.* 1972. 178: 423–26

Goldreich, P., Scoville, N. 1976. *Ap. J.* 205: 144–54

Hagen, W. 1978. *Ap. J. Suppl.* 38: 1–18

Haisch, B. M., Linsky, J. L. 1976. *Ap. J. Lett.* 205: L39–42

Heap, S. R. 1978. In preparation

Hearn, A. G. 1975a. *Astron. Astrophys.* 40: 277–83

Hearn, A. G. 1975b. *Astron. Astrophys.* 40: 355–64

Hearn, A. G. 1979. Private communication

Hearn, A. G., Mewe, R. 1976. *Astron. Astrophys.* 50: 319–21

Hollweg, J. V. 1978. *Rev. Geophys. Space Sci.* 16: 689–720

Holzer, T. E. 1977. *J. Geophys. Res.* 82: 23–35.

Holzer, T. E. 1978. In *Solar System Plasma Physics: A Twentieth Anniversary Overview*, ed. C. F. Kennel, L. J. Lanzerotti, E. N. Parker. Amsterdam: North-Holland. In press

Holzer, T. E., Axford, W. I. 1970. *Ann. Rev. Astron. Astrophys.* 8: 31–60

Hundhausen, A. 1972. *Coronal Expansion and Solar Wind.* New York: Springer

Hutchings, J. B. 1978. *Earth Extraterr. Sci.* 3: 123–34

Iben, I. 1974. *Ann. Rev. Astron. Astrophys.* 12: 215–56

Jones, T. W., Merrill, K. M. 1976. *Ap. J.* 209: 509–24

Kelch, W. L., Linsky, J. L., Basri, G. S., Chiu, H. Y., Chang, S. W., Maran, S. P., Furenlid, I. 1978. *Ap. J.* 220: 962–79

Klein, R. I., Castor, J. I. 1978. *Ap. J.* 220: 902–23

Kraft, R. 1967. *Ap. J.* 150: 551–70

Kudritzki, R. P., Reimers, D. 1978. *Astron. Astrophys.* 70: 227–76

Kuhi, L. V. 1964. *Ap. J.* 140: 1409–33

Kwok, S. 1975. *Ap. J.* 198: 583–91

Kwok, S. 1978. Private communication

Lamers, H. J. G. L. M., Morton, D. C. 1976. *Ap. J. Suppl.* 32: 715–36

Lamers, H. J. G. L. M., Rogerson, J. B. 1978. *Astron. Astrophys.* 66: 417–30

Lamers, H. J. G. L. M., Snow, T. P. 1978. *Ap. J.* 219: 504–14

Linsky, J. L., Haisch, B. M. 1979. *Ap. J. Lett.* 229: L33–38

Lucy, L. B. 1971. *Ap. J.* 163: 95–110

Lucy, L. B. 1976. *Ap. J.* 205: 482–91

Lucy, L. B., Solomon, P. M. 1970. *Ap. J.* 159: 879–93

Marlborough, J., Roy, J. R. 1970. *Ap. J.* 160: 221–24

Marlborough, J., Zamir, M. 1975. *Ap. J.* 195: 145–56

Menietti, J. D., Fix, J. D. 1978. *Ap. J.* 224: 961–68

Merrill, K. M. 1978. In *Proc. IAU Colloq. No. 42, The Interaction of Variable Stars with their Environment*, ed. R. Kippenhahn, J. Rahe, W. Strohmeier. *Publ. Bamberg Obs.* Bd. xi, No. 121, pp. 446–94

Mihalas, D. 1978. *Stellar Atmospheres*, pp. 511–66, San Francisco: Freeman. 2nd ed.

Moran, J. M. 1976. In *Frontiers of Astrophysics*, ed. E. H. Avrett, pp. 385–437. Cambridge: Harvard Univ. Press

Morton, D. C. 1967. *Ap. J.* 147: 1017–24

Morton, D. C. 1976. *Ap. J.* 203: 386–98

Morton, D. C., Jenkins, E. B., Brooks, N. 1969. *Ap. J.* 155: 875–85

Morton, D. C., Wright, A. E. 1978. *MNRAS* 182: 47P–51P

Muchmore, D. O., Böhm, K. H. 1978. *Astron. Astrophys.* 69: 113–17

Mullan, D. J. 1976. *Ap. J.* 209: 171–78

Mullan, D. J. 1978. *Ap. J.* 226: 151–66

Olson, G. L. 1978. *Ap. J.* 226: 124–37

Panagia, N., Felli, M. 1975. *Astron. Astrophys.* 39: 1–5

Parker, E. N. 1958. *Ap. J.* 128: 664–76

Parker, E. N. 1963. *Interplanetary Dynamical Processes*. New York: Interscience

Pneumann, G. W., Kopp, R. A. 1978. *Sol. Phys.* 57:49–64

Reid, M. J., Muhlman, D. O., Moran, J. M., et al. 1977. *Ap. J.* 214:60–77

Reimers, D. 1975. In *Problems in Stellar Atmospheres and Envelopes*, ed. B. Baschek, W. H. Kegel, G. Traving, pp. 229–56. Berlin: Springer

Reimers, D. 1977. *Astron. Astrophys.* 57: 395–400

Reimers, D. 1978. In *Proc. IAU Colloq. No. 42, The Interaction of Variable Stars with their Environment*, ed. R. Kippenhahn, J. Rahe, W. Strohmeier, pp. 559–76

Roberts, P. H., Soward, A. M. 1972. *Proc. Roy. Soc. London Ser. A* 328:185

Rumpl, W. M., Cassinelli, J. P. 1979. In preparation

Rybicki, G. B., Hummer, D. G. 1978. *Ap. J.* 219:654–75

Salpeter, E. E. 1974. *Ap. J.* 193:585–92

Sanner, F. 1976. *Ap. J. Suppl.* 32:115–45

Sargent, W. L. W. 1961. *Ap. J.* 134:142–50

Sargent, W. L. W., Osmer, P. S. 1969. In *Mass Loss for Stars*, ed. M. Hack, pp. 57–63. Dordrecht: Reidel

Snow, T. P., Jenkins, E. B. 1977. *Ap. J. Suppl.* 33:269–360

Snow, T. P., Morton, D. C. 1976. *Ap. J. Suppl.* 32:429–65

Sobolev, V. V. 1960. *Moving Envelopes of Stars*. Cambridge: Harvard Univ. Press

Stalio, R., Upson, W. L. 1979. In *Proc. 4th Int. Colloq. Astrophys., Trieste: High Resolution Spectrometry*, ed. M. Hack. In press

Stencel, R. E. 1978. *Ap. J. Lett.* 223:L37–39

Stothers, R. 1975. *Ap. J. Lett.* 197:L25–27

Thomas, R. N. 1978. In *Proc. IAU Symp. No. 83, Mass Loss and Evolution of O-Type Stars*. In press

Vaiana, G. S., Rosner, R. 1978. *Ann. Rev. Astron. Astrophys.* 16:393–428

Van Blerkom, D. 1978. *Ap. J.* 221:186–92

van der Hucht, K. A., Bernat, A. P., Kondo, Y. 1979. *Astron. Astrophys.* In press

Weaver, R., McCray, R., Castor, J., Shapiro, P., Moore, R. 1977. *Ap. J.* 218:377–95

Wendker, H. J., Smith, L. F., Israel, F. P., Habing, H. J., Dackel, H. P. 1975. *Astron. Astrophys.* 42:173–80

Weymann, R. J. 1978. In *Proc. IAU Colloq. No. 42, The Interaction of Variable Stars with their Environment*, ed. R. Kippenhahn, J. Rahe, W. Strohmeier, pp. 577–90

Wilson, W. J., Barrett, A. H. 1972. *Astron. Astrophys.* 17:385–402

Winnberg, A. 1978. In *Proc. IAU Symp. No. 42, The Interaction of Variable Stars with their Environment*, ed. R. Kippenhahn, J. Rahe, W. Strohmeier, pp. 495–520

Woolf, N. J., Ney, E. P. 1969. *Ap. J. Lett.* 155:L181–84

Wright, A. E., Barlow, M. J. 1975. *MNRAS* 170:41–51

York, D. G., Vidal-Madjar, A., Laurent, C., Bonnet, R. 1977. *Ap. J. Lett.* 213:L61–65

Zirin, H. 1976. *Ap. J.* 208:414–25

Ann. Rev. Astron. Astrophys. 1979. 17:309–43

ON THE NONHOMOGENEITY OF METAL ABUNDANCES IN STARS OF GLOBULAR CLUSTERS AND SATELLITE SUBSYSTEMS OF THE GALAXY

*2153

Robert P. Kraft
Lick Observatory, Board of Studies in Astronomy and Astrophysics,
University of California, Santa Cruz, California 95064

1 INTRODUCTION: SCOPE OF THE ARTICLE

Observational and theoretical evidence gathered in the score of years following the mid-1950s established conclusively that globular clusters are old; quite possibly they contain the oldest stars of the Galaxy. It was quickly realized that if chemical compositions, ages, and kinematical parameters of clusters could be obtained as a function of galactocentric distance, the early dynamical and chemical history of the Galaxy could be reconstructed. The recent literature contains a number of papers and reviews (cf e.g., Woltjer 1975, Harris 1976, Searle & Zinn 1978) in which this global problem is addressed, and we refer to it as the "classical cluster problem." On the other hand, knowledge of the chemical composition of cluster stars may also be applied to less grand, but no less interesting, problems in the evolution of low-mass stars; for example, theories of nucleosynthesis, deep convective mixing, and mass loss at different evolutionary stages may be tested. The cluster domain seems to provide a uniquely opportune setting for studies of this kind because one expects the following quite reasonable assumptions to be satisfied: *cluster stars are coeval and have the same initial chemical composition.* Moreover, since evolved cluster stars have evolutionary time scales small compared with cluster ages, they have also very nearly the same initial mass at main sequence turnoff. Thus in a cluster, one presumes that surface chemical compositions of individual stars may be compared with essentially all parameters held constant except *evolutionary state.*

0066-4146/79/0915-0309$01.00

It is true that composition studies, some quite detailed, have been carried out for bright field halo stars having galactic orbits similar to those of globular clusters, but the number of such stars is quite small and determinations of relative mass and age lack precision. Thus evolutionary connections between halo field stars having different spectroscopic characteristics necessarily remain somewhat obscure. Contrariwise, although the brightest cluster giants are comparatively faint and studies of their compositions thus necessarily rather crude and lacking in detail, the evolutionary relation between individual stars is quite clear. Thus an improvement in our knowledge of surface abundances in cluster stars would appear to be of critical importance.

In the first part of this article, we review techniques for estimating cluster metal abundances in which application to the classical cluster problem is the principal consideration. These methods, which employ a limited spectral resolution and usually a single parameter describing cluster metallicity, were first developed in the earliest period of photo-electric photometry, and may themselves be described as "classical." We also discuss in this part of the article techniques developed later on, which have similar methodology and objectives. We then turn to recent, largely higher-resolution results, many of which were reported in the 1978 Cambridge NATO Summer School on "Globular Clusters," and show how these extend, but also challenge, some of the leading ideas about cluster abundances that had been derived from classical methods. In particular, we cite evidence indicating 1. that cluster giants mix more nuclearly processed material to the surface than had been suspected on the basis of classical evolutionary theory, 2. that some clusters are chemically inhomogeneous, even when "primary" elements are considered and 3. that the chemical history of halo stars sharing the same Fe/H ratio, but belonging to different stellar subsystems, may not be the same.

The resurgence of interest in cluster abundances is attributable in large part to improvements in observational technique, viz., the application of image-tube spectrographs of echelle design, medium- to high-resolution optical scanning spectrometers, and IR broad-band spectroscopic techniques to the observation of cluster stars. In addition, results obtained with newly commissioned large telescopes in Chile and Australia, locations from which several of the nearest and brightest globular clusters can be observed, have proved especially valuable. Finally, workers in the field of synthetic spectrum analysis, by applying the technique to the low- and medium-resolution spectra available for globular cluster giants, have shown that quite accurate abundances for several *different* metallic species can be obtained.

2 METAL-ABUNDANCE DETERMINATIONS (CLASSICAL METHODS)

Unfortunately, the brightest globular cluster giants are judged to be quite faint by observers who operate conventional high-resolution spectrographs at the coude foci of large telescopes. Consequently abundance determinations based on such equipment were, until quite recently, limited to one star in each of M92 and M13 (Helfer et al. 1959, Pagel 1966). On the other hand, beginning in the early 1950s, multicolor photoelectric broad-band photometry (or photoelectrically calibrated photographic photometry) and low-resolution scanner fluxes became available for many individual stars in several clusters and from these observations, techniques for the determination of metal abundances were developed. These depend on the fact that, in the spectra of F-, G-, and K-type stars, Fe-peak metals are principally responsible for line blocking which in turn increases in effectiveness with decreasing wavelength (Figure 1). The blocking

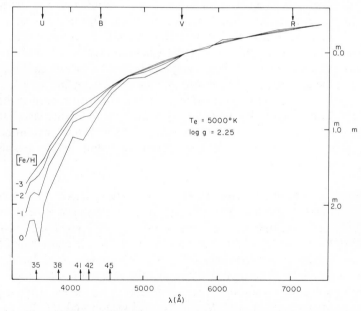

Figure 1   Theoretical scanner fluxes for giant stars of different metal content at $T_e = 5000$ K, $\log g = 2.25$ (Bell & Gustafsson 1978, 1979). Fluxes per unit frequency, converted to magnitudes and averaged over 50-Å intervals, are given. Although spectral regions having particularly strong lines have been avoided, the effect of line blanketing in the UV is apparent. Effective wavelengths of filters in the extended UBV and DDO systems are indicated.

itself is derived from a suitable comparison of broad-band filter fluxes, and has been calibrated by measurement of these fluxes in bright field halo stars in which Fe abundance had been determined from curve-of-growth or model atmosphere analysis. Attempts to calculate the metallicity directly from the blocking have encountered difficulties owing possibly to uncertainties in the solar UV-opacity (Bell & Gustafsson 1979). We describe briefly some of the principal techniques.

Line-Blocking Methods

SUBDWARFS: UBV PHOTOMETRY   Stars of the Hyades main sequence, which have very nearly solar metallicity, define a standard reference sequence in the $(U-B)$ vs $(B-V)$ diagram of the Johnson-Morgan (1953) UBV system. Owing to a general reduction in metallicity, subdwarfs are generally displaced in this diagram in a direction corresponding to larger U and B fluxes relative to V (Sandage 1964, 1969a,b). The vertical vector $\delta(U-B)$, measured at a standard fiducial $(B-V) = 0.60$, is found to be correlated with $[Fe/H] \equiv \log (Fe/H)_* - \log (Fe/H)_\odot$, the latter being derived from high-dispersion analysis of spectroscopically accessible nearby subdwarfs (cf Wallerstein et al. 1963, Wallerstein & Carlson 1960). On the assumption that the main sequence dwarfs of a globular cluster are identical with field subdwarfs having the same $\delta(U-B)$, one derives cluster abundances for those cases in which reliable U magnitudes can be measured. In Table 1 we list the abundances determined (Sandage 1970) for four well-known metal-poor clusters M92, M15, M13, and M3. It is worth adding that if the absolute magnitudes of field subdwarfs can be established from reliable trigonometric parallaxes, then application of stellar evolutionary theory permits one to derive also cluster ages and stellar masses at main sequence turnoff.

GIANTS: DDO SYSTEM, WASHINGTON SYSTEM, METHODS BASED ON IR COLORS
Cluster dwarfs being faint ($V \sim 19$ to 22), accurate measurements of U are very difficult; thus several investigators have devised methods by which UV-excesses can be measured directly in cluster giants (McClure & van den Bergh 1968, Wallerstein 1962, Wallerstein & Helfer 1966). Again a correlation is established between the excess $\delta(U-B)$ and $[Fe/H]$ derived from a high-resolution analysis of bright field giants. [Cautionary note: relations between $\delta(U-B)$ and $[Fe/H]$ established for dwarfs are inapplicable to giants because the UV flux is sensitive to surface gravity as well as abundance (cf e.g., Bell & Gustafsson 1978).] Values of $\delta(U-B)$ in giants, measured at $(B-V) = 1.0$, have recently been summarized by Kron & Guetter (1976).

A line-blocking quantity similar to $\delta(U-B)$ is defined in the widely-used David Dunlap Observatory (DDO) System (McClure & van den Bergh 1968, Osborn 1973); the most extensive application of the system to globular cluster stars is that by Hesser et al. (1977). Five basic filters, located in the blue and ultraviolet spectral regions in the vicinity of $\lambda 4800$, $\lambda 4500$, etc., and denoted by 48, 45, etc., admit stellar flux from which are formed color indices $C(45-48)$, $C(42-45)$, $C(38-41)$, and $C(41-42)$ which measure surface gravity, effective temperature, heavy element abundance, and CN-band strength, respectively. The system therefore admits of the possibility of studying abundances in the CNO group independent of the Fe peak. The reader may judge how abundances are derived in a few sample cases from inspection of Figure 2. Results are given also in Table 1. The zero point of calibration of the abundances derived from field giants remains at the moment somewhat uncertain especially in the metal-rich domain (Hesser et al. 1977).

The "Washington System" (Canterna 1976) is similar in principle to the DDO system but employs filters transmitting flux in the red and near-IR regions of the spectrum, as well as in the UV. Results based on observations of a few individual giants in several clusters of exceptional interest in the outer halo (Canterna & Schommer 1978) are given also in Table 1, and will be discussed in more detail later.

More recent work (Cohen et al. 1978) throws some doubt, however, on the usefulness of color-color diagrams involving IR vs UV fluxes in the quest for accurate [Fe/H] determinations. Plots of $(U-V)$ vs $(V-K)$, for example, show little discrimination between stars of M3 ([Fe/H] $= -1.7$) and M92 ([Fe/H] $= -2.2$). From calculated model atmosphere colors, these authors show that in passing from intermediate to low metal-abundance giants, changes in surface gravity mask the expected changes in $(U-B)$ color produced by deblanketing. Effects of this kind may lie behind the discrepancies found when [Fe/H] determinations on the Washington system are compared with those of the Searle-Zinn system (described below) for clusters observed in common. It may also account for the disappointing scatter of $(U-B)$ values with [Fe/H] (Table 1) when cluster giants are observed in the $UBVRI$ system (Eggen 1972), and compared in a plot of $(U-B)$ vs $(R-I)$.

GIANTS: SEARLE-ZINN METHOD   The method is similar in principle to those just discussed, but has the advantages of increased spectral resolution and relative freedom from uncertainties in interstellar reddening. A sample energy distribution F_v for a typical globular cluster giant, expressed in magnitudes and averages over 160 Å intervals, is shown in

Table 1 Metal abundances [Fe/H] for Galactic globular clusters

Cluster NGC	Sp	Δs	$\delta(U-B)$ MS	$\delta(U-B)$ giants	DDO	Wash. Sys.	Searle-Zinn	COG	Mod. atmos.	Syn. spec.	Adopted	Wt.
107 (47 Tuc)	G3			-0.6	-0.4					-0.8	-0.6	5
288	(F-)			-1.8							-1.8	1
362	F8			-0.8	-1.0						-1.0	2
1851	F7			-0.8							-1.0	1
2419	F5					-2.2					-2.1	1
2808	F8			-0.8							-0.8	1
Pal 3	(F-)					-2.3					-2.2	1
Pal 4	(F-)					-2.4					-2.3	1
4147	F2						-1.6				-1.6	3
5024 (M53)	F4	-1.8					-1.9				-1.9	4
5053	(F-)			-1.8	-2.0	-2.4	-2.0				-2.0	5
5139 (ω Cen)	F7	-1.1 to -2.0			-1.0 to -2.2		-0.8 to -1.6			-1.7 to -2.2	-0.8 to -2.2	5
5272 (M3)	F7	-1.6		-1.1	-1.5	-1.4	-1.7				-1.7	5
5466	(F-)		-0.9				-1.9		-1.8		-1.9	2
Pal 5	(F-)				-1.5		-1.1				-1.3	3
5904 (M5)	F5	-1.0	-1.2	-0.9	-1.0		-1.2				-1.1	5

Cluster	Sp	(1)	(2)	(3)	(4)	(5)	(6)	(7)	(8)	(9)	mean	n
6121 (M4)	F8	−1.2									−1.2	2
6171 (M107)	G0.5	−0.8									−0.7	5
6205 (M13)	F5	−1.0	−1.2	−0.5	−1.6		−0.6	−1.3	−1.6		−1.5	4
6229	F7			−0.9			−1.6				−1.4	3
6254 (M10)	F8		−1.6		−1.6	−1.7	−1.4				−1.5	5
6341 (M92)	F2	−2.2		−1.9	−2.2	−2.2	−1.3	−2.3	−2.4	−2.2	−2.2	5
6352	(G)				0.0						0.0	2
6397	F5			−1.1						−2.0	−2.0	2
6656 (M22)	F5	−1.7			?	−2.0				−2.0	−1.9	4
6712	G4	−0.4									−0.4	2
6752	F6			−1.2							−1.7	3
6779 (M56)	F5						−1.8			−1.8	−1.8	2
Pal 11	(F−)					−0.7					−0.6	1
6838 (M71)	G5			−0.2	0.0	−0.3	−0.2				−0.2	5
6934	F7	0.0					−1.4				−1.4	2
6981 (M72)	G0.5		−0.4	−1.0			−1.4				−1.4	3
7006	F3.5					−2.1					−1.5	4
7078 (M15)	F3	−2.0	−1.5	−2.3	−1.6		−1.4			−2.2	−2.0	5
7089 (M2)	F3	−1.4	−2.4	−2.3	−1.9		−1.9				−1.6	3
7099 (M30)	F3	−2.0		< −2.2	−1.4		−1.4				−2.0	2
Pal 12	(F−)					−1.7					−1.6	1
Pal 13	(F−)					−1.9					−1.8	1

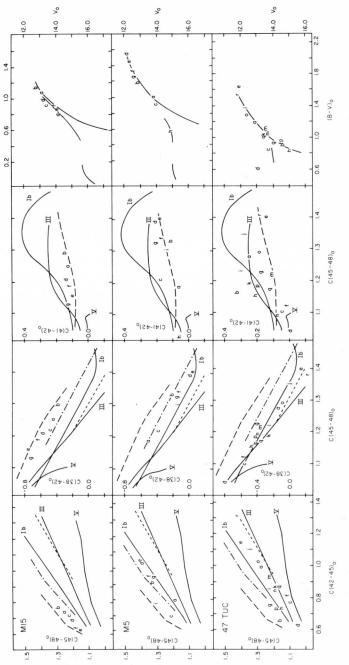

Figure 2 Color-color plots in the DDO system for three clusters of widely different metal content. Thin lines correspond to mean loci of V, III, and Ib luminosity classes in Pop I stars. Long-dash, dot-dash, and short-dash lines refer to M92, M3, and M71, for which [Fe/H] = −2.2, −1.7, and −0.2 respectively. Metal abundances for M15, M5, and 47 Tuc can be estimated by interpolation between these lines. The right-hand panel shows the location of individual cluster stars in the conventional c-m array; the upper $(B–V)_0$-scale should be used both for M15 and M5.

Figure 3, which is adapted from the paper by Searle & Zinn (1978). F_v is plotted against $\Psi(\lambda)$, where $\Psi = 1.30 \, \lambda^{-1} - 0.60$ or $0.75 \, \lambda^{-1} + 0.65$, according as $\lambda^{-1} \leqq 2.29$ or > 2.29. We note that as a function of $\Psi(\lambda)$ (or λ^{-1}), F_v is virtually a straight line between 8000 Å and 5000 Å. The slope of this line depends on the intrinsic energy distribution of the star and on interstellar reddening; however, since the transformation defining $\Psi(\lambda)$ is based on the Whitford reddening law (Miller & Mathews 1972), areas between the straight line, extrapolated shortward of $\lambda^{-1} = 2.29$ and the observed stellar flux m_λ, are reddening independent. These areas, integrated over 160 Å intervals and denoted by $Q(\lambda)$, thus measure the intrinsic spectrum of the star.

In the Searle-Zinn method, some dozen red giants are typically scanned in each cluster, the $Q(\lambda)$ measured at several wavelengths between $\lambda5160$ Å and $\lambda3880$ Å and the weighted mean of these Q's, denoted by S, is formed. For each cluster the absolute magnitude M_V is plotted as a function of S, and a fiducial value of S, denoted by $\langle S \rangle$, is read out at the convenient luminosity $M_V = -1$ for all clusters; $\langle S \rangle$ ranges over a factor of four from the most metal-poor (M92) to the most metal-rich (M71) cluster. The correlation of $\langle S \rangle$ with [Fe/H] is, however, determined

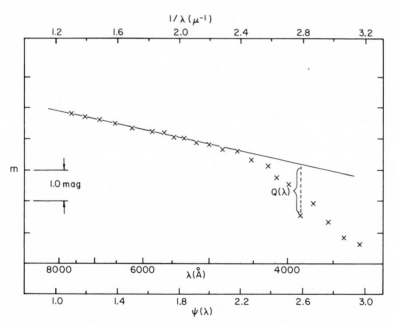

Figure 3   Sample energy distribution of a globular cluster star in the Searle-Zinn method. Units are those of Figure 1, except that the fluxes are averaged over 160-Å intervals.

from the [Fe/H]-value of RR Lyraes in the cluster (Butler 1975a) and not from direct model calculations of the UV-blanketing itself. Cluster metal abundances based on this method are also summarized in Table 1.

Other Direct Methods: Δs-Measurements for RR Lyraes

Finally, we consider a low-resolution spectroscopic method, the so-called Δs-measurements in the spectra of RR Lyrae variables, stars which span the spectral type interval A3–F5. In this temperature domain, the continuous opacity (H⁻ plus bound-free and free-free hydrogen) depends on the H abundance alone; thus hydrogen line strengths depend on temperature virtually independent of metallicity whereas the Ca II K-line strengths depend both on metallicity and temperature. In an RR Lyrae of solar metallicity, the spectral type s predicted from the K line and H_γ agree (implying $\Delta s = 0$), whereas for a metal-poor star, the K line will appear too weak and the K-line spectral type will be judged too early. The unit of Δs is tenths of spectral class and the most metal-poor RR Lyraes ([Fe/H] ~ -2) have $\Delta s \sim 11$. The method requires only modest spectroscopic resolution (~ 10 to 20 Å) and thus permits the investigator with a well-equipped medium- to large-sized telescope to reach RR Lyraes in clusters. The abundance calibration is again indirect: from curve-of-growth analysis of bright field RR Lyraes, a relation is established between [Fe/H] and Δs (Preston 1959, 1961); the most comprehensive treatment is Butler's (1975a) who derived [Fe/H] $= 0.16 \Delta s - 0.23$. It should be noted that the method makes use of the strength of the K line as an interpolation device linking [Fe/H] with Δs, and thus tacitly assumes that [Ca/H] and [Fe/H] are positively correlated. Attempts to obtain the Ca abundance from the K line directly (Rodgers 1974) lead to a different metal-abundance scaling, the significance of which will be discussed later.

The abundances derived from the preceding "classical" methods establish that the metallicity of globular cluster stars ranges from roughly solar to [Fe/H] ~ -2. But the techniques have inherent limitations: 1. each [Fe/H]-value represents an "average" often taken over only a handful of cluster stars and it is assumed that the "spread" is a result not of intrinsic fluctuations but only of observational errors; 2. the basic abundance calibrations are based almost entirely on field stars, so that one assumes overall chemical identity between cluster and field stars of a given [Fe/H]; 3. the low spectral resolution limits "abundances" to the Fe-peak isotopes considered as a group. Objections and exceptions, to be discussed later, have been raised against some or all of the assumptions and/or limitations.

Inspection of Table 1 also reveals that the methods do not yield the same value of [Fe/H] for a given cluster: M13 is a notable example.

Finally, low-resolution UV-blanketing methods have been criticized (Kraft et al. 1979) for their lack of sensitivity in the lowest metal-abundance domain. Spectral resolution higher than that employed in these techniques is needed if one is to distinguish stars with, say, $[Fe/H] = -2.0$ from those with $[Fe/H] = -2.5$, given the typical observational errors.

Metal-Abundance Determinations: Secondary Indicators

As is well known from calculations of stellar evolution, the metal abundance of a cluster is reflected in the morphology of its color-magnitude (c-m) or H-R diagram. Along the Hayashi line above the giant branch (GB) knee, an increase in $[Fe/H]$ or Z (metal abundance by mass) leads to a redward shift of the GB and the degeneration of the horizontal branch (HB) into red stub hard against the GB (Faulkner 1966, Iben 1974). In the theoretical (L, T_e) diagram, the Hayashi line moves parallel to itself and the red giant tip brightens slightly as Z increases (cf e.g., Sweigart & Gross 1978); however, in the observational $(V, B-V)$ diagram, owing to the dependence of the bolometric correction on T_e and of $(B-V)$ on both T_e and line blocking, the slope and height of the GB above the HB decline with increasing Z. The physics of the evolutionary tracks and their dependence on Y (helium abundance by mass), Z, and age are reviewed in extenso elsewhere (Iben 1974, Renzini 1977).

Several secondary abundance indicators have been developed which exploit these morphological dependences on Z. We call attention here to ΔV (Sandage & Wallerstein 1960), the height of the GB above the HB, read at (unreddened) $(B-V) = 1.40$; $(B-V)_{o,g}$, the unreddened color of the GB read out at the luminosity of the HB (Sandage & Smith 1966); the slope S of the GB above the junction of the HB and GB (Hartwick 1968); and the Dickens (1972a) type, which measures the relative population of the HB blueward and redward of the RR Lyr-star gap (cf van Albada & Baker 1973). These parameters are plotted as functions of $[Fe/H]$ (Table 1) in Figure 4; the limitations of each are discussed in turn.

ΔV The quantity is reddening dependent. In metal-poor clusters, the GB tip is near $(B-V)^\circ = 1.40$; if the stellar frequency exhibited in the c-m array is low, as for example in a cluster of small total mass like NGC 5053 (Sandage et al. 1977), ΔV may be quite difficult to estimate.

$(B-V)_{o,g}$ The quantity is also reddening dependent, and can be subject to considerable uncertainty in clusters in which photoelectrically calibrated photographic photometry alone is available. Points corresponding only to clusters with the most reliable c-m arrays and reddening determinations, as judged by Butler (1975b) and Sandage et al. (1977), are plotted in Figure

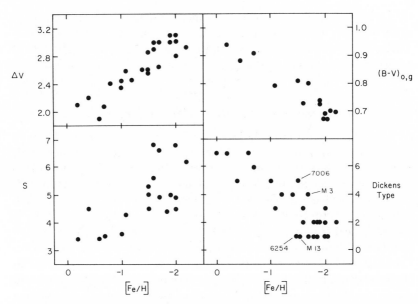

Figure 4 Secondary abundance indicators as functions of [Fe/H] (Table 1). Clusters illustrating the "second parameter problem" are indicated.

4. From a study of theoretical model atmosphere colors, Kraft et al. (1979) argue that $(B–V)_{o,g}$ is fairly insensitive to [Fe/H] when [Fe/H] $\lesssim$ -2.0; they show that $d(B–V)_{o,g}/d$[Fe/H] declines from ~ 0.15–~ 0.05 as [Fe/H] decreases from -1.6 to -2.4. Whether this is an important consideration in the "classical cluster problem" depends on whether there exist any real clusters with metal abundances significantly lower than those of M92.

HARTWICK S   The quantity is reddening independent, but unfortunately loses sensitivity when [Fe/H] < -1.4, as expected from the shapes of the evolutionary tracks.

DICKENS TYPE; SECOND PARAMETER PROBLEM   The quantity varies on the average with [Fe/H] as expected, but the spread at a given [Fe/H] (Table 1) indicates that the HB mapping is not a function of Z_{Fe} alone. The spread is particularly noticeable in clusters of intermediate metal-poorness, e.g., the HBs of NGC 7006 are red, of M3 intermediate, and of M2, M13, M10, and NGC 6752 quite blue. The effect is known as the "second parameter problem" and several explanations have been offered, among them the following:

$Z_{CNO} \neq Z_{Fe}$?   Although first explicitly suggested in connection with DDO observations of red HB stars in NGC 7006 (Hartwick & McClure

1972), the decoupling of the CNO group from the Fe peak is discussed as a possibility by Faulkner (1966) in his classical paper on HB morphology. The Z of interest is Z_{CNO}, since the CNO group governs the energy generation and opacity of HB stars. Explicit decoupling of Z_{CNO} from Z_{Fe} was considered in model calculations by Hartwick & Van den Berg (1973) and later by Castellani & Tornambè (1977), who showed that changes in Z_{CNO} by factors smaller than 10 could lead to interesting results. New evidence related to this possibility is cited later.

A range in He abundance? Attribution of the second parameter to He was first made by Sandage & Wildey (1967), further elaborated by Hartwick (1968), and discussed from the point of view of HB models by Rood (1973). Using the Sweigart & Gross (1976) HB models for $Z = 0.001$ and helium core mass $\mathcal{M}_c = 0.475\ \mathcal{M}_\odot$, we estimate that a HB star of fixed total mass can be moved across the instability strip from left to right by the required amount if Y is decreased by about 0.07; blueward of the strip, the morphology is not very dependent on Y. However, it is not sufficient to make $\mathcal{M}$ and $\mathcal{M}_c$ constant since a change in Y will change also the input parameters of stars before they reach the HB. Consider, for example, a cluster with $Y = 0.30$ and age $t_9 = 12$. Using the Rood (1972) evolutionary tracks, Renzini's (1977) interpolation equations that give t_9, turnoff mass $\mathcal{M}_{RG}$, etc. as functions of (Y, Z), and the Fusi-Pecci & Renzini (1975) theory of mass loss, one can show that in such a cluster the turnoff mass is $\mathcal{M}_{RG} = 0.75\ \mathcal{M}_\odot$, and a star now at the red giant tip has a "flash" mass $\mathcal{M}_{Fl} = 0.55\ \mathcal{M}_\odot$, and a core mass $\mathcal{M}_c = 0.475\ \mathcal{M}_\odot$. Given a stochastic mass-loss range of $\sim 0.05\ \mathcal{M}_\odot$ (Rood 1972), one finds that the HB morphology is similar to that of M13. On the other hand, a cluster with $Y = 0.275$ (only 0.025 smaller than the value above) and the same age will have parameters $\mathcal{M}_{RG} = 0.80\ \mathcal{M}_\odot$, $\mathcal{M}_{Fl} = 0.61\ \mathcal{M}_\odot$, and $\mathcal{M}_c = 0.48\ \mathcal{M}_\odot$. With a mass-loss range again of $\sim 0.05\ \mathcal{M}_\odot$, the HB stars will imitate the morphology of M3. That so small a He change ($\Delta Y = 0.025$) should induce such a large change in HB morphology results from the considerable sensitivity of HB morphology to the quantity $q = \mathcal{M}/\mathcal{M}_c$ as well as (Y, Z) (Faulkner 1966, Rood 1973).

Other changes in cluster properties induced by He variations will be discussed later in the section on He abundance.

A range in age? Because a star's HB location is sensitive to $q = \mathcal{M}/\mathcal{M}_c$, one could convert the morphology of M13 into M3 simply by increasing the mass at main sequence turnoff. Taking for example, $Y = 0.30$, $Z = 0.001$, and $t_9(M3) = 12.0$, and using the same precepts as above, we discover we must take $t_9(M13) = 13.5$; i.e., we find that an increase in age of only $\Delta t_9 = 1.5$ will convert the HB morphology of M3 into that of

M13. This time is of the same order as the present uncertainty in the determination of cluster ages (Renzini 1977, Ciardullo & Demarque 1977); it is also ten times larger than the classical time scale of galactic collapse (Eggen et al. 1962). Renzini (1977) has also noted that for $Z = 10^{-2}$ and 10^{-4}, HB morphology is rather insensitive to age variations of this sort, but shows maximum sensitivity near $Z = 10^{-3}$.

Core rotation?   The amount of mass lost by a star as it approaches the RG tip is critically dependent on L_{Fl}, and this in turn determines its subsequent location on the ZAHB. Renzini (1977) has shown that a plausible range of core rotational velocities, by affecting L_{Fl}, could produce the spread in mass loss required to obtain the observed width of cluster HBs. It is therefore plausible that a *difference* in the *mean* core rotation of stars in the pre-He core flash stage of different clusters could be responsible for the second parameter. Stars with $Z \sim 10^{-3}$ again show maximum sensitivity to the effect. Observational tests of the proposition that differences in core rotation explain the second parameter problem are, however, somewhat difficult to imagine (cf Renzini 1978).

Metal Abundances from Integrated Spectra and Photometric Indices

The determination of metal abundances for remote galactic globular clusters, for globular clusters in local group galaxies, and for related stellar populations in the nuclei of elliptical and spiral galaxies must necessarily depend on integrated spectra of colors. The subject of stellar

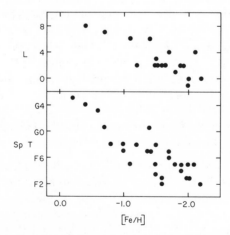

Figure 5   Spectral indices derived from integrated cluster light as functions of [Fe/H] (Table 1).

populations in galaxies is beyond the scope of this article, but has recently been given comprehensive treatment (van den Bergh 1975, Faber 1977); we call attention here only to the principal methods.

LOW-RESOLUTION SPECTROSCOPY   The most extensive lists of spectral types for clusters, based essentially on the ratio of the G-band to H_γ, are those of Kinman (1959) and Kron & Mayall (1960), which extend the seminal work of Morgan (1956). Spectral types determined from spectrograms obtained on conventional blue-sensitive plates depend on [Fe/H] not only because of direct changes in metal-line strength but also because of changes in cluster morphology. As [Fe/H] declines, clusters have hotter and brighter GBs and bluer HBs with only little change in the relative populations of GB and HB; thus the integrated spectral types become earlier. A plot of Kinman types against [Fe/H]-values from Table 1 is shown in Figure 5; a plot involving the more complexly defined, but similar, van den Bergh (1969) L-value is also shown.

Integrated pseudo-equivalent widths of spectral features measured from image-tube scanner (Robinson & Wampler 1972) spectra obtained at Lick (Burstein 1979) for Mg "b" + MgH, CH, and Na-D also show close correlations with [Fe/H], especially the first named.

PHOTOMETRIC COLOR INDICES   A comprehensive review of the problem has been given by Kron & Guetter (1976). Owing to the changes in cluster morphology with [Fe/H] cited above, Kron and Guetter argue that the effect of metal-line absorption on conventional broad-band photometric indices is much diluted in the integrated cluster light; thus a tight linear correlation between these indices and metal abundance is not to be expected. Rough correlations, not reproduced here, can certainly be shown to exist between the [Fe/H]-values of Table 1 and the reddening-free parameter $Q = (U-B) - 0.72 (B-V)$ (van den Bergh 1967), the integrated UV-index $(U-B)^\circ$ (Kron & Guetter 1976, Harris & van den Bergh 1974, Racine 1973), and the Strömgren-system metallicity parameter $[m_1]$ (Johnson & McNamara 1969).

Recent attempts to use near-IR colors have proved more successful. From model atmosphere calculations of K giants (Bell & Gustafsson 1978, 1979, Cohen et al. 1978), one finds that color indices such as $(V-R)$ and $(V-K)$ are sensitive primarily to T_e, and are almost independent of metallicity and surface gravity. The integrated V, K, and R magnitudes of clusters are all heavily dependent on light from giants and are relatively less affected by HB light than are U and B, for example. Moreover, the giant branch moves redward with increasing metallicity. Thus the integrated $(V-K)^\circ$ of clusters shows a good correlation with [Fe/H] (Aaronson et al. 1978, Frogel et al. 1978). Reasons for an earlier conclusion (Grasdalen

1974) to the contrary are discussed by Aaronson et al. and Pritchet (1977). The Aaronson et al. work includes also a narrow bandpass centered on CO at 2.4μ. The corresponding integrated CO-index scales rather slowly with [Fe/H] and there is considerable scatter, some of which may be real and associated with the second parameter problem.

3   HELIUM-ABUNDANCE DETERMINATIONS

He I lines appear in the spectra of hot HB stars, and in comparison with Pop I stars of similar temperature, are anomalously weak (Greenstein & Münch 1966). Spectroscopic analysis of the HB star S-18 in M15 (Newell 1970) gave $N(He)/N(H) \sim 0.01$. On the other hand, analysis (Auer & Norris 1974) of the bright blue giant Barnard 29 in M13 gave the "normal" He/H ratio of 0.115 ± 0.03 ($Y = 0.31$). Since the smaller value leads to quite impossible input parameters for HB models (cf Iben & Faulkner 1968, and many others), it seems likely that the weakness of the lines in M15, S-18 results from an atmospheric effect, perhaps "gravitational settling" or something even more exotic (Sargent & Searle 1967).

The most recent analysis of the M15 planetary (K648) (Hawley & Miller 1978) yields $N(He)/N(H) = 0.10$ ($Y = 0.28$), also close to the "normal" value. This object and Barnard 29 are in evolutionary stages more advanced than the HB, and might be thought to have suffered atmospheric contamination from the dredging-up of He-rich material. If so, it is surprising that the resulting He/H ratio should "turn out" to agree with the values universally found in H II regions and hot Pop I stars (Danziger 1970, Peimbert 1975).

The preceding summarizes all that is known from "direct measurement," so that in general He abundances in clusters depend on indirect arguments, among them the following.

Pulsation Theory

The relationship between the temperature at the blue edge of the instability strip and Y is given by $Y = 0.22 + 5$ (log $T_{e(BE)} - 3.863$), if $Z \approx 0.001$ (Iben 1974). Sandage (1969a) found $(B-V)^{\circ}_{BE} = 0.18$ for M3, so that in accordance with the temperature scale of Philip (1977) we have $T_{e(BE)} = 7450$ K and $Y = 0.265$. McDonald's (1977) temperature scale gives $T_e = 7380$ K and thus $Y = 0.245$. The derived value of Y is obviously rather sensitive to errors in the scaling of T_e with $(B-V)^{\circ}$, although the actual value of $(B-V)^{\circ}_{BE}$ is well known.

Deupree (1977) has shown that an increase in the He abundance decreases T_e at the red edge of the instability strip; the observed color width of the strip then implies $Y \sim 0.3$ with little difference between clusters. Considering the uncertainties this does not seem significantly different from the value of Y based on $T_{e(BE)}$.

Evolutionary Considerations

The luminosities of HB stars and RR Lyraes are dependent on Y, but they are also dependent on Z and q, so that in view of the observational uncertainties, present knowledge of L_{HB} and $M_V(RR)$ does not provide a very close constraint on Y. Of greater value is the dependence of Y on the relative population R of the GB and HB (Iben 1968, Iben & Rood 1969, Iben et al. 1969). This has been rediscussed by Renzini (1977) who allowed explicitly for semiconvection and other factors omitted in earlier discussions. He derived $Y = 0.22 \pm 0.04$, for a value of $R = 1.6$, averaged over several clusters. This agrees quite well with the result obtained from $T_{e(BE)}$, but is somewhat smaller than the values obtained by the other methods.

Cosmological Constraints

From a comparison of observed values of Y in H II regions in the Galaxy, the Magellanic Clouds, and young H II regions in compact galaxies (Searle & Sargent 1972), Peimbert (1975) suggested a "pregalactic" He abundance of $Y \approx 0.22$. Canonical "big-bang" cosmological calculations allow only a small range in pregalactic Y, viz., 0.235 to 0.240 (Gott et al. 1974). We expect Y appropriate to clusters to lie near the pregalactic value, but the scatter in the foregoing estimates precludes a definitive test of the proposition.

4 CHEMICAL INHOMOGENEITY OF STARS IN GLOBULAR CLUSTERS AND DWARF SPHEROIDAL SYSTEMS

As a result of the technological advancements discussed in Section 1 it is now possible to study the spectra of individual globular cluster stars in some detail, and also to extend classical methods to cluster and dwarf spheroidal satellites in the remote halo of the Galaxy. The new observational material presents an abundance pattern more complex than would have been thought plausible on the basis both of observations and theory developed in the classical period. We first examine abundance problems of isolated clusters considered one at a time, in particular the problems of 1. chemical homogeneity and 2. chemical uniformity relative to the sun. Afterwards we take up new observational evidence bearing on the more global questions of chemical uniformity of one cluster vs another and of cluster stars vis-à-vis halo field stars.

Are Stars in a Given Cluster Chemically Identical?

More precisely put, if s_1 and s_2 are two different stars in a cluster, do we find that $[X/H]_{s_1} = [X/H]_{s_2}$, for all X and for all pairs s_1 and s_2? Possible chemical inhomogeneity could be a reflection either of *mixing* of interior nuclear-processed material into the envelope during evolution or of *primordial abundance variations* from star to star.

If the latter hypothesis were required by the observations, then the early nuclear history of the cluster material would come under consideration, and it is conceivable that the sequence of events that had generated the isotopic ratios found in the sun and solar system material would be incompletely or differently realized in the primordial material of a cluster. Thus closely related to the question raised above is the following question: *Is $[X/H]$ independent of X for all stars of a given cluster?* We are concerned here, of course, with the initial Xs, not the values modified by possible mixing processes. Since the Fe abundance is the standard of reference and since Fe is generally deficient in clusters relative to the sun, we are for all practical purposes asking if elements such as the CNO group, Ca, etc. are deficient by the same factor as Fe.

We first explore the mixing hypothesis and inquire after the expectations from classical evolutionary theory.

MIXING AND CLASSICAL EVOLUTIONARY THEORY   The evolution of stars near 1 $\mathcal{M}_\odot$ has been reviewed by Iben (1974). Recent discussions of surface abundance changes are those by Dearborn et al. (1976) for states prior to the He core flash (pre-HCF) in metal-rich stars, and by Gingold (1974, and references therein) for states corresponding to the asymptotic giant branch (AGB), i.e., double shell-source (DSS) evolution. We list the expected effects in order of increasing atomic weight.

He   A small ($< 10\%$) increase in the envelope value of Y in pre-HCF evolution (Gingold 1977, Dearborn et al. 1976) and an even smaller increase in the DSS phase (Gingold 1974) have been predicted.

CNO group   Dearborn et al. (1976) show that the $^{12}C/^{13}C$ ratio should drop to ~ 20 to 30 (from the assumed initial value of 90) on the SGB during the pre-HCF phase, as the base of the outer convective envelope reaches layers where significant CN processing has gone on. An increase in the $^{14}N/^{12}C$ ratio by a factor of about 2 (Iben 1964, 1967) is also predicted, but O remains unchanged and there is no significant processing of O into N. However, more recent calculations of classical models (Sweigart & Gross 1978, Sweigart & Mengel 1979) suggest that the extent of the penetration of the base of the outer convection zone into the region of partial CNO processing may be even smaller than thought earlier.

Changes in the observed CNO isotope ratios are expected for very luminous giants in the AGB(DDS) phase (Gingold 1974) as a result of He-shell flashes, but details are lacking. Finally, there remains the possibility that new carbon, resulting from the triple-α reaction, might be brought up from the core during a "decentered" detonation of the He core flash (Thomas 1967, Paczýnski & Tremaine 1977).

Fe peak, and light metals between the CNO group and Fe   No changes are expected since the stellar masses and central temperatures are too small.

s-process   A slow neutron flux might be produced by the chain $^{12}C(p, \gamma)$ $^{13}N(\beta^+ \nu)^{13}C(\alpha, n)^{16}O$ if hydrogen in the outer envelope could be brought into the He-shell burning region during He-shell flashes in the DSS stage. The necessary linkage seems however not to exist (Sweigart 1974), and the one other possible neutron source, viz., the reaction $^{22}Ne(\alpha, n)$ ^{25}Mg, is found only in stars of intermediate mass (Iben 1976).

In summary, classical theory suggests that surface abundances should evolve little if at all until the most advanced DSS stages. Modest changes in the CNO isotopes and (probably undetectable) changes in the H/He ratio are expected.

OBSERVATIONAL RESULTS   Recent observations show, however, that these expectations are not fulfilled: some clusters contain Ba and CH stars; depletion of atmospheric carbon on a large scale characterizes advancing evolutionary stages in many clusters; wide variations in CN, CH, and NH band strengths are found at the same evolutionary stage, and in at least one cluster fluctuations in the strengths of lines of Fe-peak and α-process elements are found. A summary of these results now follows.

CH stars   Two CH-stars, analogous to the field CH stars first studied by Wallerstein & Greenstein (1964), have been found in ω Cen (Harding 1962, Dickens 1972b), analyzed by Bell & Dickens (1974, Dickens & Bell 1976), and reanalyzed by Mallia (1977). A third CH star was found by Bond (1975) and McClure & Norris (1977) report finding a high temperature example in M22. According to Bell and Dickens, the stars in ω Cen have $[C/Fe] \approx +0.5$, $^{12}C/^{13}C \approx 10$, and $[N/Fe] \approx +1.3$, with $[Fe/H] = [O/H] = -1.3$. The large N abundance was dictated by the relative strengths of CN and C_2, the Swan bands being relatively weak. Dickens & Bell (1976) also found two additional giants in ω Cen with enormous overabundances of N, viz., $[N/Fe] \approx +1.6$, but with no obvious additional peculiarities. It is interesting that the $^{12}C/^{13}C$ ratio approaches the equilibrium value appropriate to extensive CN cycling, but the observed [N/C] ratio is too large if the primordial ratio is taken as solar.

CH(G-band) and NH($\lambda3360$) variations in normal cluster stars In a low-resolution spectral survey of M92 giants, Zinn (1973) discovered that AGB stars had systematically weaker G-bands than subgiant branch (SGB) stars of equivalent temperature. The effect was later found in other clusters, especially those of relatively low metal abundance ([Fe/H] $\lesssim$ -0.8) (Dickens & Bell 1976, Mallia 1975, 1977, Zinn 1977, Norris & Zinn 1977). Subsequently spectral surveys of stars in M92 and M15 (Carbon et al. 1977) extended and confirmed Zinn's work, and by adding observations of the NH (0, 0) and (1, 1) bands near $\lambda3360$, explored the hypothesis that CH and NH strengths were anticorrelated. Using spectral synthesis, C abundances were obtained for brighter M92 giants by Bell et al. (1978) and C and N abundances were obtained by the same method for a larger sample of stars, many on the SGB, by Carbon et al. (1979). The behavior of the C and N abundances in relation to evolutionary state, although quite complicated, can be summarized as follows.

1. A large range in C and N abundances is observed at any given place in the c-m array even on the SGB, and C and N are not necessarily anticorrelated.
2. In general, stars on the AGB have carbon depleted by a factor of 10 or more relative to Fe, but stars exhibiting carbon depletion can show up as faint as $M_V \sim +1.5$. Thus on the average carbon shows a systematic depletion with advancing evolutionary state beginning as faint as $M_V = +0.3$, log L = 1.9 on the SGB.
3. On the *average*, the nitrogen abundance shows no change with evolutionary state and is enhanced so that $\langle[N/Fe]\rangle \sim +0.5$. Individual stars with N enhanced by factors of 10 or more can be found in all parts of the c-m array, even on the SGB.
4. Although Norris & Zinn (1977) and Zinn (1977) showed that G-bands strengthen near the RG tip, Bell et al. (1978) demonstrated that the effect is entirely due to the decline in temperature and does not require an increase in C abundance.
5. No correlation is found between the strengths of Fe and Ca II features and the abundances of either C or N, i.e., there is no evidence for variations in Fe-peak or α-process elements in either M92 or M15.

CN variations A similar survey of CN ($\sim\lambda4215$) was conducted by Hesser et al. (1977) who used the DDO system to observe 145 stars in 17 globular clusters of widely different [Fe/H]. They showed that relatively metal-rich ($[Fe/H] > -1$) clusters have wide CN-strength variations, but metal-poor clusters showed no such variations (a result understandable at least for giants in M15 and M92 which show no $\lambda 4215$ feature). A particularly large range of CN variations was found in stars of ω Cen, a cluster which also exhibits variations in Fe and Ca II (see below). However, their suggestion that M22 is a cluster similar in its anomalous properties to ω Cen, on the basis of excessively wide variations in CN strength, has become more problematical with Lloyd Evans's (1978) discovery that most of the stars found by them to have strong CN are not actually members of M22.

In the metal-rich cluster 47 Tuc, the DDO-system results were confirmed by Hesser (1978) who obtained small-scale image-tube spectra of stars deep on the SGB, virtually to the main sequence. Stars at the base of the SGB in the same part of the c-m array were found to exhibit significant differences on CN strength; thus mechanisms proposed to explain these variations must be made to operate in quite early evolutionary stages.

McClure & Norris (1977) considered the overall problem of CH and CN variations in clusters and argued plausibly that the CN variations in metal-rich, and the CH variation in metal-poor, stars could be regarded as manifestations of the same basic mechanism operating in different metal-abundance domains. At the moment, however, it is not known if CH and NH variations exist in metal-poor clusters at evolutionary stages as early as those in which CN variations are exhibited in metal-rich clusters.

CO variations   Very recently, Frogel & Persson (1978) reported the results of a CO (2.4μ) survey of individual globular cluster giants. In a plot of CO-index against $(V-K)°$, the clusters segregate by CO strength along separate lines with little dispersion; the only exceptional cluster is ω Cen, in which a wide variation in CO strength was detected at a fixed T_e [derived from $(V-K)°$]. The CO-index was found to range from M92 to essentially solar (Pop I) strength. The low metal-abundance limit corresponds well with the Fe and Ca results reported in the next section, but the metal-rich limit is higher than expected and indicates the existence of "CO-strong" stars in ω Cen.

Fe-peak variations   The location of the giant branch in the $(V, B-V)$ diagram is sensitive to $[Fe/H]$, and thus indirect evidence for a spread in Fe-peak abundances would be found if clusters had significantly wide giant branches. Renzini (1977) demonstrated qualitatively that carbon and nitrogen variations would not be expected to widen the giant branch significantly at least for metal-poor clusters since these elements do not contribute to the envelope opacity; this is confirmed by recent detailed calculations (Rood 1978). In the case of M15, Sandage & Katem (1977) argue persuasively that significant [Fe/H] variations are excluded, since the width of the GB in the $(V, B-V)$ diagram is essentially that expected from the observational errors. There is no doubt that variations in [Fe/H] of the size discussed below for ω Cen are excluded, but considerably smaller variations ($<$ factor of 2) might be hard to detect from fluctuations in $(B-V)$ alone in the low metal-abundance domain inhabited by M15 (Kraft et al. 1979).

The $(V, B-V)$ diagram by Cannon & Stobie (1973) showed conclusively that ω Cen had an unusually wide giant branch, and Dickens & Bell (1976), who favored variations in [Fe/H] as an explanation for the effect, showed that variations in stellar mass were inadequate and in interstellar reddening were nonexistent. The width of the giant branch is, however, dependent on the color system employed: Norris & Bessell (1977) found that in the V vs $R-I$ plane, the giant branch of ω Cen is considerably narrower, and is in fact only a little wider than 47 Tuc. They argued that, because the IR CN-bands lie in the R and I filters, the width of the sequence in the V vs $R-I$ plane may be entirely explicable on the basis of CN and CH variations observed by them. On the other hand, IR colors are relatively insensitive to [Fe/H] and gravity variations and depend mostly on T_e; thus even if ω Cen has Fe variations, it might be expected to have a somewhat narrower giant branch in a c-m array employing an IR color rather than $(B-V)$ as abscissa (cf Lloyd Evans 1977).

Other methods show conclusively that Fe and/or Ca abundance variations exist in ω Cen. Freeman & Rodgers (1977) discovered that the RR Lyraes had a wide range in Δs; this result was confirmed and extended to nearly half the known RR Lyraes by Butler et al. (1978), who obtained a range in [Fe/H] of -1.1 to -2.0 from the calibration of [Fe/H] vs Δs (Butler 1975a). Hesser et al. (1977) reported [Fe/H] variations based on the appropriate DDO indices. From calculations of synthetic spectra, Mallia (1977) found direct evidence that RGO 65 had a significantly lower [Fe/H] than of several other ω Cen giants, and Rodgers (1978), using the method of Searle & Zinn (1978), confirmed that a wide variation in [Fe/H] exists among the giants. Owing to uncertainties in the various abundance scales, the mean metal abundance derived from RR Lyraes and giants differs somewhat, but both agree in giving a spread in [Fe/H] of about one dex. For giants the frequency distribution over [Fe/H] tends to peak toward lower [Fe/H] with a long tail in the direction of solar metal abundance (Norris 1978). The frequency distribution for RR Lyraes, on the other hand, is bimodal, with peaks corresponding to the two Oosterhoff (1939) groups, the Type II (lower [Fe/H]) stars being the more numerous (Butler et al. 1978).

Light metals (between CNO and Fe)   The nucleosynthetic history of these elements differs from that of the Fe peak; the solar system isotope ratios are thought to be generated by some combination of α-processing and explosive carbon and/or oxygen burning (Arnett 1972, also the review by Trimble 1975). The different history of ^{40}Ca, for example, in comparison with ^{56}Fe, enters the cluster abundance problem directly and indirectly. From a detailed abundance analysis of five giants in M13 and three in M3, based on high-resolution echellograms, Cohen (1978) found that Na and Ca scatter significantly in the M3 stars, even though Fe is the same for all. (No variations were found in the abundance of these metals in the M13 stars.) For the M3 giants the [Ca/Fe] ratio ranged from $+0.2$ to -0.9; the Na and Ca abundances appeared correlated. The existence of these variations in turn introduces a new uncertainty in the application of the Δs method for RR Lyraes to the cluster abundance problem, viz., whether in RR Lyraes, Ca and Fe abundances are strictly correlated on a *star-by-star* basis. The correlation, of course, need not be flat: evidence for increases in the [Ti/Fe] and [Ca/Fe] ratios with decreasing [Fe/H] in *field* RR Lyraes has been given by Butler (1975a) and discussed (Butler et al. 1978) in connection with the derivation of Fe abundances for RR Lyraes in ω Cen. However, even if the Ca and Fe abundances were not strictly correlated star by star, it is likely that averages struck over many stars could still be used to derive mean cluster [Fe/H] abundances by the Δs method. The issue would seem to be whether the mean correlation between Ca and Fe derived from field RR Lyraes is also applicable to RR Lyraes in clusters.

s-process elements   Stars with enhanced Ba and possibly Sr seem to have been found only in ω Cen (Dickens & Bell 1976) and M22 (Mallia 1976, 1977). The most extreme Ba-star in ω Cen (RGO 371) has [Ba/Fe] = $+1.5$ and less certain Sr overabundance, but less extreme objects are found. All lie near the red edge of the giant branch not far from the red giant tip, and thus are compatible with the assumption that they have experienced He-shell flashes in the DSS (AGB) phase.

Interpretation: Mixing or Primordial Abundance Variations?

EVIDENCE FOR MIXING   It is hard to avoid the conclusion that globular cluster giants mix the ashes of internal nuclear processing to the surface on a scale larger than expected on the basis of classical evolutionary theory. The existence of CH and Ba stars in late evolutionary stages in clusters in which the vast majority of stars show no *generally* strong enhancement of carbon or s-process species suggests the correctness of the mixing hypothesis. CH stars seem to require that carbon produced by triple-α processing be mixed to the surface either at the He-core or shell flashes and the presence of Ba-stars in clusters suggests that the convective linkage necessary to operate the neutron-source chain $^{12}C(p, \gamma)^{13}N(\beta^+ \nu)^{13}C(\alpha, n)^{16}O$ in low mass stars must indeed exist.

In typical metal-poor cluster stars, as early as the mid-SGB, the average carbon abundance begins to decline presumably in response to the processing of the base of the convective envelope through the CN cycle, and continues to drop to ~ 0.1 of the original value, or lower, with advancing evolutionary state. The onset of the decline thus occurs much earlier and the magnitude of the effect is much larger than we would have expected based on the degree of CN processing and mixing found in conventional evolutionary models. Unfortunately, even with the improved observational technology cited earlier, it is not yet possible to recover the expected changes in the $^{12}C/^{13}C$ ratio in these stars. The lack of anticorrelation between $\langle [N/H] \rangle$ and $\langle [C/H] \rangle$ that one expects on the basis of the mixing hypothesis, remains, however, a puzzle: if C is destroyed, it must appear as N. And since AGB and SGB stars have the same temperature, the unexpected constancy of the *average* N abundance cannot be attributed to observational error. Equally puzzling are the large variations in C and N at a given point in the c-m array, even on the SGB, the evident lack of anticorrelation between C and N at each point, and the average *overabundance* of N. C and N abundance determinations of large numbers of giants in all evolutionary stages have been carried out only in the very metal-poor clusters M92 and M15 (Carbon et al. 1979), so the general overabundance of N has not been firmly established for other clusters. Nevertheless, the unusually strong CN features found in clusters of higher metallicity, even among stars on the SGB, the wide variations in CN strength at a given point in the c-m array, and the presumption that in the early evolutionary stages the CN band-strengths are likely to be controlled by N, all suggest that N overabundance may be a widespread phenomenon (Norris 1978).

The possibility that meridional circulation driven by internal rotation might lead to mixing having observational consequences consistent with the foregoing has been considered by Sweigart & Mengel (1979). In their picture, a red giant retains in its interior the angular velocity ($\omega \sim 10^{-4}$ rad sec^{-1}) appropriate to its main sequence phase, the outer part of the convective region having been spun down either by inward redistribution of angular momentum or a magnetically coupled stellar wind (Kraft 1970). In a rather extensive precursor region above the hydrogen-burning shell CN cycling converts C into N but closer to the shell the ON-cycle produces new N. Circulation currents driven by the differential rotation bring up this N-rich material from the O-depleted region against the effect of the μ-gradient but only if the metal abundance is low. On the other hand, in all metal abundance domains, the currents are able to bring up CN-processed material against the μ-gradient. The theory thus predicts that with increasing Z N-enhancement should disappear, but until a more extensive catalog of N and C abundance determinations is available in metal-rich clusters, it is not possible to check the value of Z at which the cutoff in N-enhancement takes place. The most that can be said is that in field Pop I giants of high ($\sim$solar) metallicity, [N/Fe] $\cong 0.0$ with little spread (Janes 1975, Deming et al. 1977, Deming 1978) whereas in clusters with [Fe/H] as high as that of 47 Tuc, wide CN variations occur even in the earliest evolutionary stages (Hesser 1978).

The Sweigart-Mengel theory predicts in a natural way both the general overabundance of N in M92 and M15 stars, and also the lack of anti-correlation between C and N abundances: the amount of material undergoing ON-processing is dependent on the internal angular velocity, which reflects the well-known spread of initial main sequence (ms) angular velocity. The theory is also consistent with the spread in ω required to explain the width of the zero-age horizontal branch (ZAHB) (Renzini

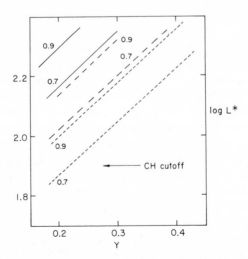

Figure 6   The luminosity L^* at which carbon shows significant depletion, as a function of helium abundance Y for turnoff masses $\mathscr{M} = 0.7$ and 0.9 $\mathscr{M}_\odot$. Dotted, dashed, and solid lines refer to metal abundance $Z = 0.001$, 0.0004, and 0.0001, respectively. The observed luminosity at which the CH abundance drops sharply in M92, according to Carbon et al. (1979), is indicated.

1977). Finally, it predicts a luminosity L^* at which carbon depletion should set in on the SGB, as a function of $(\mathcal{M}, Y, Z)$ (Figure 6). But here it seems less successful. According to Bell et al. (1978) the carbon depletion in red giants of M92 and NGC 6397 begins at log $L = 2.3$; however, Carbon et al. (1979) found log $L = 1.9$, derived from a considerably larger sample of M92 stars. For $Z = 10^{-4}$, the brighter value gives a reasonable cluster age, but log $L = 1.9$ gives $t_9 > 25$ (cf Sweigart & Gross 1978) if $Y \sim 0.2$ to 0.3. A more reasonable age is achieved if $Z \sim 10^{-3}$, which might be appropriate if Z_{CNO} were the controlling "metal" abundance and $Z_{CNO} > Z_{Fe}$.

EVIDENCE FOR PRIMORDIAL VARIATIONS   Although mixing must play a major role in determining the strengths of molecular features involving the CNO group, it cannot explain the Fe peak and light metal abundance variations seen in ω Cen and M3 and some way to produce primordial variations is required. How or when this comes about is not clear. One might imagine that 1. the original gaseous medium out of which the cluster was formed was chemically inhomogeneous, 2. the original medium was well mixed but was contaminated inhomogeneously by some external medium after the cluster gas came together (Iben 1978), or 3. the original gaseous material of the cluster was more or less homogeneous but was contaminated either by the first cluster supernovae or by the ejecta of medium-mass first generation stars, possibly both, the ejected material being retained by the cluster but poorly mixed.

Plausible extrapolation of cluster luminosity functions to higher luminosities and present notions of the types of elements to be ejected from supernovae as a function of initial stellar mass (cf Trimble 1975, Table X; Woosley 1974) suggest that interesting amounts of heavy element material could be produced by a few supernova events, especially in metal-poor clusters. Perhaps massive Fe supernovae produced the Fe fluctuations seen in ω Cen. The "overabundance" of lighter elements and their fluctuations could be produced from less massive supernovae or from secular mass loss of ordinary medium-mass first-generation stars. Norris (1978) has shown that if the initial mass-function (IMF) of ω Cen is $dN/d \log \mathcal{M} \sim \mathcal{M}^{-1.5}$, and if the 10^4 first generation stars with masses in the range 5 to 15 $\mathcal{M}_\odot$ could produce around 1000 $\mathcal{M}_\odot$ of ^{14}N, this would be sufficient to produce "primordial" N enhancements of about a factor of 10 in half the present low-mass stars of ω Cen. (The initial nitrogen abundance was assumed to be $[N/H] = -1.5$.) That 10^4 first generation stars could produce and eject the necessary ^{14}N was shown to be plausible: at each thermal pulse, a 7 $\mathcal{M}_\odot$–AGB model dredges up 0.75 $\mathcal{M}_\odot$ of ^{12}C (Iben 1975, 1976, Iben & Truran 1978) which Norris suggests may be converted to ^{14}N during interpulse phases. Extrapolating the same IMF to larger $\mathscr{M}$, we calculate the Fe enrichment (cf Trimble 1975, Table X) in ω Cen from the 200 "Fe-supernovae" expected with masses between 40 and 60 $\mathscr{M}_\odot$, and find an increase in Fe by a factor of about 3. This is a little too small to explain the observed Fe variations, but considering the uncertainties in the supernova theory and in the assumptions made, one finds that the hypothesis of a supernova origin for the Fe-fluctuations is also not completely unreasonable.

Applying these notions to metal-poor clusters less massive than ω Cen would lead to similar, but somewhat larger, fluctuations in Fe, since although the amount of contaminating mass decreases, so does also the quantity of hydrogen-rich cluster material into which the "new" Fe is deposited. Since significant Fe-peak fluctuations have been found in no other cluster, the supernova model would seem to require that lower mass clusters have steeper IMFs. In the Galaxy, there is little to guide us, but in the LMC, Freeman (1978) reported large variations in the slope of the IMF of several blue globular clusters which have very young ages ($\sim 10^7$ yr). It is not known if this is related to total cluster mass.

We have already discussed the rotationally-driven mixing theory of Sweigart & Mengel (1979) in connection with the carbon and nitrogen anomalies in M92 and M15, and we have seen that the theory predicts an onset of carbon depletion and nitrogen overabundance that is too bright for the observations. Even more difficult for the theory may be the enhancements of CN found by Hesser (1978) at the very earliest evolutionary stages of stars in 47 Tuc. Contrariwise, Carbon et al. (1979) took the view that the mean overabundance of N and fluctuations in N and C abundances resulted from primordial variations driven by medium-mass first generation supernovae; these would not be expected to produce Fe. The calculation, based on the tabulation by Trimble (1975), goes through essentially as Norris's treatment described above. They preferred a solution, however, in which an overabundance of C, by a factor of 2–3, was produced by mixing in the ejecta of supernovae in the 5 to 10 $\mathscr{M}_\odot$ domain; the subsequent overabundance of N was meant to be produced in the low-mass stars themselves by the mixing out of CN-processed material low on the SGB. In this way $\langle[N/H]\rangle$ could remain virtually constant with advancing evolution phase as required by the observations, because two thirds of the C had already been processed to N low on the SGB. However, it must be said that it is difficult to see how primordial fluctuations in N of the kind invoked by Carbon et al. could be applied to the variations in CN strength seen on the low SGB of a metal-rich cluster such as 47 Tuc. Real abundance determinations of C and N in faint 47 Tuc stars are urgently required.

So far, the competing importance of primordial fluctuations vs mixing in altering the appearance of stellar spectra in a cluster has been emphasized. Fe-peak and light-metal variations require a primordial origin. Changes in molecular features belonging to the CNO group, on the other hand, might be a result either of mixing or primordial variations. However, the possibility that the "primordial" and "mixing" indicators are in some cases "intertwined" has been emphasized by Norris (1978), who reported that in giants of ω Cen there is a strong correlation between the strength of Ca II H and K and the violet CN absorption. Norris showed that the effect cannot be attributed to atmospheric boundary cooling resulting from increased CN opacity. The abundance of N is therefore probably correlated with the general metal abundance in ω Cen. It is not known whether the CO strengths (Frogel & Persson 1978) are also correlated with Ca II and/or CN. In any case if N is *correlated* with Ca, and if the spotty in situ enrichment hypothesis for the primordial variations is correct, then the contaminating source is required to produce N along with Ca, and it is somewhat difficult to see how N could be produced by medium-mass stars while Ca is injected into the primordial medium from massive supernovae.

Against the hypothesis of in situ supernova origin for overabundances and fluctuations is the finding by Cohen (1978) that Na and Ca are significantly overabundant in one of the M3 giants compared with the other two. She argues that this would not be expected from explosive nucleosynthesis in metal-poor stars since large overabundances of Ca but not Na are predicted (Truran & Arnett 1971).

Of relevance to the question of the origin of the abundance fluctuations in clusters is the possible existence of radial abundance gradients (Freeman 1978). The case for the existence of radial *color* gradients has been put forward by several authors (Gascoigne & Burr 1956, Chun, 1976, Freeman 1978), and most investigators agree that the effect results from an excess of red giants in the cluster center; this is presumably driven by mass segregation. Color gradients are found in only one third of the clusters investigated (Chun 1976, Da Costa 1978), but these include 47 Tuc and ω Cen. In 47 Tuc the necessary mass segregation has been attributed (Freeman 1978) to an actual metal-abundance gradient but this is disputed (Da Costa 1978) on the basis of metal-sensitive narrow-band photometry, and the situation is unclear. In ω Cen the existence of a metal-abundance gradient seems more firmly established. Freeman (1978) finds that the metal-poor RR Lyraes are distributed over a wider radial domain than the metal-rich ones, as are the bluer giants at a given luminosity. Norris (1978) has given evidence that, in fact, the bluer giants of ω Cen are more metal-poor than the redder ones. The existence of such a gradient suggests that, however the abundance differences between ω Cen stars were established, the resulting radial distribution imitates that found in elliptical galaxies (cf Faber 1977); in this sense ω Cen may be thought of as a microgalaxy formed by a process similar to that acting in elliptical galaxies.

Fluctuations in Dwarf Spheroidal Systems

Little was known about metal abundances in the seven known dwarf spheroidal satellites of the Galaxy (Hodge 1971) until recently, and the latest developments have been reviewed by Zinn (1978b). The stars are faint and detailed knowledge is not extensive. Four of the five objects for which c-m arrays have been derived (Draco, Leo II, Ursa Minor, Sculptor) show a steep giant branch indicative of low metal abundance, but the HB morphology is often anomalously red. According to Demers (1978) the giant branch of Fornax is rather shallow, indicating high metallicity, and has a spread similar to that of ω Cen; a range of [Fe/H] of about 0.8 dex is indicated.

In a much more extensive and detailed study, Zinn (1978a) derived [Fe/H] values for 17 Draco giants by the method of Searle & Zinn (1978) already described. The stars divide into two metal-abundance groups separated by a factor of 2 with $\langle$[Fe/H]$\rangle = -1.86 \pm 0.09$, a value close to that of M92. The range of metal abundances in M92 is, however, almost certainly smaller than that of Draco. Zinn pointed out that the total masses of M92 and Draco are similar, but the mean stellar density of Draco is about 1000 times smaller; thus if M92 collapsed from a denser proto-cloud of gas than did Draco, it should have retained more easily the products of early supernova outbursts and therefore should have a wider abundance range than Draco, opposite to that which is observed. The results therefore suggest that M92 and Draco had different enrichment histories.

5 CHEMICAL HOMOGENEITY OF ONE CLUSTER VS ANOTHER: THE SECOND PARAMETER PROBLEM

As a consequence of observational necessity, as we have seen, metal abundances of clusters are rank-ordered by [Fe/H]. Consider two clusters M and N and the primordial, i.e., unmixed, abundance of some element X. The question is this: When $[Fe/H]_M = [Fe/H]_N$, is it also true that $[X/H]_M = [X/H]_N$ for all X? Demonstrable differences from one cluster to another would obviously imply that different overall chemical enrichment histories exist for clusters. Moreover, as discussed earlier, if X $\sim$ He or the CNO group, and $[X/H]_M \neq [X/H]_N$, the HB morphology can be strikingly changed. We now explore what recent observations say about these possibilities, and particularly whether the second parameter problem is due to CNO variations from cluster to cluster. We inquire no further about He abundances since these are virtually impossible to obtain directly from spectroscopic observation.

CNO Group

Using DDO photometry, Hartwick & McClure (1972) discovered that several giants in NGC 7006 ([Fe/H] = − 1.5, Table 1) had CN bands much stronger than corresponding giants in M3 and M13 ([Fe/H] ∼ − 1.7), and suggested that excess primordial N might be responsible for the major inconsistency in the morphology of the c-m array: ΔV and the Hartwick S suggest fairly low metals in agreement with the directly observed value of [Fe/H] in Table 1, but the HB is red (Dickens type 5).

More recently Canterna & Schommer (1978), on the basis of quite small samples of giants observed in the Washington system, found that Draco, Pal 3, 4, and 12 as well as NGC 7006 were fairly metal-poor, but had anomalously red HBs. Their data agreed with the results of a low-resolution spectral survey by Cowley et al. (1978), and agreed also with the latter's finding that of these systems only NGC 7006 and Pal 12 contain stars with enhanced CN or CH. Since only two of the four stars observed in NGC 7006 actually showed enhanced CN bands, Canterna and Schommer advanced an explanation based on mixing rather than enhanced primordial N, and thus they preferred an explanation for the red HB anomaly based on differences in He abundance or age. A new manifestation of anomalous behavior in these clusters arises when some of the secondary abundance parameters are compared. Harris (1978) found that Pal 12, despite its low [Fe/H]-value, had a giant branch virtually as red as that of 47 Tuc; the excessively red value of $(B–V)_{o,g}$ for Pal 4, 12, 13 has also been noted by Schommer (1978).

Generally speaking, these anomalous metal-poor clusters inhabit the outer regions of the galactic halo and have galactocentric distances $R > 12$ kpc (Zinn 1978b). Considerable work on distant dwarf spheroidals in addition to Draco remains in progress, but several of these dwarf spheroidals also show anomalous red HBs (Ursa Minor seems to be an exception). In these clusters and systems, CNO abundances have been obtained for only very small samples of stars; nevertheless on the basis of the available observations, it does seem rather unlikely that the HB anomalies can be explained as a result of decoupling of CNO from Fe peak elements. A preferable solution to the problem (Zinn 1978b) would seem to be a picture in which the more remote subsystems have younger evolutionary ages than the clusters of the inner halo and nuclear core of the galaxy, since we might expect on general grounds that the less dense outer halo material would take a longer time to collapse (Searle & Zinn 1978).

Turning to the nearer classical clusters, we find that when giants are rank-ordered by CO strength, the position of M3 is anomalous if its Fe abundance alone is considered (Cohen et al. 1978, Pilachowski 1978). Thus from Table 1, we see that $[Fe/H]_{M13} = [Fe/H]_{M3} + 0.2$. The c-m arrays of M3 and M13 classically exemplify the second parameter problem; thus it is interesting that the ranking of the two clusters by CO is in the expected direction and has roughly the right size to explain the anomaly (Hartwick & Van den Berg 1973).

Since M3 is not infested with carbon stars, we expect to find $[C/H]_{M3} < [O/H]_{M3}$, and the strength of CO to be governed by the abundance of C. Independent measurement of the O abundance (Cohen 1978) yields $[O/Fe] = +0.4 \pm 0.2$ in M3 and $< +0.4$ in M13, so O appears to be "overabundant" at least in M3. A similar overabundance of O is found in several metal-poor field giants (Lambert, Sneden & Ries 1974, Hartoog 1979). Worth mentioning here are the abundances of Ne and CNO group elements in the planetary nebula of M15, a cluster in which giants have a slighly lower value of $[Fe/H] = -2.0$ (Table 1) than in M3. The Hawley-Miller (1978) analysis leads to $[Ne/Fe] = +0.5$, $[O/Fe] = +0.8$ and $[N/Fe] = +1.5$. This extraordinary overabundance of N is reminiscent of the overabundance seen in M15 and M92 giant stars, which as we have seen could be primordial. As for oxygen, He-shell flashes, at least in a 7 $\mathcal{M}_\odot$ star (Iben 1975, 1976), do not seem to produce or mix significant ^{16}O to the surface; consequently the overabundance of Ne and O are also likely to be primordial.

We summarize the CNO abundances and the second parameter problem with the suggestion that, in the classical near halo clusters with $[Fe/H] < -1.5$, the CNO group may be mildly decoupled from $[Fe/H]$ and primordially somewhat overabundant on the average in certain of these clusters relative to others, particularly in M3 vis-à-vis M13, and further is inhomogeneous even within a single cluster. The situation is further confused by mixing in advanced evolutionary stages. In the outer halo clusters and dwarf spheroidals ($R > 30$ kpc), the excessive redness of the HBs probably has little to do with CNO and reflects instead the influence of some other parameter, possibly He abundance, but more probably age. If this picture is correct, and if cluster self-contamination is an important process, it may be that the near halo clusters, in comparison with the more remote stellar subsystems, retain more easily the products of first generation nucleosynthetic contamination.

Light Metals

Little is known, but Cohen (1978), in the study cited earlier, found that $[Ca/Fe]_{M13} = +0.4$ and $[Ca/Fe]_{M3} > 0$, the latter with considerable scatter. The average M13 enhancement of Ca is the same as that derived by Bell & Manduca (1978) from synthetic spectrum analysis of field RR Lyraes. This suggests that significant part of the difference between the $[Ca/H]$ and $[Fe/H]$ abundance scales for Δs in RR Lyraes (Rodgers 1974) is due to a real abundance difference between $[Ca/H]$ and $[Fe/H]$ in the same stars (Butler 1975a).

s-Process

From a spectral synthesis study of 23 giants in NGC 6397, NGC 6752, and M22, clusters with $[Fe/H] < -1.5$, Mallia (1977) found $\langle [Ba/Fe] \rangle = 0.0$ for all. In M13 Cohen (1978) reported Ba and Y slightly overdeficient, although in M3 $[Ba/Fe] \sim [Y/Fe] \sim 0.0$. Curiously, she found some evidence for an "overabundance" of the heavier s-process species in both clusters, e.g., La, Ce, and Nd, which seems at variance with expectation from theory (Seeger et al. 1965).

6 CHEMICAL HOMOGENEITY OF CLUSTER STARS VS HALO STARS

The origin of halo field stars is not clear: some or all could have been ejected from the present globular clusters, or they might have been formed in clusters that have long since dissolved. These "lost" clusters certainly would have had a different dynamical history from existing clusters, and could have had a different chemical enrichment history as well. At least one component of the halo, viz., the RR Lyraes has more or less the same (flat) abundance (i.e., $[Fe/H]$) gradient as the globular clusters (Butler et al. 1979) but the agreement in zero point of the abundances scales may be somewhat fortuitous (Kraft et al. 1979).

It is interesting therefore to inquire if the halo stars, when ordered by $[Fe/H]$, are also ordered by $[X/H]$, where $X \neq Fe$. The amount of direct evidence on this point is rather thin. In the case of α-process elements such as Ca, we have already seen that $[Ca/Fe] > 0$ in M3 and M13 by an amount more or less in agreement with that required to understand the difference in the Δs vs Fe and Δs vs Ca scales in field RR Lyraes when $[Fe/H] < -1.0$. In s-process species, Cohen (1978, 1979) reports a slight overdeficiency of Ba in M13, but $[Ba/Fe] = -0.4$ in M92, and these results agree with the magnitude of the "aging effect," as a function of $[Fe/H]$ (Spite & Spite 1978), in metal-poor halo giants.

On the other hand, in the metal-rich domain, although there is a strong correlation between CN strengths and [Fe/H] in disk giants and evidence that excessive CN strength in these stars is due to real super-metal-richness (Deming 1978), the excessive CN strengths for giants in clusters seem more likely to result from real overabundances of N. And in the metal-poor cluster domain (M92, M15), we have seen that some of the anomalies require an overabundance of N and possibly other elements in the CNO group. But Carbon et al. (1979) have noted that NH bands are not especially strong in the Sneden (1974) field giants with [Fe/H]-values similar to these clusters: the corresponding N abundances are typically 3 to 10 times smaller. If the CNO elements as a group were, in fact, anomalous in halo giants vis-à-vis cluster giants, this would be a crucial facet of any scenario of halo formation. Only a few halo giants are known: what is required is a survey technique by which halo giants can be discovered in large numbers.

ACKNOWLEDGMENTS

The author is indebted to several astronomers who communicated results in advance of publication, especially Roger Bell, Drake Deming, Dennis Butler, Judy Cohen, Bob Schommer, Bob Zinn, Allen Sweigart, Jim Hesser, Alvio Renzini, Ken Freeman, John Norris, and Dave Philip, and to the organizers of the 1978 NATO Summer School for an invitation to attend. The assistance of Jim Hesser, David Hartwick, and Bob McClure in the preparation of Figure 2 is gratefully acknowledged, as is the general assistance of Nick Suntzeff and Ed Kemper. Useful conversations were had with Ed Langer and Sandra Faber.

This review was written with partial support from NSF Grant AST 78-08612.

Literature Cited

Aaronson, M., Cohen, J. G., Mould, J., Malkan, M. 1978. *Ap. J.* 223:824
Arnett, W. D. 1972. *Ap. J.* 176:681, 699
Auer, L. H., Norris, J. 1974. *Ap. J.* 194:87
Bell, R. A., Dickens, R. J. 1974. *MNRAS.* 166:89
Bell, R. A., Gustafsson, B. 1978. *Astron. Astrophys. Suppl.* 34:229
Bell, R. A., Gustafsson, B. 1979. *Astron. Astrophys.* In press
Bell, R. A., Dickens, R. J., Gustafsson, B. 1978. In *Symp. Important Advances in 20th Century Astronomy*, Copenhagen
Bell, R. A., Manduca, A. 1978. *Ap. J.* 225:908
Bond, H. E. 1975. *Ap. J. Lett.* 202:L47
Burstein, D. 1979. In preparation
Butler, D. 1975a. *Ap. J.* 200:68

Butler, D. 1975b. *Publ. Astron. Soc. Pac.* 87:559
Butler, D., Bell, R. A., Dickens, R. J., Epps, E. 1978. In *IAU Symp. No. 80*, ed. W. B. Burton. Dordrecht: Reidel. In press
Butler, D., Kinman, T. D., Kraft, R. P. 1979. *Astron. J.* In press
Cannon, R. D., Stobie, R. S. 1973. *MNRAS.* 162:227
Canterna, R. 1976. *Astron. J.* 81:228
Canterna, R., Schommer, R. A. 1978. *Ap. J. Lett.* 219:L119
Carbon, D. F., Butler, D., Kraft, R. P., Nocar, J. L. 1977. In *CNO Isotopes in Astrophysics*, ed. J. Andouze, p. 33. Dordrecht: Reidel
Carbon, D. F., Langer, G. E., Butler, D.,

Kraft, R. P., Trefzger, Ch. F., Suntzeff, N., Kemper, E., Nocar, J. 1979. *Astron. J.* In press

Castellani, V., Tornambè, A. 1977. *Astron. Astrophys.* 61:427

Chun, M.-S. 1976. *Observational Evidence for Radial Inhomogeneities in Globular Clusters.* PhD thesis. Aust. Natl. Univ., Canberra

Ciardullo, R. B., Demarque, P. 1977. *Yale Trans.* 33:1

Cohen, J. G. 1978. *Ap. J.* 223:487

Cohen, J. G. 1979. *Ap. J.* In press

Cohen, J. G., Frogel, J. A., Persson, S. E. 1978. *Ap. J.* 222:165

Cowley, A. P., Hartwick, F. D. A., Sargent, W. L. W. 1978. *Ap. J.* 220:453

Da Costa, G. 1978. *On Radial Abundance Gradients in Globular Clusters.* Presented at NATO Cambridge Summer School on Globular Clusters

Danziger, I. J. 1970. *Ann. Rev. Astron. Astrophys.* 8:161

Dearborn, D. S. P., Eggleton, P. P., Schramm, D. N. 1976. *Ap. J.* 203:455

Demers, S. 1978. Paper presented at NATO Cambridge Summer School on Globular Clusters

Deming, D. 1978. *Ap. J.* 222:246

Deming, D., Olson, E. C., Yoss, K. M. 1977. *Astron. Astrophys.* 57:417

Deupree, R. G. 1977. *Ap. J.* 214:502

Dickens, R. J. 1972a. *MNRAS.* 157:292

Dickens, R. J. 1972b. *MNRAS.* 159:7P

Dickens, R. J., Bell, R. A. 1976. *Ap. J.* 207:506

Eggen, O. J. 1972. *Ap. J.* 172:639

Eggen, O. J., Lynden-Bell, D., Sandage, A. R. 1962. *Ap. J.* 136:748

Faber, S. M. 1977. In *The Evolution of the Galaxies and Stellar Populations*, ed. B. M. Tinsley, R. B. Larson, p. 157. New Haven: Yale Printing Service. 449 pp.

Faulkner, J. 1966. *Ap. J.* 144:978

Freeman, K. C. 1978. *Populations in Globular Clusters.* Presented at NATO Cambridge Summer School on Globular Clusters

Freeman, K. C. Rodgers, A. W. 1977. *Ap. J. Lett.* 201:L71

Frogel, J. A., Persson, S. E. 1978. *Infrared Observations of Globular Clusters.* Presented at NATO Cambridge Summer School of Globular Clusters

Frogel, J. A., Persson, S. E., Cohen, J. G. 1978. *Ap. J.* 222:165

Fusi-Pecci, F., Renzini, A. 1975. *Astron. Astrophys.* 39:413

Gascoigne, S. C. B., Burr, E. J. 1956. *MNRAS.* 116:570

Gingold, R. A. 1974. *Ap. J.* 193:177

Gingold, R. A. 1977. *MNRAS.* 178:533

Gott, J. R., Gunn, J. E., Schramm, D. N., Tinsley, B. M. 1974. *Ap. J.* 194:593

Grasdalen, G. L. 1974. *Astron. J.* 79:1047

Greenstein, J. L., Münch, G. 1966. *Ap. J.* 146:618

Harding, G. A. 1962. *Observatory.* 82:205

Harris, W. E. 1976. *Astron. J.* 81:1095

Harris, W. E. 1978. Paper presented at NATO Cambridge Summer School on Globular Clusters

Harris, W. E., van den Bergh, S. 1974. *Astron. J.* 79:31

Hartoog, M. 1979. In preparation

Hartwick, F. D. A. 1968. *Ap. J.* 154:475

Hartwick, F. D. A., McClure, R. D. 1972. *Ap. J. Lett.* 176:L57

Hartwick, F. D. A., Van den Berg, D. A. 1973. *Publ. Astron. Soc. Pac.* 85:355

Hawley, S. E., Miller, J. S. 1978. *Ap. J.* 220:609

Helfer, H. L., Wallerstein, G., Greenstein, J. L. 1959. *Ap. J.* 129:700

Hesser, J. E. 1978. *Ap. J. Lett.* 223:L117

Hesser, J. E., Hartwick, F. D. A., McClure, R. D. 1977. *Ap. J. Suppl.* 33:471

Hodge, P. W. 1971. *Ann. Rev. Astron. Astrophys.* 9:35

Iben, I. 1964. *Ap. J.* 140:1631

Iben, I. 1967. *Ann. Rev. Astron. Astrophys.* 5:571

Iben, I. 1968. *Ap. J.* 154:581

Iben, I. 1974. In *Ann. Rev. Astron. Astrophys.* 12:215

Iben, I. 1975. *Ap. J.* 196:525, 549

Iben, I. 1976. *Ap. J.* 208:165

Iben, I. 1978. *Nucleosynthesis in Evolved Globular Cluster Stars.* Presented at NATO Cambridge Summer School on Globular Clusters

Iben, I., Rood, R. T. 1969. *Nature* 223:933

Iben, I., Truran, J. W. 1978. *Ap. J.* 220:980

Iben, I., Rood, R. T., Strom, K. M., Strom, S. E. 1969. *Nature* 224:1006

Janes, K. A. 1975. *Ap. J. Suppl.* 29:161

Johnson, H. L., Morgan, W. W. 1953. *Ap. J.* 117:313

Johnson, S. L., McNamara, D. H. 1969. *Publ. Astron. Soc. Pac.* 81:415

Kinman, T. D. 1959. *MNRAS.* 119:499, 538

Kraft, R. P. 1970. In *Spectroscopic Astrophysics*, ed. G. H. Herbig, p. 385. Berkeley: Univ. Calif. Press

Kraft, R. P., Trefzger, Ch. F., Suntzeff, N. 1979. In *IAU Symposium No. 84*, ed. W. B. Burton. In press

Kron, G. E., Guetter, H. H. 1976. *Astron. J.* 81:817

Kron, G. E., Mayall, N. U. 1960. *Astron. J.* 65:581

Lambert, D. L., Sneden, C., Ries, L. M. 1974. *Ap. J.* 188:97

Lloyd Evans, T. 1977. *MNRAS.* 178:345

Lloyd Evans, T. 1978. *MNRAS.* 182:293

Mallia, E. A. 1975. *MNRAS*. 170:57P

Mallia, E. A. 1976. *MNRAS*. 177:73

Mallia, E. A. 1977. *Astron. Astrophys*. 60:195

McClure, R. D., Norris, J. 1977. *Ap. J. Lett.* 217:L101

McClure, R. D., van den Bergh, S. 1968. *Astron. J.* 73:313

McDonald, L. H. 1977. *Properties of RR Lyrae Variables*. PhD thesis. Univ. Calif., Santa Cruz

Miller, J. S., Mathews, W. G. 1972. *Ap. J.* 172:593

Morgan, W. W. 1956. *Pub. Astron. Soc. Pac.* 68:509

Newell, E. B. 1970. *Ap. J.* 159:443

Norris, J. 1978. *The Correlation of Cyanogen, Calcium, and the Heavy Elements on the Giant Branch of Omega Centauri*. Presented at NATO Cambridge Summer School on Globular Clusters

Norris, J., Bessell, M. S. 1977. *Ap. J. Lett.* 211:L91

Norris, J., Zinn, R. 1977. *Ap. J.* 215:74

Oosterhoff, P. T. 1939. *Observatory* 72:104

Osborn, W. H. 1973. *Ap. J.* 186:725

Paczyński, B., Tremaine, S. D. 1977. *Ap. J.* 216:57

Pagel, B. E. J. 1966. In *Colloquium on Late Type Stars*, ed. M. Hack, p. 133. Trieste: Osservatorio Astronomico

Peimbert, M. 1975. *Ann. Rev. Astron. Astrophys.* 13:113

Philip, A. G. D. 1977. Private communication

Pilachowski, C. 1978. *Ap. J.* 224:412

Preston, G. W. 1959. *Ap. J.* 130:507

Preston, G. W. 1961. *Ap. J.* 134:633, 651

Pritchet, C. 1977. *Astron. J.* 82:471

Racine, R. 1973. *Astron. J.* 78:180

Renzini, A. 1977. *Advanced Stages in Stellar Evolution*. Presented at 7th Advanced Course Saas-Fee, Switzerland

Renzini, A. 1978. *Influence of Core Rotation on HB Morphology*. Presented at NATO Cambridge Summer School on Globular Clusters

Robinson, L. B., Wampler, E. J. 1972. *Publ. Astron. Soc. Pac.* 84:161

Rodgers, A. W. 1974. *Ap. J.* 191:433

Rodgers, A. W. 1978. *The Metal Abundance of Giants in ω Cen*. Presented at NATO Cambridge Summer School on Globular Clusters

Rood, R. T. 1972. *Ap. J.* 177:681

Rood, R. T. 1973. *Ap. J.* 184:815

Rood, R. T. 1978. *The Effect of Variations of CNO/Fe on Globular Cluster Color Magnitude Diagrams*. Presented at NATO Cambridge Summer School on Globular Clusters

Sandage, A. R. 1964. *Ap. J.* 139:442

Sandage, A. R. 1969a. *Ap. J.* 157:515

Sandage, A. R. 1969b. *Ap. J.* 158:1115

Sandage, A. R. 1970. *Ap. J.* 162:841

Sandage, A. R., Katem, B. 1977. *Ap. J.* 215:62

Sandage, A. R., Smith, L. L. 1966. *Ap. J.* 144:886

Sandage, A. R., Wallerstein, G. 1960. *Ap. J.* 131:598

Sandage, A. R., Wildey, R. 1967. *Ap. J.* 150:469

Sandage, A. R., Katem, B., Johnson, H. L. 1977. *Astron. J.* 82:389

Sargent, W. L. W., Searle, L. V. 1967. *Ap. J. Lett.* 150:L33

Schommer, R. 1978. Paper presented at NATO Cambridge Summer School on Globular Clusters

Searle, L., Sargent, W. L. W. 1972. *Ap. J.* 173:25

Searle, L., Zinn, R. 1978. *Ap. J.* 225:357

Seeger, P. A., Fowler, W. A., Clayton, D. D. 1965. *Ap. J. Suppl.* 11:121

Sneden, C. 1974. *Ap. J.* 189:493

Spite, M., Spite, F. 1978. *Astron. Astrophys.* 67:23

Sweigart, A. V. 1974. *Ap. J.* 189:289

Sweigart, A. V., Gross, P. G. 1976. *Ap. J. Suppl.* 32:367

Sweigart, A. V., Gross, P. 1978. *Ap. J. Suppl.* 36:405

Sweigart, A. V., Mengel, J. G. 1979. *Ap. J.* 229:624

Thomas, H. C. 1967. *Z. Ap.* 67:420

Trimble, V. I. 1975. *Rev. Mod. Phys.* 47:877

Truran, J. W., Arnett, W. D. 1971. *Astrophys. Space Sci.* 11:430

van Albada, T. S., Baker, N. H. 1973. *Ap. J.* 185:477

van den Bergh, S. 1967. *Astron. J.* 72:70

van den Bergh, S. 1969. *Ap. J. Suppl.* 19:145

van den Bergh, S. 1975. *Ann. Rev. Astron. Astrophys.* 13:217

Wallerstein, G. 1962. *Ap. J. Suppl.* 6:407

Wallerstein, G., Carlson, M. 1960. *Ap. J.* 132:276

Wallerstein, G., Greenstein, J. L. 1964. *Ap. J.* 139:1163

Wallerstein, G., Helfer, H. L. 1966. *Astron. J.* 71:350

Wallerstein, G., Helfer, H. L., Greenstein, J. L. 1963. *Ap. J.* 138:97

Woltjer, L. 1975. *Astron. Astrophys.* 42:109

Woosley, S. E. 1974. Paper presented at NATO Cambridge Summer School on the Origins and Abundances of the Chemical Elements

Zinn, R. 1973. *Ap. J.* 182:183

Zinn, R. 1977. *Ap. J.* 218:96

Zinn, R. 1978a. *Ap. J.* 225:790

Zinn, R. 1978b. *Dwarf Galaxies*. Presented at NATO Cambridge Summer School on Globular Clusters

Ann. Rev. Astron. Astrophys. 1979. 17:345–85
Copyright © 1979 by Annual Reviews Inc. All rights reserved

COMPACT H II REGIONS AND OB STAR FORMATION

✻2154

H. J. Habing
Sterrewacht, Huygens Laboratorium, Leiden, The Netherlands

F. P. Israel
Owens Valley Radio Observatory, California Institute of Technology, Pasadena, California 91125

1 INTRODUCTION

1.1 Short History

In 1967 radio observations led to the discovery of very small ($d < 0.5$ pc) and dense ($n_e > 10^4$ cm^{-3}) H II regions. Ryle & Downes (1967) found that the obscured object DR 21 has a very high emission measure E.M. ($= n_e^2 d) = 5 \times 10^7$ pc cm^{-6}, while Mezger, Schraml & Terzian (1967a) deduced from the measured radio spectrum that the similarly obscured galactic source W49A contains a high density component with E.M. = 10^8 pc cm^{-6}. Subsequently, Mezger et al. (1967b) coined the term "compact H II regions," showed the close relationship to OH main-line masers, and proposed that compact H II regions/OH masers should be recognized as a new class of galactic sources. The recombination rate of the compact H II regions requires ionization by one or more early O type star; from the observed high turbulent velocities ($V \approx 25$ km s^{-1}) Mezger et al. (1967b) concluded that the objects must be in a state of rapid expansion, and must represent a very early stage in the life of an OB star. In fact, Vandervoort (1963) and W. G. Mathews (1965) had already pointed out that H II regions with such densities must of necessity undergo rapid expansion and therefore have short lifetimes.

The discovery of compact H II regions coincided with their theoretical prediction by Davidson & Harwit (1967), who also drew attention to the strong infrared emission expected from the (predicted) dust cocoon (see also Davidson 1970). The first compact H II regions to be recognized as strong sources of near-infrared emission were W 51-IRS2 and K3-50

345

0066-4146/79/0915-0345$01.00

(Neugebauer & Garmire 1970). A highlight was the study of the obscured complex region W3 at radio and infrared wavelengths (Wynn-Williams 1971, Wynn-Williams et al. 1972) that, among other things, showed the existence of a compact infrared source (W3-IRS5) *without* detectable radio emission, but with associated H_2O maser emission. A similar source (the well-known BN object) had already been discovered earlier by Becklin & Neugebauer (1967). Low & Aumann (1970) discovered that H II regions containing compact components are often strong far-infrared emitters, while several authors later observed a close association of large neutral clouds with H II regions in general.

1.2   Compact H II Regions and OB Star Formation

The study of OB associations and of associated large molecular clouds has led to the following star formation scenario (Blaauw 1964, Elmegreen & Lada 1977, Sargent 1979, Blitz 1979 and references therein): the stars appear to form progressively in large, elongated molecular clouds over a time of $\lesssim 10^8$ yr. During this time we observe the formation of subsequent groups of stars each on a very short time scale (10^5 yr) in adjacent parts of the clouds. It is plausible that OB stars form in ways different from those of lower mass. Blaauw (1964) has noted that the mass spectrum of stars in associations is much flatter than that of stars in clusters; he suggested that different mechanisms may be involved. This is to be expected; the concept of star formation implies the compression of interstellar gas with typical densities of 10^{-20} g cm^{-3} to stellar gas with densities of 1 g cm^{-3}. One suspects that several different processes and sequences can achieve this so that the end products (stars) will not always be identical.

The same compression factor of 10^{20} makes it necessary to distinguish different stages in the star formation process: the onset of the gravitational collapse, the collapse (and fragmentation) phase, the start of nuclear reactions, the separation of the star from its birthplace. For OB stars at least, the first stage can in principle be studied through molecular line observations at millimeter and centimeter wavelengths. The subsequent stage occurs within dark clouds of high densities ($\lesssim 10^3$ cm^{-3}). Observations of objects from this stage are at present not available and will become possible only through spectroscopy with high sensitivities in the far infrared. In this review we are interested in objects in the last stages.

Direct observational evidence linking OB star associations with compact H II regions is scarce, presumably because most observed OB associations represent a much later stage in the life of a star-forming region than (compact) H II regions. A very interesting exception is the situation near the OB association Cep OB3 (Figure 1), where indications

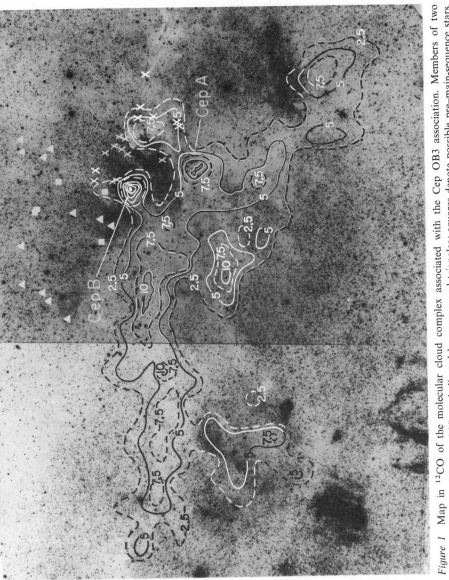

Figure 1   Map in ^{12}CO of the molecular cloud complex associated with the Cep OB3 association. Members of two different subgroups (Blaauw 1964) are indicated by crosses and triangles; squares denote possible pre–main-sequence stars. At present, star formation seems to be taking place in Cep A, and possibly in Cep B (adapted from Sargent 1979).

for ongoing star formation (compact infrared source; OH and H_2O maser sources) are found at the interface of Cep OB3 and its associated molecular cloud (Sargent 1979). Individual O stars such as exciting stars of (extended) H II regions are often associated with compact H II regions (see Section 3.4). Moreover, several groups of compact and extended H II regions can only be interpreted as OB associations in statu nascendi. Good examples are W3 (Mezger & Wink 1974), W58 (Israel 1976a), and S254/S258 (Israel 1976c). Each of these groups must contain at least half a dozen young OB stars with a separation, and in a total volume typical of that of an OB association.

The collapse of a massive star is completed in about 10^5 years (Larson 1973). The formation of a compact H II region will take up to this amount of time. Thus a compact H II region is observed at the spot where it has formed, and its surroundings will at least initially contain information on the process of formation. However, this information will soon be obliterated by the intense dynamical interaction between the newly formed star and its surroundings—it is therefore of interest to study the most compact (e.g. the smallest and densest) and therefore youngest objects in the most detail.

1.3 *Scope of This Review*

We review studies of compact objects (H II regions, IR sources, and interstellar masers) with diameters $d < 0.5$ pc and densities $n > 5 \times 10^3$ cm^{-3}, associated with stars with luminosities $L > 10^4 L_\odot$, i.e. with early-type stars of spectral types O4 through B0 and masses ranging from $\approx 60\ M_\odot$ to $\approx 10\ M_\odot$. Of course, in the places where such stars form there will probably be simultaneous formation of stars of lower mass (at least B–early A), and in that sense our discussion may be relevant also for those stars. But for a more explicit discussion of the formation of lower mass stars we refer the reader elsewhere, e.g. to the proceedings of recent conferences (Strom et al. 1975, de Jong & Maeder 1977, Solomon 1979, Gehrels 1979).

Since we want to stress the interaction of exciting stars, compact H II regions, and their neutral surroundings, we also briefly discuss objects that are not covered by the definition given above, but are clearly related to compact objects, and indirectly yield information on them. This is, for instance, the case with some observations of extended H II regions, and with observations of molecular clouds. With a few exceptions (e.g. Zuckerman 1973 and Becklin et al. 1976 for Orion A), optical observations have shed little light on the subject at hand; therefore we do not discuss the wealth of detailed optical observations on extended H II regions (see, for example, Meaburn 1975, 1977). We also exclude objects such as bipolar nebulae, since their evolutionary stage is too little understood at this moment.

2   OBSERVATIONS AND INTERPRETATIONS— GENERAL REMARKS

The observations of compact H II regions and infrared sources suffer from some important limitations, even though these are not always fundamental and may be overcome in the future. We are interested in the physical parameters of individual components (ionized gas, dust, neutral material); and these can be derived only if distance uncertainties, confusion, and radiative transfer problems are solved. In the following we discuss present problems in more detail.

2.1   Distances

Distances are sometimes poorly known. For nearby objects they can often be obtained via associated optical objects (Georgelin et al. 1973, Crampton et al. 1978). But for distant, obscured sources (W49, DR21) only a kinematic distance determination (from recombination line or foreground H I absorption lines) is possible. Apart from a fundamental ambiguity ("near" or "far" distance?) considerable errors can occur due to systematic deviations from circular galactic rotation. (See, for example, Caswell et al. 1975, Vogt & Moffat 1975). Since the ionized gas may move away from the molecular cloud (cf the "blister" model, Section 3.3) the radial velocity of the associated molecular cloud should be used for the kinematic distance, and not the radial velocity of the H II region. We expect that the distances to most H II regions are known to within a factor of two. The distance enters as a parameter in all derived quantities except in the surface brightness (e.g. the emission measure $n_e^2 d$). The derived mass of the emitting material is affected most strongly by distance uncertainties.

2.2   Resolution and Sensitivity

Resolution and sensitivity are best for the radio observations (~ 2 arcsec and ~ 5 mJy[1] for radio continuum; ~ 0.001 arcsec and ~ 100 mJy for VLB interferometric maser line observations), reasonable for near-infrared observations (typically ~ 5 arcsec and 500 mJy at 10 μm), and poor at far-infrared wavelengths (at best 20 arcsec; 5 Jy). Molecular line observations seldom have a resolution better than 1 arcmin.

[1] 1 Jy = 1 f.u. = 10^{-26} W m^{-2} Hz^{-1}.

A resolution of a few arcsec ($\sim 10^{16}$ cm at 1 kpc distance) is usually sufficient to *separate* different sources; it is usually not good enough to *resolve* the observed source significantly. As a consequence, often even the projected two-dimensional source geometry is not known. In no case is the spatial three-dimensional source structure known. Confusion problems arise with poorer resolution: sources blend together, or nonexistent extended sources appear because of projection effects (see, for example, Panagia et al. 1978). Geometrical information is, however, vital if one wants to transform observed parameters such as flux density and angular size into physical parameters such as electron density.

2.3 *Continuum Radiation*

The radiation received is often continuum emission, and spectral lines, with all they have to offer for interpretation (see Section 2.5), are lacking. In spite of this lack, some overall parameters can be derived. Suppose that the source is optically thin; optically thick radiation is discussed in Section 2.4. The measured flux density of the source is then given by $D^{-2} \int \varepsilon \, dV$, where D is the distance of the source, ε is the (volume) emission coefficient, and V the volume. If D is known and the angular diameter can be measured, one can estimate V (e.g by assuming spherical symmetry) and thus obtain an estimate of $\langle \varepsilon \rangle$.

The *radio* continuum is produced by bremsstrahlung; ε is proportional to $n_e^2 T_e^{-0.35}$, where T_e is the electron temperature. Since T_e is expected to be constant to within a factor of 1.5, one obtains a reliable estimate of $\langle n_e^2 \rangle$, and thus of the rms value of n_e. Other important parameters are the total number of recombinations inside the nebula (proportional to the optically thin radio flux density) and the mass of ionized gas, M_i, which is proportional to $\int n_e \, dV$. By replacing n_e by its rms value, one obtains an upper limit for M_i.

When the *infrared* continuum is measured at a sufficient number of wavelengths one obtains the total luminosity of the source.[2] But other parameters are difficult to obtain, even at small optical depths. At short wavelengths ($\lambda \lesssim 3.5 \ \mu m$) the received emission may be produced by a variety of mechanisms: thermal dust emission, bremsstrahlung, scattered starlight, and even direct starlight. At longer wavelengths the emission mechanism is uniquely determined (thermal dust emission), but there

---

[2] Here it should be noted that the radio and infrared emitting volumes usually are by no means coincident: near-infrared emission due to hot dust may originate from beyond the boundaries of the ionized radio emitting volume, and far-infrared emission due to cooler dust originates largely from parts of the neutral cloud adjacent to the ionized H II region; if the stellar heat source is of sufficiently late type, there may not even be an ionized region associated with the far-infrared emission.

$\varepsilon = K_v B_v(T_d)$ is poorly known (a) because the absorption coefficient K_v is poorly known (see, for example, Andriesse 1974) and (b) because the Planck function $B_v(T_d)$ depends strongly on the dust temperature T_d. As a result, it is difficult to disentangle density and temperature information from the observed value of $\langle \varepsilon \rangle$. Fundamentally important data such as the total mass of emitting dust and the dust temperature are often only very poorly determined from infrared continuum measurements. However, at very long infrared wavelengths ($\sim 1000 \ \mu$m) where the optical depth is certainly very small, reliable estimates of the surface density of the dust and the total dust mass can be obtained if one can guess a reliable value of T_d (cf Werner et al. 1975, Righini-Cohen & Simon 1977).

2.4   Effects of Significant Optical Depths

Considerable optical depths τ can occur in the dense objects under consideration. (a) The bremsstrahlung absorption coefficient is proportional to $n_e^2 \lambda^2$, both at radio wavelengths and in the infrared (λ is the wavelength of observation) and is therefore smallest at the shortest observable wavelength (in practice 2–6 cm for aperture synthesis instruments). However, even then $\tau \gg 1$ may occur through the densest parts. Under those conditions, the derivation of useful source parameters is often difficult and sometimes impossible. They are most frequently obtained from the spectrum, using idealized models (spherical, cylinder, Gaussian, or homogeneous density distribution; see, for example, Mezger et al. 1967a, Mezger & Henderson 1967, Rubin 1968). Although the results are useful for the comparison of one source of bremsstrahlung to another, they may be completely unrealistic in a physical sense. So the comparison with parameters derived, for example, for models from infrared observations can be hazardous. In particular the effects of density clumping and density gradients are usually ignored. In addition, if the radius of the ionized sphere where $\tau \geq 1$ becomes less than 10^{16} cm, the surface area is so small that the object is undetectable at a typical distance of 1 kpc. The smallest (that is, youngest) compact H II regions thus cannot be detected by radiotelescopes.[3] (b) In the infrared, large optical depths sometimes also occur. The dust absorption coefficient is a decreasing function of wavelength ($\propto \lambda^{-1}$ or λ^{-2}) with, however, a sharp maximum at 9.7 μm due to silicates. Large optical depths (due to material outside the ionized region) are occasionally present at shorter wavelengths ($\lambda < 20 \ \mu$m). This may make a source undetectably weak; in other cases it renders the interpretation of the short wavelength infrared continuum very difficult.

---

[3] Thus not all compact H II regions will be radio sources; some may only show up as infrared sources with Brackett-line emission (cf Section 4.2).

2.5   *Spectral Lines*

Spectral lines are the richest source of information on the physical condition and composition of both the ionized and neutral material. In the *infrared*, several lines have been discovered (see Section 4.2), but identification is still a problem for some. Moreover, the measured line intensities in most cases are too low to enable easy detection or mapping with present-day detectors. Clearly, the field has only started to open up.

At *radio* wavelengths the spectral lines of importance are the molecular lines and recombination lines. Molecular lines from material around the compact sources (see also Section 6) suffer from lack of spatial resolution and often from large optical depths and/or lack of knowledge of the excitation conditions. The recombination lines are influenced by nasty transfer effects caused by negative absorption. Since compact H II regions are often close to more diffuse regions of ionized gas, the line profile is

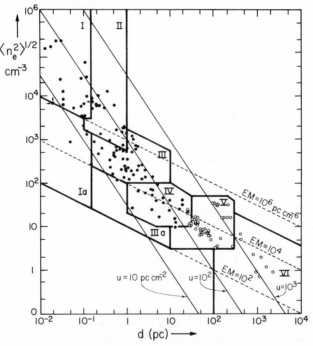

Figure 2   Plot of rms electron density (n_e) versus linear size (d) for individual H II region components. Lines of constant excitation parameter (u) and constant emission measure (E.M.) are marked. Filled circles are galactic H II region components, open circles are extragalactic components, all observed with the Westerbork Synthesis Radio Telescope. Heavy lines and Roman numerals indicate the classification scheme outlined in Table 1. From Israel (1976b).

difficult to interpret even at low optical depth (see Brown et al. 1978). The radiative transfer effects may be overcome, at least partially, when the observations are made with sufficient spatial resolution and at sufficiently high frequencies. With present-day aperture synthesis instruments at least sufficient resolution is attainable for the stronger recombination lines (H 109α; Wellington et al. 1976, 1977, J. H. van Gorkom, private communication).

2.6   Sky Surveys

Systematic, complete, and reliable sky surveys at high sensitivity and high angular resolution do not exist. *Radio continuum* surveys always have angular resolutions of more than 2 arcmin and miss several interesting sources (e.g. ON1). The reason for this is that at such resolutions, extended H II regions dominate the emission from a complex, while the very few isolated compact H II regions (cf Section 3.4) have flux densities too low to detect. Nevertheless, high-emission measure sources found in single-dish surveys are usually good hunting grounds for searches with aperture synthesis instruments (cf the observations of W33 by Goss et al. 1978). No systematic and complete survey has yet been published in the *maser lines*. In the *infrared* the only relevant catalogue, the AFGL (Price & Walker 1976), is statistically unreliable at the lower flux density levels, but not at the higher flux density levels (Harris & Rowan-Robinson 1977). A most worthwhile contribution may come from the complete and deep infrared sky survey that is to be carried out with the IRAS satellite in 1981.

3   RADIO OBSERVATIONS

In this section we make particular use of the theses by Israel (1976b) and Gilmore (1978), who have observed a large number of H II regions.

3.1   Classification of H II Regions

A convenient way to characterize a compact H II region is by two physical parameters: its size (d) and the rms electron density (n_e). The values of d and n_e can be derived directly from the observations, provided that the source is optically thin, that it is resolved by the telescope beam, and that the distance is known (see Section 2.3). Figure 2 shows the distribution of n_e and d of a large sample of H II regions (Israel 1976b). (This diagram is strictly meant as an illustration of the values encountered. It should not be used for statistical purposes, because the observed sample is not well defined.) Figure 2 leads naturally to an empirical scheme for classifying compact H II regions according to their n_e and d

Table 1 Classification of H II regions[a]

	Definition			Properties						
Class (1)	n_e(cm^{-3}) (2)	d(pc) (3)	E.M. (pc cm^6) (4)	Nature (5)	Examples (6)	Optical appearance (7)	Group properties (8)	OH/H$_2$O masers (9)	S(20μm)/ S(6cm) (10)	Ionized mass ($M_\odot$) (11)
I	>3000	<0.15	>10^6	Ultra-compact	W3OH, S157B	Obscured	Never isolated	+	<10^4	≈10^{-2}
II	>1000	0.1–1.0	>10^5	Compact	W3A, K3-50	Strongly reddened	Never isolated	–	<10^3	≈1
III	100–3000	0.15–10	1×10^4– 3×10^6	Dense	S158, S90	Partially obscured	Sometimes isolated	–	<5×10^2	≈10
IV	100–100	1–30	5×10^2– 1×10^5	Classical	S104, S162	Globules, bright rims	Sometimes in groups	–	<10^2	≈10–5×10^2
V	3–50	10–300	<5×10^5	Giant	Core, NGC5461	Complex or diffuse	–	–	–	≈500–5×10^6
VI	10	>100	<1×10^5	Super-giant	Envelope, NGC5461	Structureless	–	–	–	≈10^6–10^8

[a] From Israel (1976b).

values (see Table 1). It turns out that other properties of compact H II regions correlate well with the position of (n_e, d) in Figure 2 (see Columns 5–11). The classification is roughly equivalent with an evolutionary sequence; for complicating factors see Israel (1976b) and Section 3.4.

H II regions in Class I (ultracompact) and Class II (compact) are the ones of interest to us, especially in connection with two additional requirements: (*a*) the requirement that the star ionizing the region is inside the densest part of the region; that is, bright rims, elephant trunks, and externally ionized globules are excluded; and (*b*) the requirement that sufficient circumstantial evidence exists that the object is young (i.e. other "signposts" of star formation should occur nearby). Ultracompact H II regions are never (optically) visible; compact H II regions sometimes.

3.2   The Sample of Known Ultracompact and Compact H II Regions

Over a hundred ultracompact and compact H II regions have now been detected at radio wavelengths. The precise number is hard to determine because of instrumental resolution effects (cf Section 2), confusion with objects as bright rims, and incomplete observations. Most (ultra)compact H II regions have been discovered by aperture synthesis observations of thermal radio sources with high surface brightness that were found in low-resolution, single-dish surveys such as those by Altenhoff et al. (1970), Kazès et al. (1975), and Goss & Shaver (1970). Recent aperture synthesis surveys of optically visible H II regions have yielded a surprisingly small number of new (ultra)compact H II regions (Israel 1977b, Felli et al. 1978, Graf 1978). A similar result was obtained for dark clouds in surveys by Myers (1977) and by Gilmore (1978). Brown & Zuckerman (1975) and Falgarone et al. (1978) found only two ultracompact H II regions in the active ρ Oph dark cloud. In Table 2 we list all regions where to our knowledge ultracompact or compact components have been found.

3.3   Structure of Compact H II Regions; "Blisters"

Well-resolved and more extended H II regions often show a structure that is best described as a bright, ridgelike core surrounded by a less bright envelope (Israel 1978). This kind of structure is very common and requires an explanation. Following the proposal originally made for Orion A (Zuckerman 1973, Balick et al. 1974) and Orion B (Grasdalen 1974), Israel (1976b, 1978) explains the core-envelope structure by assuming that H II regions generally have a "blister" type structure. The ionizing star is embedded in a half-open cavity in a molecular cloud; inside the cavity, the ionized gas has a very high density near the ioniza-

Table 2 Star Forming regions in the Galaxy

Name (1)	α(1950.0) (2)	δ(1950.0) (3)	l (4)	b (5)	Class I/II compact H II region[a] (6)	Extended H II region[a] (7)	Compact near-IR source[a] (8)	OH maser[a] (9)	H₂O maser[a] (10)	D (kpc)[b] (11)	Reference[c] (12)
S184	00 49.8	+56 15	123.2	−6.5	+	+	0	0	+	2.0	1,2
S187	01 19.9	+61 33	126.7	−1.0	+	+	0	0	+	3.0	3,4
W3 Main	02 22.0	+61 53	133.7	+1.2	+	+	+	−	+	3.0	5–13
W3 OH	02 21.9	+61 52	133.7	+1.2	+	+	+	+	+	3.0	6,9,14–19
IC1848	02 57.6	+60 17	138.3	+1.5	0	+	+	0	0	2.5	20,21,136
LKHα101	04 26.9	+35 10	165.4	−9.1	+	−	+	−	−	0.8	22–25
S228	05 10.0	+37 24	169.2	−1.0	−	+	+	0	−	2.6	1,26
Orion A	05 32.8	−05 26	208.9	−19.3	+	+	+	+	+	0.5	9,11,27–31
S235	05 37.5	+35 40	173.7	+2.7	+	+	+	−	+	2.5	9,20,32–34
Orion B	05 39.2	−01 58	206.5	−16.4	+	+	+	0	+	0.4	9,31,35–37
NGC2071	05 44.5	+00 21	205.1	−14.1	−	+	+	+	+	1.3	31,38–40
MonR2 (NGC2170)	06 05.3	−06 23	213.7	−12.6	+	+	+	+	+	1.0	9,31,41,42,135
S247	06 05.6	+21 36	189.8	+0.8	0	+	0	+	0	2.5	43
S252	06 06.0	+20 40	189.9	+0.5	+	+	0	0	+	2.0	44,45
S255/7	06 10.0	+18 01	192.6	−0.0	+	+	+	+	+	1.5	9,20,31,46,47, 49
S269	06 11.8	+13 51	196.5	−1.7	+	+	+	+	+	2.0	9,20,31,47
NGC2264	06 38.4	+09 32	203.4	+2.0	0	+	+	+	0	0.9	9,50
AFGL1074	07 05.5	−10 39	224.3	−1.3	0	0	+	0	+	0.1?	51,52
OH0739	07 40.0	−14 36	231.8	+4.2	−	−	0	+	0	1.3	9,31,34,53,54
RCW48	10 19.8	−57 50	284.2	−0.8	0	0	0	+	0	5.4	31
G285.2	10 29.5	−56 47	285.2	+0.0	0	+	0	+	+	0.2 (5.1)	31,55,56

Name	α(1950)	δ(1950)	l	b							Ref
NGC3603	11 12.9	−60 53	291.6	−0.4	+	+	○	+	○	3.5	31,55,57,58
RCW65	12 32.0	−61 23	301.1	+1.1	−	+	○	○	○	4.9	31
G305.2	12 07.9	−62 19	305.2	+0.2	+	+	○	+	○	2.4	59
G305.4	13 09.3	−62 21	305.4	+0.2	+	+	○	+	○	8.0	31,58
G308.9	13 39.6	−61 54	308.9	+0.1	−	+	○	○	○	3.7 (8.9)	31
G309.9	13 47.2	−61 20	309.9	+0.5	−	+	○	○	○	5.8 (7.1)	31
G316.6	14 40.0	−59 42	316.6	−0.1	+	+	○	+	○	12.1	59
G320.3	15 06.0	−58 16	320.2	−0.3	−	+	○	+	○	5.0 (10.4)	31
G324.2	15 29.0	−55 45	324.2	+0.1	−	+	○	○	○	6.9 (9.3)	31
RCW97	15 49.2	−54 28	327.3	−0.6	+	+	○	+	○	3.5	31,55,58
G328.2	15 54.0	−53 50	328.2	−0.5	−	+	○	○	○	3.0	31
G329.0	15 56.7	−53 04	329.0	−0.2	○	+	○	○	○	3.7	31
G329.4	15 59.7	−53 02	329.4	−0.5	○	+	○	○	○	5.1 (12.2)	31
G330.9	16 06.5	−51 58	330.9	−0.4	−	+	○	+	○	4.7	31
G331.0	16 06.1	−51 48	331.0	−0.2	+	+	○	+	○	6.6	31,57,58
G331.5	16 08.3	−51 21	331.5	−0.1	+	+	○	+	○	6.7	31,57
G331.4	16 08.6	−51 38	331.4	−0.3	+	+	○	+	○	4.8	31
G332.7	16 16.2	−50 54	332.7	−0.6	○	+	○	+	○	3.9	59
G333.2	16 17.2	−50 28	333.2	−0.5	+	+	○	+	○	6.7	31,55,58
G333.6	16 18.4	−49 59	333.6	−0.2	+	+	○	+	○	5.2	21,55,58
ρ Oph	16 22.0	−24 25	352.8	+17.1	○	○	+	−	+	0.2	60–64,138
G337.7	16 34.8	−46 55	337.7	−0.1	−	+	○	+	○	12.7	31
G338.9	16 36.9	−45 36	338.9	+0.6	−	+	○	+	○	5.1	31
G337.9	16 37.5	−47 02	337.9	−0.5	+	+	○	+	○	4.2	31,57,58
G338.9	16 39.5	−46 03	338.9	−0.1	−	+	○	+	○	3.6 (15.1)	31

Table 2 (*continued*)

Name (1)	α(1950.0) (2)	δ(1950.0) (3)	l (4)	b (5)	Class I/II compact H II region[a] (6)	Extended H II region[a] (7)	Compact near-IR source[a] (8)	OH maser[a] (9)	H₂O maser[a] (10)	D (kpc)[b] (11)	Reference[c] (12)
G339.6	16 42.5	−45 32	339.6	−0.1	0	0	0	+	0	2.9 (15.8)	31
G340.1	16 44.7	−45 16	340.1	−0.2	0	+	0	+	0	4.0 (14.8)	31
G345.5	17 00.9	−40 40	345.5	−0.3	0	0	0	+	−	1.5 (17.9)	31
G345.7	17 03.3	−40 47	345.7	−0.1	0	0	0	+	+	0.7 (18.9)	31,58
H 2-3	17 06.1	−41 32	354.4	−0.9	+	0	+	−	+	4.0	41,64
G347.6	17 08.4	−30 05	347.6	+0.2	0	0	0	+	−	7.9 (11.6)	31
G348.2	17 08.9	−38 26	348.2	+0.5	0	0	+	+	0	1.7	57
G349.1	17 13.0	−37 58	349.1	+0.0	0	0	0	+	+	7.5 (12.5)	42,65
G348.6	17 15.9	−39 01	348.6	−1.0	0	+	0	+	+	2.4 (17.2)	57,59
NGC6334	17 16.5	−35 56	351.1	+0.7	+	+	+	+	+	1.7	31,66–69
G351.6	17 25.9	−36 37	351.6	−1.3	+	0	0	0	0	5.0	57,70
G353.4	17 27.1	−34 39	353.4	−0.3	+	0	0	+	0	4.3	31,38
G355.2	17 30.2	−32 46	355.2	+0.1	+	0	0	+	0	1.4	31,38
Sgr B2	17 44.2	−28 22	0.7	+0.0	+	0	+	+	+	10.0	9,71–74
G 0.6	17 47.1	−28 53	0.6	−0.9	+	0	0	+	0	10.0	1,73
W 28	17 58.0	−24 00	5.9	−0.4	+	+	0	+	+	3.5	9,31,35
W 29(M8)	18 00.0	−24 23	6.0	−1.2	+	+	0	−	0	1.4	35,75–77

Source	RA	Dec	l	b						Size	References
W 31	18 06.3	−20 19	10.2	−0.4	+	+	o	+	+	5.1	9,35,57,78
G 10.6	18 07.5	−19 57	10.6	−0.4	+	+	o	+	+	6.0	9,31,79−82
G 12.4	18 07.9	−17 57	12.4	+0.5	−	o	+	o	o	2.7	83
G 12.2	18 09.7	−18 25	12.2	−0.1	+	+	o	+	+	3.7	9,38,84
W 33	18 11.0	−18 02	12.7	−0.2	+	o	o	+	+	4.0	9,11,31,55,81,85
M 17	18 17.0	−16 16	15.0	−0.7	o	+	+	+	+	2.2	9,37,86−89,139
G 19.6	18 24.8	−11 58	19.6	−0.2	+	+	o	+	+	4.0	9,31,38,80,82
G 20.1	18 25.4	−11 31	20.1	−0.1	+	+	o	+	−	7.7	31,80,81
W 42	18 32.6	−07 38	24.3	+0.1	o	+	o	+	o	9.1	31,82
W 43	18 45.0	−01 59	30.8	−0.0	+	+	o	+	+	7.0	9,35,41,45
G 33.1	18 49.6	+00 04	33.1	+0.1	+	o	o	o	+	5.4	9,31
										(11.1)	
G 34.3	18 50.7	+01 11	34.3	+0.1	+	+	+	+	+	3.8	9,31,35,38
G 35.6	18 53.9	+02 17	35.6	−0.0	+	+	+	+	+	4.0	9,31,38
										(12.3)	
W 48	18 59.2	+01 09	35.2	−1.7	−	o	o	o	+	3.3	9,38
RCorAustr.	18 58.0	−37 02	66.2	−17.8	+	o	o	o	−	0.2	41,48,60,90
G 40.6	19 03.6	+06 42	40.6	−0.1	−	+	+	o	+	2.1	38
										(13.1)	
W 49	19 07.0	+09 00	43.1	+0.2	+	+	+	+	+	14.0	9,31,76,91−94
G 43.8	19 09.5	+09 31	43.8	−0.1	+	+	−	+	+	2.7	9,31,38,95
										(11.8)	
G 45.1	19 11.0	+10 46	45.1	+0.1	+	+	+	+	+	9.7	9,11,31,80,81,96
G 45.5	19 12.0	+11 04	45.5	+0.1	+	+	+	+	+	9.7	9,31,80,81,97
G 48.5	19 18.2	+13 46	48.5	−0.0	+	+	−	+	+	11.8	9,31,38,95
W 51	19 21.0	+14 21	49.5	−0.3	+	+	+	+	+	7.0	9,11,18,37,38, 98−102
S 88	19 44.0	+25 05	61.5	+0.1	o	o	+	o	o	3.0	11,103,104,137
S 98	19 57.0	+31 14	68.1	+0.9	o	o	o	o	o	3.0	3
W 58	20 00.0	+33 24	70.9	+1.6	+	+	+	+	+	9.0	9,11,18,99, 105−109

Table 2 (continued)

Name (1)	α(1950.0) (2)	δ(1950.0) (3)	l (4)	b (5)	Class I/II compact H II region[a] (6)	Extended H II region[a] (7)	Compact near-IR source[a] (8)	OH maser[a] (9)	H₂O maser[a] (10)	D (kpc)[b] (11)	Reference[c] (12)
ON-1	20 08.1	+31 23	69.5	+1.0	+	–	–	+	+	3.5	9,81
S 104	20 16.0	+36 41	74.9	+0.7	+	+	0	0	0	5.0	1,110
G 75.8	20 19.0	+37 22	75.8	+0.4	+	+	+	+	+	5.5	9,25,35,80,81, 113
S 106	20 25.4	+37 12.6	76.4	–0.6	+	+	+	–	+	3.6	11,32,51,52,111, 114,115,134
AFGL2591	20 27.6	+40 01	78.9	+0.7	+	0	+	0	+	1.5	11,116–120
W 75N	20 36.9	+42 27	81.9	+0.8	+	+	–	+	+	1.0	9,18,38,71,80, 81,121,122
DR 21	20 37.2	+42 09	81.7	+0.5	+	+	+	+	+	4.0	9,18,71,100,121, 122
S 140	22 17.6	+63 04	106.8	+5.3	+	+	+	–	+	0.9	9,11,20,123,124
S 135	22 19.7	+58 35	104.6	+1.4	+	+	0	0	0	1.9	3
Cep A	22 54.3	+61 45	109.9	+2.0	–	+	+	+	+	0.7	51,52,125
S 152	22 55.6	+58 33	108.7	–0.9	+	+	+	0	+	4.0	3,126,127
S 157	23 13.9	+59 46	111.3	–0.7	+	+	+	–	+	3.5	4,9,18,128–130
S 158	23 11.4	+61 14	111.5	+0.8	+	+	+	+	+	3.5	8,9,11,100,128, 131–133
S 159	23 13.4	+60 51	111.6	+0.4	+	+	0	–	–	3.5	34,128,129

[a] The meaning of the symbols is as follows: + detected; − not detected; 0 not observed.

[b] Distances are taken from different sources and should only be taken as indicative. In case of distance ambiguity, both possible values are given.

[c] References are only given for recent and representative literature. 1. Israel (1976c); 2. Elmegreen & Lada (1978); 3. Israel (1977b); 4. Lo & Burke (1973); 5. Harris & Wynn-Williams (1976); 6. Wynn-Williams et al. (1972); 7. Mezger & Wink (1974); 8. Beetz et al. (1976); 9. Genzel & Downes (1977a), Genzel et al. (1978); 10. Willner (1977); 11. Dyck & Capps (1978); 12. Hackwell et al. (1978); 13. Wellington et al. (1976, 1977); 14. Harten (1976); 15. Genzel et al. (1978); 16. Harris & Scott (1976); 17. Hughes & Viner (1975); 18. Thronson & Harper (1979); 19. Mader et al. (1978); 20. Beichman et al. (1979); 21. Lada et al. (1979); 22. Altenhoff et al. (1976); 23. Brown et al. (1976); 24. Herbig (1971); 25. Knapp et al. (1976); 26. Frogel & Persson (1973); 27. Martin & Gull (1976); 28. Becklin et al. (1976); 29. Zuckerman (1973); 30. Zuckerman & Palmer (1974); 31. Knowles et al. (1976); 32. Israel & Felli (1978); 33. Giushkov et al. (1975); 34. Sibille et al. (1976); 35. Turner et al. (1974); 36. Grasdalen (1974); 37. Löbert & Goss (1978); 38. Evans et al. (1979); 39. Campbell (1978); 40. Pankonin et al. (1977); 41. Knapp & Morris (1976); 42. Beckwith et al. (1976); 43. Turner (1971); 44. Felli et al. (1977); 45. Pipher et al. (1976a); 46. Pipher & Soifer (1976); 47. Israel (1976c); 47. Vrba et al. (1976); 49. Evans et al. (1977); 50. Harvey et al. (1977b); Thompson & Tokunaga (1978); 51. Blitz & Lada (1979); 52. Blitz (1979); 53. Goss et al. (1973b); 54. Gillett & Soifer (1976); 55. Frogel & Persson (1974); 56. Johnston et al. (1971); 57. Frogel et al. (1977); 58. Caswell et al. (1974); 59. Caswell et al. (1977); 60. Brown & Zuckerman (1975); 61. Falgarone et al. (1978); 62. Encrenaz et al. (1975); 63. Elias (1978); 64. Becklin et al. (1974); 65. Turner & Rubin (1971); 66. Cheung et al. (1979); 67. Johnston et al. (1973); 68. Dickel et al. (1977); 69. Emerson et al. (1973); 70. Brown & Broderick (1973); 71. Harvey et al. (1977a); 72. Balick & Sanders (1974); 73. Downes et al. (1978); 74. Gatley et al. (1978); 75. Wright et al. (1977a); 76. Wink et al. (1975); 77. Woolf et al. (1973); 78. Wright et al. (1977b); 79. Fazio et al. (1978); 80. Matthews et al. (1977); 81. Habing et al. (1974); 82. Goss et al. (1973a); 83. Wright et al. (1979); 84. Shaver & Danks (1978); 85. Goss et al. (1978); 86. Wilson et al. (1979); 87. Lada (1976); 88. Gatley et al. (1979); 89. Lemke (1975); 90. Loren (1978); 91. Wynn-Williams (1971); 92. Wilson (1975); 93. Becklin et al. (1973); 94. Mufson & Liszt (1977); 95. Matthews et al. (1978); 96. Hefele & Schulte in den Bäumen (1978); 97. Baud (1977); 98. Martin (1972); 99. Puetter et al. (1979); 100. Wynn-Williams et al. (1974); 101. Scott (1978); 102. Bieging (1975); 103. Pipher et al. (1977); 104. Blair et al. (1978b); 105. Israel (1976a); 106. Harris (1975a); 107. Colley & Scott (1977); 108. Wynn-Williams et al. (1978a); 109. Persson & Frogel (1974); 110. Weiler & Shaver (1978); 111. Pipher et al. (1976b); 112. Cesarsky et al. (1978); 113. Pipher et al. (1979); 114. Lucas et al. (1978); 115. Maucherat (1975); 116. Brown (1974); 117. Wendker & Baars (1974); 118. White et al. (1975); 119. Rieke et al. (1973); 120. Evans et al. (1976); 121. Dickel et al. (1978); 122. Harris (1975b); 123. Harvey et al. (1978); 124. Blair et al. (1978a); 125. Sargent (1977, 1979); 126. Lo et al. (1975b); 127. Frogel & Persson (1972); 128. Israel et al. (1973), Israel (1977a); 129. Birkinshaw (1978); 130. Zeilik (1976); 131. Werner et al. (1979); 132. Willner (1976); 133. Downes & Wilson (1974); 134. Sibille et al. (1975); 135. Loren (1977); 136. Loren & Wootten (1978); 137. Deharveng & Maucherat (1978); 138. Fazio et al. (1976); 139. Matthews et al. (1979).

tion fronts (the core). Ionized gas flows out of the cavity, forming a low density envelope, like a blister on the skin of a molecular cloud.

Since 1974 more cases of similar H II regions have been reported. The observational evidence for the blister model has been strengthened by optical information on the systematic gas flows and the strong interaction between the ionizing stars and molecular clouds (Meaburn 1975, 1977). In addition, carbon recombination lines (Brown et al. 1978) can pinpoint the location of the interface of molecular cloud and ionized gas. They are a very important tool in unravelling the three-dimensional geometry of an H II region/molecular cloud complex. The convincing cases for blisters are all H II regions of Classes III and IV. There exists some observational evidence that (ultra)compact H II regions may also have blister-type structures. First, very high resolution observations (~ 1 arcsec) with the Cambridge Five-Kilometer synthesis radio telescope often show structures consistent with the blister model (cf the compact objects S158-G2, Martin 1973; K3-50, Harris 1975a and Colley & Scott 1977; and the ultracompact H II regions S157 and S159, Birkinshaw 1978). Second, (ultra)compact H II regions are often found at the edges of molecular clouds and the H109α radial velocities indicate outflow from the molecular cloud (Israel 1978).

3.4 Group Properties

An interesting property of compact H II regions is their gregarious nature: they are rarely isolated, but are usually found in groups or in the immediate vicinity of more evolved H II regions, or close to near-IR continuum/OH maser sources. A good example is W3(OH), where at least six ultracompact components associated with OH/H_2O masers fill a volume of about 0.5 pc in diameter (Harten 1976). Another example is NGC 7538 where a group of three obscured compact and ultracompact H II regions (Martin 1973, Wynn-Williams et al. 1974) with a cluster diameter of about 0.5 pc is found at the edge of a much larger, optically visible H II region. About one parsec south, a group of H_2O and OH masers indicate a different center of star-forming activity, while yet another, somewhat mysterious compact IR source is found about 2.5 pc to the east (Werner et al. 1979; see Figure 3). This whole configuration can exist only for a relatively short time. With continuing expansion, the three close H II regions will have merged into a single H II region in about 10^5 years; at about the same time this object will have been in turn engulfed by the extended H II region. How complex a source may be is demonstrated by the compact object(s) K3-50 in W58 (compare, for instance, the discussion of observations by Persson & Frogel 1974, by Israel 1976a, and by Wynn-Williams et al. 1978a). Although an impressive amount of optical, infrared, and radio data is available, it is in this case impossible to draw a firm conclusion about whether the source structure is one component with a density gradient and varying amounts of extinction and scattered light, or two or three components with different extinction.

A well-studied individualist among the compact H II regions is the ultracompact source associated with the OH maser ON1, (H. E. Matthews et al. 1977). It is located in an extensive CO cloud (Baud 1977, F. P. Israel and H. A. Wootten, in preparation), and appears to be the only young object in this cloud. Possibly ON1 is the first significant star born in the CO cloud.

These examples indicate that OB stars form in close groups, in space about a parsec or less in diameter (we assume that ON1 is the exception that proves the rule). These groups should not be confused with Blaauw's (1964) subgroups, which have much larger diameters (≈ 50 pc). The groups of (ultra)compact H II regions are, rather, subgroups within Blaauw's subgroups. It is remarkable that H II regions often appear to be lined up so as to suggest sequential star formation: the most evolved objects are on one side, the most compact objects appear on the other side. A very clear example is NGC 7538 (Habing et al. 1972, Martin 1973, Israel 1977a, Werner et al. 1979) on a total scale of about 10 pc. On a vastly larger scale (200 pc) the whole IC1848/IC1805/IC1795/W3 complex (Wendker &

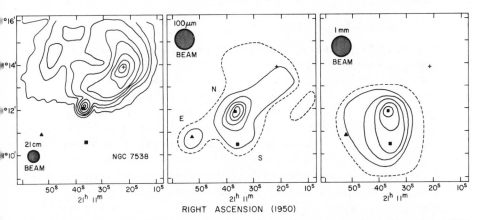

Figure 3   Maps of the NGC 7538 region at 21 cm, 100 μm, and 1 mm, made with the Westerbork SRT, the Kuiper Airborne Observatory, and the Hale 5-m telescope respectively. The cross marks the position of IRS 5, the presumed exciting star of the optical nebula; the two squares show the position of the OH/H_2O masers, and the triangle shows the position of the remarkable source IRS 9. With higher resolution, the compact object associated with the northern maser source is shown to consist of three (ultra)compact objects. The three maps indicate the position of ionized gas, warm dust, and dust with a high column density respectively. From Israel et al. (1973) and Werner et al. (1979).

Altenhoff 1977, Rohlfs et al. 1977, Lada et al. 1978) shows a similar phenomenon. The observed large-scale morphology ties in very nicely with the concept of sequential star formation as formulated by Elmegreen & Lada (1977) (see Section 1).

4 INFRARED OBSERVATIONS

4.1 What Sources Do We Know?

If one assumes that newborn stars are embedded in dust clouds, all the energy liberated during the formation process and in the nuclear-burning stage, and transformed into photons, will end up in the infrared. Therefore unbiased infrared surveys are the best way of detecting all regions of star formation. Unfortunately, unbiased and reliable infrared sky surveys at sufficiently long wavelengths (say $\lambda > 10$ μm) do not exist. Infrared observations have therefore often been made in the direction of regions selected because of other criteria: near (ultra)compact H II regions, near OH and H_2O masers, and in hot and dense regions of molecular clouds. Often such observations yielded infrared sources not only at the position of the H II region or the maser, but also at neighboring positions. For a specific review of IR sources we refer to Panagia (1977) and to Werner et al. (1977). The most important questions for this moment appear to be: what is the nature of compact IR sources, what powers them, and in what state of evolution are they?

4.2 Near-Infrared Observations

Historically, most infrared observations have been broad-band continuum observations (see, for instance, the review by Wynn-Williams & Becklin 1974), but spectral line observations are rapidly becoming more important. The 9.7-μm spectral band, usually ascribed to silicates (see, for example, Day & Donn 1978; however, for a controversial view, see Hoyle & Wickramasinghe 1977) has been studied intensively. This band may appear in emission (e.g. in the Orion nebula itself) or in absorption (Aitken & Jones 1973, Gillett et al. 1975, Soifer & Pipher 1975, Willner 1976, 1977). It has frequently and successfully been analyzed using a model consisting of a one-dimensional, two-layer distribution of material, where a hot emitting layer is observed through a cold layer of dust (Gillett et al. 1975). The accuracy with which the observed line can be predicted is reassuring and lends support to the main contention, namely that in all cases considered the optical depth of the emitting layer is very low ($\tau \ll 1$) and that of the absorbing layer is large ($\tau \gtrsim 1$). A different interesting explanation for the 9.7-μm feature was recently given by Sarazin (1978) who considers the effect of multiple grain sizes. In particular, he explains the observed relationship between τ_{Si} and τ_{H_2CO}.

Several other emission features have been measured between 2 and 13 μm, especially in the spectrum of the planetary nebulae NGC 7027. A few have also been found in compact H II regions (e.g. K3-50). For a list of emission features see, for example, Allamandola et al. (1978). With increasingly sophisticated infrared techniques, it is becoming possible to detect and map compact objects in infrared atomic and molecular lines. Recombination lines such as (hydrogen) Brackett α and Brackett γ have already been measured (see, for example, Grasdalen 1976, Joyce et al. 1978), and also fine structure lines of other atomic species [O III], [S III] (Baluteau et al. 1976, Greenberg et al. 1977), [OI] (Melnick et al. 1979), [Ar III] and [S IV] (Willner 1977), and [Ne II] (Aitken & Jones 1974). Vibrational-rotational lines of H_2 have been detected by Gautier et al. (1976) and Beckwith et al. (1978) in Orion, probably coming from a shock front. CO vibrational-rotational lines have been seen in absorption against the BN object by Hall et al. (1978).

Mainly on the basis of infrared continuum and radio continuum observations, we distinguish three categories of near-infrared sources. These three categories may imply an evolution in the sequence B,C,A, although C is probably not a necessary evolutionary step.

The *first* category (A) consists of those sources coinciding with compact H II regions found at radio wavelengths; most known compact infrared objects belong to this category, but that may be due to selection effects. The main properties of the infrared emission are well known: the flux density increases with wavelength (and reaches a maximum in the far infrared between 50 and 120 μm); at $\lambda \lesssim 3.5$ μm the flux density drops below that expected from free-free emission; the contours of near-infrared brightness compare generally very well with those of the radio brightness. At short wavelengths ($\lambda \lesssim 2$ μm) the ionizing stars are sometimes seen (e.g. Beetz et al. 1976). Elaborate models have been made to explain the infrared emission and there is some debate (e.g. Petrosian & Dana 1975, Wright 1973, Natta & Panagia 1976, Tielens & de Jong 1979) over the question of whether or not the ratio of dust density to gas density is the same in the ionized gas and in the neutral interstellar medium. If it is the same, then the dust is a major source of opacity for Lyman continuum photons (direct heating); if the ratio is much smaller in the ionized gas (dust depletion) then the gas will absorb a major fraction of the Lyman continuum photons, convert them partially into Lyman α photons, which will then heat the dust. To obtain the same amount of observable ionization one requires a much larger energy production in the first case than in the second case. Far-infrared measurements (see below) indicate such a larger energy production, but, unfortunately, the interpretation of the far-infrared observations is not unique. Lyman α heating would provide a more uniform heating and thus would explain well the absence of color gradients in the near-infrared measurements and the great similarity between the radio and near-infrared contours of equal brightness. A detailed model calculation for W3A by Tielens & de Jong (1979) supports the assumption of severe dust depletion; dust depletion also follows from an analysis of the ratio between the Ne line and dust continuum around 12.8 μm in G333.6-0.2 by Aitken et al. (1977); see also Wynn-Williams et al. (1978b) and Rank et al. (1978). However, in the analysis by Aitken et al. only the grains radiating at 13 μm are taken into account. Cooler grains, if they exist, would have escaped detection. Nevertheless, we find the arguments for dust depletion, at least in compact H II regions, rather convincing, and the counterevidence from far-infrared measurements rather weak, but not all investigators will agree on this point.

The *second* category (B) of near-infrared sources is radioquiet. The best examples in this category are the BN object in Orion and W3-IRS5, but several more are known (see, for example, Beichman et al. 1979). They appear to have very large optical depths (from the analysis of the 9.7-μm silicate feature), and their infrared spectrum can be well explained by assuming a cool central source, surrounded by a very dense shell ($A_v \approx$ 50–100 mag); see, for example, Finn & Simon (1977) and Bedijn et al. (1978). However, such models are certainly not unique and additional spectroscopic observations are required. For example, Brackett-line emission has been observed in the BN object (Grasdalen 1976, Joyce et al. 1978) consistent with the presence of a compact H II region of such a small diameter ($< 10^{15}$ cm) that it is not observable at radio wavelengths. Although several radioquiet sources have a spectrum similar to BN or to

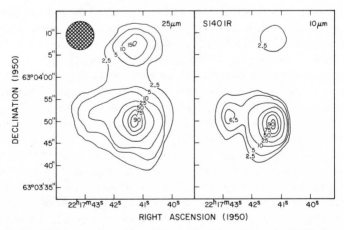

Figure 4   Near-infrared maps at 10' and 25 μm of compact sources in S140 indicate rather different extinction values for close objects. From Beichman et al. (1979).

W3-IRS5 not all of them share this property: point sources exist (e.g. those in the Kleinmann-Low nebula) that only became visible at $\lambda > 25$ μm instead of $\lambda = 10$ μm (cf Figure 4). We presume that many such sources remain to be discovered. They could be very similar to sources like BN, but have a surrounding dust shell with an optical depth larger by a factor of two.

The BN object possibly contains a prototype of an early B star (B0 or B1) still on its way to or very recently arrived at the main sequence. Very promising for the future are several techniques until now only applied to the BN object. For instance, recent spectroscopic observations with 10 km s^{-1} resolution between 2 and 4 μm by Hall et al. (1978) of Brackett and Pfund recombination lines of hydrogen show considerable systematic motions. Measurements by the same authors of vibrational-rotational lines of CO indicate significant motions in the molecular cloud, confirming millimeter observations of CO lines (e.g. Zuckerman et al. 1976). Hall et al. argue that it is more likely that the central star of the BN object is ejecting matter, rather than accreting. Hall's work shows that very informative spectroscopic observations can now be made of objects like the BN object. The discovery by Dyck and associates (Dyck et al. 1973, Loer et al. 1973) that the near-infrared emission from the BN object is considerably polarized appears to indicate that strong magnetic fields occur around the object (but for a different interpretation, see Elsässer & Staude 1978). Because of the poorly known grain properties it is difficult to estimate the strength of the magnetic fields. Other similar objects also show polarization, with the amount of polarization at 2.2 μm increasing for decreasing source size (Dyck & Capps 1978; see also Section 5.2).

The *third* category (C) of objects consists of radio compact H II regions without any associated near-infrared source. ON1 is an example, but more are known (Hefele et al. 1977). The existence of the H II region indicates the presence of a hot, ionizing source of radiation. The absence of near-infrared emission is most easily explained by assuming that a cool circumstellar dust shell absorbs all near-infrared emission; i.e. the object is even more extreme than the objects in the Kleinmann-Low nebula mentioned before. That such an explanation is not at all unlikely is shown by far-infrared measurements.

4.3   Far-Infrared Observations

Far-infrared fluxes from H II regions correlate well with the radio fluxes (for a compilation see Wynn-Williams & Becklin 1974, Emerson 1978) and the standard interpretation is that the stars providing the energy for the infrared radiation (taken as thermal dust emission well in excess of the expected free-free emission due to ionized components) also provide the ionizing photons. This has led to a debate on the amount of dust inside an H II region (see also Section 4.2). Imagine an H II region around an ionizing star both surrounded by a dense dust layer absorbing all radiation from star and H II region. The total radiation from the object summed over all infrared wavelengths equals the stellar luminosity L_*. The observed radio flux density provides a determination of L_{Lyc}, the luminosity of the star below $\lambda = 91.2$ nm. L_{Lyc} and L_* are related through the theory of stellar atmospheres. However, for a given L_* the observed value of L_{Lyc} is systematically smaller by a factor of two to four than is predicted, a discrepancy that can be explained in two different ways (see, for example, Johnson 1973a,b). The first explanation is that, although L_{Lyc} is predicted correctly, the observed value of L_{Lyc} is smaller because a fraction of the Lyman continuum photons is absorbed by dust particles inside the H II region, and therefore does not contribute to the ionization. Dust at an abundance normal for the interstellar medium can provide enough absorption inside an H II region. The second explanation says that L_{Lyc} (predicted) is too large because the heating and ionization balance is determined not by a single star, but by a cluster. For a cluster of stars the ratio L_{Lyc}/L_* is smaller than for the brightest member of the cluster, because member stars of lower luminosity are also cooler. The inclusion of other cluster members lowers the predicted value of L_{Lyc} enough to explain a large part of the discrepancy (Panagia 1977). This last explanation seems to have the necessary consequence that the dust abundance inside the H II region is less than normal (depletion), but geometrical and filling factors may play a significant role.

Each of these two explanations has its proponents. One of the factors impeding a solution was that the determination of L_* was made through far-infrared measurements obtained with wide beams, so that it was impossible to exclude nonrelated sources. This situation has been remedied recently by airborne observations (Thronson & Harper 1979); in a number of reasonably well-resolved cases the observed value of L_* is smaller, thus leading to a smaller predicted value of L_{Lyc} and reducing the discrepancy. However, a definitive choice between the two explanations cannot yet be made.

A considerable number of airborne observations of compact H II regions with resolution between 20 arcsec and 60 arcsec have recently been published (P. M. Harvey et al. 1975, 1976, 1977a,b, Thronson & Harper 1979). In a number of complex sources it has been possible to determine the contributions from each component to the total far-infrared flux densities. The major new result is that several infrared sources, including some compact H II regions, are found to be surrounded by very dense dust clouds. All observers give estimates of very large extinction values ($A_v \approx 100$ mag is mentioned). These values are uncertain, but by no more than a factor of two. The quoted values are often in agreement with the analysis of the deep 9.7-μm absorption features mentioned above and with molecular observations; they imply fairly high extinction values even at near-infrared wavelengths up to 20 μm.

Observations of continuum emission at very long infrared wavelengths (350 μm–3 mm) have been made by several groups (e.g. Werner et al. 1975, Westbrook et al. 1976, Righini et al. 1976, Hudson & Soifer 1976). The emission sources are optically thin at these wavelengths and the surface brightness distribution is determined largely by the surface density of the dust. The observations have been made with angular resolutions of 1–2 arcmin and are directly comparable with molecular line observations. They show conclusively that compact H II regions are associated with the densest parts of molecular clouds, but are not in disagreement with the statement (Section 3.3) that the compact H II regions occur near the edges of the dense parts.

Of great interest is the measurement of considerable far-infrared luminosities for sources having no radio emission (W75-S-OH, DR 21-IRS 1, NGC 7538-S-OH). The inference that the stars involved are pre-main-sequence stars (by Harvey et al. 1977a, and by Thronson & Harper 1979) is not necessarily correct. The absence of a detectable H II region may indeed imply that the star does not (yet) produce an appropriate amount of Lyman continuum photons. However, it is equally possible that the size of the H II region is too small for detection at radio wavelengths (see the discussion on the BN object in Section 3.2). Whatever the explanation for the absence of a radio source, the object certainly has to be very young.

5   MASERS AND COMPACT H II REGIONS

5.1   General Remarks

The first maser source was detected in 1965 in Orion when clear non-LTE emission was measured from four hyperfine transitions in the rotational ground state of OH (Gundermann 1965, Weaver et al. 1965). Somewhat later maser sources of H_2O and SiO lines were found, usually close to OH maser sources. Since 1965 several hundred maser sources have been detected and it is now clear that, with very few exceptions, the maser sources fall into one of two categories (see, for example, Moran 1976):
(*a*) satellite-line maser sources coinciding with cool, well-evolved stars;
(*b*) main-line maser sources primarily associated with compact H II regions, but also with other signposts of star formation ("young masers").

Obviously we are concerned here only with maser sources in the second category.

Considerable differences exist between the two categories. First, among the young masers much higher luminosities occur (by factors of 10^3) than among the masers around evolved stars. Only in the 1612-MHz line of OH do evolved-star masers reach luminosities comparable to the strongest 1612-MHz emission of young masers. Nevertheless, very weak maser sources occur in both categories. Second, the young masers often show strong circular (and sometimes linear) polarization in the OH lines. Third, SiO masers occur only around evolved stars, with the one exception of the SiO maser in the Kleinmann-Low nebula in Orion (Snyder & Buhl 1974, Moran et al. 1977). Orion is an exception in any case (perhaps due to its relatively small distance from the Sun) because at almost the same spot the only known methanol maser (CH_3OH) is found (Buxton et al. 1977). The situation now appears as follows: masers around evolved stars are much more numerous, but generally weaker and therefore closer than young masers; masering conditions around young objects are more extreme than around well-evolved stars: longer path lengths, higher densities, stronger infrared pumps, stronger magnetic fields.

The value of maser observations is two-fold. First they permit the detection of regions of peculiar excitation conditions, such as occur around stars or starlike objects with thick circumstellar shells. Second, because of the high intensity of the radiation, the source structure can be mapped with very high precision and the geometry and kinematics of the masing region can be studied; maser components with apparent sizes of 1 AU were accurately located in recent VLBI experiments (see Section 5.3).

5.2 Hydroxyl Masers

Some 100 young masers have been detected. Most of these have their strongest emission at 1665 or at 1667 MHz ("main-line masers") but for a few the 1720-MHz line is strongest [NGC 7538 OH(N); ON3]. Unique were the properties of the temporary 1720-MHz OH maser associated with V1057 Cyg (Lo & Bechis 1974); since this is a young object of only 8 $M_\odot$ (Grasdalen 1973), we do not discuss it further. Hydroxyl masers show time variation on a scale of 10^6 to 10^7 seconds, but time variation has not been sufficiently well monitored to lead to any firm conclusion on their nature (Sullivan & Kerstholt 1976). OH maser emission is not only found in the transitions at 1612, 1665, 1667, 1720 MHz belonging to the $2\Pi_{\frac{3}{2}}$, $J = \frac{3}{2}$ ground state, but also in microwave transitions occurring in other rotational levels; for a (no longer complete) listing see ter Haar & Pelling (1974). There is no satisfactory explanation for the OH maser emission associated with young objects; we refer to Elitzur (1979) for a thorough discussion of the problem. A good, published compilation of all OH masers sources is lacking (but see Table 2). There exists an extensive, unpublished survey for the northern hemisphere by B. E. Turner. Turner's work, as well as more incidental searches in the southern hemisphere by Caswell et al. (1977), has yielded a number of very weak young masers. This indicates that a large spread in OH luminosities exists; many weak undiscovered maser sources will exist.

VLBI observations of OH masers have shown conclusively (Moran et al. 1968) that maser sources consist of groups of intense "spots" of radiation. A "spot" has a size $< 10^{15}$ cm, and the spots are distributed over an area of order 10^{16} cm in diameter. VLBI observations in different lines show that there is no correlation between maser spots in one line with those in another line, nor between maser spots belonging to different peaks in one line (with an important exception to be discussed). This property led Cook (1975) to suggest that within each maser source a selection effect works: the combination of Doppler shift, due to gas-dynamic motions, and Zeeman splitting provides sufficient path lengths only along certain (almost accidental) lines of sight. The exception here is that emission line peaks of opposite circular polarization and of slightly different frequency often come from the same spot within a very high degree of positional agreement (< 0.005 arcsec; P. J. Harvey et al. 1974, Lo et al. 1975a, Moran et al. 1978). This could imply that masers occur in areas where highly symmetrical configurations of gas flow and magnetic fields occur, as near neutral sheets (H. J. Habing and C. A. Norman, in preparation). Magnetic fields of a few milligauss are probably involved (Harvey et al. 1974). In Section 4.2 the occurence of such magnetic fields was discussed in connection with the strong polarization of the near-infrared continuum from small sources.

The relation between compact H II regions and young OH masers was established by Mezger et al. (1967b) and subsequently elaborated upon by several investigators (Habing et al. 1972, 1974, H. E. Matthews et al. 1977, Evans et al. 1979). The following conclusions can be drawn. 1. A few young OH masers show no associated compact H II region at all; very low upper limits have been measured to the radio flux density [e.g. NGC 7538-OH(S)], with a radio flux density $\leqq 5$ mJy at 6 cm (Israel 1977a). This situation is reminiscent of and probably similar to that of the radioquiet infrared sources (see Section 4); in fact, some of them are the same source (e.g. S255IR/OH; Israel 1976c). For the others, the associated H II region is almost always within the error box of the OH maser position (see Figure 5). 2. In at least one case (W3-OH) a very accurate positional coincidence (within one arcsec) has been found between OH maser sources in two different lines and an ultracompact H II region (cf Moran et al. 1978). Several other very good candidates for such an accurate positional coincidence are known, but the OH maser positions still have slightly uncomfortable errors ($\lesssim 5$ arcsec). All candidates are very small H II regions, with diameters $\leqq 3 \times 10^{17}$ cm (Habing et al. 1974). 3. Young OH masers are associated only with Class I ultracompact regions (see Section 3.1); about 20% of all Class I regions are associated with an OH maser (Israel 1976b).

These three conclusions lead to a straightforward, qualitative explanation: OH maser sources reside in dense circumstellar shells around newly formed OB stars. Some stars are still so young that the associated H II region is sufficiently small to remain undetected. As the ultracompact H II region expands, it becomes detectable, but when the expanding diameter exceeds $\approx 3 \times 10^{17}$ cm the OH maser conditions disappear.

5.3   Water Masers

The most important observed properties of the H_2O masers are the following (Moran 1976): they show complex spectra without much common structure, consisting often of narrow peaks, some with weak

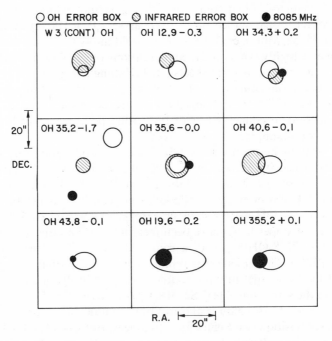

Figure 5   Relative positions of OH masers, compact infrared sources, and compact radio sources. From Evans et al. (1979).

linear polarization (Bologna et al. 1975); some are highly variable on short time scales (days or weeks). The variations in average intensity of a maser source indicate that the pump energy is derived from a common central object (Little et al. 1977). Emission peaks are found even at large Doppler shifts (up to 200 km s^{-1}); the high velocity peaks are on the average somewhat weaker than the low velocity peaks. Several H_2O maser mechanisms have been proposed, but none is generally accepted. Theoretical advance has been very slow in recent years: with the addition of a paper by Goldreich & Kwan (1974), ter Haar & Pelling's review (1974) is probably still up to date.

On the observational side progress has been very fast. 1. Many new maser sources have been found. In view of the variability of the sources a nondetection is hardly meaningful; many H_2O masers have popped up where, say, a few months before nothing could be found. An extensive, but not exhaustive list of young H_2O maser sources in the Galaxy is given by Genzel & Downes (1977a,b). 2. VLBI mapping of some 12 sources (Genzel et al. 1978, Johnston et al. 1977, Walker et al. 1978) has shown conclusively that in each maser source the low velocity components form a few elongated groups of maser spots (see, for example, Figure 6). Such groups have a typical diameter of 3×10^{16} cm, individual spots having a diameter $\approx 10^{14}$ cm. The spots within a group may show signs of systematic

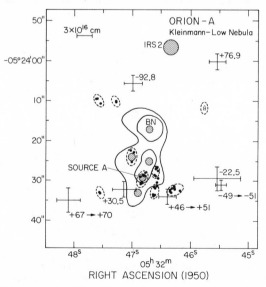

Figure 6 H_2O masers in the Orion KL nebula are shown as black dots and crosses; the (near-)infrared emission is indicated by shaded circles (compact sources) and 21-μm contour lines (extended KL nebula). From Genzel et al. (1978).

motions (expansion, rotation). 3. For a few masers very accurate absolute positions have been determined (Forster et al. 1977, 1978, Mader et al. 1978). These show that H_2O maser sources do not coincide with compact, or ultracompact, H II regions, i.e. in all cases the nearest H II region is well outside the error box of the H_2O position (exceptions: ON1, NGC7538/IRS1), although the two in many cases lie very close together. However, sometimes the groups of H_2O maser spots do coincide with (radioquiet) infrared sources and with OH masers. 4. High velocity emission peaks have been found in a considerable number of sources (Genzel & Downes 1977a, Goss et al. 1976). They are situated outside the groups of low velocity components (Genzel & Downes 1977a,b, Walker et al. 1977). An appealing, qualitative interpretation of these conclusions was summarized by Genzel & Downes (1977a, 1979). First, they suggest that each group of low velocity maser spots contains a recently formed star — it is even possible that the H_2O luminosity is correlated with the stellar luminosity. This implies that each group should coincide with an infrared point source. Searches at short wavelengths ($\lambda < 20$ μm) have sometimes failed to detect such a point source but, as discussed earlier, this may be due to the presence of obscuring dust with a considerable optical depth even at 20 μm. Second, groups of H_2O maser spots do not coincide with (detectable) compact H II regions, indicating that the star is very young indeed. Third, the occurrence of high velocity peaks is explained as being

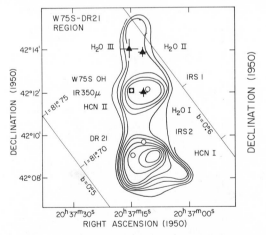

Figure 7   Map in HCN of the densest parts of the molecular cloud complex associated with W75S-DR 21. Open circles indicate compact near-infrared sources (DR 21 itself is also a compact radio source); H_2O masers are indicated by triangles and the OH maser source/ far-infrared source is marked by a square. The situation, like the one in NGC 7538, is suggestive of sequential star formation, here in a northern direction. From Cato et al. (1975).

due to blobs of material accelerated by a stellar wind. This interpretation was first put forward by Strelnitskii & Sunyayev (1973) and is supported in analyses by Heckman & Sullivan (1976) and by Morris (1976). The high velocity peaks, but also the velocities of the low velocity peaks, suggest strongly that the H_2O masering material is no longer collapsing onto the star, but is already expanding.

In a recent paper Genzel & Downes (1979) estimate from somewhat uncertain statistical arguments that the H_2O maser phase lasts $\gtrsim 10^5$ yr. This is quite a long time, as we shall see in Section 7. Individual spots are often active over only a short time. The well-observed rise and decline, within a few months in 1977, of an intense linearly polarized peak in the W3(OH) water maser profile (Burke et al. 1978) led the authors to propose a model in which a burst of energy diffuses through a "maser prone" medium. The model explains very well the time behavior of the flux and indicates a release of about 10^{40} erg in a very short time.

6 MOLECULAR EMISSION ASSOCIATED WITH COMPACT H II REGIONS

Most if not all H II regions, and therefore also compact H II regions, are associated with molecular clouds, as is clear from several CO surveys (Wilson et al. 1974, Blair et al. 1975, Gillespie et al. 1977). The first studies showed that the molecular clouds are usually much larger than the associated H II regions, but that the compact H II regions coincide with the densest parts (e.g. observations of HCN by Gottlieb et al. 1975; of HCN and CS by Morris et al. 1974; and millimeter and submillimeter continuum measurements by Westbrook et al. 1976; see Figure 7). Recent detailed mapping of large areas has made it clear that the molecular clouds often form very large elongated complexes with sizes of typically 10×40 pc (Lada 1976, Kutner et al. 1977, Sargent 1977, Thaddeus 1977). These facts fit very well within the framework of recent ideas on the formation of OB associations, in which the star formation process is initiated on one side of an elongated molecular cloud complex, and autocatalytically proceeds into the cloud complex [Elmegreen & Lada 1977; a good example is also the Cep OB3 association, Sargent 1979; see also the review by Thaddeus 1977 and the proceedings of the Gregynog (Solomon 1979) and Tucson (Gehrels 1979) workshops].

An important discovery is that compact H II regions have always been found near the edge of dark or molecular clouds and almost never deep inside them (Myers 1977, Israel 1976b, 1978, Gilmore 1978). As a consequence, the evolution of an H II region cannot be spherically symmetric. At a very early stage the expanding compact H II region breaks through the skin of the molecular cloud, leading to the so-called blister model (see Section 3). Also of interest is a recent survey of star-forming regions in NH_3 line emission by Ho (1978). Ho finds small NH_3 emission regions associated with every compact H II region observed. With typical densities of 10^4 to 10^5 cm^{-3}, these NH_3 clouds have total masses of 10 to 1000 $M_\odot$. The estimated kinetic temperatures are intermediate between those of "hot" molecular clouds near H II regions and "cool" dark clouds without H II regions, strengthening support for the evolutionary sequence mentioned in Section 1.2. Ho remarks that in the ten cases he observed, the association between the compact H II region and the dense molecular cloud is quite reminiscent of Orion where a rather compact H II region is adjacent to and in contact with a dense (part of a) molecular cloud— sizes, densities, and relative distances are all similar.

Because of its small distance the molecular cloud surrounding the infrared KL nebula is certainly the best studied example of a dense (part of a) cloud directly producing new stars. The results have already been reviewed many times (see, for example, Zuckerman & Palmer 1974) and are not repeated here. Some new important facts have been discovered, that are possibly immediately related to the star formation process, but whose role in this process is not yet understood. 1. H_2 emission has been discovered through vibrational-rotational lines (Gautier et al. 1976, Beckwith et al. 1978, Joyce et al. 1978); the lines are produced in an excited gas behind a shock wave (Kwan 1977, Hollenbach & Shull 1977, London et al 1977), but the origin of the shock is unknown. 2. Very high velocities in the CO line profile of the KL nebula have been discovered (Zuckerman et al. 1976, Phillips et al. 1977), probably indicating large expansion velocities. The extent of this phenomenon is not very clear at present. 3. Hyperfine lines of NH_3 have been measured (Ho & Barrett 1978) leading to two separate but very tentative conclusions: (a) the KL region is part of a dense layer that forms the contact between two clouds at slightly different velocities; (b) and/or the NH_3 gas around the KL nebula shows a strong clumping. 4. Emission in the 6-cm H_2CO line has been discovered indicating very large densities directly associated with compact H II regions (Orion: Kutner & Thaddeus 1971, Zuckerman et al. 1975; NGC 7538: Downes & Wilson 1974).

It is a commonplace statement that we need observations of higher spatial resolution in the molecular lines and development of submillimeter receivers. Only when both goals have been achieved will it be possible to study molecular clouds in sufficient detail. This is of fundamental importance, because observations of compact objects yield only post hoc information, while molecular line observations will enable us to catch the process of star formation in the act.

7   REDISCUSSION OF OB STAR FORMATION

Here we summarize the main results from Sections 2–6 and discuss their relevance to the problem of OB star formation. We distinguish two different aspects: 1. the tendency of OB stars to form in groups, and 2. the formation of individual objects.

7.1   The Formation of OB Stars in Groups

We have noted that H II regions of Class I and Class II are rare and short-lived phenomena. They share this characteristic with compact infrared and OH/H_2O maser sources. From this point of view it is remarkable that despite their rareness (their lifetime is less than 1% of that of a main sequence O star, see Section 7.2) they are not isolated. Indeed, if one of them is found, there is a very large probability of finding another within about 0.5 pc—for a typical distance of 1 kpc this corresponds to about 1.5 arcmin or 1.5 mm on the scale of the Palomar Sky Survey. OB stars also appear to form in small groups of less than 0.5 pc in size, within a time of approximately 10^5 yr from each other.

A second point to notice is the tendency of the compact H II regions to form near the outer edge of molecular clouds and not in the central part (see Section 3.3). This result may be related to the finding (Burki 1978) that the most massive stars in very young clusters have a tendency to be at the outer edge of the cluster. Such behavior is explained by the self-supporting mechanism of Elmegreen & Lada (1977), but these authors still need an external influence to start the process (see the review by Lada et al. 1978). A new and intriguing suggestion was recently made by Icke (1979); he explains the observed behavior by assuming that the molecular cloud is embedded in and accreting material from a larger cloud complex. He supposes that in the accretion flow fragments occur that are gravitationally stable, but become unstable as soon as they enter the much denser (and hotter?) cloud. Immediate collapse follows, so that the fragment does not have the time to penetrate deeper into the cloud.

7.2   Formation of Individual Stars

Numerical simulation of the spherically symmetric collapse of small dense clouds has led to the conclusion (for details see the reviews by Larson 1973, and Woodward 1978) that in the center of the cloud a proto-stellar core is formed, which continues to grow by accretion. Larson & Starrfield (1971) and Kahn (1974) suggested that the accretion time for massive stars is long compared to the Kelvin-Helmholtz time of the protostar, so that nuclear processes will be ignited already during accretion. Kahn's analytical results are supported in their main conclusion by a numerical study by Yorke & Krügel (1977). The large energy production by nuclear fusion will affect the accretion flow. First, grains will melt at some distance from the star. Second, the radiation pressure of direct starlight, coupled with the infrared radiation field farther outward, will decelerate the accretion flow and hence build up a layer of enhanced dust and gas densities ("cocoon"), which is limited on the inner side by the melting radius. During accretion any H II region is probably limited to the dust-free sphere inside the melting radius. As the stellar mass increases, the radiation pressure of the star will invert the flow of the cocoon. From that moment on, the H II region will start expanding very rapidly (Cochran & Ostriker 1977). A shock wave will be set up outside the dust cocoon; both will move outward very quickly. The situation is now very reminiscent of the original "cocoon star" proposal by Davidson & Harwit (1967).

Ultimately the ionization front will reach the outer edge of the part of the cloud where the star formed, the H II region will break open, and a "blister type" H II region follows. This nonspherical process has been studied theoretically only very recently (Icke 1979, Tenorio-Tagle 1979, Kandel & Sibille 1978); an interesting conclusion is the prediction of significant expansion velocities of up to 30 km s^{-1} (Mach 3) in the ionized gas speeding away from the molecular clouds. Optical observations (reviewed by Meaburn 1977) indeed show such velocities to be present at least in more extended H II regions.

7.3   Limitations of the Theory of Individual Star Formation

Before we compare the theoretical models in Section 7.1 and 7.2 with the observations, we emphasize a few basic limitations. First, all models are based on the assumption of spherical symmetry, except for the recent models that attempt to explain the "blister-type" H II regions. Observations clearly indicate large deviations from spherical symmetry and suggest axial symmetry as a better approximation: groups of H_2O maser spots show a flattened distribution (Genzel et al. 1978) and in at least one far-infrared source the circumstellar dust appears to have an opening at the poles (Harvey et al. 1977b), while NGC 7538(E) seems to show similar characteristics (Werner et al. 1979). Second, in calculating the accretion flow, and especially the expansion of the very small H II region, a possible stellar wind is ignored. Direct evidence for the existence of such a wind comes from the high velocity components of H_2O masers. Third, the influence of a magnetic field in the neutral circumstellar shell has been ignored. Nevertheless, VLBI observations of OH maser emission indicate that fields of up to $B = 5$ mG occur. Hence, $B^2/8\pi = 10^{-6}$ erg cm^{-3}, which corresponds to a gas pressure $nT = 10^{10}$ cm^{-3} K. It thus follows that the magnetic field has some influence on the expansion.

7.4 Confrontation of Theory and Observations

Quantitatively, it is not yet clear that theory and observations agree. An important check is to compare theoretically predicted relative time scales with observed relative numbers of sources. Here, in our opinion, a problem probably emerges with the present data: one of the striking facts discovered in observing "signposts" of star formation is that several phenomena appear to occur simultaneously. This suggests that these phenomena have comparable lifetimes, differing by at most factors of two or three. Of these signposts, ultracompact H II regions have a well-predicted lifetime of the order of 10^4 yr to reach a diameter of 3×10^{17} cm.

However, this lifetime is shorter than expected from two other considerations. 1. Comparing the number of masers with the stellar birth rate function, Genzel & Downes (1979) estimate that H_2O masers associated with OB stars last at least 6×10^4 yr. 2. The collapse of fragments to form OB stars lasts at least 10^5 yr (Yorke & Krügel 1977). Imagine that several stars form simultaneously. If at the end of each collapse a short-lived ($\approx 10^4$ yr) phenomenon occurs, we do not expect to observe these phenomena simultaneously. Nevertheless, we observe ultracompact H II regions and H_2O masers simultaneously. This suggests that each phenomenon lasts a significant fraction of the collapse time.

Both considerations are suggestive, neither is conclusive. If super-compact H II regions expand more slowly than nondynamical models indicate, one has to look for a cause. Such a cause could be the strong magnetic field in the cocoon (see Section 5.2). If so, then the OH masers obtain a special significance: they inform us that a newborn H II region wants to expand, but is restrained by the magnetic field.

Another confrontation between theory and observations emerges in the discussion of the energetics of compact H II regions, IR sources, masers, and molecular clouds. Such discussions have been presented by several investigators. Two recent papers (Evans et al. 1977, Blair et al. 1978) analyzed relatively simple configurations. The conclusion is that the energetics can be understood at least approximately: outside stars are the main heating sources; their radiation flux is converted into infrared photons that heat the dust; the dust cools largely by infrared radiation and partially by heating the molecular gas through inelastic collisions.

Qualitatively there is good agreement between theory and observations. From the observations discussed in the previous sections we obtain the following evolutionary scheme. 1. The youngest objects are infrared sources without any H II regions, associated, if at all, only with H_2O

masers having simple line profiles. These objects may be identified with accreting stars. Possible examples include some stars in the KL nebula. 2. Slightly older are those objects which are similar, but which in addition show high velocity peaks in the H_2O maser line profile; these peaks indicate the existence of a significant stellar wind, implying the end of the accretion phase. Possible examples include the BN object, W3-IRS 5. 3. Next are the infrared sources associated with the smallest H II regions (3×10^{16} cm diameter). Possible examples are ON1 and NGC7538/IRS 1. Apparently the H II region has grown sufficiently in size to become visible. Surrounding the H II region is a dense circumstellar shell that sometimes houses an OH maser source. 4. When the H II region has expanded beyond a diameter of 3×10^{17} cm, the OH maser disappears and a blister-type H II region will soon appear.

ACKNOWLEDGMENTS

We thank those colleagues who sent us preprints and various other pieces of information. We are also grateful to S. Harris, H. C. van de Hulst, A. I. Sargent, M. W. Werner, and several other colleagues for discussions and critical comments. Last, but not least, we thank L. de Leeuw, M. van Haaster, J. Ramsey, and F. De Windt for their excellent typing. FPI acknowledges the support of NSF grant AST77-00247 and a travel grant of the Leidsch Kerkhoven-Bosscha Fonds that enabled him to complete this review.

Literature Cited

Aitken, D. K., Jones, B. 1973. *Ap. J.* 184:127

Aitken, D. K., Jones, B. 1974. *MNRAS* 167: 11p

Aitken, D. K., Griffiths, J., Jones, B. 1977. *MNRAS* 179:179

Allamandola, L. J., Greenberg, J. M., Norman, C. A. 1978. *Astron. Astrophys.* 66:129

Altenhoff, W. J., Braes, L. L. E., Olnon, F. M., Wendker, H. J. 1976. *Astron. Astrophys.* 46:11

Altenhoff, W. J., Downes, D., Goad, L., Maxwell, A., Rinehart, R. 1970. *Astron. Astrophys. Suppl.* 1:319

Andriesse, C. D. 1974. *Astron. Astrophys.* 37:257

Balick, B., Sanders, R. H. 1974. *Ap. J.* 192:325

Balick, B., Gammon, R. H., Doherty, L. H. 1974. *Ap. J.* 188:45

Baluteau, J. P., Bussoletti, E., Anderegg, M., Moorwood, A. F. M., Caron, N. 1976. *Ap. J. Lett.* 210:L45

Baud, B. 1977. *Astron. Astrophys.* 57:443

Becklin, E. E., Neugebauer, G. 1967. *Ap. J.* 147:799

Becklin, E. E., Neugebauer, G., Wynn-Williams, C. G. 1973. *Astrophys. Lett.* 13:147

Becklin, E. E., Frogel, J. A., Kleinmann, D. E., Neugebauer, G., Persson, S. E., Wynn-Williams, C. G. 1974. *Ap. J.* 187:487

Becklin, E. E., Beckwith, S., Gatley, I., Matthews, K., Neugebauer, G., Sarazin, C., Werner, M. W. 1976. *Ap. J.* 207:770

Beckwith, S., Evans, N. J., Becklin, E. E., Neugebauer, G. 1976. *Ap. J.* 208:390

Beckwith, S., Persson, S. E., Neugebauer, G., Becklin, E. E. 1978. *Ap. J.* 223:464

Bedijn, P. J., Habing, H. J., de Jong, T. 1978. *Astron. Astrophys.* 69:73

Beetz, M., Elsässer, H., Poulakos, C., Weinberger, R. 1976. *Astron. Astrophys.* 50:41

Beichman, C. A., Becklin, E. E., Dyck, H. M., Capps, R. W. 1979. *Ap. J.* Submitted

Bieging, J. 1975. In *H II Regions and Related Topics*, ed. T. Ł. Wilson, D. Downes, p. 443. Heidelberg: Springer

Birkinshaw, M. 1978. *MNRAS* 182:401

Blaauw, A. 1964. *Ann. Rev. Astron. Astrophys.* 2:213

Blair, G. N., Peters, W. L., Vanden Bout, P. A. 1975. *Ap. J. Lett.* 200:L161

Blair, G. N., Evans, N. J., Vanden Bout, P. A., Peters, W. L. 1978a. *Ap. J.* 219:896

Blair, G. N., Davis, J. H., Dickinson, D. F. 1978b. *Ap. J.* 226:435

Blitz, L. 1979. In Proceedings Gregynog conference, ed. P. Solomon. In press

Blitz, L., Lada, C. J. 1979. *Ap. J.* 227:152

Bologna, J. M., Johnston, K. J., Knowles, S. H., Mango, S. A., Sloanaker, R. M. 1975. *Ap. J.* 199:86

Brown, R. L. 1974. *Ap. J. Lett.* 194:L9

Brown, R. L., Broderick, J. J. 1973. *Ap. J.* 181:125

Brown, R. L., Zuckerman, B. 1975. *Ap. J. Lett.* 202:L125

Brown, R. L., Broderick, J. J., Knapp, G. R. 1976. *MNRAS* 175:87P

Brown, R. L., Lockman, F. J., Knapp, G. R. 1978. *Ann. Rev. Astron. Astrophys.* 16:445

Burke, B. F., Giuffrida, T. S., Haschick, A. D. 1978. *Ap. J. Lett.* 226:L21

Burki, G. 1978. *Astron. Astrophys.* 62:159

Buxton, R. B., Barrett, A. H., Ho, P. T. P., Schneps, M. H. 1977. *Astron. J.* 82:985

Campbell, P. D. 1978. *Publ. Astron. Soc. Pac.* 90:262

Caswell, J. L., Batchelor, R. A., Haynes, R. F., Huchtmeier, W. K. 1974. *Aust. J. Phys.* 27:417

Caswell, J. L., Murray, J. D., Roger, R. S., Cole, D. J., Cooke, D. J. 1975. *Astron. Astrophys.* 45:239

Caswell, J. L., Haynes, R. F., Goss, W. M. 1977. *MNRAS* 181:427

Cato, B. T., Rönnäng, B. O., Lewin, P. G., Rydbeck, O. E. H., Yngvesson, K. S., Cardiasmenos, A. G., Shanley, J. F. 1975. *Res. Rep. No. 123.* Chalmers University, Sweden

Cesarsky, C. J., Cesarsky, D. A., Churchwell, E., Lequeux, J. 1978. *Astron. Astrophys.* 68:33

Cheung, L., Frogel, J. A., Gezari, D. Y., Hauser, M. G. 1979. *Ap. J. Lett.* 226:L149

Cochran, W. D., Ostriker, J. P. 1977. *Ap. J.* 211:392

Colley, D., Scott, P. F. 1977. *MNRAS* 181:703

Cook, A. H. 1975. *MNRAS* 171:605

Crampton, D., Georgelin, Y. M., Georgelin, Y. P. 1978. *Astron. Astrophys.* 66:1

Davidson, K. 1970. *Astrophys. Space Sci.* 6:422

Davidson, K., Harwit, M. 1967. *Ap. J.* 148:443

Day, K. L., Donn, B. 1978. *Ap. J. Lett.* 222:L45

Deharveng, L., Maucherat, M. 1978. *Astron. Astrophys.* 70:19

de Jong, T., Maeder, A., eds. 1977. *I.A.U. Symp. No. 75, Star Formation.* Dordrecht: Reidel

Dickel, H. R., Dickel, J. R., Wilson, W. J. 1977. *Ap. J.* 217:56

Dickel, J. R., Dickel, H. R., Wilson, W. J. 1978. *Ap. J.* 223:840

Downes, D., Wilson, T. L. 1974. *Ap. J. Lett.* 191:L77

Downes, D., Goss, W. M., Schwarz, U. J., Wouterloot, J. G. A. 1978. *Astron. Astrophys. Suppl.* 35:270

Dyck, H. M., Capps, R. W. 1978. *Ap. J. Lett.* 220:L49

Dyck, H. M., Capps, R. W., Forrest, W. J., Gillett, F. C. 1973. *Ap. J. Lett.* 183:L99

Elias, J. H. 1978. *Ap. J.* 224:453

Elitzur, M. 1979. *Astron. Astrophys.* 73:322

Elmegreen, B. G., Lada, C. J. 1977. *Ap. J.* 214:725

Elmegreen, B. G., Lada, C. J. 1978. *Ap. J.* 219:467

Elsässer, H., Staude, M.-J. 1978. *Astron. Astrophys.* 70:L3

Emerson, J. P. 1978. PhD thesis. University College, London

Emerson, J. P., Jennings, R. E., Moorwood, A. F. M. 1973. *Ap. J.* 184:401

Encrenaz, P. J., Falgarone, E., Lucas, R. 1975. *Astron. Astrophys.* 44:73

Evans, N. J., Crutcher, R. M., Wilson, W. J. 1976. *Ap. J.* 206:440

Evans, N. J., Blair, G. N., Beckwith, S. 1977. *Ap. J.* 217:448

Evans, N. J., Beckwith, S., Brown, R. L., Gilmore, W. 1979. *Ap. J.* 227:450

Falgarone, E., Cesarsky, D. A., Encrenaz, P. J., Lucas, R. 1978. *Astron. Astrophys.* 65:L13

Fazio, G. G., Zeilik, M., Low, F. J. 1976. *Ap. J. Lett.* 206:L165

Fazio, G. G., Lada, C. J., Kleinmann, D. E., Wright, E. L., Ho, P. T. P., Low, F. J. 1978. *Ap. J. Lett.* 221:L77

Felli, M., Habing, H. J., Israel, F. P. 1977. *Astron. Astrophys.* 59:43

Felli, M., Harten, R. H., Habing, H. J., Israel, F. P. 1978. *Astron. Astrophys. Suppl.* 32:423

Finn, G. D., Simon, Th. 1977. *Ap. J.* 212:472

Forster, J. R., Welch, W. J., Wright, M. C. H. 1977. *Ap. J. Lett.* 215:L121

Forster, J. R., Welch, W. J., Wright, M. C. H., Baudry, A. 1978. *Ap. J.* 221:137

Frogel, J. A., Persson, S. E. 1972. *Ap. J.* 178:667

Frogel, J. A., Persson, S. E. 1973. *Ap. J.* 186:207

Frogel, J. A., Persson, S. E. 1974. *Ap. J.* 192:35

Frogel, J. A., Persson, S. E., Aaronson, M. 1977. *Ap. J.* 213:723

Gatley, I., Becklin, E. E., Werner, M. W., Harper, D. A. 1978. *Ap. J.* 220:822

Gatley, I., Becklin, E. E., Werner, M. W., Sellgren, K. 1979. *Ap. J.* Submitted

Gautier, T. N., Fink, U., Treffers, R. R., Larson, H. P. 1976. *Ap. J. Lett.* 207:L129

Gehrels, T., ed. 1979. *Proc. Conf. Protostars Planets.* Tucson, Arizona, Jan. 1978

Genzel, R., Downes, D. 1977a. *Astron. Astrophys. Suppl.* 30:145

Genzel, R., Downes, D. 1977b. *Astron. Astrophys.* 61:117

Genzel, R., Downes, D. 1979. *Astron. Astrophys.* 72:234

Genzel, R., Downes, D., Moran, J. M., Johnston, K. J., Spencer, J. H., Walker, R. C., Haschick, A., Matveyenko, L. I., Kogan, L. R., Kostenko, V. I., Rönnäng, B., Rydbeck, O. E. H., Moiseev, I. G. 1978. *Astron. Astrophys.* 66:13

Georgelin, Y. M., Georgelin, Y. P., Roux, S. 1973. *Astron. Astrophys.* 25:337

Gillespie, A. R., Huggins, P. J., Sollner, T. C. L. G., Phillips, T. G., Gardner, F. F., Knowles, S. H. 1977. *Astron. Astrophys.* 60:221

Gillett, F. C., Soifer, B. T. 1976. *Ap. J.* 207:780

Gillett, F. C., Forrest, W. J., Merrill, K. M., Capps, R. W., Soifer, B. T. 1975. *Ap. J.* 200:609

Gilmore, W. S. 1978. PhD thesis. Univ. Maryland

Glushkov, Yu. I., Denisyuk, E. K., Karyagina, S. V. 1975. *Astron. Astrophys.* 39:481

Goldreich, P., Kwan, J. 1974. *Ap. J.* 191:93

Goss, W. M., Shaver, P. A. 1970. *Aust. J. Phys. Astrophys. Suppl.* 14:1

Goss, W. M., Lockhart, I. A., Fomalont, E. B., Hardebeck, E. G. 1973a. *Ap. J.* 183:843

Goss, W. M., Nguyen-Quang-Rieu, Winnberg, A. 1973b. *Astron. Astrophys.* 29:435

Goss, W. M., Knowles, S. H., Balister, M., Batchelor, R. A., Wellington, K. J. 1976. *MNRAS* 174:541

Goss, W. M., Matthews, H. E., Winnberg, A. 1978. *Astron. Astrophys.* 65:307

Gottlieb, C. A., Lada, C. J., Webster Gottlieb, E., Lilley, A. E., Litvak, M. M. 1975. *Ap. J.* 202:655

Graf, W. 1978. *Natl. Space Sci. Data Cent. Rep. No. 78-05,* p. 66. *Astron. J.* In press

Grasdalen, G. L. 1973. *Ap. J.* 182:781

Grasdalen, G. L. 1974. *Ap. J.* 193:373

Grasdalen, G. L. 1976. *Ap. J. Lett.* 205:L83

Greenberg, L. T., Dyal, P., Geballe, T. R. 1977. *Ap. J. Lett.* 213:L71

Gundermann, E. J. 1965. PhD thesis. Harvard Univ.

Habing, H. J., Israel, F. P., de Jong, T. 1972. *Astron. Astrophys.* 17:329

Habing, H. J., Goss, W. M., Matthews, H. E., Winnberg, A. 1974. *Astron. Astrophys.* 35:1

Hackwell, J. A., Gehrz, R. D., Smith, J. R., Briotta, D. A. 1978. *Ap. J.* 221:737

Hall, D. N. B., Kleinmann, S. G., Ridgway, S. T., Gillett, F. C. 1978. *Ap. J. Lett.* 223:L47

Harris, S. 1975a. *MNRAS* 170:139

Harris, S. 1975b. In *H II Regions and Related Topics,* ed. T. L. Wilson, D. Downes, p. 393. Heidelberg: Springer

Harris, S., Scott, P. F. 1976. *MNRAS* 175:371

Harris, S., Wynn-Williams, C. G. 1976. *MNRAS* 174:649

Harris, S., Rowan-Robinson, M. 1977. *Astron. Astrophys.* 60:405

Harten, R. 1976. *Astron. Astrophys.* 46:109

Harvey, P. J., Booth, R. S., Davies, R. D., Whittet, D. C. B., McLaughlin, W. 1974. *MNRAS* 169:545

Harvey, P. M., Hoffman, W. F., Campbell, M. F. 1975. *Ap. J. Lett.* 196:L31

Harvey, P. M., Campbell, M. F., Hoffman, W. F. 1976. *Ap. J. Lett.* 205:L69

Harvey, P. M., Campbell, M. F., Hoffman, W. F. 1977a. *Ap. J.* 211:786

Harvey, P. M., Campbell, M. F., Hoffman, W. F. 1977b. *Ap. J.* 215:151

Harvey, P. M., Campbell, M. F., Hoffman, W. F. 1978. *Ap. J.* 219:891

Heckman, T. M., Sullivan, W. T. 1976. *Astrophys. Lett.* 17:105

Hefele, H., Schulte in den Bäumen, J. 1978. *Astron. Astrophys.* 66:465

Hefele, H., Wacker, W., Weinberger, R. 1977. *Astron. Astrophys.* 56:407

Herbig, G. H. 1971. *Ap. J.* 169:537

Ho, P. T. P. 1978. PhD thesis. Mass. Inst. Tech.

Ho, P. T. P., Barrett, A. H. 1978. *Ap. J. Lett.* 224:L23

Hollenbach, D. J., Shull, J. M. 1977. *Ap. J.* 216:419

Hoyle, F., Wickramasinghe, N. C. 1977. *Nature* 268:610

Hudson, H. S., Soifer, B. T. 1976. *Ap. J.* 206:100

Hughes, V. A., Viner, M. R. 1975. *MNRAS* 173:73P

Icke, V. 1979. *Ap. J.* In press

Israel, F. P. 1976a. *Astron. Astrophys.* 48:193

Israel, F. P. 1976b. PhD thesis. Univ. Leiden

Israel, F. P. 1976c. *Astron. Astrophys.* 52:175

Israel, F. P. 1977a. *Astron. Astrophys.* 59:27

Israel, F. P. 1977b. *Astron. Astrophys.* 61:377
Israel, F. P. 1977c. *Astron. Astrophys.* 60:233
Israel, F. P. 1978. *Astron. Astrophys.* 70:769
Israel, F. P., Felli, M. 1978. *Astron. Astrophys.* 63:325
Israel, F. P., Habing, H. J., de Jong, T. 1973. *Astron. Astrophys.* 27:143
Johnston, K. J., Knowles, S. H., Sullivan, W. T. 1971. *Ap. J. Lett.* 167:L93
Johnston, K. J., Sloanaker, R. M., Bologna, J. M. 1973. *Ap. J.* 182:67
Johnston, K. J., Knowles, S. H., Moran, J. M., Burke, B. F., Lo, K. Y., Pappadopoulos, G. D., Read, R. B., Hardebeck, E. G. 1977. *Astron. J.* 82:403
Johnson, H. M. 1973a. *Ap. J. Lett.* 180:L7
Johnson, H. M. 1973b. *Ap. J.* 182:497
Joyce, R. R., Simon, M., Simon, Th. 1978. *Ap. J.* 220:156
Kahn, F. D. 1974. *Astron. Astrophys.* 36:149
Kandel, R. S., Sibille, F. 1978. *Astron. Astrophys.* 68:217
Kazès, I., Le Squéren, A. M., Gadéa, F. 1975. *Astron. Astrophys.* 42:9
Knapp, G. R., Morris, M. 1976. *Ap. J.* 206:713
Knapp, G. R., Kuiper, T. B. H., Knapp, S. L., Brown, R. L. 1976. *Ap. J.* 206:443
Knowles, S. H., Caswell, J. L., Goss, W. M. 1976. *MNRAS* 175:537
Kutner, M., Thaddeus, P. 1971. *Ap. J. Lett.* 168:L67
Kutner, M., Tucker, K. D., Chin, G., Thaddeus, P. 1977. *Ap. J.* 215:521
Kwan, J. 1977. *Ap. J.* 217:771
Lada, C. J. 1976. *Ap. J. Suppl.* 32:603
Lada, C. J., Elmegreen, B. G., Blitz, L. 1978. Preprint
Lada, C. J., Elmegreen, B. G., Cong, H.-l., Thaddeus, P. 1978. *Ap. J. Lett.* 226:L39
Larson, R. B. 1973. *Ann. Rev. Astron. Astrophys.* 11:219
Larson, R. B., Starrfield, S. 1971. *Astron. Astrophys.* 13:190
Lemke, D. 1975. In *H II Regions and Related Topics*, ed. T. L. Wilson, D. Downes, p. 372. Heidelberg: Springer
Little, L. T., White, G. J., Riley, P. W. 1977. *MNRAS* 180:639
Lo, K. Y., Bechis, K. P. 1974. *Ap. J. Lett.* 190:L125
Lo, K. Y., Burke, B. F. 1973. *Astron. Astrophys.* 26:487
Lo, K. Y., Walker, R. C., Burke, B. F., Moran, J. M., Johnston, K. J., Ewing, M. S. 1975a. *Ap. J.* 202:650
Lo, K. Y., Burke, B. F., Haschick, A. D. 1975b. *Ap. J.* 202:81
Löbert, W., Goss, W. M. 1978. *MNRAS* 183:119
Loer, S. J., Allen, D. A., Dyck, H. M. 1973. *Ap. J. Lett.* 183:L97

London, R., McCray, R., Chu, S. I. 1977. *Ap. J.* 217:442
Loren, R. B. 1977. *Ap. J.* 215:129
Loren, R. B. 1979. *Ap. J.* 227:832
Loren, R. B., Wootten, H. A. 1978. *Ap. J. Lett.* 225:81
Low, F. J., Aumann, H. H. 1970. *Ap. J. Lett.* 162:L79
Lucas, R., Le Squéren, A. M., Kazès, I., Encrenaz, P. J. 1978. *Astron. Astrophys.* 66:155
Mader, G. L., Johnston, K. J., Moran, J. M. 1978. *Ap. J.* 224:115
Martin, A. H. M. 1972. *MNRAS* 157:31
Martin, A. H. M. 1973. *MNRAS* 163:141
Martin, A. H. M., Gull, S. P. 1976. *MNRAS* 175:235
Mathews, W. G. 1965. *Ap. J.* 142:1120
Matthews, H. E., Goss, W. M., Winnberg, A., Habing, H. J. 1977. *Astron. Astrophys.* 61:261
Matthews, H. E., Shaver, P. A., Goss, W. M., Habing, H. J. 1978. *Astron. Astrophys.* 63:307
Matthews, H. E., Harten, R. H., Goss, W. M. 1979. *Astron. Astrophys.* 72:224
Maucherat, M. 1975. *Astron. Astrophys.* 45:193
Meaburn, J. 1975. *H II Regions and Related Topics*, ed. by T. L. Wilson, D. Downes, p. 22. Heidelberg: Springer
Meaburn, J. 1977. In *Topics in Interstellar Matter*, ed. H. van Woerden, p. 81. Dordrecht: Reidel
Melnick, G., Gull, G. E., Harwit, M. 1979. *Ap. J. Lett.* 227:L29
Mezger, P. G., Henderson, A. P. 1967. *Ap. J.* 147:471
Mezger, P. G., Wink, J. 1974. *Mem. Soc. Astron. Italiana* 45:315
Mezger, P. G., Schraml, J., Terzian, Y. 1967a. *A. J.* 150:807
Mezger, P. G., Altenhoff, W., Schraml, J., Burke, B. F., Reifenstein, E. C., Wilson, T. L. 1967b. *Ap. J. Lett.* 150:L137
Moran, J. M. 1976. In *Frontiers of Astrophysics*, ed. E. H. Aurett, p. 385. Cambridge, Mass.: Harvard Univ. Press
Moran, J. M., Burke, B. F., Barrett, A. H., Rogers, A. E. E., Ball, J. A., Carter, J. C., Cudaback, D. D. 1968. *Ap. J. Lett.* 152:L97
Moran, J. M., Johnston, K. J., Spencer, J. H., Schwartz, P. R. 1977. *Ap. J.* 217:434
Moran, J. M., Reid, M. J., Lada, C. J., Yen, J. L., Johnston, K. J., Spencer, J. H. 1978. *Ap. J. Lett.* 224:L67
Morris, M. 1976. *Ap. J.* 210:100
Morris, M., Palmer, P., Turner, B. E., Zuckerman, B. 1974. *Ap. J.* 191:349
Mufson, S. L., Liszt, H. S. 1977. *Ap. J.* 212:664

Myers, P. C. 1977. *Ap. J.* 211:737
Natta, A., Panagia, N. 1976. *Astron. Astrophys.* 50:191
Neugebauer, G., Garmire, G. 1970. *Ap. J. Lett.* 161:L91
Panagia, N. 1977. In *Infrared and Submillimeter Astronomy*, ed. G. G. Fazio, p. 43. Dordrecht: Reidel
Panagia, N., Natta, A., Preite-Martinez, A. 1978. *Astron. Astrophys.* 68:265
Pankonin, V., Winnberg, A., Booth, R. S. 1977. *Astron. Astrophys.* 58:L25
Petrosian, V., Dana, R. A. 1975. *Ap. J.* 196:733
Persson, S. E., Frogel, J. A. 1974. *Ap. J.* 188:523
Phillips, T. G., Huggins, P. J., Neugebauer, G., Werner, M. W. 1977. *Ap. J. Lett.* 217:L161
Pipher, J. L., Soifer, B. T. 1976. *Astron. Astrophys.* 46:153
Pipher, J. L., Grasdalen, G. L., Soifer, B. T. 1976a. *Ap. J.* 193:283
Pipher, J. L., Sharpless, S., Savedoff, M., Kerridge, S. J., Krassner, J., Schurmann, S., Soifer, B. T., Merrill, K. M. 1976b. *Astron. Astrophys.* 51:255
Pipher, J. L., Sharpless, S., Savedoff, M. P., Krassner, J., Varbese, S., Soifer, B. T., Zeilik, M. 1977. *Astron. Astrophys.* 59:215
Pipher, J. L., Soifer, B. T., Krassner, J. 1979. *Astron. Astrophys.* Submitted
Price, S. D., Walker, R. G. 1976. *Air Force Geophys. Lab. Rep.* AFGL-TR-76-0208
Puetter, R. C., Russell, R. W., Soifer, B. T., Willner, S. P. 1979. *Astrophys. J.* 228:118
Rank, D. M., Dinerstein, H. L., Lester, D. F., Bregman, J. D., Aitken, D. K., Jones, B. 1978. *MNRAS* 185:179
Rieke, G. H., Harper, D. A., Low, F. J., Armstrong, K. R. 1973. *Ap. J. Lett.* 183:L67
Righini, G., Simon, M., Joyce, R. R. 1976. *Ap. J.* 207:119
Righini-Cohen, G., Simon, M. 1977. *Ap. J.* 213:390
Rohlfs, K., Braunsfurth, E., Hills, D. L. 1977. *Astron. Astrophys.* 30:369
Rubin, R. H. 1968. *Ap. J.* 153:761
Ryle, M., Downes, D. 1967. *Ap. J. Lett.* 148:L17
Sarazin, C. L. 1978. *Ap. J.* 220:165
Sargent, A. 1977. *Ap. J.* 218:736
Sargent, A. 1979. *Ap. J.* In press
Scott, P. F. 1978. *MNRAS* 183:435
Shaver, P. A., Danks, A. C. 1978. *Astron. Astrophys.* 65:323
Sibille, F., Bergeat, J., Lunel, M., Kandel, R. 1975. *Astron. Astrophys.* 40:441
Sibille, F., Lunel, M., Bergeat, J. 1976. *Astron. Astrophys.* 47:161
Snyder, L. E., Buhl, D. 1974. *Ap. J. Lett.* 189:L31
Soifer, B. T., Pipher, J. L. 1975. *Ap. J.* 199:663
Solomon, P. M., ed. 1979. *Proc. Gregynog Workshop, Giant Molecular Clouds.* In press
Strelnitskii, V. S., Sunyayev, R. A. 1973. *Sov. Astron. AJ* 16:579. *Astron. Zh.* 49:704 (1972)
Strom, S. E., Strom, K. M., Grasdalen, G. L. 1975. *Ann. Rev. Astron. Astrophys.* 13:187
Sullivan, W. T., Kerstholt, J. H. 1976. *Astron. Astrophys.* 51:427
Tenorio-Tagle, G. 1979. *Astron. Astrophys.* 71:59
ter Haar, D., Pelling, M. A. 1974. *Rep. Prog. Phys.* 37:487
Thaddeus, P. 1977. In *I.A.U. Symp. No. 75, Star Formation*, ed. T. de Jong, A. Maeder. Dordrecht: Reidel
Thompson, R. I., Tokunaga, A. T. 1978. *Ap. J.* 226:119
Thronson, H. A., Harper, D. A. 1979. *Ap. J.* 230:133
Tielens, A. G. G. M., de Jong, T. 1979. *Astron. Astrophys.* Submitted
Turner, B. E. 1971. *Astrophys. Lett.* 8:73
Turner, B. E., Rubin, R. H. 1971. *Ap. J. Lett.* 170:L13
Turner, B. E., Balick, B., Cudaback, D. D., Heiles, C., Boyle, R. J. 1974. *Ap. J.* 194:279
Vandervoort, P. O. 1963. *Ap. J.* 139:869
Vogt, N., Moffat, A. F. J. 1975. *Astron. Astrophys.* 45:405
Vrba, F. J., Strom, S. E., Strom, K. M. 1976. *Astron. J.* 81:317
Walker, R. C., Johnston, K. J., Burke, B. F., Spencer, J. H. 1977. *Ap. J. Lett.* 211:L135
Walker, R. C., Burke, B. F., Haschick, A. D., Crane, P. C., Moran, J. M., Johnston, K. J., Lo, K. Y., Yen, J. L., Broten, N. W., Legg, T. H., Greisen, E. W., Hansen, S. S. 1978. *Ap. J.* 226:95
Weaver, H. F., Williams, D. R. W., Dieter, N. H., Lum, W. T. 1965. *Nature* 208:29
Weiler, K. W., Shaver, P. A. 1978. *Astron. Astrophys.* 65:305
Wellington, K. J., Sullivan, W. T., Goss, W. M., Matthews, H. E. 1976. *Astron. Astrophys.* 47:351
Wellington, K. J., Sullivan, W. T., Goss, W. M., Matthews, H. E. 1977. *Astron. Astrophys.* 54:319
Wendker, H. J., Altenhoff, W. J. 1977. *Astron. Astrophys.* 54:301
Wendker, H. J., Baars, J. W. M. 1974. *Astron. Astrophys.* 33:157
Werner, M. W., Elias, J. H., Gezari, D. Y., Hauser, M. G., Westbrook, W. E. 1975. *Ap. J. Lett.* 199:L185
Werner, M. W., Becklin, E. E., Neugebauer, G. 1977. *Science* 197:723
Werner, M. W., Becklin, E. E., Gatley, I.,

Matthews, K., Neugebauer, G., Wynn-Williams, C. G. 1979. *Ap. J.* Submitted

Westbrook, W. E., Werner, M. W., Elias, J. H., Gezari, D. Y., Hauser, M. G., Lo, K. Y., Neugebauer, G. 1976. *Ap. J.* 209:94

White, G. J., Little, L. T., Parker, E. A., Nicholson, P. S., MacDonald, G. H., Bale, F. 1975. *MNRAS* 170:32P

Willner, S. P. 1976. *Ap. J.* 206:728

Willner, S. P. 1977. *Ap. J.* 214:706

Wilson, T. L. 1975. In *H II Regions and Related Topics*, ed. T. L. Wilson, D. Downes, p. 372. Heidelberg: Springer

Wilson, T. L., Fazio, G. G., Jaffe, D., Kleinmann, D. E., Wright, E., Low, F. J. 1979. *Astron. Astrophys.* In press

Wilson, W. J., Schwartz, P. R., Epstein, E. E., Johnson, W. A., Etcheverry, R. D., Mori, T. T., Berry, G. G., Dyson, H. B. 1974. *Ap. J.* 191:357

Wink, J. E., Altenhoff, W. J., Webster, W. J. 1975. *Astron. Astrophys.* 38:109

Woodward, P. R. 1978. *Ann. Rev. Astron. Astrophys.* 16:555

Woolf, N. J., Stein, W. A., Gillett, F. C., Merrill, K. M., Becklin, E. E., Neugebauer, G., Pepin, T. J. 1973. *Ap. J. Lett.* 179:L111

Wright, E. L. 1973. *Ap. J.* 185:569

Wright, E. L., Lada, C. J., Fazio, G. G., Kleinmann, D. E., Low, F. J. 1977a. *Astron. J.* 82:132

Wright, E. L., Fazio, G. G., Low, F. J. 1977b. *Ap. J.* 217:724

Wright, E. L., de Campli, W., Fazio, G. G., Kleinmann, D. E., Lada, C. J., Low, F. J. 1979. *Ap. J.* 228:439

Wynn-Williams, C. G. 1971. *MNRAS* 151:397

Wynn-Williams, C. G., Becklin, E. E. 1974. *Publ. Astron. Soc. Pac.* 86:5

Wynn-Williams, C. G., Becklin, E. E., Neugebauer, G. 1972. *MNRAS* 160:1

Wynn-Williams, C. G., Becklin, E. E., Neugebauer, G. 1974. *Ap. J.* 187:473

Wynn-Williams, C. G., Becklin, E. E., Matthews, K., Neugebauer, G., Werner, M. W. 1978a. *MNRAS* 179:255

Wynn-Williams, C. G., Becklin, E. E., Matthews, K., Neugebauer, G. 1978b. *MNRAS* 183:237

Woodward, P. R. 1978. *Ann. Rev. Astron. Astrophys.* 16:555

Yorke, H. W., Krügel, E. 1977. *Astron. Astrophys.* 54:183

Zeilik, M. 1976. *Astron. Astrophys.* 46:319

Zuckerman, B. 1973. *Ap. J.* 183:863

Zuckerman, B., Palmer, P. 1974. *Ann. Rev. Astron. Astrophys.* 12:279

Zuckerman, B., Palmer, P., Rickard, L. J. 1975. *Ap. J.* 197:571

Zuckerman, B., Kuiper, T. B. H., Rodriguez-Kuiper, E. N. 1976. *Ap. J. Lett.* 209:L137

Ann. Rev. Astron. Astrophys. 1979. 17 : 387–413
Copyright © 1979 by Annual Reviews Inc. All rights reserved

MARTIAN METEOROLOGY ✻2155

Conway B. Leovy
Department of Atmospheric Sciences and Geophysics Group,
University of Washington, Seattle, Washington 98195

INTRODUCTION

The atmosphere of Mars has been an object of fascination and often erroneous speculation for a very long time (see Glasstone 1968 for a survey of early work), but only since the beginning of the era of space flight has an accurate picture begun to emerge. Much of our information concerning the atmosphere and its meteorology has been obtained from three spectacularly successful spaceprobes: Mariner 9 (1971–1972), and Viking 1 and 2 (1976–present), and many of the early results of these missions were published in two special issues of the *Journal of Geophysical Research* (Volume 78, 1973, and Volume 82, 1977). Reviews of Martian meteorology include those of Mintz (1961) and Leovy (1969). Studies of the Martian atmosphere were reviewed prior to Mariner 9 by Ingersoll & Leovy (1971), and after Mariner 9 by Barth (1974).

The Viking missions were the first to explore the atmosphere with entry probes and landers as well as from orbit, and in so doing they provided a detailed picture of structure and dynamical processes, sufficiently detailed that valuable comparisons can now be made between the Martian atmosphere, the atmosphere of the earth, and other rotating differentially heated fluid systems. This review is an attempt to describe that picture.

For reference, Table 1 lists some of the principle parameters that govern the meteorological processes on Mars and the earth. Many parameters are similar, especially when contrasted with values for the other planets. It is particularly significant that neither planet has an opaque atmosphere. As a result, the two meteorologies have much in common, and it is relatively easy to understand Martian meteorology and to draw useful comparisons between the dynamics of the two atmospheres.

Table 1 shows that the Martian solar day is slightly longer than the terrestrial day, and I shall make use of the term "sol," coined during

387

Table 1 Atmospheric parameters

	Mars	Earth
Planetary mass (kg)	5×10^{22}	6×10^{23}
Planetary radius (km)	3394	6369
Acceleration of gravity (m/s^2)	3.72	9.81
Mean solar constant (W/m^2)	591	1373[a]
Orbital eccentricity	0.093	0.017
Axial inclination ($^\circ$)	25	23.5
Length of year (earth days)	687	365
Length of solar day (s)	88775	86400
Bolometric albedo	0.24[b]	0.29[c]
Atmospheric visible opacity	0.1 to $\sim 10^d$	0.2 to $\sim 100^e$
Atmospheric infrared emissivity	0.15 to 0.9[d]	0.4 to 1[e]
Principle constituents	$CO_2(0.95), {}^{40}A(0.016),$ $N_2(0.027)^f$	$N_2(0.78), O_2(0.21),$ ${}^{40}A(0.01), H_2O(<0.03)$
Surface pressure (mb)	6	1013
Gas constant (m^2/s^2–K)	188	287
Temperature range (near-surface)	145–245[g]	220–310
Mean scale height (km)	10	7.8
Adiabatic lapse rate (K/km)	4.5	9.8

[a] Fröhlich 1977.
[b] Kieffer et al. 1973, Irvine et al. 1968.
[c] Vonder Haar & Suomi 1969.
[d] Variability due mainly to suspended dust.
[e] Variability due mainly to water clouds.
[f] Owen et al. 1977.
[g] Based on Viking 1 and 2 measurements, Ryan & Henry 1979.

preparations for Viking, to refer to it. Martian seasons are referenced to the longitude of the sun in Mars-centered coordinates, L_s; during the present epoch $L_s = 0$ corresponds to northern vernal equinox.

PRESSURE, TEMPERATURE, AND ZONALLY AVERAGED WIND

In this section, the large-scale variability of some of the principle quantities is described as background for discussion of more detailed processes.

Time Variations of Pressure

Martian seasonal variability can be keyed to the seasonal pressure patterns observed at the two Viking sites (Figure 1; see Hess et al. 1979). The most obvious feature is the large-amplitude seasonal oscillation at both sites. This is due to the variation in atmospheric mass as CO_2 condenses out on the winter polar caps, a process originally predicted by Leighton

& Murray (1966) after Mariner 4 had confirmed earlier spectroscopic evidence that the mean surface pressure is very low (Kliore et al. 1965, Kaplan et al. 1964), too low to prevent polar temperatures from falling to the CO_2 frost point. The deep pressure minimum near $L_s = 150$ corresponds to maximum CO_2 accumulation in the south polar cap at the end of winter; the secondary minimum near $L_s = 350$ corresponds to maximum accumulation in the northern cap. Much of the asymmetry is due to the orbital eccentricity which causes a long southern winter and a short northern winter. More recent data for the Martian year end, not plotted in Figure 1, show the expected return of pressure at both sites to the values measured at the beginning of the year.

Superposed on this low-frequency annual oscillation is a much higher frequency component of variability, especially at the Lander 2 site, which increased during autumn and winter until $L_s = 280$, when there was a sudden rise in Lander 2 pressure followed by a decrease in the high-frequency variability. A period of even more intense activity began about $L_s = 325$ and lasted about 100 sols before fading into the more quiescent pattern of late spring and summer. These variations are due to traveling storm systems, similar to terrestrial storms, and they are described in detail below.

The pressure jump at Lander 2 near $L_s = 280$ was associated with the second of two planetwide dust storms. The first of these began near $L_s = 205$ and does not show such a prominent signature. These spectacular global storms are also discussed below.

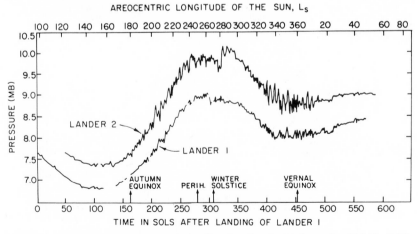

Figure 1 Daily average pressure at the landing sites. Lander 1 is located at 22.5 N, 48 W; Lander 2 is at 48.0 N, 230 W. Data from the Viking meteorology experiments (see Hess et al. 1979).

Temperature Distributions

Figure 2 shows the vertical temperature profiles measured by Viking (Seiff & Kirk 1977). They resemble reasonably well a less detailed profile derived from Mars 5 entry data (Kerzhanovich 1977). In the lowest 15 km, temperature falls more slowly with height than predicted by radiative-convective equilibrium models, which neglect suspended dust and the stabilizing effect of circulation (Gierasch & Goody 1968). Absorption of solar radiation by suspended dust can account for this discrepancy (Gierasch & Goody 1972, Conrath 1975, Zurek 1978, Pollack et al. 1977, 1979a). Above 40 km, the temperature averages about 140 K, but there are large-amplitude, large-scale irregularities. Similar irregularities were detected by means of the stellar occultation technique and were interpreted as tidally forced oscillations (Elliot et al. 1977). The diurnal tide, forced by heating in the lower atmosphere, can drive vertical motions in the upper atmosphere (Chapman & Lindzen 1970). The corresponding temperature variations are expected to have amplitudes and vertical scales like those in Figure 2 (Zurek 1976).

Distributed temperature profiles have been obtained by the radio occultation and infrared sounding techniques (Kliore et al. 1973, Fjeldbo et al. 1977, Hanel et al. 1972, Conrath et al. 1973, Conrath 1975). A temperature cross section derived from a composite of Mariner 9 infrared data obtained between $L_s = 43$ and $L_s = 54$ is shown in Figure 3. In the

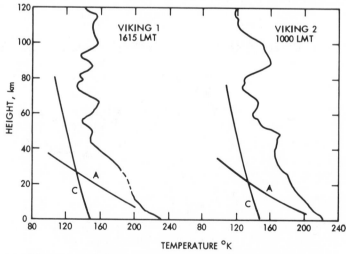

Figure 2   Viking entry temperature profiles. Curves marked C and A correspond to CO_2 condensation and an adiabat respectively. Reproduced from Seiff & Kirk (1977), *Journal of Geophysical Research*, Volume 82, copyrighted by American Geophysical Union.

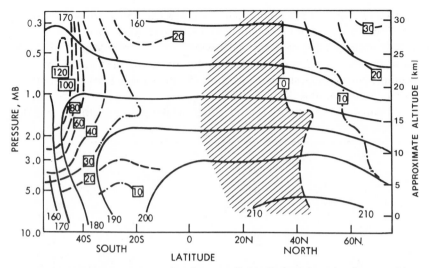

Figure 3 Temperature cross section for $L_s = 43$–54 obtained from Mariner 9 orbiter infrared measurements. Corresponding eastward zonal wind shown by curves $(-\cdot-)$, westward winds shaded (m/s). Equatorial winds are uncertain. Data were provided by B. Conrath, Goddard Space Flight Center, and the figure is reproduced from Pollack et al. (1979b), *Journal of Atmospheric Sciences*, Volume 36, copyrighted by American Meteorological Society.

tropics, subtropics, and late spring (northern) midlatitudes, temperature is nearly uniform horizontally with a lapse rate equal to about half the adiabatic rate. At high latitudes of the autumn hemisphere and below about 15 km, there is a strong horizontal temperature gradient associated with the edge of the newly forming seasonal CO_2 polar cap. Above 15 km in the same region, there is a temperature maximum with temperatures far above their radiative equilibrium values. Similar temperature maxima are found above regions of intense dynamical activity in the earth's stratosphere and mesosphere. They are indicative of the dynamical activity, and are produced by compressional heating, presumably associated with large-scale subsidence. This Martian temperature maximum is most intense during midwinter, and during the most active phases of planetwide dust storms (Figure 4).

Zonal Mean Winds

Very few direct observations of winds are available. Cloud motions in the summer tropics indicate westward winds of 30–55 m/s at altitudes of 15–25 km (Briggs et al. 1977). The Viking landers measured diurnally variable surface winds having small daily mean values (Hess et al. 1977). During winter, prior to the onset of the global dust storm at $L_s = 280$,

Lander 2 winds were predominantly eastward, but during the storm itself, winds were predominantly toward southwest. Briggs & Leovy (1974) indirectly inferred eastward winds in winter poleward of 45 N from the morphology of clouds. In addition, some dust cloud motions and structures suggest wind direction (Gifford 1964, Leovy et al. 1972, 1973a, Peterfreund & Kieffer 1979).

Fortunately, the thermal wind equation is applicable to the zonally

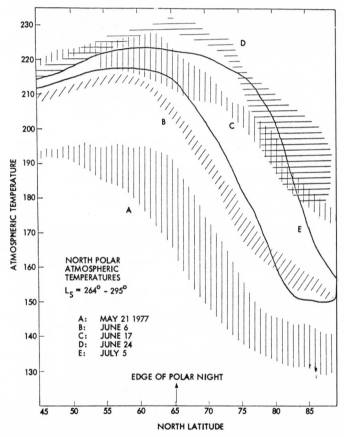

Figure 4 Atmospheric temperature profiles from Viking orbiter infrared measurements (T. Martin & Kieffer 1979). Shaded areas and enclosed area *E* correspond to ranges on the indicated dates (*A*: $L_s = 264$, *B*: $L_s = 274$, *C*: $L_s = 281$, *D*: $L_s = 285$, *E*: $L_s = 292$). Temperatures correspond to a broad weighting function centered near 25 km altitude, and show the enhancement of the high altitude subpolar temperature maximum during the most intense phase of a planetwide dust storm. Reproduced from Martin & Kieffer (1979), *Journal of Geophysical Research*, Volume 84, copyrighted by American Geophysical Union.

averaged zonal winds on Mars (see Holton 1975, p. 48). Thus the mean zonal wind at height z and latitude ϕ is given with good accuracy by

$$\bar{u}(z,\phi) = \bar{u}(0,\phi) + g(fa)^{-1} \int_0^z \bar{T}^{-1} (\partial \bar{T}/\partial \phi) \, dz, \tag{1}$$

where a is the planetary radius, g is the acceleration of gravity, $f = 2\Omega$ sin ϕ is the Coriolis parameter for rotation rate Ω, and $\bar{T}$ is zonally averaged temperature. Because of surface friction, it is reasonable to expect that $\bar{u}(0,\phi)$ is small, and the Viking measurements do show relatively weak mean zonal winds at two sites. Contours of $\bar{u}(z,\phi)$ appear in Figure 3 based on the assumptions that $\bar{u}(0,\phi) = 0$ everywhere and the temperature distribution is representative of the zonal mean. The most significant feature is the intense eastward jet at 20 km, directly above the region of strong horizontal temperature gradient in the south.

Winds calculated with a general circulation model (GCM; Pollack et al. 1976, 1979b) show zonally averaged temperatures and zonal winds similar to those in Figure 3 except that the calculated atmosphere is considerably less statically stable (Figure 5). This defect is a consequence of the neglect of dust heating. The model predicts distributions of surface winds as well as winds aloft, and shows eastward surface winds in winter high latitudes and summer subtropics, and generally westward winds elsewhere.

Mean meridional winds are less certain, since the thermal wind equation is not applicable. The GCM predicts a strong thermally driven circulation at the solstices: flow toward the summer pole at low levels, rising in the summer subtropics, return flow toward the winter pole aloft, and descending flow in winter midlatitudes (Figure 5). This is analogous to the weaker thermally driven circulation in the earth's tropics, which was first interpreted by Hadley (1735). An earlier GCM calculation indicated that the equinoctial meridional circulation would be much weaker and symmetric about the equator, while the zonal wind would be characterized by weak high latitude eastward jets, like the one shown on the right side of Figure 3 (Leovy & Mintz 1969).

There is some indirect evidence that the inferred pattern of meridional circulation at solsticial seasons is correct. (a) Surface streaks and surface drifts, believed to have formed during planetwide dust storms, fit the inferred pattern of southern summer surface winds (Sagan et al. 1973, 1977). (b) The high level temperature maximum above the region of strong subpolar temperature gradient indicates a subsidence pattern consistent with such a circulation. (c) During the second global dust storm, the pressure difference between the Viking landers and the winds measured at the landers were consistent with an intensification of the thermally driven circulation. (*d*) Ascending motions in the summer subtropics comparable to those in the rising branch of the calculated circulation are required to explain the observed midsummer temperature distribution over Lander 1 (Pollack et al. 1979a).

SMALL SCALE PHENOMENA

The Boundary Layer

Heat, momentum, and mass are exchanged between the surface and atmosphere across a boundary layer whose depth varies over the diurnal cycle. During the night, radiative heat loss produces a strong inversion

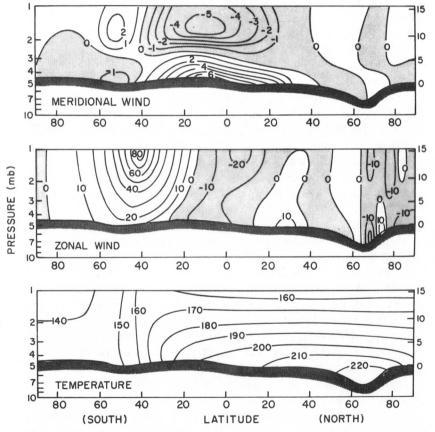

Figure 5 Zonally averaged cross sections of temperature (K, *lower*), mean zonal wind (m/s, *middle*, westward wind shaded), and mean meridional wind (m/s, *upper*, southward wind shaded), $L_s \sim 105°$. Results generated by a general circulation model (Pollack et al. 1976, 1979b).

in the lowest 1 or 2 km. As the surface warms during the morning, the inversion is eventually broken, and a convective layer grows upward until late afternoon. As heating ceases, the convective layer collapses rapidly, and is replaced again by the nighttime inversion (Gierasch & Goody 1968, Flasar & Goody 1976, Hess 1976).

Because almost all exchange between the surface and atmosphere takes place during the daytime when the atmosphere near the surface is unstable and the convective layer is active, I discuss only the daytime convective case in further detail. This layer can be divided into three regions. (a) Below heights of about 1 cm, depending on surface stress, mass and heat transfer are by molecular diffusion; momentum transfer may or may not be molecular, depending on whether the surface roughness length is greater than or less than this height. (b) Between this height and the height at which the Richardson Number Ri is of order unity, turbulence generated by shear stress is the main agent of transfer (Ri measures the ratio of buoyancy accelerations to shear stresses). (c) At higher levels transfer is by intermittent convective plumes driven by buoyancy forces. This uppermost portion of the boundary layer is the "convective layer," and it may extend upward for several kilometers as the plumelike convective elements increase in scale. Through most of the convective layer, heat flux is upward and mixing is efficient, even though the mean lapse rate may be slightly less than the adiabatic lapse rate. Near the top of the convective layer, the heat flux changes sign as the convective elements "overshoot." The Viking entry temperature profiles indicate that the midafternoon convective layer does not extend above about 4 km (Figure 2). This height is consistent with the sizes and inferred heights of cumuliform clouds observed in the tropics during early afternoon (Briggs et al. 1977), and with the height inferred from the duration of afternoon gusts at the Viking landers (Tillman 1976), but it is much less than that predicted by radiative-convective equilibrium models (Gierasch & Goody 1968). Heating by suspended dust, which stabilizes the lower atmosphere, is also responsible for suppressing the growth of the convective layer (Pollack et al. 1979a).

Sutton et al. (1978) used the Viking lander data to evaluate the surface heat flux F_H (positive for upward flux) and friction velocity u_* [$u_* = (\tau_s/\rho_s)^{1/2}$ where τ_s is the stress exerted on the surface by the atmosphere, and ρ_s is surface atmospheric density]. They found that F_H ranged up to about 15 W/m^2, approximately 3% of the available solar flux. This contrasts with the corresponding terrestrial ratio of about 30%. Since the density at the surface of Mars is about 1% as great as at the earth's surface, the Martian atmosphere may be about 10 times as efficient in transporting heat by convection. The most interesting comparison for u_* is with the threshold value of about 2.5 m/s believed to be required to initiate saltation (Greeley et al. 1976). During summer the maximum values of u_* at the sites are estimated to be 0.4–0.6 m/s. Values approached, but probably did not reach, 2.5 m/s at the Viking 1 site during the second global dust storm. Apparently little or no saltation took place at the sites, although a regional scale dust cloud (distinct from the planetwide dust storm clouds) was observed to pass directly over Lander 1 during late winter. Sutton et al. also compared the relationships between gustiness and temperature variance and the bulk parameters: mean surface wind and temperature drop across the unstable near-surface layer, with the corresponding terrestrial relationships. They found a satisfying similarity, justifying the extrapolation of the semiempirical body of terrestrial boundary layer theory to Mars.

In the upper part of the boundary layer, the wind profile is influenced by the Coriolis parameter f, and the simplest model of this layer is that introduced by Ekman (1902) in which nonlinear terms in the equation of motion are neglected and turbulent transfer is assumed to be accomplished by an ad hoc constant eddy viscosity coefficient, v_e. Leovy & Zurek (1979) applied this theory to the semidiurnal wind and pressure variations measured by Lander 2. The pressure gradient required by the theory was estimated by assuming that the semidiurnal tide is a westward propagating wave with longitudinal wavenumber 2 and broad scale meridional pressure variations determined by semidiurnal pressure differences between the landers. The theory contains a single free parameter, β, the ratio of the Ekman layer depth, $d_e = (2v_e/f)^{1/2}$, to the surface layer length scale (u_*/f). They were able to obtain a good fit to the data for a wide range of amplitudes of the semidiurnal tide using a value for β in good agreement with the corresponding value for the terrestrial neutrally stable boundary layer (Businger & Arya 1973).

The Ekman layer model can also be used to fit the diurnally averaged component of the wind and pressure. The pressure difference between landers shown in Figure 1 is due mainly to difference in elevation of the sites, but there is also a meteorological component that varies with time. If this component is assumed to vanish during the inactive midsummer period ($L_s \sim 100$), it can be evaluated from the data in Figure 1. During the second global dust storm the mean wind shifted, blowing from northeast at the same time that the pressure at Lander 2 jumped. If both of these effects are assumed to be due to enhancement of the zonally symmetric circulation, and if the pressure and low level temperature difference between Lander 1 and Lander 2 are assumed to be representative of conditions at Lander 2, the steady winds can be calculated. Table 2 summarizes the comparison of observed semidiurnal and steady winds at Lander 2 with those calculated using the Ekman layer theory. The

Table 2   Lander 2 wind comparisons (m/s)[a]

	Semidiurnal[b] meridional		Semidiurnal[b] zonal		Steady[c]	
	Amplitude	Phase[d]	Amplitude	Phase[d]	Zonal	Meridional
Observed	2.75	0.73	3.06	1.21	−2	−6
Calculated	2.64	0.19	3.31	1.09	−4.4	−3.2

[a] Calculated using $\beta = 0.11$, compared with $\beta = 0.19$ obtained from Businger & Arya's (1973) neutral boundary layer model.
[b] Average over eight 6-sol intervals between $L_s = 200$ and $L_s = 300$.
[c] Average over 10-sol interval around $L_s = 285$.
[d] Phase lag in radians of northward or eastward maximum relative to maximum pressure.

general agreement between observed and calculated values in Table 2 suggests several conclusions. (*a*) The semidiurnal wind and pressure oscillations at the landers are due to a planetary scale westward propagating tide of zonal wavenumber 2. (*b*) There is a substantial enhancement of the mean meridional circulation in the northern hemisphere at the beginning of the second planetary scale dust storm. (*c*) The characteristics of the Martian planetary boundary layer are similar to those of the terrestrial planetary boundary layer; in particular, the values derived from the analysis of Lander 2 data imply effective values $d_e \sim 500$ m, $v_e \sim 10$ m^2/s.

Large-scale slopes, which are ubiquitous on Mars, affect the diurnal wind pattern in the earth's boundary layer (Holton 1967, Blackadar 1957) and on Mars (Blumsack 1971a,b, Blumsack et al. 1973). Buoyancy forces produce downslope acceleration in the cold nighttime boundary layer and upslope acceleration in the warm daytime boundary layer so that maximum upslope winds occur in the evening, maximum downslope winds occur in the morning. This simple pattern is modified by the Coriolis acceleration and by diurnally varying turbulent momentum exchange in the boundary layer. Effects of each of these processes can be seen in the Viking lander data (Hess et al. 1977), and the diurnal pattern of midsummer winds at Lander 1 was generally consistent with the pattern expected for a strongly heated boundary layer on a large-scale slope.

Internal Gravity Waves

These are ubiquitous in stratified fluid flows on earth, and they are widespread on Mars as well. Internal gravity waves forced by topographic variations are easily seen in cloud patterns near the edges of the winter polar caps (Leovy et al. 1972, Briggs & Leovy 1974). Since they form downstream of the generating topographic obstacle, they are wind direction indicators, and their patterns may also be indicative of vertical wind profiles (Pirraglia 1976). For example, resonant gravity waves can arise when weak winds occur beneath much stronger winds from the same direction (Eliassen & Palm 1960). Such resonant waves are common in the winter subpolar zone (45–60° latitude), and their pattern has been used to estimate the magnitude of the eastward surface winds (Briggs & Leovy 1974).

If dissipation and internal reflection are not too severe, internal gravity waves generated near the surface can propagate into the upper atmosphere with amplitude increasing approximately as $\exp(z/2H)$ where H is the scale height. This rate corresponds to constant wave energy density. Such waves would break at sufficiently great heights, generating turbulence and acting as effective mixing agents in the upper atmosphere (Hines 1963). It is not known whether gravity wave mixing is an important effect in the upper Martian atmosphere; but long internal gravity waves have been observed as high as 70 km (Anderson & Leovy 1978).

Internal gravity wave solitary waves may also occur in planetary oceans and atmospheres. A truncated wave train resembling a group of solitary waves was observed near the large Martian volcano Pavonis Mons (Briggs et al. 1977). This feature may have been produced by the same mechanism suspected of producing solitary waves in the terrestrial ocean: a strong tidal flow over a topographic ledge or obstacle produces an internal gravity wave train in the lee; as the tidal flow ebbs, the large amplitude wave train may begin to propagate in the form of a group of solitary waves.

PLANETARY WAVES

Transient Waves, Frequency $< \Omega$

Transient planetary scale waves arising from baroclinic instability were predicted for Martian winter subpolar regions by Mintz (1961), and they have been generated in GCM calculations for Mars (Leovy & Mintz 1969, Pollack et al. 1979b). The Viking landers detected the signatures of such waves at the surface (Ryan et al. 1978); they are responsible for the high-frequency pressure variations seen in Figure 1. A time series of meridional wind component and pressure for $L_s = 145$–205 is shown in Figure 6. There is a remarkably regular oscillation of the meridional wind with an average period of about 3.1 sols. The pattern falters for a few sols following VL-2 sol 170, near the start of the first planetwide dust storm, but otherwise it continues uninterrupted. Pressure shows the 3.1-sol periodicity clearly, but it also contains other low-frequency components. Since wind is associated with pressure gradient, the additional components of pressure variability apparently have a larger horizontal extent than the 3.1-sol component.

The nature of the 3.1-sol disturbances can be inferred from the relationship between wind and pressure. The measured surface winds reflect the wind near the top of the boundary layer, which is approximately geostrophic. Poleward meridional winds always occur with falling pressure and conversely (Figure 6), indicating eastward propagating disturbances (the same relationship holds for terrestrial mid-latitude disturbances). The phase speed and longitudinal wavelength can be inferred as follows. The geostrophic relation gives

$$v_g = (fa \cos \phi)^{-1} RT \frac{\partial}{\partial \lambda} (\delta p/p_0) = -(fC)^{-1} RT \frac{\partial}{\partial t} (\delta p/p_0) \qquad (2)$$

where v_g is the meridional component of geostrophic wind, R is the gas constant, $\delta p/p_0$ is the fractional pressure perturbation, and λ is longitude. The second relationship in (2) follows from the assumption that the waves are propagating eastward at constant phase speed C. The value of v_g is not known, but the approximate ratio $v_0/v_g \approx 0.5$ follows from the boundary layer analysis of the preceding section (v_0 is the measured surface meridional wind). Application of this ratio to the data shown in Figure 6 yields $C \sim 15$ m/s, and longitudinal wavenumber $n \sim 4$ (Ryan et al. 1978). These values for the phase speed, wavenumber, and period are consistent with those anticipated for baroclinic waves under Martian conditions (Mintz 1961, Leovy 1969) and with Martian waves simulated

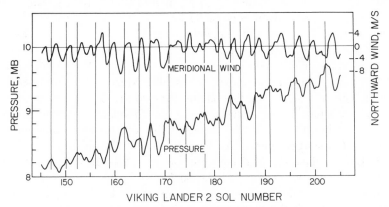

Figure 6   Time series of meridional wind (v_0, *upper*) and surface pressure (*lower*) at Viking Lander 2. Vertical lines are drawn through pressure maxima. A 1-sol running mean has been applied to remove the tides. See Ryan et al. (1978).

Table 3   Dynamical parameters for midlatitude winter

	$(fa)^2(gH)^{-1}$	$H\,\overline{(\ln\theta)}_z$	R	n	$\overline{(\ln T)}_\phi$	Ro_T	Ri
Mars	3.1	0.40	0.13	3	0.2	0.065	31
Earth	5.4	0.12	0.023	7	0.1	0.019	64

by GCM calculations (Leovy & Mintz 1969); the amplitudes also correspond closely to those calculated by GCMs (Pollack et al. 1979b). Detailed analysis of individual disturbances and images of wave-related cloud features also suggest that the disturbances are baroclinically unstable waves with fairly sharp cold fronts in some cases (Briggs & Leovy 1974, Tillman et al. 1979).

An interesting aspect of these waves is their regularity. They are more regular than baroclinic waves in the earth's atmosphere, and resemble the regular wave regime of rotating differentially heated tank experiments (Spence & Fultz 1977). The more regular regimes in the tank experiments are characterized by relatively high static stability as measured by the parameter R: $R \equiv gH^2\,\overline{(\ln\theta)}_z/(fa)^2$, where θ is potential temperature and H is the depth scale. The longitudinal derivative length scale of baroclinic waves tends to be proportional to $R^{1/2}$. Winter midlatitude values of R and the corresponding longitudinal wavenumber n are given in Table 3. Values of n based on parameters evaluated near latitude 45° are in reasonable agreement with observed terrestrial values and inferred Martian values. In general, the smaller the value of n based on R, the more stable is the system in the sense of the likelihood of regular rather than irregular wave regimes. Also given in Table 3 are corresponding values of the Richardson Number $[Ri \equiv g\,\overline{(\ln\theta_z)}/(\bar{u}_z)^2 = (f^2a^2/g)\overline{(\ln\theta)}_z/\overline{(\ln T)}_\phi^2$, from (1)], and the planetary thermal Rossby Number, $Ro_T = gH(fa)^{-2}\,\overline{(\ln T)}_\phi$. Note that $R = Ri(Ro_T)^2$. The values of Ri are comparable on Mars and earth during midlatitude winter, but Ro_T (earth) $\ll Ro_T$ (Mars); this relationship implies the smaller Martian wavenumber and more regular regime. Another factor may contribute to the relatively regular Martian regime. Spence and Fultz found that regular regimes occurred when their tank was heated at the top, thereby increasing the stability in the upper part of the fluid. Because of the absorption of solar radiation by dust, the Martian atmosphere is also "heated at the top," at least to a greater degree than the earth's atmosphere.

If the Martian and laboratory regimes are indeed analogous, Mars may help to link these experiments and the earth's atmosphere. The zonally averaged thermal state of the Martian midlatitude winter atmosphere varies because of the varying atmospheric dust load. There are corresponding variations of the wave properties (Ryan et al. 1978).

The temperature distributions inferred from Viking orbiter infrared measurements, such as those of Martin & Kieffer (1979), can be related to wave property variations inferred from the surface wind and pressure measurements. In this way, Mars may be able to act as a sort of "controlled experiment" on the relationship between wave properties and mean flow. Mars is a cleaner object of study than earth from this point of view because the terrestrial waves are subject to the complications of latent heat release and differential heating between continents and oceans, and the earth's atmosphere is nearly always in the irregular wave regime. Such an analysis of the Martian data has yet to be carried out.

Forced Quasi-Stationary Waves

The large-amplitude variations in the global scale topography of Mars should force quasi-stationary long waves, just as topography and differential heating do on earth. Martian quasi-stationary waves have been modeled by Webster (1977), Mass & Sagan (1976), Pollack et al. (1976), and Moriyama et al. (1978). Properties of the waves developed in the linearized model of Webster are similar to those found in the nonlinear models of Pollack et al. (1976) and Mass & Sagan (1976). Waves are forced in two different ways. (a) In the eastward mean zonal flow of the winter subpolar region, waves are forced kinematically by flow over topography. Barotropic pressure troughs tend to form in the basins such as Hellas and Argyre, with pressure ridges over the large-scale topographic ridges. These waves transport eastward momentum toward the zonal jet, and thus may play a role in maintenance of the jet. (b) In the weak eastward zonal wind regime of the summer subtropics, waves are forced by the differential heating influence of topography; uplands act as heat sources, lowlands as heat sinks. The resulting waves are baroclinic with warm upper level high pressure areas overlying surface lows over the topographic ridges and cool upper level lows and surface highs overlying the large-scale valleys. In regions of westward surface wind there is little wave response, the directional asymmetry arising from the character of Rossby wave propagation (Dickinson 1978). The distinction between kinematically forced waves and thermally forced waves may be clearer on Mars than on the earth because of the absence of oceans, but none of the data available now bear directly on the Martian quasi-stationary waves. Either synoptic coverage from an orbiter or a distributed network of surface weather stations would be required.

Thermal Tides

Prior to Viking, theoretical studies of the thermally generated diurnal tide had been carried out by Zurek (1976) and Conrath (1976). Since the convective thermal drive on Mars is about ten times as effective as the diurnal forcing on earth (per unit mass), the tidal response may be expected to scale up accordingly. In addition, the Martian atmosphere is strongly heated internally because of suspended dust, leading to a further enhancement of the tides, the enhancement varying with the dust load. An index of this enhancement is the daily pressure variation, $\delta p/p_0$, which is largest in the tropics. The maximum terrestrial value of $\delta p/p_0$ is about 10^{-3}; in the Martian tropics, it has been observed to range from 10^{-2} to 6×10^{-2}, depending on dust loading (Hess et al. 1977, Leovy & Zurek 1979).

Four main tidal components can be distinguished. (a) *Vertically propagating diurnal components.* These are confined to low latitudes, have vertical wavelengths of order 30 km or less, and, in the absence of damping, tend to amplify upwards as $\exp(z/2H)$ until they reach an altitude at which they begin to produce static instabilities. Above this level, their behavior is somewhat speculative, but they probably begin to break, and their amplitudes may remain roughly constant with height as tidal energy is dissipated into smaller scale waves and turbulence (Lindzen 1968). (b) *Vertically trapped diurnal components.* These predominate at high latitudes; their vertical structure is not wavelike, and they do not amplify upward above the forcing region. (c) *Semidiurnal tidal components.* These are wavelike at all latitudes, although the dominant mode has a vertical wavelength of order 200 km. Higher order modes have shorter vertical wavelengths and amplify upward until they too reach a possible "breaking level." (d) *Tidal components produced by topographic and nonlinear coupling.* Interaction of these three types of primary tidal components with large-scale topography produces secondary components, both trapped and vertically propagating. Additional components can also be produced by interaction between tidal components and the background wind, and between large-amplitude primary tidal components.

The tidal intensity is of particular interest because breaking tides may act as effective stirring agents for the upper atmosphere (Lindzen 1968, 1970). Zurek has shown that Martian tides may break as low as 35 km, or even lower during global dust storms. This is in contrast to a breaking level of about 90 km for terrestrial tides. It is tempting to attribute the relatively low pressure of the Martian homopause[1] to the strong Martian tides. The Martian homopause occurs at about 1/50th of the pressure of the terrestrial homopause (Nier & McElroy 1977), and this is about the same as the average enhancement of Martian tides relative to the earth.

The interaction of the westward propagating diurnal tide with large-

[1] The "homopause," sometimes referred to as the "turbopause," is the transition region separating the lower part of an atmosphere in which turbulent mixing dominates over molecular diffusion from the upper part in which molecular diffusion dominates.

amplitude topography is probably important and has been investigated by both Conrath and Zurek. This interaction produces an eastward traveling secondary component of zonal wavenumber 2 (this is the dominant longitudinal wavenumber of Martian equatorial topography), enhancing the primary tide over uplands and decreasing it over lowlands. Thus, Zurek finds the strongest diurnal tidal winds over upland longitudes near 120 W and 300 W, and near latitudes 25 N and 25 S.

DUST AND ATMOSPHERIC DYNAMICS

The most spectacular Martian meteorological phenomena are the dust storms that regularly reach planetary scale, sometimes obscuring almost all surface features on the planet (Masursky et al. 1972, Briggs et al. 1979, Capen 1974). These planetwide storms occur systematically when Mars is near perihelion, which corresponds closely to southern summer solstice during the present epoch. During recent years, when earth-based observations have been most numerous, these storms have been observed to originate from large but localized dust clouds in two general regions: near 110 W, 25 S and near 310 W, 30 S very close to the favored regions suggested by Zurek's theory (L. Martin 1974, 1976, Gierasch 1974). In the following discussion, dust storm properties deduced from spacecraft observations are emphasized.

Properties and Distribution of the Dust

One of the more surprising of the Viking atmospheric discoveries was the high atmospheric dust content even during the clearest parts of the year. Optical depth remained above 0.4 at the Viking sites throughout the first year of observations. Properties of the suspended dust have been inferred from spacecraft optical and infrared data (Pollack et al. 1977, 1979a, Toon et al. 1977). The mean particle radius is found to be near 0.4 μm; the particles contain silicates (Hanel et al. 1972) as well as a highly absorbing component and their infrared spectra suggest clay minerals, montmorillonite, for example. The absorbing component may be magnetite or maghemite (Pollack et al. 1977, Hargraves et al. 1977). For the purpose of radiative calculations, the most important aerosol properties are the single scattering albedo $\overline{\omega_0}$, the asymmetry factor g, the infrared opacity, and the optical depth at visible wavelength. Pollack et al. (1979a) estimate $\overline{\omega_0} \simeq 0.75$, $g \simeq 0.5$. The infrared opacity is largest near 9 μm, and the wavelength average infrared opacity is comparable to the visible opacity (Peterfreund & Kieffer 1979, T. Martin et al. 1979).

During planetwide dust storms the aerosol is distributed globally and fairly uniformly, except that opacity is lower in polar regions (Leovy et al. 1972, Briggs et al. 1979). Between such storms there is evidence for a latitude-dependent component of variability as well as one related to topography (Thorpe 1977). The mixing ratio in the vertical is rather uniform with little or no evidence for layering. During the planetwide storms, the aerosol extends to above 50 km, and the aerosol tends to remain high longer in the tropics than in high latitudes, as it would if the propagating component of the diurnal tide plays an important role in mixing the dust upward (Pollack et al. 1977, Anderson & Leovy 1978).

Time Evolution of Planetwide Dust Storms

The two planetwide dust storms observed by Viking have been described in papers by Pollack et al. (1979a), Briggs et al. (1979), Ryan & Henry (1979), Leovy & Zurek (1979), T. Martin et al. (1979), and Peterfreund & Kieffer (1979). The first storm was observed originally as a large but distinctly bounded dust cloud in the Claritas Fossae (110 W, 25 S). This dust cloud may have grown singly or combined with others to become a global storm whose subsequent evolution can be traced in the variation patterns of opacity and pressure at the lander sites (Figures 7 and 8). At

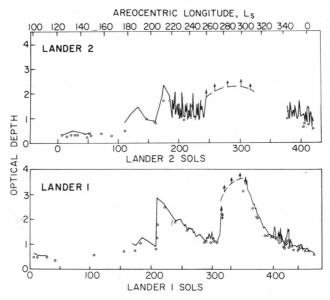

Figure 7 Normal incidence visible opacities deduced by Pollack et al. (1979a) from solar extinction measurements taken by the Viking landers. Solid lines are morning measurements, circles are afternoon measurements, and dashed segments with arrows indicate periods where only lower limits could be determined. Reproduced from Leovy & Zurek (1979), *Journal of Geophysical Research*, Volume 84, copyrighted by American Geophysical Union.

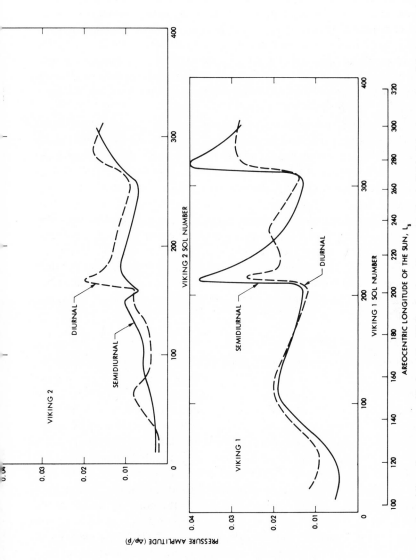

Figure 8 Diurnal and semidiurnal pressure amplitudes at the Viking landers. Data points for individual 6-sol averages of the pressure harmonics are shown. Arrows indicate onset times of the global storms. Reproduced from Leovy & Zurek (1979), *Journal of Geophysical Research*, Volume 84, copyrighted by American Geophysical Union.

$L_s = 208$ there was a rapid jump in opacity at both sites and an almost simultaneous jump in amplitude of the diurnal and semidiurnal tides. The global storm apparently carried the seed of its own demise because opacity and tidal pressure amplitudes began to decay immediately after this onset phase. Behavior of the second storm was similar, but peak opacities and pressure amplitudes apparently exceeded those during the first storm. Wind speeds at the sites increased and decreased along with the tidal amplitudes, and showed a predominantly semidiurnal component of variation. The maximum wind speed measured by Viking 1 shortly after the onset of the second storm was 31 m/s.

It is particularly noteworthy that the semidiurnal tide became dominant at Viking 1 during the peak phases of both storms. The pressure variations indicate an enhancement of both the semidiurnal tide and the trapped components of the diurnal tide relative to the vertically propagating diurnal components. Temperature distributions measured by the Viking orbiters also suggest enhancement of the trapped diurnal components. These features could be brought about by a combination of a shift in the heating to high levels and high latitudes with a heating decrease near the surface (Leovy & Zurek 1979). The shift to high levels can occur if the opacity becomes very large (optical depth $\gtrsim 5$) and the dust very deep. Shift of the heating to high latitudes would occur if the maximum dust load shifts to high latitudes. Both of these shifts did occur (T. Martin & Kieffer 1979, Pollack et al. 1979a).

Dust Storm Genesis and Decay

Several mechanisms have been proposed for dust storm genesis, including topographic winds (Mass & Sagan 1976), positive feedback between large cyclonic wind systems, dust raising, and differential heating of dust-laden and dust-free air (Gierasch & Goody 1973), and positive feedback between dust heating, dust raising, and global scale winds: tides and mean meridional circulation (Leovy et al. 1973b). The latter mechanism requires a buildup of dust load prior to the onset of the global dust storm itself, and Leovy and co-workers speculated that such a buildup could arise from local dust storms generated by the diurnal tide in the favored subtropical uplands, and by strong winds along the edge of the subliming south polar cap. Factors in each of these mechanisms appear to play a role, but the mechanisms proposed by Leovy et al. (1973b) appear particularly relevant. There is evidence of intimate coupling between the tides and the meridional circulation and dust load (Conrath et al. 1973). Increase in atmospheric opacity seems to occur prior to the onset of the first global storm, and there is a marked tendency for local storms to originate near the edge of the south polar cap and in the southern subtropical uplands; all but one of 20 local storms observed in the period $L_s = 170-270$ originated in one of these two regions (Peterfreund & Kieffer 1979). Pirraglia (1975), Burk (1976), Haberle et al. (1979), Haberle (1979), and French & Gierasch (1979) have modeled the polar winds. These can be strong enough to raise dust, but only during a brief part of the year: the period near perihelion when the rate of sublimation of the south polar cap is maximum. The sublimation itself produces an outflow that, deflected toward the west by the Coriolis acceleration, contributes significantly to surface wind and u_*. Winds capable of producing saltation may also occur during midwinter when the polar cap regions are covered by CO_2 frost or snow. Thus, the coincidence between perihelion and the maximum rate of sublimation of the polar cap during the present orbital epoch appears to be an important factor in dust storm genesis.

A mechanism for dust storm decay has been proposed by Pollack et al. (1979a), who noted the commencement of storm decay immediately after the onset phase. The vertically averaged atmospheric temperature change due to dust heating maximizes when the optical depth is a little greater than one and begins to decrease for larger values. Thus some components of circulation may be suppressed by the rapid increase in opacity that accompanies global dust storm onset. The propagating components of the diurnal tide are particularly suspect since they are important factors in storm generation that are suppressed by very large opacity. In any case, the data in Figures 7 and 8 strongly suggest that global dust storm generation is favored only when the optical depth is near unity.

Pollack and collaborators also noticed that the rate of opacity decrease is greater than can be accounted for by particle fallout alone. They suggest that mass flow into the condensing north polar cap region carries dust with it; the dust then serves to nucleate condensation of H_2O and CO_2 ice to form snow flakes, predominantly of CO_2, which grow large enough to precipitate. In this way, dust and water ice are removed from the atmosphere and are deposited in semipermanent layers in the polar region. The deposited CO_2 and some of the water are recycled seasonally with the spring and summer sublimation, but layered deposits of dust and some water ice remain. Both polar regions are covered with relatively recent layered deposits, forming "laminated terrain" (Cutts 1973, Cutts et al. 1976), but since the global storms occur only during northern winter, according to this model, deposition can only be taking place to a significant degree in the north during the current epoch. Laminated terrain in the south must have formed during earlier orbital epochs. Several lines of evidence support this conclusion. (a) The thickness of individual polar laminae (a few tens of meters) is quite consistent with the amount of dust that could be deposited from the atmosphere during the precessional or orbital obliquity cycles ($\sim 10^6$ earth years) if the current depositional rate is maintained for a significant fraction of such a cycle (Pollack et al. 1979a). (b) Seasonal exchange of water vapor takes place between the atmosphere and the north polar cap but not the south polar cap (Farmer et al. 1977, Davies et al. 1979). (c) Extensive fresh dune fields in the north polar region are indicative of eastward surface winds, while dunes and other indicators near the south pole are indicative only of northwestward winds (A. Ward 1978). Midwinter snow drifts in the seasonal south polar condensate deposit, indicative of eastward drift, are an exception (Thomas et al. 1979). The wind direction associated with the northern dunes suggests dune formation during winter as a mixture of dust and CO_2 snowflakes (Haberle et al. 1979).

This picture of the connection between dust storms, the seasonal condensation-sublimation cycle, and Martian orbital characteristics may prove to be an important key to unraveling both the history of the laminated terrains and the nature of past Martian climates. The occurrence of global dust storms and their phasing with respect to polar cap condensation are likely to depend sensitively on long-term variations in orbital ellipticity and in the precessional cycle (Pollack 1979).

PAST CLIMATES, VOLATILE RESERVOIRS, AND ESCAPE

One of the most tantalizing problems in Martian planetology is the nature of past climates. Pollack (1979) reviewed in some detail the evidence for warmer and wetter climates in the past, and discussed various possible causes. This evidence consists of orbital views of channels apparently formed by running water. The channels are widespread and show a wide variety of morphologies. Many may have been caused or modified by volcanic or aeolian activity, or by ground ice sapping rather than by runoff of liquid water. Others may have been produced by runoff from subsurface ice masses heated by volcanism or meteoritic impact, but widely distributed small channels or gullies cannot easily be explained in this way, and these features are often cited as evidence for a climate warm enough and wet enough for H_2O precipitation and runoff. The age of the gullies is not known, but may be several billion years (Pieri 1976, Masursky et al. 1977).

Two mechanisms have been invoked to account for sufficiently high temperatures for H_2O precipitation and runoff: (a) increased atmospheric emissivity due to surface CO_2 and water vapor pressures, (b) increased emissivity due to additional infrared absorbing gases, particularly ammonia. The viability of these mechanisms depends on several subsidiary issues. How much CO_2, H_2O, or NH_3 would have to be added to the atmosphere to bring temperatures up to the H_2O melting point? What are the sizes and compositions of any volatile reservoirs? Under what circumstances could these reservoirs be tapped? What would be the lifetime of NH_3 against photodissociation or other chemical loss processes in the Martian environment?

The first question has been addressed by Pollack (1979) using radiative equilibrium models that allow a variety of possible lapse rate modifications due to convection and large-scale dynamics. In order to obtain ground temperatures sufficiently high to melt H_2O ice in low and middle latitudes with CO_2 and H_2O atmospheres only, surface pressures in excess of one bar are required. If NH_3 is present, the required surface pressure is much less, ~ 0.1 bar for a fully reducing atmosphere, or a few tenths of a bar for mixtures of a few percent ammonia with CO_2.

It is very difficult to estimate the possible sizes of Martian volatile reservoirs. Anders & Owen (1977) used the Viking measurements of ^{36}A together with estimates of the surface ^{40}K abundance buttressed by Mars 5 observations, and the Viking measurements of the ratio of Cl to S in the surface material to construct a model of the Martian volatile inventory. They conclude that Mars began as a volatile-poor planet relative to the earth and Venus, and it has outgassed less efficiently. They estimate a total mass of outgassed CO_2 equivalent to 0.14 bars, acknowledging that values larger by a factor of two or three could be accommodated by the available data. A tenfold increase could not be accommodated, however.[2]

On the other hand, the C, O, and N isotopic data provide strong indications that the volatile reservoir is much larger than the current atmospheric mass (Nier & McElroy 1977, McElroy et al. 1977). The ratio $^{15}N/^{14}N$ is enhanced by 70% relative to the terrestrial value, while there is no detectable enhancement of ^{13}C relative to ^{12}C or ^{18}O relative to ^{16}O ($<5\%$ enhancement). Since N, C, and O can all escape from the Martian exosphere by similar nonthermal processes, these results suggest that ^{15}N has been enriched by selective escape of the light isotope (Brinkmann 1971, McElroy 1972, McElroy & Yung 1976). The lack of enrichment in C and O is then traced to the existence of a large reservoir, presumably of CO_2 as well as H_2O, which is in at least occasional contact with the atmosphere. It is possible to estimate the minimum size of the C reservoir from estimates by McElroy and his associates of the escape rate. If the reservoir is in the form of CO_2, it is equivalent to a surface

[2] The recent Pioneer Venus measurements have been reported to show a much higher ratio of ^{36}A to CO_2 and N_2 than assumed by Anders and Owen so that attempts to scale the Martian volatile inventory that are heavily dependent on the ^{36}A abundance should now be regarded with caution.

pressure of 0.3 bars or more. The ^{15}N enrichment value is so large that several further conclusions are possible, based on the nonthermal escape model. (a) The average outgassing over 4.5 Gyr must have been less than 0.3 times the present escape rate so that early outgassing is suggested. (b) The initial abundance of N_2 must have been at least 3 mb, consistent with a CO_2 reservoir of a few tenths of a bar. (c) If the homopause pressure and escape mechanism have remained constant, the average surface pressure over 4.5 Gyr could not have been greater than a few times the present surface pressure. If it were greater by more than a factor of four, the exospheric nitrogen concentration would have been too low to give the observed enrichment. This is an interesting constraint in that it limits the total length of time a given pressure enhancement is likely to have persisted. However, if the homopause is maintained by the tides, it is possible that homopause pressure would increase with a decrease in tidal amplitude and hence with increased atmospheric mass. Such a variation would tend to increase the exospheric nitrogen concentration and the corresponding enrichment rate.

The above arguments suggest a total evolved CO_2 reservoir in the range 0.3–0.5 bars, too small to warm the equatorial surface to the ice melting point, but large enough to have major effects on circulation and on the rates of surface erosion and deposition. Based on analogy with the earth, there is probably an even larger reservoir of stored H_2O, but it cannot enter the atmosphere without a substantial rise in temperature.

The stored CO_2 is most likely adsorbed in the regolith, primarily in the polar regions (Fanale & Cannon 1971, Fanale 1976, Pollack 1979). Carbonate minerals may also constitute a reservoir for CO_2 (Toulmin et al. 1977). After Mariner 9, it was believed that a reservoir of CO_2 might exist in the residual north polar cap (Murray & Malin 1973). Viking observations showed that this was not the case (Kieffer et al. 1976, Farmer et al. 1976), but recent Viking measurements indicate that the late summer residual south polar ice cap may contain CO_2 (Kieffer 1977). This cap occupies about 0.05% of the planetary surface, and even if 400 m deep, it could only contain as much CO_2 as is in the atmosphere now.

Seasonal variations of the CO_2 ice caps and the atmospheric mass (Figure 1) are in general agreement with models invoking radiative control of the condensation and sublimation rates (Leighton & Murray 1966, Cross 1971, Briggs 1974, Davies et al. 1979), but the stability of a residual ice cap is uncertain at present because of uncertainties in albedo, atmospheric radiation to the surface, and emission from the cap (Ingersoll 1974, Kieffer et al. 1977). The stability of polar CO_2 ice and absorbed polar CO_2 must vary widely over periods from 10^5–10^6 earth years as a result of orbital variations, particularly in the obliquity (Murray et al. 1973, W. Ward 1974, Gierasch & Toon 1973, Pollack 1979). Since the total CO_2 reservoir is apparently inadequate to account for surface temperatures near the melting point, NH_3 remains the viable candidate for raising temperatures to H_2O melting. However NH_3 would photodissociate readily in a primitive Martian atmosphere, so that without special replenishment mechanisms, its lifetime would be quite short (Abelson 1966). As a result of these difficulties, it is worthwhile to look closely at the possibility of mechanisms other than climate change as possible causes of the Martian channels (e.g. Yung 1978).

CONCLUDING REMARKS

In any survey of this kind, it is necessary to omit major topics of interest. This selection reflects my own particular biases that the most currently exciting Martian meteorological problems include (a) transient waves, their relationship to the zonal mean flow, to laboratory experiments, and to terrestrial transient waves, (b) tides and dust storms, their interconnection, and their relationship to homopause pressure, (c) the processes controlling CO_2, dust, and water ice deposition and removal in the polar cap regions, (d) the composition of the early Martian atmosphere and the influence of composition on climate.

ACKNOWLEDGMENTS

My work in this field has been supported over a number of years by grants from the NASA Planetary Programs Office, Atmospheric Sciences Division, and the Mariner 9 and Viking projects. I am indebted to many generous colleagues who have been immeasurably helpful, in particular to G. A. Briggs, R. W. Zurek, J. B. Pollack, and J. E. Tillman, and to Y. Mintz, R. B. Leighton, and B. C. Murray who first stimulated my interest in the problems of Martian meteorology. Y. Mintz and F. Fanale made valuable comments on an earlier version of this paper.

Literature Cited

Abelson, P. H. 1966. *Proc. Natl. Acad. Sci. USA* 55:1365
Anders, E., Owen, T. 1977. *Science* 198:453
Anderson, E. H., Leovy, C. B. 1978. *J. Atmos. Sci.* 35:723
Barth, C. A. 1974. *Ann. Rev. Earth Planet. Sci.* 2:333
Blackadar, A. K. 1957. *Bull. Am. Meteorol. Soc.* 38:283
Blumsack, S. L. 1971a. *J. Atmos. Sci.* 28:1134
Blumsack, S. L. 1971b. *Icarus* 15:429

Blumsack, S. L., Gierasch, P. J., Wessel, W. R. 1973. *J. Atmos. Sci.* 30:66
Briggs, G. A. 1974. *Icarus* 23:167
Briggs, G. A., Leovy, C. B. 1974. *Bull. Am. Meteorol. Soc.* 55:278
Briggs, G. A., Klaasen, K., Thorpe, T. E., Wellman, J., Baum, W. 1977. *J. Geophys. Res.* 82:4121
Briggs, G. A., Baum, W., Barnes, J. 1979. *J. Geophys. Res.* 84: In press
Brinkmann, R. T. 1971. *Science* 174:944

412 LEOVY

Burk, S. D. 1976. *J. Atmos. Sci.* 33:923

Businger, J. A., Arya, P. A. L. 1973. *J. Geophys. Res.* 78:7092

Capen, C. F. 1974. *Icarus* 22:345

Chapman, S., Lindzen, R. S. 1970. *Atmospheric Tides.* New York: Gordon & Breach

Conrath, B. J. 1975. *Icarus* 24:36

Conrath, B. J. 1976. *J. Atmos Sci.* 33:2430

Conrath, B. J., Curran, R., Hanel, R. A., Kunde, V., Maguire, J., Pearl, J., Pirraglia, J., Welker, J., Burke, T. 1973. *J. Geophys. Res.* 78:4267

Cross, C. A. 1971. *Icarus* 15:110

Cutts, J. A. 1973. *J. Geophys. Res.* 78:4231

Cutts, J. A., Blasius, K. R., Briggs, G. A., Carr, M. H., Greeley, R., Masursky, H. 1976. *Science* 194:1329

Davies, D. W., Farmer, C. B., LaPorte, D. D. 1979. *J. Geophys. Res.* 84: In press

Dickinson, R. E. 1978. *Ann. Rev. Fluid Mech.* 10:159

Ekman, V. W. 1902. *Nyt Mag. Naturv.* 40:1

Eliassen, A., Palm, E. 1960. *Geofys. Publ.* 22:1

Elliot, J., French, R. G., Dunham, E., Gierasch, P. J., Veverka, J., Church, C., Sagan, C. 1977. *Science* 195:485

Fanale, F. P. 1976. *Icarus* 28:179

Fanale, F. P., Cannon, W. 1971. *Nature* 230:502

Farmer, C. B., Davies, D. W., LaPorte, D. D. 1976. *Science* 193:1339

Farmer, C. B., Davies, D. W., Holland, A. L., LaPorte, D. D., Doms, P. E. 1977. *J. Geophys. Res.* 82:4225

Fjeldbo, G., Sweetnam, D., Brenkle, J., Christensen, E., Farless, D., Mehta, J., Seidel, B., Michael, W. Jr., Wallio, A., Grossi, M. 1977. *J. Geophys. Res.* 82:4317

Flasar, F. M., Goody, R. M. 1976. *Planet. Space Sci.* 24:161

French, R. G., Gierasch, P. J. 1979. *J. Geophys. Res.* In press

Fröhlich, C. 1977. *The Solar Output and Its Variation,* Chap. 3. Boulder, Colo.: Colorado University Press

Gierasch, P. J. 1974. *Rev. Geophys. Space Phys.* 12:730

Gierasch, P. J., Goody, R. M. 1968. *Planet. Space Sci.* 24:161

Gierasch, P. J., Goody, R. M. 1972. *J. Atmos. Sci.* 29:400

Gierasch, P. J., Goody, R. M. 1973. *J. Atmos. Sci.* 30:169

Gierasch, P. J., Toon, O. B. 1973. *J. Atmos. Sci.* 30:1502

Gifford, F. A. 1964. *Mon. Weather Rev.* 92:435

Glasstone, S. 1968. *The Book of Mars.* Washington DC: US GPO

Greeley, R., White, B., Leach, R., Iversen, J., Pollack, J. B. 1976. *Geophys. Res. Lett.* 3:417

Haberle, R. M. 1979. *Icarus.* In press

Haberle, R. M., Leovy, C. B., Pollack, J. B. 1979. *Icarus.* In press

Hadley, G. 1735. *Philos. Trans. R. Soc. London* 39:58

Hanel, R. A., Conrath, B. J., Hovis, W. A., Kunde, V. G., Lowman, P. D., Pearl, J. C., Prabhakara, C., Schlachman, B., Levin, G. V. 1972. *Science* 175:305

Hargraves, R. B., Collinson, D. W., Arvidson, R. E., Spitzer, C. R. 1977. *J. Geophys. Res.* 82:4547

Hess, S. L. 1976. *Icarus* 28:269

Hess, S. L., Henry, R. M., Leovy, C. B., Ryan, J. A., Tillman, J. E. 1977. *J. Geophys. Res.* 82:4559

Hess, S. L., Henry, R. M., Tillman, J. E. 1979. *J. Geophys. Res.* 84: In press

Hines, C. O. 1963. *Q. J. R. Meteorol. Soc.* 89:1

Holton, J. R. 1967. *Tellus* 19:199

Holton, J. R. 1975. *Introduction to Dynamic Meteorology.* New York: Academic

Ingersoll, A. P. 1974. *J. Geophys. Res.* 79:3403

Ingersoll, A. P., Leovy, C. B. 1971. *Ann. Rev. Astron. Astrophys.* 9:147

Irvine, W. M., Simon, T., Menzel, D. H., Pikoos, C., Young, A. T. 1968. *Astron. J.* 73:807

Kaplan, L. D., Münch, G., Spinrad, H. 1964. *Ap. J.* 139:1

Kerzhanovich, V. V. 1977. *Icarus* 30:1

Kieffer, H. H. 1977. *Bull. Am. Astron. Soc.* 9:540 (Abstr.)

Kieffer, H. H., Chase, S. C., Miner, E., Münch, G., Neugebauer, G. 1973. *J. Geophys. Res.* 78:4291

Kieffer, H. H., Chase, S. C., Martin, T. Z., Miner, E. D., Palluconi, F. D. 1976. *Science* 194:1341

Kieffer, H. H., Martin, T. Z., Peterfreund, A. R., Jakosky, B. 1977. *J. Geophys. Res.* 82:4249

Kliore, A. J., Cain, D., Levy, G., Eshleman, V., Fjeldbo, G., Drake, F. 1965. *Science* 149:1243

Kliore, A. J., Fjeldbo, G., Seidel, B. L., Sykes, M. J., Woiceshyn, P. M. 1973. *J. Geophys. Res.* 78:4331

Leighton, R. B., Murray, B. C. 1966. *Science* 153:136

Leovy, C. B. 1969. *Appl. Optics* 8:1278

Leovy, C. B., Mintz, Y. 1969. *J. Atmos. Sci.* 26:1167

Leovy, C. B., Zurek, R. W. 1979. *J. Geophys. Res.* 84: In press

Leovy, C. B., Briggs, G. A., Young, A. T., Smith, B. A., Pollack, J. B., Shipley, E. N., Wildey, R. L. 1972. *Icarus* 17:373

Leovy, C. B., Briggs, G. A., Smith, B. A. 1973a. *J. Geophys. Res.* 78:4252

Leovy, C. B., Zurek, R. W., Pollack, J. B. 1973b. *J. Atmos. Sci.* 30:749

Lindzen, R. S. 1968. *Proc. R. Soc. London Ser. A* 303:299

Lindzen, R. S. 1970. *J. Atmos. Sci.* 27:536

Martin, L. J. 1974. *Icarus* 23:108

Martin, L. J. 1976. *Icarus* 29:363

Martin, T. Z., Kieffer, H. H. 1979. *J. Geophys. Res.* 84: In press

Martin, T. Z., Peterfreund, A. R., Miner, E. D., Kieffer, H. H., Hunt, G. 1979. *J. Geophys. Res.* 84: In press

Mass, C., Sagan, C. 1976. *J. Atmos. Sci.* 33:1418

Masursky, H., Batson, R. M., McCauley, J. F., Soderblom, L. A., Wildey, R. L., Carr, M. H., Milton, D. J., Wilhelms, D. E., Smith, B. A., Kirby, T. B., Robinson, J., Leovy, C. B., Briggs, G. A., Young, A. T., Duxbury, T., Acton, C., Murray, B. C., Cutts, J. A., Sharp, R., Smith, S., Leighton, R., Sagan, C., Veverka, J., Noland, M., Lederberg, J., Levinthal, E., Pollack, J. B., Moore, J., Hartmann, W., Shipley, E. N., deVaucouleurs, G., Davies, M. 1972. *Science* 175:294

Masursky, H., Boyce, J. M., Dial, A. L., Schaber, G. G., Strobell, M. E. 1977. *J. Geophys. Res.* 82:4016

McElroy, M. B. 1972. *Sciene* 175:443

McElroy, M. B., Yung, Y. L. 1976. *Planet. Space Sci.* 24:1107

McElroy, M. B., Kong, T. Y., Yung, Y. L. 1977. *J. Geophys. Res.* 82:4379

Mintz, Y. 1961. In *The Atmospheres of Mars and Venus.* NAS-NRC Pub. 944:107

Moriyama, S., Iwashima, T., Yamamoto, R. 1978. In *Proc. Lunar Planet. Symp., 11th,* Univ. Tokyo, Japan, p. 23

Murray, B. C., Malin, M. 1973. *Science* 183:437

Murray, B. C., Ward, W. R., Yeung, S. 1973. *Science* 180:638

Nier, A. O., McElroy, M. B. 1977. *J. Geophys Res.* 82:4341

Owen, T., Biemann, K., Rushneck, D. R., Biller, J. E., Howarth, D. W., LaFleur, A. L. 1977. *J. Geophys. Res.* 82:4635

Peterfreund, A. R., Kieffer, H. H. 1979. *J. Geophys. Res.* 84: In press

Pieri, D. 1976. *Icarus* 27:25

Pirraglia, J. A. 1975. *J. Atmos. Sci.* 32:60

Pirraglia, J. A. 1976. *Icarus* 27:517

Pollack, J. B. 1979. *Icarus* 36: In press

Pollack, J. B., Leovy, C. B., Mintz, Y. H., Van Camp, W. 1976. *Geophys. Res. Lett.* 3:479

Pollack, J. B., Colburn, D., Kahn, R., Hunter, J., Van Camp, W., Carlston, C. E., Wolf, M. R. 1977. *J. Geophys. Res.* 82:4479

Pollack, J. B., Colburn, D., Flasar, F., Carlston, C. E., Pidek, D., Kahn, R. 1979a. *J. Geophys. Res.* 84: In press

Pollack, J. B., Leovy, C. B., Greiman, P., Mintz, Y. 1979b. *J. Atmos. Sci.* Submitted

Ryan, J. A., Hess, S. L., Henry, R. M., Leovy, C. B., Tillman, J. E., Walcek, C. 1978. *Geophys. Res. Lett.* 5:715, 815

Ryan, J. A., Henry, R. M. 1979. *J. Geophys. Res.* 84: In press

Sagan, C., Veverka, J., Fox, P., Dubisch, R., French, R., Gierasch, P. J., Quam, L., Lederberg, J., Levinthal, E., Tucker, R., Eross, B. 1973. *J. Geophys. Res.* 78:4163

Sagan, C., Pieri, D., Fox, P., Arvidson, R., Guinness, E. 1977. *J. Geophys. Res.* 82:4430

Seiff, A., Kirk, D. B. 1977. *J. Geophys. Res.* 82:4364

Spence, T. W., Fultz, D. 1977. *J. Atmos. Sci.* 34:1261

Sutton, J. L., Leovy, C. B., Tillman, J. E. 1978. *J. Atmos. Sci.* 35:2346

Thomas, P., Veverka, J., Campos-Marquetti, R. 1979. *J. Geophys. Res.* Submitted

Thorpe, T. E. 1977. *J. Geophys. Res.* 82:4151

Tillman, J. E. 1976. In *Proc. Symp. Planet. Atmos.,* p. 145. Ottawa, Ontario: Royal Society of Canada

Tillman, J. E., Henry, R. M., Hess, S. L. 1979. *J. Geophys. Res.* 84: In press

Toon, O. B., Pollack, J. B., Sagan, C. 1977. *Icarus* 30:633

Toulmin, P., Baird, A., Clark, B., Keil, K., Rose, H., Christian, R., Evans, P., Kelliher, W. 1977. *J. Geophys. Res.* 82:4625

Vonder Haar, T. M., Suomi, V. E. 1969. *Science* 163:667

Ward, A. W. 1978. PhD thesis. Dept. Geol. Sci., Univ. of Washington, Seattle

Ward, W. R. 1974. *J. Geophys. Res.* 79:3375

Webster, P. J. 1977. *Icarus* 30:626

Yung, Y. L. 1978. *Nature* 273:730

Zurek, R. W. 1976. *J. Atmos. Sci.* 33:321

Zurek, R. W. 1978. *Icarus* 35:196

Ann. Rev. Astron. Astrophys. 1979. 17:415–43
Copyright © 1979 by Annual Reviews Inc. All rights reserved

PHYSICS OF NEUTRON STARS[1]

×2156

Gordon Baym

Department of Physics, University of Illinois at Urbana-Champaign,
Urbana, Illinois 61801

Christopher Pethick

Department of Physics, University of Illinois at Urbana-Champaign,
Urbana, Illinois 61801, and NORDITA, DK-2100 Copenhagen Ø, Denmark

1 INTRODUCTION

Since the first identification of neutron stars, in pulsars, a decade ago, theoretical and observational knowledge of these unusual objects has grown at a rapid rate. In this article we describe developments that have taken place since our 1975 review article on neutron stars (Baym & Pethick 1975, referred to hereafter as BP), as well as review several of their more astrophysical aspects not discussed there.

The most striking observational fact about neutron stars is their existence: at present 321 pulsars, which are generally accepted to be rapidly rotating neutron stars, have been observed in our galaxy (Manchester et al. 1978, Taylor & Manchester 1977, Manchester & Taylor 1977, Smith 1976). In addition, most of the 16 pulsating compact X-ray sources so far discovered are likely to be accreting neutron stars in close binaries (for a review see Lamb 1977). The association of the Crab and Vela pulsars with supernova remnants provides evidence for the formation of neutron stars in supernovae, a picture supported to a limited extent by comparison of pulsar populations and lifetimes with estimated supernova rates (reviewed in Manchester & Taylor 1977). Optical and X-ray observations of binary X-ray sources provide the possibility of determining the masses of the neutron stars in these objects

[1] Supported in part by National Science Foundation Grants DMR75-22241 and PHY78-04404.

415

(reviewed in Bahcall 1978); the results are consistent with present theories of neutron star structure and formation in supernovae. Quoted results, with statistical errors, include, for example, $M_{\text{Her X-1}} = 1.33 \pm 0.2\ M_\odot$ (Middleditch & Nelson 1976) and $M_{\text{Vela X-1}} = 1.5 \pm 0.2\ M_\odot$ (van Paradijs et al. 1976, Rappaport et al. 1976); however as Bahcall (1978) emphasizes, systematic errors in determination of these masses could lead to significantly greater true uncertainties.

Measurements of the surface thermal luminosity of a neutron star (in the soft X-ray, for expected surface temperatures) can allow one to deduce its surface temperature, T_e. By this method Wolff et al. (1975) have placed an upper bound, $T_e \lesssim 4.7 \times 10^6$ K, for the neutron star in the Crab Nebula. Additional bounds have been reported by Greenstein et al. (1977).

Surface magnetic fields of neutron stars in active pulsars and binary X-ray sources are inferred, from models of these systems, to be $\sim 10^{12}$ G. The principal observational inputs are, for pulsars, the rates of energy loss (reviewed in Ruderman 1972), and, for X-ray sources, the structure of the radiation and spinup rates (Lamb 1977, Ghosh & Lamb 1979). Trümper et al. (1978) observed a feature in the hard X-ray spectrum of Her X-1, which if correctly interpreted as electron cyclotron absorption at ~ 42 keV, would imply a neutron star field of 4×10^{12} G (or if emission at 58 keV, a field $\sim 6 \times 10^{12}$ G).

Information on moments of inertia of neutron stars may be obtained from observations of the secular rates of change of their spin periods. Comparison of the slowdown rate of the Crab pulsar with the luminosity of the Crab Nebula provides a lower bound on its moment of inertia ($> 1.5 \times 10^{44}$ g cm^2) (Ruderman 1972) while observed speedups, on time scales $\sim 10^2$–10^5 yr, of pulsating X-ray sources, combined with model descriptions of accretion torques, indicate moments of inertia and radii consistent with the compact objects being neutron stars (Elsner & Lamb 1976, Rappaport & Joss 1977, Ghosh & Lamb 1979). The wealth of detailed observations of short term variations of pulse arrival times, for both pulsars and pulsating X-ray sources, offers the prospect of enabling one to deduce, within the framework of theoretical models, information about the internal structure of neutron stars. For descriptions of such work see, for example, Lamb (1977), Lamb et al. (1978), and Pines et al. (1974), as well as Section 4.3.

We remind the reader of the general structure of neutron stars. Typical radii are ~ 10 km, masses $\sim M_\odot$, and central densities exceed that of nuclear matter, $\rho_0 \equiv 2.8 \times 10^{14}$ g cm^{-3}. Neutron stars have a solid crust, ~ 1 km thick, beneath which is a liquid interior, likely superfluid in part, beginning at density $\sim \rho_0$. A number of possible phases of matter at densities $\rho \gtrsim 2\rho_0$ have been investigated, but which actually occur remains somewhat uncertain. We now review recent developments on the equation of state. In subsequent sections, we discuss models of neutron stars and dynamical properties.

2   RECENT DEVELOPMENTS ON THE EQUATION OF STATE

The structure of neutron star matter is reasonably well understood (see BP) up to about nuclear matter density ρ_0 where the crust dissolves. In the very low density regime principal progress has been a better description of the properties of matter near the surface in strong magnetic fields. Recent work in the theory of nuclear matter has called into question our understanding of the properties of the liquid regime in the neighborhood of ρ_0, and brought out the sensitivity of the stellar radius and crust thickness to microscopic details of the equation of state in this regime. States of higher density matter that have received considerable recent attention are pion condensation, quark matter, and "abnormal matter" in which the nucleons become essentially massless entities. For detailed reviews of the physics of higher density matter see Baym (1977a,b, 1978).

2.1   *The Liquid Regime*

As in ordinary nuclear matter theory, calculation of the properties of neutron rich matter in the liquid regime is presently beset by a number of uncertainties: the choice of the two-body interaction and how to calculate with it, the role of the internally excited state of the nucleon— the isobar $\Delta(1236\,\text{MeV})$, or N^*—in intermediate states in nucleon-nucleon scattering, and the proper inclusion of tensor correlation effects. Until recently, the phenomenological Reid soft-core nucleon-nucleon potential, fit to phase shifts, was generally felt to be satisfactory for use in calculation of nuclear and neutron star matter. The calculational methods used have been Brueckner-Bethe-Goldstone "nuclear matter theory", which in lowest order sums contributions from two-body scattering processes, and variational techniques based on trial wave functions. See Bethe (1971) and BP for reviews of such calculations.

In the past several years the inadequacies of the conventional calculations have become apparent (Pandharipande et al. 1975, Negele 1976, Bäckman et al. 1972), and more accurate calculational techniques are currently being developed (see reviews by Clark 1978 and Day 1978). With these improved calculations, the indications are that when one uses common phenomenological nucleon-nucleon interactions, such as the Reid, the calculated binding energy and saturation density (i.e. equilibrium zero-pressure density) of symmetric nuclear matter (equal number of neutrons and protons) are too large; the conclusion, given the correctness of these calculations, is that the Reid soft-core potential is too soft. Whether better two-body interactions, derived theoretically from dispersion theory, will give more accurate results for symmetric nuclear matter than the Reid potential remains to be seen. Present indications are that for Reid type interaction potentials the Brueckner method is adequate for neutron matter up to $\rho \sim 2\rho_0$; however, improved potentials can be expected to modify previous results, e.g. the Baym-Bethe-Pethick-Pandharipande and Bethe-Johnson equations of state (surveyed by Canuto 1974).

A second important question, which has great consequences for the structure of neutron stars, is the effect of the medium on the interaction potential itself. The important attractive components of the nucleon-nucleon interaction are believed to arise from processes (analogous to the atomic van der Waals interaction) in which the two nucleons scatter, via pion exchange, to virtual intermediate states in which, for example, one or both nucleons are excited to a Δ state. Since such intermediate states generally have higher energy than the initial states, these processes produce (through the usual second order perturbation formula) a net attraction. In the two-nucleon scattering problem the energies of the intermediate states in these processes have their free-space values, and any potential that fits nucleon-nucleon phase shifts implicitly takes the intermediate states to be in free space. However, as Green and Haapakoski (see Green 1976) pointed out, the nuclear medium will have two important effects on the intermediate range attraction. First, the Pauli exclusion principle forbids processes in which one of the intermediate nucleon states is already occupied; this effect eliminates some of the attraction. Second, because the particles are not in free space the intermediate state energy denominators in such processes will also be modified; this "dispersion correction" also tends to reduce the attraction. The net result is a decrease in the intermediate range attraction in the medium, which becomes more important with increasing density. This effective repulsion is not taken into account in calculations that use a phenomenological two-body interaction.

The process most strongly affected is that in which just one of the nucleons is excited to a Δ state, a process particularly important in neutron matter. [The reason is that two neutrons, or two protons, must have total isospin $T = 1$, while a neutron and proton can have total isospin $T = 1$ or 0. In the isobar process the nucleon plus Δ in the intermediate state can only have $T = 1$ or 2 (the isospin of the Δ is $\frac{3}{2}$) and thus this process occurs only in $T = 1$ states.] To compute the effects of the medium one must solve a coupled channel problem, treating the Δ as an elementary particle that can be present in the medium. Detailed calculations (Holinde & Machleidt 1977, Green 1976, and references therein) show that with Δ's explicitly included, one finds that at higher densities (above ρ_0) the $T = 1$ interactions can change from attractive to repulsive. The effect in symmetric nuclear matter is to lower the saturation density (since in the conventional calculations only interactions in $T = 0$ states became repulsive at higher density; the $T = 1$ interactions remained attractive).

In neutron stars, this effect implies a stiffening, for ρ around ρ_0, of the equation of state, i.e. an increase of the pressure for given density (Smith & Pandharipande 1976), and thus it tends to make a neutron star of a given mass larger in size and lower in density, as well as making the crust increase in mass and volume (Pandharipande et al. 1976). Effects on models are reviewed in Section 3 below.

Another effect on neutron stars is an increase, compared with the predictions of the Reid potential, in the proton fraction in the matter at higher densities. Because in the Reid calculations the $T = 0$ interactions, which are effective between protons and neutrons, become repulsive at high density, while the $T = 1$ do not, it is expensive to have protons at higher densities. However if the $T = 1$ interactions also become repulsive, it is then more favorable to have a fraction of the nucleons be protons.

Calculations for symmetric nuclear matter that include Δ's have been carried out in Brueckner theory and by variational methods only in low order, and do not produce the correct saturation density and binding energy. Also, the matrix elements describing the transitions from the initial neutron-neutron state to the intermediate nucleon-Δ state are uncertain, especially at higher momentum transfer. Thus, the implications for neutron stars should be regarded as tentative. In particular, tensor correlation effects (see, for example, Friman & Nyman 1978), which tend to soften the matter (as well as lead to pion condensation), have not yet been adequately included in either nuclear matter or neutron star matter calculations. Until one has a satisfactory theory of symmetric nuclear matter, the equation of state of neutron star matter must be regarded as uncertain.

The neutron liquid both in the crust and interior as well as the proton liquid in the interior are believed to be superfluid. Pairing calculations, reviewed in BP, indicate that at lower densities the neutrons are paired in $^1S_0$ states, as are the protons in the interior, while at densities $\gtrsim \rho_0$, the neutrons are instead paired in $^3P_2$ states. The calculated energy gaps, a measure of the strength of the superfluid pairing, depend sensitively, however, on detailed assumptions about the interactions between nucleons. Clark et al. (1976) have examined the sensitivity of the neutron $^1S_0$ gaps to interactions between neutrons induced by particle and spin density fluctuation effects, and conclude that at low densities ($\lesssim 0.1 \rho_0$) such effects reduce the effective interactions between neutrons that are responsible for pairing by $\sim 30\%$, and reduce the gap and corresponding transition temperature by a factor of ~ 3. Further explorations of effects on both neutron and proton gaps of polarization-induced interactions would be useful. Variations in nucleon effective masses assumed in the calculation of pairing strengths can also lead to modifications of the gaps of similar magnitude.

Sauls & Serene (1978) have estimated, via the Ginzburg-Landau approach, corrections to the weak coupling BCS calculations of $^3P_2$ neutron pairing, and conclude that such corrections should not produce a qualitative change in the properties of the superfluid. Effects of superfluidity on cooling and dynamics are discussed in Section 4.

2.2   Pion Condensation

A pion-condensed state of matter is one in which the pi meson field, which normally fluctuates about nucleons, develops a nonzero expectation value. In general, pion condensation in matter tends to soften the equation of state, countering the stiffening effect of Δ isobars. It is also of astrophysical interest for its important enhancement of neutrino cooling of neutron stars (described in Section 4.2). Furthermore, pion condensation might lead to possible solidification of high density matter.

Present calculations (reviewed in Brown & Weise 1976, Migdal 1978) indicate the onset of condensation of the charged-pion field in neutron matter at a density $\sim 2\rho_0$. In this charged pion-condensed state the neutrons become rotated in isospin space into coherent superpositions of neutrons and protons, with the microscopic pion field carrying a compensating negative charge density. Methods for describing the properties of the condensed state based on the chiral symmetry of low energy pion-nucleon physics, including effects of isobars and nuclear correlations, have been developed by Campbell, Dashen, and Manassah, and Baym, Au, and Flowers (see Baym & Campbell 1979 for a detailed review and list of references to earlier work). Brown & Weise (1976) have calculated equations of state for spatially uniform charged pion-condensed neutron matter, and Au (1976) has extended this work to include effects of beta equilibrium. The calculations of Migdal and collaborators on the properties of the condensed state are reviewed in Migdal (1978). Calculations of a spatially uniform neutral π^0-condensed state of neutron matter are given by Dautry & Nyman (1979).

As an example of the effect on the equation of state that can be produced by pion condensation we note that Au (1976) finds at $3\rho_0$ a reduction $\sim 75\%$ in the pressure from its value in the noncondensed state. The detailed modification of the equation of state by pion condensation is quite sensitive, however, to the magnitude of the effective nucleon-nucleon interactions (the Landau Fermi-liquid parameter g') assumed, a quantity somewhat uncertain, both theoretically and experimentally, for high density neutron matter; extrapolation of pion-nucleon scattering amplitudes to the "off-shell" regime of pion condensation results in further uncertainties. Thus, present estimates of the modification of the equation of state by pion condensation should be regarded as preliminary but illustrative. A full reliable calculation of the equation of state including the stiffening effects of isobars described in Section 2.1 together with the softening effects of pion condensation has yet to be carried out.

A question of substantial interest is whether neutron matter in the deep interior of a neutron star can solidify. Calculations based on conventional two-body forces acting between neutrons (reviewed in BP) indicate that solidification of neutron matter would not take place. However, as Pandharipande & Smith (1975a) described, π^0 condensation offers a possible mechanism for producing a solid state. Takatsuka et al. (1978) have shown similarly that π^0 condensation might lead to a one-dimensional "solidification" of the matter, analogous to a liquid crystal. The answer to whether such states can actually occur in neutron star matter must await a fuller understanding of the nuclear matter problem.

2.3   Field Theoretic Models of High Density Matter and the Abnormal State

At densities much greater than ρ_0 the meson clouds surrounding the nucleons in matter become strongly overlapping, and one does not expect a description in terms of distinct particles—neutrons, protons, etc.—interacting via two-body forces to remain valid. One approach that has been explored is to describe high density matter in terms of relativistic "bare" nucleons interacting via explicit meson fields. The basic type of model is that given by Walecka (1974) in which the nucleons interact attractively via coupling to a scalar meson field σ, and repulsively through coupling to a more massive vector field ω. The meson fields are assumed to be linear, i.e. not coupled to themselves. Chin & Walecka (1974) fitted the coupling constants and meson masses in the theory to reproduce, in the mean field approximation, the properties of symmetric nuclear matter and then derived a neutron matter equation of state. Similar calculations on this model were carried out by Bowers et al. (1975), and by Pandharipande & Smith

(1975b) who include pion exchanges as well. The physics of the model has been explored in further detail by Chin (1977).

Walecka's model predicts that at high densities the energy density $\varepsilon \equiv \rho c^2$ equals $\frac{1}{2}(g_\omega^2/m_\omega^2)n^2$, where n is the baryon density, g_ω is the ω meson-nucleon coupling constant, and m_ω is the ω mass. The behavior $\rho \sim n^2$, which is also characteristic of calculations of matter interacting via finite range two-body potentials, predicts a limiting pressure $P = \rho c^2$. However, in similar models in which the mass of the ω meson is generated dynamically, $\rho \sim n^{4/3}$ and thus $P = \frac{1}{3}\rho c^2$ at high densities (see, for example, Harrington & Yildiz 1974, Krive & Chudnovskii 1976, Källman 1978). In another variation of the mean field theory model, Canuto et al. (1978) include attractive nucleon-nucleon interactions via exchange of massive spin-2 f° mesons (Bodmer 1971, 1973); in this model, however, matter is unstable under collapse to arbitrarily high density.

Lee & Wick (1974) have proposed a possible high density "abnormal state" of matter in which the nucleons become nearly massless. Such a state can arise in a field-theoretic model of matter as follows. The scalar field σ couples to the nucleons as an addition to the nucleon rest mass, i.e. it appears in the energy in the form $[m_n + g\sigma(x)]\bar{\psi}(x)\psi(x)$, where m_n is the usual nucleon mass, g is the coupling constant, and $\psi(x)$ is the nucleon field. Thus the σ field acts as a dynamical modification of the nucleon mass. The energy also contains a potential energy density $V(\sigma(x))$, which is minimum (and $=0$) when $\sigma(x) \equiv 0$, and guarantees that in the vacuum the mean value $\langle \sigma \rangle$ of the σ field vanishes and the effective nucleon mass equals m_n. When the density is finite $\langle \sigma \rangle$ will be nonzero in the ground state and the effective nucleon mass will be $m^* = m_n + g\langle \sigma \rangle$.

Now at high densities it becomes favorable for $\langle \sigma \rangle$ to become $\simeq -m_n/g$, and thus $m^* \simeq 0$, since then the nucleon rest energy is lowered by $\sim m_n n$, while the cost of having $\langle \sigma \rangle \simeq -m_n/g$ is essentially the field energy $V(\langle \sigma \rangle = -m_n/g)$, a term independent of the density. Thus, for $n \gtrsim V(-m_n/g)/m_n$ one might expect the system to undergo a transition to an abnormal state in which $m^* \simeq 0$. In Walecka's model the σ field is linear, $V(\sigma) = \frac{1}{2}m_\sigma^2\sigma^2 c^5/\hbar^3$, where m_σ is the mass of the quantum of the σ field, and m^* decreases smoothly with increasing density. On the other hand, in more complicated models such as the σ-model (see Baym 1977a, 1978 for a description in this context) in which $V(\sigma) = \frac{1}{2}m_\sigma^2\sigma^2(1 + g\sigma/2m_n)^2 c^5/\hbar^3$ the abnormal state can appear via a sharp phase transition.

The possibility of an abnormal state in pure neutron matter was first considered by Källman (1975) and Källman & Moszkowski (1975).

They included a mean ω field in the σ-model, and predicted that pure neutron matter would have an abnormal state; however, since their model was not fit to normal symmetric nuclear matter, their prediction of an abnormal state was inconclusive. Pandharipande & Smith (1975b) have given a detailed analysis of the problem of fitting the σ-model to the binding energy, saturation density, and symmetry energy of symmetric nuclear matter, and conclude that if the model is made to fit these properties it will not have an abnormal state. Moszkowski & Källman (1977) have also fit mean field models to symmetric nuclear matter and the symmetry energy, through inclusion of a mean ρ field in Walecka's model, or through adjustment in the σ-model of the terms in $V(\sigma)$ cubic and quartic in σ; they no longer find an abnormal state in neutron star matter.

Given the difficulties of fitting the properties of the normal state within the σ-model, one can adopt the point of view that even though the details of the normal symmetric nuclear matter state, which, being on a scale of tens of MeV, are too subtle to be fit by simple models, such models might still give a reasonable description of the larger scale energy changes in the abnormal state. In this spirit Nyman & Rho (1977) have given a phenomenological calculation in the σ-model of a (first order) transition in neutron matter to an abnormal state, but conclude that while the energy of the abnormal phase lies below that of the normal phase at sufficiently high density (the abnormal phase is not self-bound there), the transition to the abnormal phase always occurs at too high a density for abnormal matter to be present in neutron stars. See also Migdal (1978).

Model calculations illustrating the relation of pion-condensed and abnormal states in matter have been given by Chanowitz & Siemens (1977) and by Akhiezer et al. (1979).

2.4   Quark Matter

The picture that quarks are the basic constituents of strongly interacting elementary particles (such as nucleons, Δ's, hyperons, and π, ρ and ω mesons) has by now gained wide acceptance, and suggests that a more fundamental description of matter at very high densities is in terms of quarks. In particular, one expects that when matter is sufficiently compressed, the nucleons will merge together and undergo a phase transition to quark matter, a degenerate Fermi liquid, in which the basic constituents are the quarks of which the nucleons were composed. In addition to its possible occurrence in neutron stars, quark matter is of interest in the description of the early universe when the baryon density greatly exceeded that of nuclear matter (Chapline 1976). One can very roughly estimate the density at which the transition to quark matter might occur by noting that nucleons begin to touch at a particle density $\sim (4\pi r_N^3/3)^{-1}$, where $r_N \lesssim 1\,\text{fm} \equiv 10^{-13}\,\text{cm}$ is an effective nucleon radius; this density is of the order of a few times ρ_0.

In the basic quark model, the quarks are spin-$\frac{1}{2}$ fermions, of baryon number $\frac{1}{3}$, which come in at least four "flavors," u, d, s, and c (up, down, strange and charmed). The electrical charges of these four flavors are $\frac{2}{3}$, $-\frac{1}{3}$, $-\frac{1}{3}$, and $\frac{2}{3}$ respectively; all have strangeness zero, except s which has strangeness -1. For each quark q there exists a corresponding anti-quark $\bar{q}$ with opposite quantum numbers. Mesons are composed of a quark and anti-quark, and baryons (of unit baryon number) of three quarks. For example, a proton is a *uud* bound state, while a neutron is *udd*. In addition quarks have an internal degree of freedom, color, originally introduced to enable quarks to obey the Pauli principle. In the fundamental model of quark-quark interactions—quantum chromodynamics (or non-Abelian Yang-Mills SU(3) gauge theory), in which colored quarks interact via exchange of eight massless vector gluons (analogues of photons in ordinary electrodynamics)—color functions effectively as a charge for gluon interactions. Loosely speaking, two quarks of the same color "repel," while two quarks of different color "attract" with half the strength. Thus a combination of three quarks each of different color (more correctly, a color singlet) acts as a neutral object, producing no long-range gluon "Coulomb" field.

The u and d quarks are believed to have a fairly small mass, m_u, $m_d \sim 10\,\text{MeV}$; the strange quark is heavier, with m_s perhaps ~ 100–$300\,\text{MeV}$, while the charmed quark is much heavier ($m_c > 1\,\text{GeV}$). Because of its high mass the charmed quark (as well as newer high mass quarks) is not expected to be present in quark matter that could occur in neutron stars.

The quark-gluon theory has the remarkable property that quark interactions at sufficiently short distances become arbitrarily weak. Furthermore in quark matter that is in a color singlet (or color neutral) state, the interactions between quarks at distances large compared to the interparticle spacing will be screened out, analogous to screening of long range Coulomb fields in a plasma in equilibrium. Thus, as Collins & Perry (1975) pointed out, at high densities the net quark interactions in quark matter in an overall color singlet state should be sufficiently weak that the matter can to a first approximation be taken as a non-interacting relativistic Fermi gas. Quark matter formed from compression of pure neutrons will have twice as many down quarks as up quarks (and will have 12 Fermi seas, two for u and d, times 2 for spin, times 3 for color). Similarly high density quark matter consisting of u, d, and s quarks in beta equilibrium can be shown to have equal densities of these three flavors, and no electrons or muons present (Collins & Perry 1975). In both cases the energy density is $\propto n^{4/3}$ at high baryon density n. Because quark-quark interactions become weak at high density (equivalently, the effective quark-gluon coupling constant g decreases with increasing density), one can calculate the high density equation of state as a perturbation expansion in the effective fine-structure constant $\alpha_c = g^2/4\pi$. Terms up to order α_c^2 have been calculated by Baluni (1978a,b) and Freedman & McLerran (1977, 1978). As the density decreases the interactions become more and more important, leading eventually to confinement of the quarks in hadrons, a phenomenon not described by such a perturbation expansion.

One simple phenomenological picture of quark confinement is the MIT bag model (Chodos et al. 1974), in which the quarks in a nucleon are assumed confined to a finite region of space, the "bag," whose volume is limited by the introduction of a term in the nucleon energy equal to the volume of the bag times a constant $B > 0$. With $m_u = m_d = 0$, the parameters $B \simeq 55\,\mathrm{MeV\,fm^{-3}}$ and a constant $\alpha_c \simeq 2.2$ give a reasonable fit to observed masses of strongly interacting baryons and mesons. To calculate the energy of quark matter in this model one simply adds a term B to the free particle plus interaction energies $\propto n^{4/3}$. The result for the energy density is qualitatively similar to that found from an exact perturbation expansion, and both may be used to produce a first estimate of the transition to quark matter by seeing, at a given baryon density, which phase, quark or nucleon, has a lower energy density.

Such calculations within the framework of the bag model of the phase transition from pure neutron matter to quark matter have been given by Baym & Chin (1976), Chapline & Nauenberg (1976), and Keister & Kisslinger (1976). The conclusion of these calculations is that, for all neutron matter equations of state examined, the phase transition to quark matter takes place at too high a density for quark matter to be found in neutron stars. For example, with the Reid pure neutron equation of state, the density jumps from 14–$40 \times 10^{15}\,\mathrm{g\,cm^{-3}}$ at the transition, while the maximum central density found in neutron stars described by the Reid equation of state is $4.1 \times 10^{15}\,\mathrm{g\,cm^{-3}}$. On the other hand the bag model calculations are quite phenomenological and neither they nor the nuclear matter calculations ought to be regarded as conclusive for densities well above that within nucleons, $\sim 1.4\,\rho_0$. When one compares perturbation expansions of the quark matter energy, using a density-dependent coupling constant, with pure neutron equations of state (Chapline & Nauenberg 1977a, Baluni 1978a,b, Freedman & McLerran 1978, Kislinger & Morley 1978, Baym 1977b) one finds that, depending on the assumed coupling strength, the transition to quark matter could well occur at densities sufficiently low that neutron stars could have quark matter cores. At present, the strength of quark-gluon interactions is not well enough established experimentally, nor is the quark confinement problem at low densities adequately understood for one to say more precisely whether quark matter is present in neutron stars. (Possible existence of a class of dense "quark stars" is discussed in Section 3.) The quark matter calculations described here are reviewed in detail in Baym (1977a, 1978).

2.5   Finite Temperature Equations of State

The equation of state of dense matter at finite temperature is crucial to understanding formation of neutron stars in supernovae, and also the early moments of their life. In the final stages of stellar collapse in supernovae, matter passes through a range of densities from $\sim 10^9 \, \mathrm{g \, cm^{-3}}$ to about several times ρ_0; temperatures in the collapse may reach a few times 10^{11} K. During the collapse, whose time scale is $\lesssim 1$ sec, substantial electron capture takes place; because of neutral-current weak interactions, the neutrinos produced can be trapped in the matter for time scales up to $\sim$ seconds, sufficiently long to reach approximate thermal equilibrium. The primary constituents of the matter undergoing collapse are nuclei (undergoing various transformations), "free" neutrons and protons, electrons and neutrinos, both of which behave as free Fermi gases, and photons. The initial collapse is reversed in a bounce at densities ($\gtrsim \rho_0$) at which the adiabatic index of the matter, $\Gamma = \partial \ln P / \partial \ln n$ (where P is the pressure), rises well above $\frac{4}{3}$.

The simplest approximation to the equation of state at subnuclear densities treats the matter as a mixture of free neutron, proton, electron, and photon gases (see, for example, Van Riper & Bludman 1977). While this approach is valid at temperatures high compared with that required to dissociate nuclei, $\sim 20 \, \mathrm{MeV}$, it is necessary in collapse to take the nuclei present into account. In most collapse calculations to date (e.g. Arnett 1977, Mazurek 1979, see also Nadyozhin 1977) this is done by use of a semiempirical mass formula extrapolated from laboratory nuclei. As in the case of zero-temperature matter at subnuclear densities in neutron stars, a more accurate description requires that one go beyond the semiempirical mass formula and include the following important physical effects: (a) the effect of nuclear excited states, (b) interactions between nucleons outside nuclei, (c) the reduction of the nuclear surface energy due to finite temperature, and (d) the Coulomb interaction between nuclei. Sato (1975) took (a) and (b) into account by employing the $T = 0$ result of Baym et al. (1971a), which includes nucleon-nucleon interactions, for the bulk energies of matter both inside and outside nuclei, plus a finite temperature correction calculated assuming that the kinetic energy of the nucleons is that of a free gas (see also Neatrour 1979). One of the important conclusions of this work was that at high densities nuclei survive to relatively high temperatures. Mackie (1976) took (d) into account and used an improved nuclear mass formula (Mackie & Baym 1977), which allowed for the reduction of the nuclear surface energy due to the rather large neutron excess in the nuclei. Mazurek et al. (1979) also estimated the effects of (a), (b), and (d).

Bethe et al. (1979) have recently suggested a possible restriction on the range of parameters over which one needs to know the equation of state in collapse. They argue that as the entropy of matter in collapse is likely small, effects of nuclear excited states cause most nucleons to be confined to nuclei, with the pressure of matter being provided chiefly by the degenerate electrons. Consequently, under these circumstances the adiabatic index, Γ, of the matter below $\rho \lesssim \rho_0$ is close to $\frac{4}{3}$, and bounce will occur at $\rho > \rho_0$.

The presently most detailed calculation of the finite-temperature equation of state at subnuclear densities is that of D. Q. Lamb et al. (1978), who include effects (a)–(d) through a finite-temperature generalization of the work of Baym et al. (1971a). The bulk free energies of matter both inside and outside nuclei are those calculated by Lattimer & Ravenhall (1978) using a Skyrme effective nucleon-nucleon interaction, which was also used to calculate the surface free energy (Ravenhall & Lattimer 1979). The internucleus Coulomb energy is taken into account using the Coulomb liquid calculations of Hansen (1973). An important conclusion of this work is that Γ at subnuclear densities remains close to $\frac{4}{3}$ for a wide range of entropies and lepton fractions, implying that bounce will occur at densities $> \rho_0$ over a large range of initial entropies during the collapse. Calculation of the finite-temperature equation of state in the neighborhood of nuclear matter density, a quantity needed for understanding details of the bounce and subsequent shock formation, contains at least as many difficulties as the zero-temperature calculations (Section 2.1).

Buchler & Coon (1977) and Buchler & Datta (1979) have calculated the equation of state of a hot neutron gas from a more microscopic approach employing various two-body scattering approximations (in the sense of Brueckner theory) for the free energy. One can conclude from these papers that neglect of the temperature dependence of the effective nucleon-nucleon interaction, as when using the Skyrme interaction, is a reasonable approximation in the temperature—density regimes where the interactions produce a significant contribution to the free energy.

Further calculations of hot neutron star matter, using the finite-temperature Thomas-Fermi method, are given by El Eid & Hilf (1977).

An equation of state of finite-temperature neutron matter has been computed by Walecka (1975) from his relativistic mean field theory model (Walecka 1974). (See also Bowers et al. 1977b,c, and Freedman 1977.) While such a calculation omits too much of the physics at low densities to be relevant to the supernova problem, it is possibly applicable at densities well above ρ_0. The "liquid-gas" phase transition at subnuclear densities present in this calculation (for which there is no evidence in potential theory models, expected to be more reliable at such densities) is most likely an artifact of the model and the choice of parameters.

An initial calculation of finite-temperature quark matter within the framework of the MIT bag model has been carried out by Chin (1978), who concludes that the density of the transition to uniform quark matter falls with increasing temperature. See also Shuryak (1978).

2.6 Matter in High Magnetic Fields

In the strong magnetic fields expected in neutron stars, the properties of matter at relatively low densities are very different from those in the absence of the field. Individual atoms become compressed, and relatively elongated in the direction of the magnetic field, and can bind together covalently to form polymerized chains in which the electrons are free to move along the length of the chain. The chains may then be bound together by electrostatic forces to form a solid, which at zero pressure has a density much greater than that of terrestrial solids. During the past few years there has been a considerable amount of work on the properties of matter in strong magnetic fields, and while the basic picture remains the same as that described in BP, a number of the quantitive results have changed.

An astrophysically important quantity in descriptions of neutron star surfaces and hence in theories of pulsar emission (Ruderman & Sutherland 1975) is the cohesive energy, i.e. the energy of the condensed polymerized matter in a strong field, compared to that of isolated atoms in the same field. Hillebrandt & Müller (1976) pointed out an error in the earlier calculations of Chen et al. (1974) that led to an overestimate of the cohesive energy. The most reliable calculation to date is that of Flowers et al. (1977), who used a variational approach including electron exchange, the importance of which was stressed by Glasser & Kaplan (1975), and allowed for buildup of the electronic charge in the vicinity of nuclei. They find cohesive energies ranging from 2.6 keV/atom at field $B = 10^{12}$ G to 8.0 keV/atom at $B = 5 \times 10^{12}$ G, compared with Chen et al's (1974) results of 10 keV/atom and 20 keV/atom. The zero-pressure densities are not significantly modified from previous results.

The new calculated cohesive energies are small compared with the binding energy of an individual iron atom, ~ 50 keV. Given the fact that the cohesive energy is the difference of two large energies, each calculated variationally, one cannot conclude with certainty that at zero pressure the condensed state formed of polymerized chains is energetically favorable compared with a more ordinary state of elongated atoms with weaker metallic binding. (Because of the compression of atoms in strong fields, this latter state would at zero pressure still be at density much higher than terrestrial solids, but perhaps an order of magnitude below that of the magnetically polymerized solid.) To improve upon the estimates of Flowers et al. will be difficult, and the extent to which models of pulsar radio emission will be affected by a reduced cohesive energy is an open question.

The spectra of atoms in strong fields have been reviewed by Garstang (1977), with particular emphasis on H and He. Also, Sarazin & Bahcall (1977) have calculated Zeeman splitting of X-ray emission lines from Fe XXV and XXVI and other ions, and suggest that observations of these line profiles may enable one to determine both the direction and magnitude of neutron star fields.

3   NEUTRON STAR MODELS

Construction of neutron star models is of interest not only for learning their physical properties—radius, moment of inertia, density profile—for a given mass (and equation of state), but also for determining the allowed range of neutron star masses. In particular, knowledge of the maximum allowable neutron star mass is an important ingredient in attempts to identify black holes from measurement of masses of compact objects.

Neutron star models are constructed from a given equation of state $P(\rho)$ by integration of the equation of hydrostatic balance, which for general relativity is the Tolman-Oppenheimer-Volkoff equation. See Baym et al. (1971b) for a discussion of numerical integration methods, and Arnett & Bowers (1977) for an extensive comparison of models.

Figure 1 shows (for general relativistic stars) the resultant gravitational masses M as a function of the central mass density ρ_c for a representative sample of equations of state: MF is the Pandharipande-Smith (1975b) mean field theory calculation, similar to that of Chin & Walecka (1974); TI is the Pandharipande-Smith (1975a) tensor-interaction model, which incorporates approximately stiffening effects on the equation of state due to isobars (Section 2.1); BJ is for the Bethe-Johnson (1974) equation of state I, which includes small effects of hyperons (see also Malone et al. 1975); R is for the pure neutron equation of state with the Reid potential (Pandharipande 1971); π is for the Reid equation of state as modified by charged-pion condensation of a strength that may reasonably be expected, while π' shows the results of somewhat stronger pion condensation, in which there is a large first order phase transition at the onset of condensation (Maxwell & Weise 1976). The estimates of Hartle et al. (1975) on effects of pion condensation (their case n') are quite similar to the case π'. The maximum possible neutron star mass, given by the maximum of $M(\rho_c)$, is fairly sensitive to the equation of state, tending to increase as the equation of state becomes stiffer, but remaining, as we see, $\lesssim 3\,M_\odot$, and above the masses $\sim 1.4\,M_\odot$ observed in Her X-1 and Vela X-1. The central density for a given mass decreases as the equation of state becomes stiffer.

The corresponding radii of the stars, shown in Figure 2, are also sensitive to the stiffness of the equation of state, and in particular to the detailed behavior in the neighborhood of nuclear matter density. (The moments of inertia of these models are discussed in the respective references.) One sees from the TI curve how inclusion of explicit effects

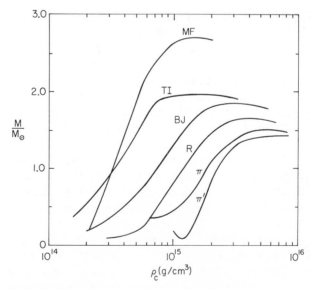

Figure 1   Gravitational mass (in solar units) versus central density for a variety of equations of state. The rising portions of the curves represent stable neutron stars. See text for identification of curves.

of isobars can increase the radius of a neutron star of a given mass. Stars with stiffer equations of state will also have thicker crusts since, for a given mass liquid core, the stiffer the equation of state is around ρ_0, the weaker is gravity at the core-crust interface. For example the crust of a 1.33 $M_\odot$ TI star reaches from a radius of 11.3 to 16.1 km, while in the same mass R star the crust exists between 9.1 and 9.9 km. (See Pandharipande et al. 1976 for more detailed discussions.) On the other hand, as we described earlier, pion condensation can significantly soften the equation of state, an effect not included in the stiffer MF or TI models; it tends to contract neutron stars of a given mass, as well as decrease M_{max}. Because there do not yet exist equations of state reliably including effects of both isobars and pion condensation, one must regard the models shown in Figures 1 and 2 as illustrating the range of possibilities.

A further interesting possibility could occur were neutron star matter to develop a self-bound state, due, for example, to pion condensation or an abnormal state. Then, as illustrated by Hartle et al's (1975) case C′, stable neutron stars could exist with arbitrarily small mass and radius ("golf balls"), and surface density equalling that of the self-bound state.

As discussed in Section 2.4, the transition to uniform quark matter may occur at densities above the maximum central density ρ_{max} in neutron stars. An interesting question then is whether there perhaps

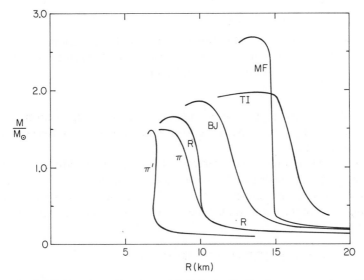

Figure 2   Gravitational mass (in solar units) versus radius for the same equations of state described in Figure 1.

exists a third stable branch of cold stars (after white dwarfs and neutron stars), "quark stars," whose central densities lie beyond ρ_{max}, and which are supported against gravitational collapse primarily by the degeneracy pressure of unconfined quarks (Itoh 1970). The general relativistic stability condition for a star of mass M and radius R—that the mean adiabatic index Γ exceed $\frac{4}{3}(1 + KR_s/R)$, where $R_s = 2MG/c^2$ is the Schwarzschild radius and K is of order unity—places a strong constraint on those equations of state of quark matter that can yield a stable third branch (see Gerlach 1968). The stability requirement on Γ tends to grow with increasing ρ_c, and is ~ 2 at M_{max} for neutron stars. Since in the limit of large density, $P \propto n^{4/3}$ in quark matter, and thus $\Gamma \propto \frac{4}{3}$, such a condition is unlikely to be fulfilled for a significant range of densities of quark matter (see Bowers et al. 1977a). We note though that, by appropriate choice of parameters, one can construct model equations of state of quark matter that yield a stable quark star branch (Chapline & Nauenberg 1977b, Fechner & Joss 1978).

Substantial effort has been devoted to determination of exact bounds on masses and moments of inertia of neutron stars. Since this subject is lucidly reviewed by Hartle (1978, see also Sabbadini & Hartle 1977), we quote only a few major results. If the equation of state is regarded as known below a fiducial density ρ_f, assumed to be not far above ρ_0 (so that $P/\rho c^2$ is small at ρ_f), then the bound on the maximum mass is given to a good approximation by

$$M_{bound}/M_\odot = 6.75(\rho_0/\rho_f)^{1/2}, \tag{1}$$

assuming only that the equation of state above ρ_f obeys $\rho > 0$ and $\partial P/\partial\rho > 0$. If one furthermore imposes the "causality condition" $\partial P/\partial\rho < c^2$, then the bound is sharpened to

$$M_{bound}/M_\odot = 4.0(\rho_0/\rho_f)^{1/2}. \tag{2}$$

It should be noted that these bounds lie, for $\rho_f = \rho_0$, significantly above the various M_{max} that one finds from equations of state derived from microscopic theory.

The above bounds were derived assuming the validity of general relativity. Other theories of gravity, however, such as the Rosen-Rosen bimetric theory and some versions of Ni's theory, permit significantly greater maximum neutron star masses than does general relativity (see Hartle 1978).

4 NONEQUILIBRIUM PROCESSES

In this section we review aspects of the physics underlying non-equilibrium processes in neutron stars that determine their evolution and dynamical behavior, concentrating in particular on transport and hydro-dynamic properties, cooling processes, and nonequilibrium effects of superfluidity.

4.1 Transport Properties and Hydrodynamics

Transport properties of dense matter have an important influence on neutron star magnetic fields and cooling, as well as on other features of their behavior. Detailed calculation of the electrical and thermal conductivities of matter at densities $\lesssim \rho_0$ have been given by Flowers & Itoh (1976), who have also discussed transport properties of the liquid regime, assuming it to be a mixture of normal (i.e. nonsuperconducting and nonsuperfluid) Fermi liquids (Flowers & Itoh, 1979). Ewart et al. (1975) have discussed the electrical conductivity of the crust, and conclude that decay of large-scale magnetic fields in the crust during a characteristic pulsar lifetime will be small unless either the temperature is higher than estimated, or the crust is very impure or composed of microcrystallites.

The high magnetic fields expected in neutron stars can have a drastic effect on transport properties, since electrons can move more easily along field lines than perpendicular to them. Consequently, when $\omega_c \tau > 1$, where ω_c is the cyclotron frequency and τ is the relevant transport relaxation time, the electrical and thermal conductivities, and the viscosity, are much larger along the field than perpendicular to it. At densities such that the electron Larmor radius is large compared with the electron spacing $[\rho \gg 2 \times 10^4 (B/10^{12})^{3/2} \, \mathrm{g \, cm^{-3}}]$ the structure of the matter, and consequently the scattering rates, are little affected by the magnetic field, and transport properties in the presence of the field may be calculated straightforwardly in terms of zero-field quantities (see, for example, Flowers & Itoh 1976, Easson & Pethick 1979). At lower densities, which are of particular interest for determining the properties of neutron star surfaces, the problem of calculating transport coefficients is much more difficult. Itoh (1975) has discussed the electrical conductivity due to electron-phonon scattering, and has also discussed the refractive index. See also the review by Canuto & Ventura (1977). The reflection of X rays by a neutron star surface assumed to be composed of the magnetically condensed matter discussed in Section 2.6 has been considered by Lenzen & Trümper (1978); their results are applicable only if the neutron star surface may be treated as smooth on the scale of $\sim 10 \, \mathrm{keV}$ X-ray wavelengths and scattering outside the surface may be neglected.

Sources of photon opacity, especially in the relatively low densities in the outer parts of a neutron star, have an important effect on neutron star surface temperatures. These processes have been studied extensively for zero magnetic field (see, for example, Carson 1976). Lodenquai et al. (1974) have studied free-free, bound-free, and Thomson scattering in the presence of a field; they find that for photons of frequency ω having an electric field vector perpendicular to the magnetic field the opacity is reduced, for $\omega \ll \omega_c$, by a factor $\sim (\omega/\omega_c)^2$.

The rather unusual conditions in neutron star interiors lead to magnetohydrodynamic behavior different from that in more common situations. Hide (1971) has pointed out that, for rapid stellar rotation, the elementary magnetohydrodynamic modes will have frequency spectra modified from the soundlike dispersion relations in the absence of rotation. For a review see Acheson & Hide (1973). Time scales to establish beta equilibrium, allowing for magnetohydrodynamic flows, have been calculated by Baym et al. (1979); they conclude that the possible force-free configuration of a magnetic field in a neutron star in magnetohydrostatic and beta equilibrium suggested by Easson (1976) can never be realized in practice. The basis of the magnetohydrodynamic equations has been examined by Easson & Pethick (1979), who also point out that when the neutrons are superfluid, the strong interactions will couple the motion of the charged particles to that of the neutrons, even though collisional effects are unimportant. Jones (1975) pointed out that proton superconductivity could drastically influence the magnetic contribution to the stress tensor, which is of importance for the static deformation of neutron stars, and for the dynamics. Detailed calculations of the effect were carried out by Easson & Pethick (1977), who showed that the components of the stress tensor are $\sim BH_{c1}/4\pi$ (where H_{c1} is the lower critical field), a result typically a few orders of magnitude greater than the normal result $B^2/8\pi$.

4.2   Cooling of Neutron Stars

Measurement of surface temperatures of neutron stars, through observation of thermal black body emission, can in principle yield substantial information about the interior structure of the stars. One may ask, for example, what are the internal states of matter and processes that could have enabled the Crab pulsar, made in 1054 A.D. with an interior temperature $\sim 10^{11}$ K, to cool to its present state with a surface temperature $< 4.7 \times 10^6$ K (Wolff et al. 1975). Knowledge of the temporal evolution of the internal temperatures of neutron stars is also important for estimating temperature-dependent properties such as transport coefficients, transitions to superfluid states, and solidification of the crust.

After formation the predominant cooling mechanism is neutrino emission. Photon emission begins to play a role only when the internal temperature falls a thousandfold to $\sim 10^8$ K, with a corresponding surface temperature about two orders of magnitude smaller.

In degenerate matter in neutron star cores cooling by the URCA process

$$n \rightarrow p + e^- + \bar{\nu}_e, \quad e^- + p \rightarrow n + \nu_e \tag{3}$$

becomes ineffective unless momentum can be transferred, in the reactions, to the rest of the system. We can see this by considering the neutron beta decay. In beta equilibrium, $\mu_n = \mu_p + \mu_e$ to within terms of order $(\kappa T)^2/\mu_p$, where μ_n, μ_p and μ_e are the neutron, proton, and electron chemical potentials. The neutrons capable of decaying lie within $\sim \kappa T$ of the neutron Fermi surface, and thus the final proton and electron must lie within κT of their Fermi surfaces; the neutrino energy is also $\sim \kappa T$. Because the electron and proton Fermi momenta are small compared with the neutron Fermi momentum p_n, the final proton, electron, and neutrino, as well, must have small momenta. But the initial neutron must have momentum $\sim p_n$, and thus the decay cannot conserve momentum if it conserves energy. In order for the process to work, a bystander particle must absorb momentum, as in the "modified URCA process" (Chiu & Salpeter 1964), $n + n \rightarrow n + p + e^- + \bar{\nu}_e$, $n + p + e^- \rightarrow n + n + \nu_e$. The neutrino luminosity due to this process has been calculated by Bahcall & Wolf (1965), Itoh & Tsuneto (1972), and most recently by Friman & Maxwell (1979), who use more realistic expressions for the nucleon-nucleon interactions, and find a stellar luminosity

$$L_\nu \sim (6 \times 10^{39}\,\mathrm{erg/s})(M/M_\odot)(\rho_0/\rho)^{1/3}T_9^8 \tag{4}$$

(where T_9 is the interior temperature in units of 10^9 K) an order of magnitude larger than that calculated by Bahcall & Wolf. Due to the available phase space for the bystander neutron this rate is down by a factor $\sim(\kappa T/\mu_n)^2$ from the rate for (3), were that process allowed by energy and momentum conservation. The calculated rate (4) is sensitive to the effective masses assumed for the nucleons.

The experimental discovery of weak neutral currents in 1974 suggested further cooling processes. The most important in the interior of neutron stars are the nucleon pair bremsstrahlung processes $n + n \rightarrow n + n + \nu + \bar{\nu}$ and $n + p \rightarrow n + p + \nu + \bar{\nu}$, first considered by Flowers et al. (1975), and re-examined by Friman & Maxwell (1979). The luminosity from these processes also varies as T^8, but is less than $\frac{1}{30}$ the magnitude of that from the modified URCA process (Friman & Maxwell 1979).

All the above calculations assumed the nucleons to be normal. If instead they are superfluid, the rates are reduced by factors $\sim \exp(-\Delta_{n,p}/\kappa T)$ due to reduction of the number of thermal excitations

(Wolf 1966, Itoh & Tsuneto 1972). (Here $\Delta_{n,p}$ are the neutron and proton superfluid gaps.) Under these circumstances the neutrino pair bremsstrahlung process,

$$e^- + (Z,A) \to e^- + (Z,A) + \nu + \bar{\nu} \tag{5}$$

from nuclei in the crust, can be important. Initially estimated by Festa & Ruderman (1969), who considered only the weak charged-current contribution, it has more recently been discussed by Flowers (1973, 1974), who also allowed for the finite nuclear size, and by Dicus (1972), Dicus et al. (1976), and Soyeur & Brown (1979), who calculated neutral-current contributions. The luminosity from this process varies as T^6, and therefore decreases less rapidly than the modified URCA process with decreasing temperature. Maxwell (1979) estimates the total luminosity due to this process as

$$L_\nu \sim (5 \times 10^{39}\,\text{erg/s})(M_{cr}/M_\odot)T_9^6, \tag{6}$$

where M_{cr} is the mass of the crust. In the inner crust, where free neutrons coexist with nuclei, neutrino pairs can also be produced by bremsstrahlung in the scattering of neutrons from nuclei (Flowers & Sutherland 1977); the rate of this process is of order $\eta(\varepsilon_f/30\,\text{MeV})^{3/2}$ $(A/200)^{1/3}(30/Z)^2$ times that for electron bremsstrahlung, where $\eta \sim \frac{1}{4}\text{–}\frac{1}{2}$, and ε_f is the neutron Fermi energy; this factor is generally <1.

Flowers et al. (1976) have pointed out that if neutrons are superfluid, two neutron-like excitations can annihilate to produce neutrino pairs. This process is most important just below the transition temperature, and it can dominate the pair bremsstrahlung process under some circumstances.

Pion condensation can significantly enhance the cooling rate of neutron stars in their hot early period, since it permits the analogue of (3) to occur conserving energy and momentum. (The condensed pion field itself does not beta decay.) Essentially the excess momentum in (3) is absorbed by an Umklapp process involving the condensed pion field. In a weak condensate one can think of the neutron decay process as occurring by first $n \to p + e^- + \bar{\nu}_e$, with the proton after the beta decay existing only in a virtual intermediate state before scattering from the condensed pion field, changing into a neutron and absorbing energy $\mu_\pi = \mu_n - \mu_p$ and momentum $\mathbf{k}$, the wave vector of the condensed pion field. Thus as long as the initial and final neutron states are separated by $\sim\mathbf{k}$ across the Fermi surface, the process is allowed by energy and momentum conservation. More precisely, because the nucleon eigenstates are linear combinations of neutrons and protons, nucleons in these states (call them f) can beta decay into themselves: $f \to f + e^- + \bar{\nu}_e$,

$f + e^- \rightarrow f + v_e$. In the first process, for example, the neutron component of the initial f decays to a proton state, which has nonzero overlap with the final f. In the beta decay the particle f moves a momentum $\sim k$ across the f Fermi surface. The net result is a luminosity (Maxwell et al. 1977)

$$L_v^\pi \sim (1.5 \times 10^{46} \, \text{erg/s})\theta^2 (M/M_\odot)(\rho_0/\rho)T_9^6, \tag{7}$$

where $\theta \sim 0.3$ is an angle measuring the degree of pion condensation. This rate is $\sim 2.5 \times 10^6 (\rho_0/\rho)^{2/3}\theta^2/T_9^2$ larger than the modified URCA rate (4), and will, if there is pion condensation, dominate at all temperatures of interest. [The result (7) is quite close to the estimate of Bahcall & Wolf (1965) for cooling via decay of free pions, were they to exist, in the medium.]

The cooling time of a neutron star may be estimated using the fact that the thermal energy U, which resides almost exclusively in the degenerate neutrons, is $U \simeq (10^{47} \, \text{erg})(M/M_\odot)(\rho/\rho_0)^{-2/3}T_9^2$. Equating dU/dt to the neutrino luminosity, one finds that the star cools from an initial temperature $T(i)$ to a final temperature $T(f)$ in a time

$$\Delta t \simeq (0.2 \, \text{yr})(\rho/\rho_0)^{-1/3}[T_9(f)^{-6} - T_9(i)^{-6}] \tag{8}$$

for the modified URCA process (4), and

$$\Delta t \simeq (3 \, \text{sec})\theta^{-2}(\rho/\rho_0)^{1/3}[T_9(f)^{-4} - T_9(i)^{-4}] \tag{9}$$

for the pion condensation process (7).

The temperature that determines the thermal emission from a neutron star is that at the surface T_e, rather than the interior temperature T. Neutron star interiors are to a good approximation isothermal, but near the surface the temperature drops rapidly, and $T_e/T \sim 10^{-2}$–10^{-3}. Detailed calculations of T_e/T have been made by Tsuruta (1974) and by Malone (1974).

The relative importance of various cooling processes on the interior temperature of a neutron star of mass $\sim M_\odot$, as a function of time, may be assessed from the schematic Figure 3; each line in this figure shows $T(t)$, assuming only a single process, that labelling the line, to be operant. Results are shown for the modified URCA process (4), for pion condensation cooling (7), for cooling from crust bremsstrahlung (6), and for emission of photons from the surface, assuming $T_e = (10T)^{2/3}$, an approximate fit to Tsuruta's (1979) calculations. The formulae were evaluated for $\rho = \rho_0$ and $\theta^2 = 0.1$. At any given time the most effective process will be the one with the lowest $T(t)$, and the temperature of the star will be roughly that T.

The curves were plotted assuming the neutrons and protons to be

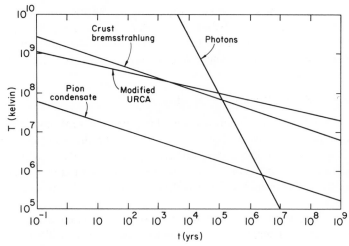

Figure 3   Schematic neutron star cooling curves of interior temperature versus time for various processes, were they to operate alone.

normal. Superfluidity has two effects; first it modifies the specific heat, which at the transition temperature jumps discontinuously to a value greater than that in the normal state, and then falls off exponentially at lower temperatures. This tends to decrease the cooling rate immediately below T_c, and to increase it at lower temperatures. A second effect is to suppress neutrino processes in the interior, and thereby decrease the cooling rate. Which of these two effects dominates depends on how important a role the neutrino processes in the interior play in the cooling.

While magnetic fields do not affect the neutrino processes, they can, as discussed in Section 4.1, reduce the opacity of the surface of a neutron star appreciably, and thus, for a given internal temperature, increase the surface temperature, and hence increase the photon luminosity (Tsuruta et al. 1972).

Detailed studies of neutron star cooling have been made by Tsuruta (1975) and Malone (1974) and reviewed by Tsuruta (1979). See also Brown (1977). More recently Maxwell (1979) has calculated the cooling curve for some simplified models, and concludes that the present upper bound on the Crab pulsar temperature is compatible with any cooling scenario involving neutrinos; however, a bound as low as 2×10^6 K would be more difficult to understand without pion condensation.

The neutrino emissivity from a neutron star immediately after its formation may be detectable on earth. At formation the neutrinos in the star are degenerate and the neutrino mean free paths are short, due to scattering via neutral current interaction with nucleons, as well as absorption of neutrinos by neutrons (Sawyer & Soni 1979). As Sawyer & Soni (1977) pointed out, even when the neutrinos are no longer degenerate, but the neutron star is still very hot, neutrino mean free paths are small compared with the stellar radius, and neutrino luminosities and stellar cooling rates will be reduced from those given above. These effects will not significantly affect the temperature of neutron stars more than a few hours old. Considerable work is in progress on neutrino processes in very young neutron stars, and more detailed calculations of the early life of a neutron star should be available in the near future.

4.3 Dynamical Effects of Superfluidity

The irregular fluctuations in neutron star rotational periods observed in both pulsars and pulsating X-ray sources, as well as the behavior of pulsars after sudden speedup (Manchester & Taylor 1977, Lamb 1977), may be governed in part by superfluidity of the internal regions of the stars. The rotational dynamics of the two-component model of a neutron star with a superfluid interior weakly coupled to a normal crust, introduced by Baym et al. (1969), provides a plausible explanation of the long relaxation times associated with speedups of the Crab and Vela pulsars (see Pines et al. 1974); however, relation of the observed relaxation times to possible microscopic coupling processes requires detailed study. Aspects of the dynamics were discussed by Ruderman & Sutherland (1974), and by Greenstein (1975, 1976, 1977), who pointed out possible internal heating of neutron stars due to the superfluid-crust coupling; Harding et al. (1978) have recently calculated relaxation processes between superfluid neutron vortex lines and the crust. Effects of pinning of the vortices to nuclei in the crust were considered by Anderson & Itoh (1975), by Ruderman (1976), who suggested possible breaking of the crust due to pinning forces, by Alpar (1977), and by Shaham (1977).

An interesting exploration of possible relations to relaxation after a sudden pulsar speedup has been the experimental simulation of such behavior by Tsakadze & Tsakadze (1975), using rotating superfluid helium. Theoretical interpretation of these experiments for neutron stars has been given by Anderson et al. (1978) and Alpar (1978).

5 CONCLUSION

As can be judged from this review, knowledge of the properties of neutron stars has advanced remarkably over the past decade. One is now much more aware of the gaps and uncertainties in the picture, and the important problems for further research. Particularly crucial is an improved description of nuclear matter and forces, from which we will gain a better understanding of the equation of state, of the effects of pion condensation and superfluidity, and of neutron star models. A second outstanding problem, which we have only touched on peripherally, is the formation of neutron stars in supernovae, a problem on which the discovery of weak neutral currents has had a strong recent impact. For an overview of problems in this area see, for example, Arnett (1977) and Freedman et al. (1977).

Much work has been done on many areas of neutron star behavior that we unfortunately have not had adequate space to discuss. Among these are: internal dynamical processes, including models of short term variability of neutron star rotation periods and neutron star wobble (Lamb 1975, 1977, F. K. Lamb et al. 1978); processes outside neutron stars, including pulsar emission mechanisms (Manchester & Taylor 1977), accretion and X-ray emission (Sunyayev 1978, Lightman et al. 1978), surface nuclear burning (Woosley & Taam 1976, Joss 1978, Lamb & Lamb 1978, Taam & Picklum 1978), and neutron star models of X-ray (Lamb et al. 1977) and γ-ray burst sources (Lamb et al. 1973, Ruderman 1975); interactions of neutron stars with other stars, including binary pulsars (Manchester & Taylor 1977), interactions with black holes (Lattimer & Schramm 1976, Lattimer et al. 1977), and neutron stars as cores of red giants and supergiants (Thorne & Żytkow 1977).

We are grateful to our colleagues in Urbana for numerous helpful discussions during the preparation of this review.

Literature Cited

Acheson, D. J., Hide, R. 1973. *Rep. Prog. Phys.* 36:159
Akhiezer, A. I., Chudnovskii, E. M., Krive, I. V. 1979. *Ann. Phys. N.Y.* In press
Alpar, M. A. 1977. *Ap. J.* 213:527
Alpar, M. A. 1978. *J. Low Temp. Phys.* 31:803
Anderson, P. W., Itoh, N. 1975. *Nature* 256:25
Anderson, P. W., Pines, D., Ruderman, M. A., Shaham, J. 1978. *J. Low Temp. Phys.* 30:839
Arnett, W. D. 1977. *Ap. J.* 218:815
Arnett, W. D., Bowers, R. L. 1977. *Ap. J. Suppl.* 33:411
Au, C.-K. 1976. *Phys. Lett.* 61B:300
Bäckman, S.-O., Clark, J. W., Ter Louw, W. J., Chakkalakal, D. A., Ristig, M. L. 1972. *Phys. Lett.* 41B:247
Bahcall, J. N. 1978. *Ann. Rev. Astron. Astrophys.* 16:241
Bahcall, J. N., Wolf, R. A. 1965. *Phys. Rev.* 140B:1445, 1452
Baluni, V. 1978a. *Phys. Lett.* 72B:381
Baluni, V. 1978b. *Phys. Rev.* D17:2092

Baym, G. 1977a. *Neutron Stars and the Properties of Matter at High Density.* Copenhagen: Nordita
Baym, G. 1977b. *Proc. Int. Conf. High Energy Phys. Nucl. Struct., 7th,* ed. M. P. Locher, p. 309. Zurich: Birkhäuser
Baym, G. 1978. *Proc. 1977 Les Houches Summer School,* ed. R. Balian, G. Ripka, p. 745. Amsterdam: North-Holland
Baym, G., Bethe, H. A., Pethick, C. J. 1971a. *Nucl. Phys.* A175:225
Baym, G., Campbell, D. 1979. In *Mesons and Fields in Nuclei,* ed. M. Rho, D. Wilkinson. Amsterdam: North-Holland
Baym, G., Chin, S. 1976. *Phys. Lett.* 62B:241
Baym, G., Easson, I., Pethick, C. J. 1979. To be published
Baym, G., Pethick, C. J. 1975. *Ann. Rev. Nucl. Sci.* 25:27
Baym, G., Pethick, C. J., Pines, D., Ruderman, M. 1969. *Nature* 224:872
Baym, G., Pethick, C. J., Sutherland, P. G. 1971b. *Ap. J.* 170:299
Bethe, H. A. 1971. *Ann. Rev. Nucl. Sci.* 21:93
Bethe, H. A., Brown, G. E., Applegate, J.,

Lattimer, J. M. 1979. *Nucl. Phys. A.* In press

Bethe, H. A., Johnson, M. B. 1974. *Nucl. Phys.* A230:1

Bodmer, A. R. 1971. *Phys. Rev.* D 4:1601

Bodmer, A. R. 1973. *The Nuclear Many-Body Problem*, ed. F. Calogero, C. Ciofi degli Atti. Bologna: Editrice Compositori

Bowers, R. L., Gleeson, A. M., Pedigo, R. D. 1975. *Phys. Rev.* D12:3043, 3056

Bowers, R. L., Gleeson, A. M., Pedigo, R. D. 1977a. *Ap. J.* 213:840

Bowers, R. L., Gleeson, A. M., Pedigo, R. D., Wheeler, J. W. 1977b. *Phys. Rev.* D15:2125

Bowers, R. L., Gleeson, A. M., Wheeler, J. W. 1977c. *Ap. J.* 213:531

Brown, G. E. 1977. *Comments Astrophys. Space Phys.* 7:67

Brown, G. E., Weise, W. 1976. *Phys. Rep.* 27C:1

Buchler, J.-R., Coon, S. A. 1977. *Ap. J.* 212:807

Buchler, J.-R., Datta, B. 1979. *Phys. Rev.* C19:494

Canuto, V. 1974. *Ann. Rev. Astron. Astrophys.* 12:167

Canuto, V., Datta, B., Kalman, G. 1978. *Ap. J.* 221:274

Canuto, V., Ventura, J. 1977. *Fundam. Cosmic Phys.* 2:203

Carson, T. R. 1976. *Ann. Rev. Astron. Astrophys.* 14:95

Chanowitz, M., Siemens, P. 1977. *Phys. Lett.* 70B:175

Chapline, G. 1976. *Nature* 261:550

Chapline, G., Nauenberg, M. 1976. *Nature* 264:235

Chapline, G., Nauenberg, M. 1977a. *Phys. Rev.* D16:450

Chapline, G., Nauenberg, M. 1977b. *Ann. N.Y. Acad. Sci.* 302:191

Chen, H.-H., Ruderman, M. A., Sutherland, P. G. 1974. *Ap. J.* 191:473

Chin, S. A. 1977. *Ann. Phys. NY* 108:301

Chin, S. A. 1978. *Phys. Lett.* 78B:552

Chin, S. A., Walecka, J. D. 1974. *Phys. Lett.* 52B:24

Chiu, H.-Y., Salpeter, E. E. 1964. *Phys. Rev. Lett.* 12:413

Chodos, A., Jaffe, R. L., Johnson, K., Thorn, C. B., Weisskopf, V. F. 1974. *Phys. Rev.* D9:3471

Clark, J. W. 1978. *Prog. Part. Nucl. Phys.* Oxford: Pergamon. In press

Clark, J. W., Källman, C.-G., Yang, C.-H., Chakkalakal, D. A. 1976. *Phys. Lett.* 61B:331

Collins, J. C., Perry, M. J. 1975. *Phys. Rev. Lett.* 34:1353

Dautry, F., Nyman, E. 1979. *Nucl. Phys. A.* 319:323

Day, B. 1978. *Rev. Mod. Phys.* 50:495

Dicus, D. A. 1972. *Phys. Rev.* D6:941

Dicus, D. A., Kolb, E. W., Schramm, D. N., Tubbs, D. L. 1976. *Ap. J.* 210:481

Easson, I. 1976. *Nature* 263:486

Easson, I., Pethick, C. J. 1977. *Phys. Rev.* D16:275

Easson, I., Pethick, C. J. 1979. *Ap. J.* 227:995

El Eid, M. F., Hilf, E. R. 1977. *Astron. Astrophys.* 57:243

Elsner, R. F., Lamb, F. K. 1976. *Nature* 262:356

Ewart, G. M., Guyer, R. A., Greenstein, G. 1975. *Ap. J.* 202:238

Fechner, W. B., Joss, P. C. 1978. *Nature* 274:347

Festa, G. G., Ruderman, M. A. 1969. *Phys. Rev.* 180:1227

Flowers, E. G. 1973. *Ap. J.* 180:911

Flowers, E. G. 1974. *Ap. J.* 190:381

Flowers, E. G., Itoh, N. 1976. *Ap. J.* 206:218

Flowers, E. G., Itoh, N. 1979. To be published

Flowers, E. G., Lee, J.-F., Ruderman, M. A., Sutherland, P. G., Hillebrandt, W., Müller, E. 1977. *Ap. J.* 215:291

Flowers, E. G., Ruderman, M. A., Sutherland, P. G. 1976. *Ap. J.* 205:541

Flowers, E. G., Sutherland, P. G. 1977. *Astrophys. Space Sci.* 48:159

Flowers, E. G., Sutherland, P. G., Bond, J. R. 1975. *Phys. Rev.* D12:315

Freedman, B., McLerran, L. 1977. *Phys. Rev.* D16:1130, 1147, 1169

Freedman, B., McLerran, L. 1978. *Phys. Rev.* D17:1109

Freedman, D. Z., Schramm, D. N., Tubbs, D. L. 1977. *Ann. Rev. Nucl. Sci.* 27:167

Freedman, R. A. 1977. *Phys. Lett.* 71B:369

Friman, B. L., Maxwell, O. V. 1979. *Nordita preprint 78/15*

Friman, B. L., Nyman, E. 1978. *Nucl. Phys.* A302:365

Garstang, R. H. 1977. *Rep. Prog. Phys.* 40:105

Gerlach, U. 1968. *Phys. Rev.* 172:1325

Ghosh, P., Lamb, F. K. 1979. *Ap. J.* 232. In press

Glasser, M. L., Kaplan, J. I. 1975. *Ap. J.* 199:208

Green, A. M. 1976. *Rep. Prog. Phys.* 39:1109

Greenstein, G. 1975. *Ap. J.* 200:281

Greenstein, G. 1976. *Ap. J.* 208:836

Greenstein, G. 1977. *Ap. J.* 211:308

Greenstein, G., Margon, B., Bowyer, S., Lampton, M., Paresce, F., Stern, R., Gordon, K. 1977. *Astron. Astrophys.* 54:623

Hansen, J. P. 1973. *Phys. Rev.* A8:3096

Harding, D., Guyer, R. A., Greenstein, G. 1978. *Ap. J.* 222:991

Harrington, B. J., Yildiz, A. 1974. *Phys. Rev. Lett.* 33:324

Hartle, J. B. 1978. *Phys. Rep.* 46:201

442   BAYM & PETHICK

Hartle, J. B., Sawyer, R. F., Scalapino, D. J. 1975. *Ap. J.* 199:471
Hide, R. 1971. *Nature Phys. Sci.* 221:114
Hillebrandt, W., Müller, E. 1976. *Ap. J.* 207:589
Holinde, K., Machleidt, R. 1977. *Nucl. Phys.* A280:429
Itoh, N. 1970. *Prog. Theor. Phys.* 44:291
Itoh, N. 1975. *MNRAS* 173:1P
Itoh, N., Tsuneto, T. 1972. *Prog. Theor. Phys.* 48:1849
Jones, P. B. 1975. *Astrophys. Space Sci.* 33:215
Joss, P. 1978. *Ap. J. Lett.* 225:L123
Källman, C.-G. 1975. *Phys. Lett.* 55B:178
Källman, C.-G. 1978. *Phys. Lett.* 76B:377
Källman, C.-G., Moszkowski, S. A. 1975. *Phys. Lett.* 57B:183
Keister, B. D., Kisslinger, L. S. 1976. *Phys. Lett.* 64B:117
Kislinger, M. B., Morley, P. D. 1978. *Ap. J.* 219:1017
Krive, I. V., Chudnovskii, E. M. 1976. *Pisma Zh.E.T.F.* 23:531. Transl. 1976. *JETP Lett.* 23:485
Lamb, D. Q., Lamb, F. K. 1978. *Ap. J.* 220:291
Lamb, D. Q., Lamb, F. K., Pines, D. 1973. *Nature Phys. Sci.* 246:52
Lamb, D. Q., Lattimer, J. M., Pethick, C. J., Ravenhall, D. G. 1978. *Phys. Rev. Lett.* 41:1623
Lamb, F. K. 1975. *Ann. NY Acad. Sci.* 262:331
Lamb, F. K. 1977. *Ann. NY Acad. Sci.* 302:482
Lamb, F. K., Fabian, A. C., Pringle, J. E., Lamb, D. Q. 1977. *Ap. J.* 217:197
Lamb, F. K., Pines, D., Shaham, J. 1978. *Ap. J.* 224:969
Lattimer, J. M., Mackie, F., Ravenhall, D. G., Schramm, D. N. 1977. *Ap. J.* 213:225
Lattimer, J. M., Ravenhall, D. G. 1978. *Ap. J.* 223:314
Lattimer, J. M., Schramm, D. N. 1976. *Ap. J.* 210:549
Lee, T. D., Wick, G. C. 1974. *Phys. Rev.* D9:2291
Lenzen, R., Trümper, J. 1978. *Nature* 271:216
Lightman, A. P., Rees, M. J., Shapiro, S. L. 1978. *Proc. Varenna Summer School*, LXV, p. 786. Amsterdam: North-Holland
Lodenquai, J., Canuto, V., Ruderman, M. A., Tsuruta, S. 1974. *Ap. J.* 190:141
Mackie, F. 1976. PhD thesis. Univ. Ill., Urbana
Mackie, F., Baym, G. 1977. *Nucl. Phys.* A285:332
Malone, R. C. 1974. PhD thesis. Cornell Univ., Ithaca
Malone, R. C., Johnson, M. B., Bethe, H. A. 1975. *Ap. J.* 199:741

Manchester, R. N., Lyne, A. G., Taylor, J. H., Durdin, J. M., Large, M. I., Little, A. G. 1978. *MNRAS* 185:409
Manchester, R. N., Taylor, J. H. 1977. *Pulsars*. San Francisco: Freeman
Maxwell, O. 1979. *Ap. J.* In press
Maxwell, O., Brown, G. E., Campbell, D. K., Dashen, R. F., Manassah, J. T. 1977. *Ap. J.* 216:77
Maxwell, O., Weise, W. 1976. *Phys. Lett.* 62B:159
Mazurek, T. J. 1979. To be published
Mazurek, T. J., Lattimer, J. M., Brown, G. E. 1979. *Ap. J.* 229:713
Middleditch, J., Nelson, J. 1976. *Ap. J.* 208:567
Migdal, A. B. 1978. *Rev. Mod. Phys.* 50:107
Moszkowski, S. A., Källman, C.-G. 1977. *Nucl. Phys.* A287:495
Nadyozhin, D. K. 1977. *Astrophys. Space Sci.* 49:399; 51:283
Neatrour, J. 1979. To be published
Negele, J. W. 1976. *Comments Nucl. Part. Phys.* 5:149
Nyman, E., Rho, M. 1977. *Nucl. Phys.* A290:493
Pandharipande, V. R. 1971. *Nucl. Phys.* A178:123
Pandharipande, V. R., Pines, D., Smith, R. A. 1976. *Ap. J.* 208:550
Pandharipande, V. R., Smith, R. A. 1975a. *Nucl. Phys.* A237:507
Pandharipande, V. R., Smith, R. A. 1975b. *Phys. Lett.* 59B:15
Pandharipande, V. R., Wiringa, R. B., Day, B. D. 1975. *Phys. Lett.* 57B:205
Pines, D., Shaham, J., Ruderman, M. A. 1974. In *Physics of Dense Matter, Proc. IAU Symp. No. 53*, ed. C. J. Hansen, p. 189. Dordrecht: Reidel
Rappaport, S., Joss, P. 1977. *Nature* 266:123
Rappaport, S., Joss, P., McClintock, J. E. 1976. *Ap. J.* 206:L103
Ravenhall, D. G., Lattimer, J. M. 1979. To be published
Ruderman, M. A. 1972. *Ann. Rev. Astron. Astrophys.* 10:427
Ruderman, M. A. 1975. *Ann. NY Acad. Sci.* 262:164
Ruderman, M. A. 1976. *Ap. J.* 203:213
Ruderman, M. A., Sutherland, P. G. 1974. *Ap. J.* 190:137
Ruderman, M. A., Sutherland, P. G. 1975. *Ap. J.* 196:51
Sabbadini, A. G., Hartle, J. B. 1977. *Ann. Phys. NY* 104:95
Sato, K. 1975. *Prog. Theor. Phys.* 54:1325
Sarazin, C. L., Bahcall, J. N. 1977. *Ap. J. Lett.* 216:L67
Sauls, J. A., Serene, J. W. 1978. *Phys. Rev.* D17:1524
Sawyer, R. F., Soni, A. 1977. *Ap. J.* 216:73

Sawyer, R. F., Soni, A. 1979. *Ap. J.* 230:859
Shaham, J. 1977. *Ap. J.* 214:251
Shuryak, E. V. 1978. *Zh.E.T.F.* 74:408. Transl. 1978. *Sov. Phys. JETP* 47:212
Smith, F. G. 1976. *Pulsars.* NY: Cambridge Univ. Press
Smith, R. A., Pandharipande, V. R. 1976. *Nucl. Phys.* A256:327
Soyeur, M., Brown, G. E. 1979. *Nucl. Phys.* A. In press
Sunyayev, R. A. 1978. *Proc. Varenna Summer School,* LXV, p. 697. Amsterdam: North-Holland
Taam, R. E., Picklum, R. E. 1978. *Ap. J.* 224:210
Takatsuka, T., Tamiya, K., Tatsumi, T., Tamagaki, R. 1978. *Prog. Theor. Phys.* 59:1933
Taylor, J. H., Manchester, R. N. 1977. *Ann. Rev. Astron. Astrophys.* 15:19
Thorne, K. S., Zytkow, A. N. 1977. *Ap. J.* 212:832
Trümper, J., Pietsch, W., Reppin, C., Voges, W., Staubert, R., Kendziorra, E. 1978. *Ap. J. Lett.* 219:L105

Tsakadze, J. S., Tsakadze, S. J. 1975. *Usp. Fiz. Nauk* 115:503. Transl. 1975. *Sov. Phys. (Usp.)* 18:242
Tsuruta, S. 1974. In *Physics of Dense Matter, Proc. IAU Symp. No. 53,* ed. C. J. Hansen, p. 209. Dordrecht: Reidel
Tsuruta, S. 1975. *Astrophys. Space Sci.* 34:199
Tsuruta, S. 1979. *Phys. Rep.* In press
Tsuruta, S., Canuto, V., Lodenquai, J., Ruderman, M. A. 1972. *Ap. J.* 176:739
van Paradijs, J. A., Hammerschlag-Hensberge, G., van den Heuvel, E. P. J., Takens, R. J., Zuiderwijk, E. J. 1976. *Nature* 259:547
Van Riper, K. A., Bludman, S. A. 1977. *Ap. J.* 213:239
Walecka, J. D. 1974. *Ann. Phys.* 83:491
Walecka, J. D. 1975. *Phys. Lett.* 59B:109
Wolf, R. A. 1966. PhD thesis. Caltech, Pasadena
Wolff, R. S., Kestenbaum, H. L., Ku, W., Novick, R. 1975. *Ap. J.* 202:L22
Woosley, S. E., Taam, R. E. 1976. *Nature* 263:101

Ann. Rev. Astron. Astrophys. 1979. 17 : 445–75

STELLAR OCCULTATION ×2157
STUDIES OF THE SOLAR SYSTEM

James L. Elliot[1]

Laboratory for Planetary Studies, Cornell University, Ithaca, New York 14853

1 INTRODUCTION

Although the discovery of the Uranian rings is perhaps the most widely known result of occultation observations, stellar occultations by planets, satellites, and asteroids have been used to obtain precise dimensions of these bodies and to learn the temperature structure of planetary upper atmospheres. This review covers the principles, observational procedures, and results relating to occultations of stars by solar system bodies other than the moon. Discussion of similar phenomena, not treated in this review, are given in the following sources: (*a*) lunar occultations of stars (Barnes et al. 1978); (*b*) lunar occultations of solar system bodies (Elliot et al. 1975b, Vilas et al. 1977, Buarque 1978); (*c*) eclipses of satellites by Jupiter and Saturn and its rings (Smith et al. 1977, Barnard 1890, Harris 1978); (*d*) mutual occultations and eclipses by the Galilean satellites (Brinkmann & Millis 1973, Aksnes & Franklin 1976) and Saturn's satellites (Aksnes & Franklin 1978); (*e*) occultations of spacecraft radio signals by planets (Kloire & Woiceshyn 1976, Fjeldbo et al. 1976, Hunten & Veverka 1976); (*f*) occultations of planets by planets (Albers 1979).

The word *occultation* comes from the Latin word *occultare*, meaning "to hide," while *eclipse* comes directly from the Greek word ἔκλειψις, referring to a disappearance of the sun or moon. The difference between an occultation and an eclipse is that an occultation occurs when one body blocks our view of another, while an eclipse occurs when a body disappears because the light from the sun has been prevented from reaching it by a second body (see Figure 1 from Brinkmann & Millis 1973). Hence what is commonly referred to as a solar eclipse is, according to the above definition, an occultation of the sun by the moon. The term

[1] Present address: Department of Earth and Planetary Sciences, Massachusetts Institute of Technology, Cambridge, Massachusetts 02139.

0066-4146/79/0915-0445$01.00

immersion refers to the disappearance of a body when it is occulted and *emersion* refers to its reappearance.

Occultations of relatively bright stars by planets, satellites, and asteroids are rare phenomena: observations of less than fifty of these events have been recorded in the astronomical literature from historical times to the present (Sander 1973). Long ago Pannekoek (1904) recognized that when a planet with an atmosphere occults a star, the main process causing the starlight to dim is the progressively larger refraction by the planetary atmosphere as the starlight passes through deeper and deeper levels. From a recording of the intensity of the starlight vs time, the temperature of an isothermal atmosphere of known composition can be obtained. However, it was not until photoelectric photometry and high speed data recording techniques were employed that this information could be gleaned from observational data. The first photoelectric observations of an occultation were made in 1952 (σ Areitis by Jupiter) by Baum & Code (1953), but nearly all useful information about solar system bodies learned from occultations has been obtained during the past decade.

What can be learned about the occulting body will depend on the circumstances of a given occultation and how it is observed. The following information about the occulting body is potentially obtainable from the data: (*a*) an accurate position relative to the occulted star, (*b*) its radius and/or apparent elliptical figure, (*c*) temperature, pressure, and number density profiles of its upper atmosphere (if certain assumptions are satisfied—see Section 5), (*d*) information about the composition of its atmosphere, (*e*) the extinction of its atmosphere, (*f*) the optical depth structure and precise dimensions of any ring material, and (*g*) the angular diameter of the occulted star.

2   PHYSICAL PROCESSES INVOLVED IN OCCULTATIONS

The physical processes responsible for dimming the starlight during an occultation are (*a*) extinction by ring material, (*b*) differential refraction by a planetary atmosphere, (*c*) extinction by the planetary atmosphere, and (*d*) Fresnel diffraction by sharp edges, such as the limbs of satellites and asteroids and the narrow rings of Uranus. A schematic light curve that would be observed for a central occultation by a ringed planet is shown in Figure 1. The lower portion of the figure shows the planet, its atmosphere and rings, while the upper portion of the figure shows the light intensity as a function of the apparent position of the star, as it appears to move behind the planet (due to the relative motion of the Earth and the planet). First, extinction of the starlight by the ring material is encountered. Analysis of this portion of the light curve yields the optical depth of the rings vs distance from the planet. Then differential refraction by the atmosphere is encountered, which causes the starlight to dim, followed by an abrupt increase in intensity near the center of the shadow. Then the preceding processes are repeated in reverse order.

The process of differential refraction is illustrated by the ray diagram in Figure 2. Starlight is incident from the left and is refracted by an isothermal atmosphere having a scale height, H, shown in the figure. The scale height is the e-folding distance for the density vs altitude. Because of this exponential density gradient in the atmosphere, the starlight is refracted by an ever increasing angle as deeper regions of the atmosphere are probed. As the Earth travels through the pattern of refracted starlight, the intensity decreases as shown by the graph on the right-hand portion of Figure 2. When the light intensity is exactly half of the unocculted

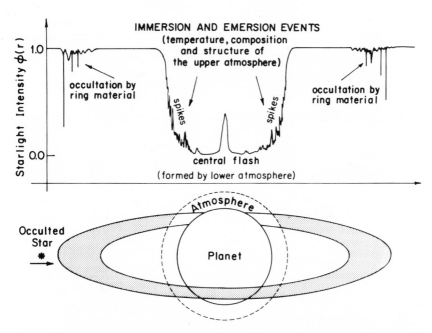

Figure 1   Schematic light curve for a central occultation by a ringed planet. The upper portion of the figure shows the light intensity of the star that would be observed from Earth as a function of the position of the star behind the planet. The first attenuation of starlight is due to the extinction by the ring material. The occultation of the star by the atmosphere occurs through the process of differential refraction (see Figure 2), with irregular variations (spikes) caused by atmospheric structure that deviates from being isothermal. The central flash is observed when the star is directly behind the center of the planet and can yield information about the extinction of the lower atmosphere.

intensity, the starlight has been refracted through an angle $\theta = H/D$ by the atmosphere, where D is the distance between the planet and the Earth. The original treatment of differential refraction by a planetary atmosphere was given by Pannekoek (1904) and Fabry (1929). Baum & Code (1953) have derived the equation of the light intensity vs time that is shown on the right-hand side of Figure 2. These authors show that for the large distances involved in planetary occultations, the differential refraction exceeds the effect of Rayleigh scattering by a factor of 10^5. Occultations by an atmosphere with a temperature gradient were discussed by Goldsmith (1963).

Extinction by the planetary atmosphere becomes an important process only when the observer is near the atmosphere (i.e. the setting sun) or near the center of the shadow of the occulting planet. In the latter case the rays converge from the entire circumference of the planetary limb,

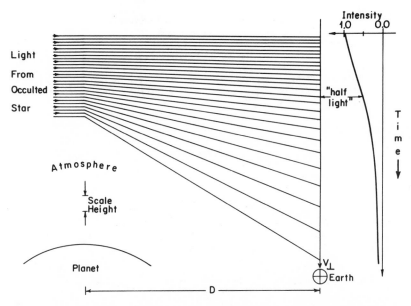

Figure 2 Stellar occultation geometry and optics. In this ray diagram the light from the occulted star is incident from the left and is refracted by the refractivity (density) gradient in the atmosphere. Since the density gradient becomes progressively greater, deep into the atmosphere, the rays are spread by an ever increasing amount by the process of differential refraction. As the Earth travels through the pattern of refracted starlight, the stellar intensity decreases as shown by the graph at the right-hand side of the figure. The equation describing the starlight vs time for the isothermal atmosphere illustrated here is given by Baum & Code (1953).

producing a large increase in intensity known as the "central flash" (Elliot et al. 1977e). Extinction by the atmosphere reduces the amplitude of the central flash (see Section 5.3) from that expected for a clear atmosphere.

Occultations by real planetary atmospheres are usually not smooth like the ideal isothermal curve of Figure 2, but contain abrupt intensity variations known as "spikes," shown in Figure 3 (Freeman & Lyngå 1970, Veverka et al. 1974a, Young 1976, Elliot & Veverka 1976). These abrupt increases in intensity may be produced by density variations of only a few percent and may be caused by density waves, layers, or turbulence

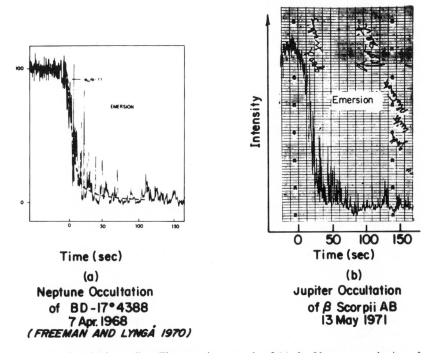

Time (sec)

(a)

**Neptune Occultation
of BD-17°4388
7 Apr. 1968**
(FREEMAN AND LYNGÅ 1970)

Time (sec)

(b)

**Jupiter Occultation
of β Scorpii AB
13 May 1971**

Figure 3   Occultation spikes. The emersion records of (*a*) the Neptune occultation of BD-17°4388 and (*b*) Jupiter occultation of β Scorpii record are the output of the monitor chart recorder used by Veverka et al. (1974a). The two chart records shown here have nearly the same time resolution and nearly equal time scales. The curves bear a striking resemblance. One reason is that the ratio of the scale height H of the atmosphere to the occultation velocity $v_\perp$ is nearly equal for the two occultations ($H/v_\perp \sim 3$ sec for each). The spikes in the two curves are of similar amplitude and number, suggesting that the mechanism causing the spikes is at work in the atmospheres of the two planets (after Elliot & Veverka 1976).

in the atmosphere of the occulting planet. This topic is discussed further in Sections 5.1 and 5.2.

For bodies without atmospheres, the occultation process is analogous to lunar occultations by stars (Nather & Evans 1970), except for differences in scale. The Fresnel scale is larger because of the increased distance of the occulting body. However, the angle subtended by the Fresnel scale is over an order of magnitude smaller, which allows the angular diameter of much smaller stars to be measured. These scales for all the planets are given in Table 1.

For bodies with very thin atmospheres, both differential refraction by the atmosphere and Fresnel diffraction by the limb must be considered in the analysis (French & Gierasch 1976).

3   PREDICTION AND OBSERVATIONAL TECHNIQUES

Before an occultation can be observed, one must know the path of the shadow cast by the occulting body. Predicting the shadow track is a two-step process. First, the ephemerides of the planets, satellites, and asteroids are compared with the positions of stars in order to determine what occultations might occur. Next, when a potential occultation is found, precise astrometric measurements of the relative position of the star and occulting body are required to establish the final prediction for the shadow track.

One method of searching for potential occultations, practiced by Taylor (1962) since 1952, is to use a computer to compare the ephemerides with the star positions in catalogues, such as the SAO and AGK3. One limitation of the catalogue search technique is that the current epoch positions of the stars can be in error due to erroneous proper motions if the catalogue is based on older data. Another limitation of the catalogue search technique is that useful occultations of stars fainter than the catalogue limit are overlooked. This is an important problem for the fainter objects, such as Uranus, Neptune, Pluto, satellites, and asteroids.

Another method of searching for potential occultations is accomplished by comparing the ephemerides of planets, satellites, and asteroids with the positions of stars on current epoch plates. Hence, proper motions are not a problem and the magnitude limit of the search can be made as faint as desired. This plate-scanning technique was used by Klemola & Marsden (1977) to predict occultations by Uranus for the period 1977–1980 and by Klemola et al. (1978) to predict occultations by Neptune for the same period. Shelus & Benedict (1978) also used this technique

Table 1 Length, angle, and time scales

Body	Mean opposition or conjunction distance (au)	Length subtended by 0.1 arc sec (km)	Fresnel scale, $\sqrt{(\lambda D)}$ (km)	Angle subtended by Fresnel scale, $\sqrt{\left(\dfrac{\lambda}{D}\right)}$ (milli-arc sec)	Time scale, $\Delta t = \dfrac{\sqrt{(\lambda D)}}{20 \text{ km/sec}}$ (sec)
Moon	0.0026	0.2	0.015	7.8	0.015[a]
Mercury	0.61	44	0.22	0.51	0.011
Venus	0.28	20	0.15	0.75	0.008
Mars	0.52	38	0.21	0.55	0.010
Ceres	1.8	131	0.38	0.29	0.019
Jupiter	4.2	310	0.59	0.19	0.029
Saturn	8.5	620	0.84	0.14	0.042
Uranus	18	1300	1.2	0.093	0.061
Neptune	29	2100	1.6	0.073	0.077
Pluto	38	2800	1.8	0.064	0.16

[a] This timescale was calculated for a lunar shadow velocity of 1.0 km sec^{-1} in place of the 20 km sec^{-1} used for the other entries in this column.

to search for potential occultations by outer solar system bodies through the early 1980s.

The next step in the process of predicting an occultation is to obtain an accurate relative position of the star and the occulting body. Usually the refinement of positions is done with photographic astrometry (Wasserman et al. 1979), however, transit circle observations can also be used for brighter objects. The accuracy of predictions based on photographic astrometry improves when images of the two bodies can be obtained on the same plate, and further improvement is realized as the distance between the two bodies decreases. The accuracy achievable at present appears to be about $\pm 0''.1$ in the relative position of the occulting body and the star. The linear distances corresponding to an angle of $0''.1$ are given in Table 1 for solar system bodies at various distances from the Earth. From this table we see that the probable prediction error nearly equals or exceeds the radii of the smaller satellites of Jupiter, all satellites (except Titan) further away, the asteroids, and Pluto. Hence for these bodies the reliability of the occultation predictions is crucial.

A complete set of observations of an occultation would consist of a two-dimensional photometric record of the entire occultation shadow pattern, but what can be obtained from a single observatory is only a single "line scan" across the shadow as it moves over the observer due to the relative motion of the Earth and the occulting body (see Figure 1). The two-dimensional intensity pattern can be partially reconstructed if line scans are obtained from several observatories, separated in the direction perpendicular to the motion of the shadow. The data from different observatories can be readily compared with an accuracy of a few milliseconds through the use of radio broadcast time signals of coordinated universal time (UTC). The UTC scale can be converted to the UT1 and ET timescales required for data reduction (Mulholland 1972, Duncombe & Seidelmann 1977).

Photometry of occultations has been done by visual, photographic, and photoelectric methods. For abrupt occultations of bright stars, visual timing data can be of value. However, discordant observations are common, and in many cases it is difficult to know which are the valid observations. Extensive visual observations were made of the occultation of Regulus by Venus on July 7, 1959 (see de Vaucouleurs & Menzel 1960, Taylor 1963 and references therein) and for several occultations by asteroids (O'Leary et al. 1976, Taylor & Dunham 1978, Bowell et al. 1978). Photographic observations of the β Scorpii occultation by Jupiter were reported by Fairall (1972) and Larson (1972).

Photoelectric photometry is of course the highly preferred technique because it is far superior in photometric and timing accuracy. The goal of occultation photometry is to record the intensity of the occulted star as a function of time with the maximum possible signal-to-noise ratio. The signal is the light from the occulted star, while the noise in the light curve comes from several sources: (a) photon noise from the star and the background; (b) scintillation noise from the star and the background; (c) variations in the background light caused by seeing and/or poor telescope tracking. The time resolution required for recording the photometric data depends on several factors, but is approximately the time during which the occultation shadow moves one Fresnel scale (Table 1).

In order to obtain high quality observations of an occultation, several factors must be considered: (a) the dominant source of background light—usually this is the occulting body itself, but it could be moonlight or twilight; (b) the relative spectra of the background light and the occulted star (Elliot et al. 1977g, Nicholson et al. 1978); (c) the size and location of telescopes available within the path of the occultation shadow; (d) the altitude above the horizon of the occulted star at each potential observing site; and (e) the expected levels of photon and scintillation noise (Elliot 1977). Several techniques for reducing the contribution of background light from a planet of large angular diameter have been reviewed by Hunten & Veverka (1976). Simultaneous observations at different wavelengths have proven valuable for several reasons: (a) obtaining information about the composition of an atmosphere (Section 5.2); (b) allowing correction for a variable amount of background light (Elliot et al. 1975b); and (c) providing redundant information in case of equipment malfunction.

Another important factor for occultation observations has been the Kuiper Airborne Observatory (Cameron 1976), which offers the advantages of optimum location, reduced scintillation noise, and clear skies (Elliot et al. 1976a, 1977a).

4 RADII

From a sufficient number of immersion and emersion timings of a stellar occultation, the radius and ellipticity of the occulting body can be accurately determined. For bodies without atmospheres, the occultation time is defined to be that of geometrical occultation (the 1/4 intensity point on the Fresnel diffraction pattern; Nather & Evans 1970). For bodies with atmospheres, the occultation time usually used is that of "1/2 light," determined from fitting a model occultation curve to the data (see Figure 2, and Baum & Code 1953). The data reduction procedures followed by most authors closely resemble that derived by Bessel and described

by Smart (1962). In this method the coordinates of the observed occultation points are projected onto a common plane (*fundamental plane*) that
passes through the center of the Earth and lies perpendicular to the
line between the centers of the Earth and the occulting body.

After the occultation points are projected onto the fundamental plane,
the figure of the occulting body is obtained by fitting a model limb
(or atmospheric) profile. The simplest of these is a circle, which requires
three occultation points; an ellipse requires a minimum of five points.
An example of an elliptical figure fitted to the 14 points (seven chords)
observed for the occultation of SAO 085009 by Pallas is shown in Figure 4
(Wasserman et al. 1979). Unless nearly complete coverage of the limb
profile can be obtained, the ultimate limitation of the occultation method
appears to be whatever departures exist of the actual limb profile from
an ellipse.

It is also possible to obtain the diameter of the occulting body, for an

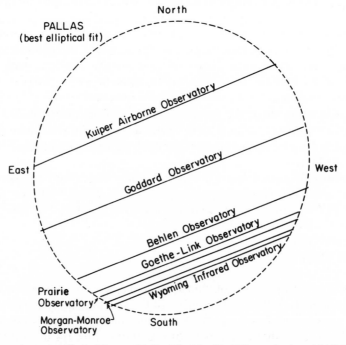

Figure 4   Apparent elliptical figure of Pallas. The May 29, 1978, occultation of SAO 085009
by Pallas was successfully observed at seven stations, illustrated by the chords in the figure.
A mean diameter of 538 ± 12 km for Pallas was determined from the fitted ellipse and
previously obtained photometric data, which was used to estimate the dimension of Pallas
perpendicular to the line of sight at the time of the occultation (after Wasserman et al.
1979).

assumed circular figure, from a single occultation chord combined with (a) precise astrometric measurements near the time of occultation (Bixby & Van Flandern 1969) or (b) the scale of the Fresnel diffraction fringes, if the occulting body has no significant atmosphere.

However, the precision of these single chord methods is much less than can be obtained with two or more chords. In order to obtain the dimension of the occulting body along the line of sight, about which the occultation profile gives no information, we must assume that the occulting body has an ellipsoidal figure and know the spatial orientation of the ellipsoid from another method. If two of the axes of the ellipsoid are equal, as is the case for a rotating planet, we can determine its polar and equatorial radii from the occultation figure and the known direction of the polar axis, unless the occulting body is seen nearly pole on. If the occulting body has three unequal radii, additional information is needed about the ratios of the radii, which may be obtained from a theoretical model (O'Leary & Van Flandern 1972) or other observational data (Wasserman et al. 1979).

The occultation radii that have been obtained for planets are given in Table 2a and for asteroids and satellites in Table 2b. Each radius has been determined from data obtained during a single occultation, so that multiple entries for a given body represent analyses by different authors, usually with different subsets of data. For planets, the radii and ellipticities refer to the number density level of 1/2 light, as given in Table 2a. The 1/2 light level is well-defined by the data and approximates an isonumber density level in the atmosphere. The number density of 1/2 light was calculated from a corrected version[2] of Equation (3) in Elliot et al. (1977g). The uncertainties in the number density levels arise primarily from imprecise knowledge of the atmospheric composition (v_{STP}) and the scale height (H). If our assumptions about the compositions are correct (see footnote a of Table 4), then the number densities typically have uncertainties of $\pm 20\%$, but could be as large as $\pm 50\%$ in the case of Venus (Hunten & McElroy 1968). Corrections to the 1/2 light radii for refraction and the deflection of light have been added (as noted) if they were not included in the original analysis.

The remaining columns of Table 2 give the number of different points

---

[2] Equation (3) of Elliot et al. (1977g) is incorrect and should read

$$n = (L/v_{STP})H^{3/2}D^{-1}(2\pi R_p)^{-1/2},$$

where n is the number density (molecules cm^{-3}) at 1/2 light; L is Loschmidt's number (molecules cm^{-3}); v_{STP} is the refractivity of the atmosphere at STP; H (km) is the scale height (see Table 3), D (km) is the Earth-planet distance, and R_p (km) is the radius of the planet at the 1/2 light level.

Table 2a   Stellar occultation radii of planets

Body	Occulted star(s) and date	Equatorial radius[a] (R_{eq}, km)	Ellipticity $\dfrac{R_{eq} - R_p}{R_{eq}}$	Number density level[b] (molecules cm^{-3})	Number of observations at different points on the limb		Fraction of diameter between extreme chords	References
					Photo-electric	Visual		
Venus	Regulus, 7 July 1959	6,169 ± 2	0[c]	6 × 10^{13}	2	39	0.58	de Vaucouleurs & Menzel 1960
Mars	ε Gem 8 April 1976	3,472[d] ± 2	0.013 ± 0.001	4 × 10^{13}	6	0	0.21	Taylor 1976
Jupiter	β Sco A and C 13 May 1971	71,880 ± 30	0.060 ± 0.001	6 × 10^{13}	7	0	0.36	Hubbard & Van Flandern 1972
		71,802 ± 55	0.060[e]	6 × 10^{13}	4	0	0.31	Lecacheux et al. 1973
		71,900 ± 20	0.0595 ± 0.0015	6 × 10^{13}	8	0	0.36	Taylor 1974
Uranus	SAO 158687 10 March 1977	26,200 ± 150	0.035 ± 0.015	6 × 10^{13}	4	0	0.01	Elliot et al. 1979
Neptune	BD-17°4399 7 April 1968	25,225 ± 30	0.021 ± 0.004	4 × 10^{13}	6	0	0.13	Kovalevsky & Link 1969
		25,275[f]	0.020[e]	4 × 10^{13}	13	0	0.14	Taylor 1970

[a] All equatorial radii refer to the level of 1/2 light.
[b] The number density levels for 1/2 light (if not given in the original analysis) were calculated as described in the text for the assumed compositions given in footnote a of Table 3.
[c] This analysis does not describe any attempt to fit an elliptical figure to the data, although in a later publication Taylor (1974) mentions that "there was no evidence of any oblateness."
[d] A refraction correction of 8 km was added to the result of the original analysis.
[e] Fixed value.
[f] A refraction correction of 50 km and a correction of 55 km to account for the relativistic deflection of light were added to the result of the original analysis.

Table 2b Stellar occultation radii of satellites and asteroids

Body	Occulted star and date	Adopted mean radius (km)	Radius from circular solution (km)	Tri-axial radii (km) Equatorial a	b	Polar c	Number of observations at different points on the limb Photo-electric	Visual	Fraction of diameter between extreme chords	References
Io	β Sco C 14 May 1971	1818 ± 5	1829.5 ± 2.0	1829	1815	1810	6	0	0.24	O'Leary & Van Flandern 1972
Ganymede	SAO 186800 7 June 1972	2635 ± 25	2635^{+15}_{-100}				4	0	0.37	Taylor 1972 Carlson et al. 1973
2 Pallas	SAO 085009 29 May 1978	269 ± 6	273 ± 3	279 ± 3	263 ± 5	266 ± 15	14	0	0.56	Wasserman et al. 1979
6 Hebe	γ Ceti 5 March 1977	93[a]	93 ± 5	98 ± 3		85 ± 3	0	4	0.37	Taylor & Dunham 1978
433 Eros	κ Gem 24 January 1975	~5[b]	11.6 ± 0.3	8.7 ± 0.3		3.4 ± 0.2	0	16	~1.0	O'Leary et al. 1976
532 Herculina	SAO 120774 7 June 1978	109 ± 8	121.5 ± 0.7				2	4	0.14	Bowell et al. 1978

[a] The uncertainty is difficult to estimate because of discrepancies of the observations with a circular solution, yet an insufficient number of distributed observations to determine an elliptical or irregular figure.

[b] This value, preferred by O'Leary et al. (1976), was obtained from a "graphical solution"; see that paper for details.

on the limb for which independent photoelectric and visual timing observations were obtained, and the fraction of the occulting body's diameter
lying between the outermost chords. As can be appreciated from Figure
4, more reliable results can be expected for solutions based on a large
number of photoelectric observations that are well distributed over the
occulting body.

Several comments should be made about some of the entries in Table 2.
These are necessarily brief, but can be pursued in greater detail in the
cited literature. Two other solutions exist for the radius of Venus
(Martynov 1961, Taylor 1963) and one for the radius of Neptune
(Freeman & Lyngå 1970) that were not included in Table 2 because they
do not refer to the 1/2 light level. For Jupiter, the astrometric solution
depends not only on the occultation times, but also on a measurement
of the relative position of β Sco A and C. The radius and ellipticity of
Uranus were obtained under the assumption that the apparent center of
Uranus coincides with the apparent center of its ring system; hence these
values should be more reliable than would normally be expected for two
chords spaced by only 0.01 of the planet's diameter.

Ganymede was reported to have a thin atmosphere on the basis of the
occultation data (Carlson et al. 1973), but the data appear equivocal on
this point. Data from Voyager should soon settle the issue. The adopted
error for the radius of Ganymede follows the assessment of Morrison et
al. (1977). Bowell et al. (1978) report that 532 Herculina may possess a
companion (about 46 km in diameter), based on two mutually corroborat-
ing observations (one photoelectric and one visual) of its apparent
occultation of SAO 120774, which occurred about two minutes before
Herculina itself occulted the star. There have been other reports of brief
disappearances of stars near asteroids (Binzel & Van Flandern 1979),
indicating a preponderance of small satellites surrounding asteroids. At
the time of this writing, the evidence for the existence of such satellites,
including the possible companion of Herculina, leaves some unconvinced.

5 ATMOSPHERES

From an occultation by a planet with an atmosphere we can, if certain
conditions are satisfied, (a) obtain temperature, pressure, and number
density profiles over several scale heights of the upper atmosphere, (b)
obtain information about the composition of the atmosphere, and (c)
learn about the extinction.

5.1 Scale Heights and Temperatures

The scale height, H, of an atmosphere is the quantity obtained from the
occultation light curve, and the temperature, T, is then determined through

the relation $T = \bar{\mu}gH/R$, where $\bar{\mu}$ is the mean molecular weight of the atmosphere, g is the gravitational acceleration, and R is the universal gas constant. The scale height can be determined either through numerical inversion of the light curve (which gives H as a function of altitude—H would be constant if the atmosphere is isothermal) or by fitting a model to the light curve (which gives a single "average" value for H).

Numerical inversion yields temperature, pressure, and number density profiles and is a valid procedure for analysis if two main assumptions are satisfied. First, the density gradients in the atmosphere must be parallel to the local gravity gradient, and second, the amount of "ray crossing" must not be severe. These assumptions are discussed by Elliot & Veverka (1976). Examples of processes in planetary atmospheres that would satisfy these assumptions are tides (Zurek 1976), photochemical layering, and vertically propagating density waves (French & Gierasch 1974).

The inversion technique was developed through its application to Neptune occultation data by Kovalevsky & Link (1969). Wasserman & Veverka (1973b) showed that the presence of noise in the occultation light curve would not invalidate the procedure, and French et al. (1977) presented several improvements to the procedure, including a quantitative treatment of the effects of light curve noise on the profiles obtained through inversion. Further discussion of the application of the inversion method is given by Hunten & Veverka (1976) and a comparison of results obtained by inversion and fitting a model light curve to several sets of data is given by French & Elliot (1979).

Examples of temperature profiles obtained by the inversion method are given in Figure 5, where the profiles extend about three scale heights. The larger errors at the top of the profiles are due to uncertainties in the initial conditions for the inversion calculation, and the termination of the profile at lower altitudes occurs at a level where the flux from the star has dropped to a few percent of its unocculted value.

The gravity gradient assumption, required for numerical inversion of the light curves, would not be correct if the spikes (see Figure 3) were due to turbulence in planetary atmospheres, as proposed by Young (1976). The turbulence model was developed further by Hubbard & Jokipii (1977), Jokipii & Hubbard (1977), the Texas-Arizona Occultation Group (1977), Hubbard et al. (1978), and Hubbard (1979); somewhat different conclusions on what effects turbulence would produce in occultation light curves were reached by Eshleman & Haugstad (1978). For the Jovian atmosphere, Elliot & Veverka (1976) feel that inertia-gravity waves (French & Gierasch 1974) or layering by photochemical or other processes (Dütsch 1971) to be as acceptable as the turbulence model, and for Mars, French & Elliot (1979) feel that the data indicate the presence

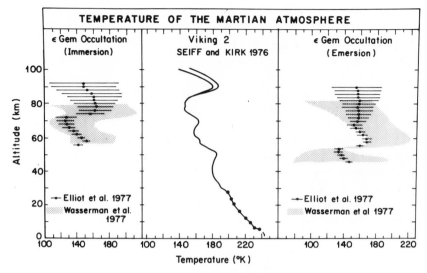

Figure 5   Temperature profiles of the Martian atmosphere. Temperature profiles obtained for immersion and emersion of the ε Gem occultation are compared with the profile obtained by the Viking 2 entry experiments. Although different regions of the Martian atmosphere were probed at different times, similar mean temperatures and wavelike temperature variations are observed. The wavelike temperature variations are probably due to tides in the Martian atmosphere (Zurek 1976, Seiff & Kirk 1977, Elliot et al. 1977f).

of tides (Zurek 1976, Elliot et al. 1977e,f), rather than turbulence. More theoretical work and/or definitive data will be needed to determine what dynamical processes are occurring in the region of planetary atmospheres probed by stellar occultations.

A summary of the scale heights and temperatures obtained from stellar occultations is given in Table 3. These results apply to an imprecisely defined region of the atmosphere, which turns out to be approximately the same number density range for all planets. Different authors have employed different methods for obtaining mean scale heights: (*a*) fitting a model isothermal light curve (e.g. Hubbard et al. 1972, Texas-Arizona Occultation Group 1977); (*b*) a linear fit to the logarithm of the refractivity (number density) vs altitude, obtained from numerical inversion (e.g. Veverka et al. 1974a); and (*c*) a mean over a specified altitude range of the scale height obtained from numerical inversion (e.g. Elliot et al. 1977e, French & Elliot 1979). The errors in the scale heights are not known precisely, due to a variety of factors (nonisothermal features, different analysis methods, noise in the data); they are probably ±10%, but could be as large as ±20%. The temperature variations given in Table 3 refer to analyses by the inversion method and would not be correct if the turbulence model applied to the data.

Table 3 Scale heights and temperatures of upper atmospheres

Planet	Number density range[a] (molecules cm^{-3})	Approximate altitude above cloud (c) or solid surface (s) (km)	Mean scale height (km)	Mean temperature[a] (K)	Temperature variations[a] (K, peak-to-peak)	Proposed processes to account for observed nonisothermal effects
Venus	10^{14}–10^{15}	100(s)[d]	(6.8)[e,b]	(308)[e,b]	—	—
Mars	10^{14}–10^{15}	70(s)[f]	7.8[g]	146 ± 10[g]	35[f,g]	tides, excited by diurnal heating of the surface[g,h]; turbulence[i,j]
Jupiter	10^{14}–10^{15}	~300(c)	25[k]	170[k]	2–10[l]	inertia-gravity waves[l,m]; layering by photochemical processes; turbulence[n,o]
Uranus	10^{14}–10^{15}	~500(c)	50[p]	100[p]	10–30[p,q]	inertia-gravity waves photochemical layering[q]
Neptune	10^{14}–10^{15}	~500(c)	50[r–v]	140[r–v]	5–10[t]	inertia-gravity waves; turbulence[n]

[a] The number density range and the mean temperature have been calculated for the following assumed atmospheric compositions: for Venus, pure CO_2; for Mars, the composition measured by the Viking entry experiments (Owen & Biemann 1976); for Jupiter, Uranus, and Neptune, the solar abundance for H_2 and $He([He]/[H_2] \simeq 0.1)$.
[b] See text for discussion.
[c] Definite evidence for other than an isothermal atmosphere is not readily apparent from either the light curve (de Vaucouleurs & Menzel 1960) or the temperature profile (Veverka & Wasserman 1974).

References
[d] Veverka & Wasserman 1974
[e] de Vaucouleurs & Menzel 1960
[f] Elliot et al. 1977e
[g] French & Elliot 1979
[h] Elliot et al. 1977f

[i] Texas-Arizona Occultation Group 1977
[j] Hubbard 1979
[k] Hunten & Veverka 1976
[l] Veverka et al. 1974a

[m] French & Gierasch 1974
[n] Young 1976
[o] Jokipii & Hubbard 1977
[p] Elliot & Dunham 1979
[q] Dunham et al. 1979

[r] Osawa et al. 1968
[s] Kovalevsky & Link 1969
[t] Veverka et al. 1974b
[u] Rages et al. 1974
[v] Wallace 1975

The results entered in Table 3 represent the consensus of the literature cited, and additional information, as well as discrepant results, are given in following discussions for the individual planets. The discrepant results are most likely due to "low-quality observations combined with in-adequate analysis," to borrow the phrasing of Hunten & Veverka (1976).

VENUS   The occultation of Regulus by Venus on July 7, 1959, was widely observed visually, but only two photoelectric traces were obtained (de Vaucouleurs & Menzel 1960). By fitting a model curve to the data, de Vaucouleurs & Menzel (1960) find a scale height of 6.8 ± 0.2 km, which corresponds to a temperature of 308 K for the pure CO_2 atmosphere. However, Hunten & McElroy (1968) argue that such a high temperature is unlikely because a model atmosphere, based on the Venus 4 results (McElroy 1968), gives a temperature of 200–210 K at the altitude probed by the occultation. Furthermore, they feel that a scale height of 4.4 km is consistent with the occultation data. A temperature profile was obtained with numerical inversion by Veverka & Wasserman (1974). They confirm that the data indicate the high temperatures found by the original analysis, but agree with Hunten & McElroy (1968) that such high temperatures cannot be correct.

MARS   The occultation of ε Geminorum by Mars occurred just three months before the Viking 1 entry experiments and provided the first opportunity to compare occultation results with measurements in situ. This event was widely observed in the eastern half of the United States; references to analyzed data are: Texas-Arizona Occultation Group (1977), Wasserman et al. (1977), de Vaucouleurs et al. (1976), Elliot et al. (1977e,f), French et al. (1978), Groth et al. (1978), and Liller et al. (1978). Discussions of these results have been given by French & Elliot (1979) and Hubbard (1978, 1979).

The mean temperature of 146 ± 10 K (French & Elliot 1979) agrees well with the mean temperatures of 144 K (Nier & McElroy 1977) and 142 K (Seiff & Kirk 1977) obtained by Vikings 1 and 2 over the same altitude range. The mean temperature of 165 $K \pm 25$ K obtained by Hubbard (1978) is somewhat higher, but also consistent with the Viking results. The Viking temperature profiles also show evidence for tides in the upper atmosphere (see Figure 5 and French & Elliot 1979), but as mentioned previously the nonisothermal effects in the occultation have been alternatively interpreted as evidence for turbulence (Hubbard 1979). The relative merits of the tidal and turbulence interpretations are discussed by French & Elliot (1979) and Hubbard (1979).

JUPITER   Jupiter occulted σ Arietis in 1952, and from their observations of this event Baum & Code (1953) found a scale height of 8 km for

Jupiter's atmosphere. This result proved too low by a factor of three, as shown by results of the Jovian occultations of β Sco A and C in 1971. Observations were made at six sites: Bhattacharyya (1972), Freeman & Stokes (1972), Hubbard et al. (1972), Larson (1972), Vapillon et al. (1973), and Veverka et al. (1974a). The mean scale height obtained was about 25 km, with some discordant results (Hunten & Veverka 1976).

The three groups that pursued analyses of their data most extensively are the Texas Group (Hubbard et al. 1972), the Meudon Group (Vapillon et al. 1973), and the Cornell-Harvard Group (Veverka et al. 1974a). These groups generally agree on the scale height at the 5×10^{14} cm^{-3} number density level, but several differences should be noted. The Meudon Group finds a large positive temperature gradient at number density levels above 10^{14} cm^{-3}; this result is not supported by the results of the other two groups. The high temperature ($T > 300$ K) thermosphere reported by the Texas Group is apparently an artifact of their inversion procedure (Combes et al. 1975). The Texas Group finds a variation of scale height with Jovian latitude, although this result is not supported by the Cornell-Harvard results (Elliot et al. 1975c). Reviews of the β Scorpii results have been presented by Hunten & Veverka (1976), Jokipii & Hubbard (1977), and Combes et al. (1975).

URANUS   From the occultation of SAO 158687 by Uranus on March 10, 1977, Elliot & Dunham (1979) obtained a scale height of 50 km. High time resolution data is "spiky," resembling that for Jupiter and Neptune.

NEPTUNE   The occultation of BD-17°4388 was observed from several observatories in Australia and Japan. All analyses obtained scale heights near 50 km, except for a value of 28.9 ± 2.6 km obtained by Freeman & Lyngå (1970).

5.2 Composition

Since the refractivity of most gases increases at ultraviolet wavelengths, the starlight becomes dispersed in color during an occultation. The effect becomes more pronounced deeper in the atmosphere, since the dispersion is proportional to the amount of refraction. The dispersion effect causing time delays in the spikes is illustrated in Figure 6. Brinkmann (1971) pointed out that this effect could be used to obtain the composition of a hydrogen-helium atmosphere, since the light curve spikes at different wavelengths could be used to accurately measure the relative dispersion. The original formulation of the method was improved upon by Wasserman & Veverka (1973a) and Elliot et al. (1974).

This method for obtaining the composition is limited to atmospheres consisting of two major constituents whose identities are known. The method has been applied twice. For Jupiter, Elliot et al. (1974) found the helium fraction by number $f(He) = 0.16^{+0.19}_{-0.16}$. The main reason for the large error is the low refractivity of helium; however, the accuracy is comparable to other remote sensing methods for determining the helium abundance in the Jovian atmosphere (Hunten & Veverka 1976). For Mars, Elliot et al. (1977e) found the argon fraction (by number) to be $f(Ar) = 0.10^{+0.20}_{-0.10}$. This result was obtained before Viking results (Owen & Biemann 1976) were available and proved to be reliable to within its stated error. The occultation method for measuring composition becomes more accurate if occultation observations can be made in the far uv (outside the Earth's atmosphere) to obtain more leverage on the refractivity. It may prove of value for Saturn, Titan, Uranus, Neptune, and Pluto if appropriate occultations occur before in situ measurements are made.

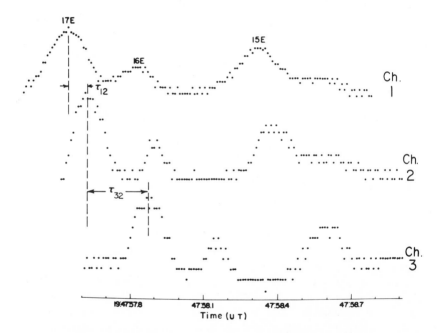

Figure 6 High time resolution records of spikes observed during the β Scorpii occultation by Jupiter. The light curves were recorded at 3535Å (Ch. 1), 3934Å (Ch. 2), and 6201Å (Ch. 3). The time delays (τ_{12} and τ_{32}) between the spike arrival in the different wavelength channels are caused by the variation of refractivity of the Jovian atmosphere with wavelength. The time between successive data points is 0.01 sec (after Elliot et al. 1974). Reprinted courtesy of *The Astrophysical Journal*, published by the University of Chicago Press; © 1974, the American Astronomical Society.

5.3   *Extinction*

Near the center of a planet's shadow, the refracted starlight from all
around the limb arrives at the same region, producing an abrupt increase

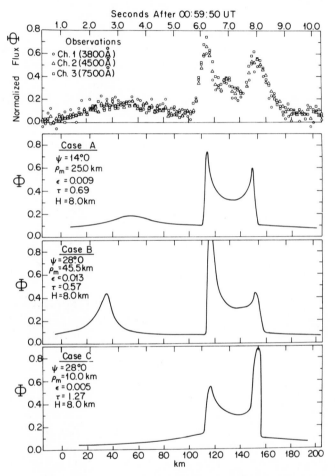

Figure 7   High time resolution data of the central flash and a model central flash profile.
The points correspond to 0.1 sec averages of the data and successive tick marks on the lower
scale correspond to a distance of 20 km. The relative areas under the profiles in the top
frame provide information about the wavelength dependence of the atmospheric extinction,
and the model profiles provide an estimate of the global atmospheric extinction along the
slant path probed by the light that forms the central flash. The quantities ψ, ρ_m, ε, τ, and
H are model parameters (after Elliot et al. 1977e). Reprinted courtesy of *The Astrophysical
Journal*, published by the University of Chicago Press; © 1977, the American Astronomical
Society.

in intensity known as the central flash (Elliot et al. 1976a, 1977e). A high-resolution trace of the central flash is shown in Figure 7. The multiple peaked structure can be largely explained by the oblate figure of the atmosphere. This model for the central flash also requires a substantial amount of extinction, and a mean optical depth of 0.23 ± 0.12 per Martian air mass most closely matches the observations (Elliot et al. 1977e). The region of the atmosphere probed by the light forming the central flash lies about 25 km above the mean surface. The derived optical depth is much larger than can be accounted for by Rayleigh scattering alone and generally agrees with values obtained by Viking observations (Thorpe 1977). Some wave-optical properties of the central flash are discussed by Hubbard (1977).

6 RINGS

To date there exist no photoelectric data for a stellar occultation by the rings of Saturn, but extensive occultation data exist for the rings of Uranus. In fact the rings of Uranus were discovered while observations of a predicted occultation of the star SAO 158687 by Uranus were being attempted. An account of the discovery was given by Elliot et al. (1977a) and *IAU Circulars 3047, 3048*, and *3051* unfold the information, as it was reported. Three groups independently concluded that material near Uranus caused brief occultations to appear in their data: a group from Cornell University observing with the Kuiper Airborne Observatory (Elliot et al. 1977b,c), a group from Lowell Observatory observing at Perth (Millis et al. 1977a,b), and observers at Kavalur (Bhattacharyya & Kupposwamy 1977). The Cornell and Lowell Groups first concluded that the occultations were caused by a belt of satellites (*IAUC 3048*). A few days later the Cornell Group noticed that five of their pre-immersion occultations aligned with five of their post-emersion occultations. After calculating the corresponding distances of the ring occultation points from the center of Uranus, assuming them to lie in the plane of the satellites of Uranus, they concluded that Uranus was surrounded by at least five narrow rings, which they called α, β, γ, δ, and ε (*IAUC 3048*). Data from the other groups confirmed their conclusion (*IAUC 3048*). The "new satellite," reported by Bhattacharyya & Kupposwamy (1977) on the basis of an occultation appearing in their data was caused by the ε ring and not a new satellite as they had originally concluded (Bhattacharyya & Bappu 1977).

   Since the confirmation of rings α–ε, the existence of four more rings has been established, raising the total to nine. Two of these are rings 4 and 5, originally reported by Millis et al. (1977b). These two rings and ring 6 were not immediately recognized as rings due to an error in the calculation of their radii from the KAO data (Elliot et al. 1978). The η ring was reported by the Cornell Group (Elliot et al. 1977d).

At the time of this writing the above nine rings (in order of increasing distance from Uranus: 6, 5, 4, α, β, η, γ, δ, and ε) are the only confirmed structures in the Uranian ring plane. Photometric traces from the KAO

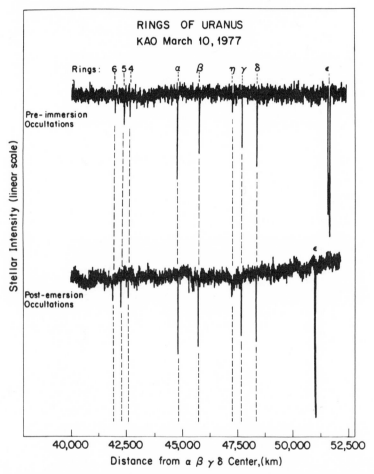

Figure 8 Occultations by the rings of Uranus. The pre-immersion and post-emersion occultations by the rings of Uranus observed with the Kuiper Airborne Observatory (Elliot et al. 1977c) have been plotted on the common scale of distance from the center of Uranus in the ring plane. Occultations corresponding to the nine confirmed rings are easily seen. Other possible occultation events are visible as shallow dips on the individual traces. Much (if not all) of the low frequency variations in the light curves are due to a variable amount of scattered moonlight on the telescope mirror.

Table 4 Occultations by the rings of Uranus

| Star | Date | Magnitudes | | | Occultations observed number west/number east | | | | | | | | | References |
| | | V | I[b] | K | Ring | | | | | | | | | |
					6	5	4	α	β	η	γ	δ	ε	
SAO 158687	10 March 1977	+8.8	+7.2		3/1	3/1	3/1	4/3	4/3	3/1	4/2	3/2	4/2	Elliot et al. 1978
BD-15°3969	23 December 1977	+10.4	+8.4		0/0	0/0	0/0	0/1	0/0?	0/0	0/1	0/1	0/1	Millis & Wasserman 1978
KM 4[a]	4 April 1978	+13.4	+12.6	+11.9	0/0	0/0	0/0	0/0	0/0	0/0	0/0	1/0	1/0	Nicholson et al. 1978
KM 5[a]	10 April 1978	+11.6	+10.7	+10.1	1/1	1/1	1/1	1/1	1/1	1/1	1/1	1/1	1/1	Nicholson et al. 1978

[a] The KM number refers to the list of Klemola & Marsden (1977).
[b] As estimated by Elliot (1977).

Table 5 Properties of the Uranian rings

| Quantity | Ring[a] | | | | | | | | |
	6	5	4	α	β	η	γ	δ	ε
Width[b] (km)	~5	~5	~5	9	14–16	~50	7	~5	20–100
Semi-major axis[c] (km)	41,980	42,360	42,663	44,839 ± 1	45,799	47,323	47,746	48,423	51,284 ± 6
Eccentricity[c] ($e \times 10^3$)				0.63 ± 0.03	~0.4?	<1.0	<1.0	<1.6	7.80 ± 0.12
Inclination[c] (deg)				0?	~0.06?		<0.08	<0.13	0
Precession rate[c] (deg/day)			2.61?	2.20?	2.04?				1.374 ± 0.006
Azimuth of pericenter[c] (deg)				325 ± 2					212 ± 2

[a] Following the notation of Elliot et al. (1978).
[b] The widths are from Nicholson et al. (1978), except for the η ring (see text).
[c] The orbital parameters for the ε and α rings and the precession rates for rings 4 and β are from the models of Nicholson et al. (1978); all other data are from Elliot et al. (1978). The semi-major axes could have systematic errors as large as ±100 km; see discussion in Elliot et al. (1978) and Nicholson et al. (1978).

data showing the occultations by these rings are shown in Figure 8. The broad rings described by Bhattacharyya & Bappu (1977) remain unconfirmed (Millis & Wasserman 1978, Nicholson et al. 1978).

After the discovery of the Uranian rings, Klemola & Marsden (1977) predicted further occultations by the rings of Uranus through 1980. One of these occultations was observed by Millis & Wasserman (1978) and two others by Nicholson et al. (1978). A summary of observed occultations by the rings of Uranus is given in Table 4. The widths of the rings obtained by Nicholson et al. (1978), who observed an occultation by the star with the smallest apparent radius at the distance of Uranus, are given in Table 5. They give the width of the η ring as ~ 5 km; however, as seen in Figure 8, the width of the η ring observed during the occultation of SAO 158687 was ~ 50 km. There is some evidence that the η ring could be a broad ring with a narrow (~ 5 km) core. Thus it could be in the process of coalescing or dispersing.

Early attempts to detect the rings by their reflected light were unsuccessful (Sinton 1977, Baum et al. 1977, Smith 1977). These attempts led to upper limits on the albedo of the ring material of ~ 0.05. A later report of the detection of the rings establishes their albedo as 0.025 at 2.2 μ (Nicholson et al. 1978, Matthews et al. 1978).

To determine the radii of the rings, Elliot et al. (1978) combined all available timing data for the ring occultations of SAO 158687. A schematic diagram of these data is shown in Figure 9. They fit a model to the occultation data for rings α, β, γ, and δ that had six free parameters: the radii of these four rings and two coordinates that specified the center of the ring system. The pole of the ring plane was assumed to be that of the satellite plane (Dunham 1971). Their resulting radii (except for rings α and ε) are given in Table 5. From their analysis they concluded that rings γ, δ, and η are nearly circular and in the satellite plane, while the others are elliptical (or inclined) to varying degrees.

Nicholson et al. (1978) established that the ε ring is elliptical rather than inclined and fit a model of a precessing ellipse to the ε ring. The parameters they derived are given in Table 5. From the precession rate they obtained a value $(3.43 \pm 0.02) \times 10^{-3}$ for J_2, assuming that the ε ring precession is due entirely to the dipole term in the gravitational potential of Uranus. It is interesting to note that the error in the value for J_2 is comparable to that obtained for Jupiter from the Pioneer flyby (Anderson 1976). Assuming the precession rate dictated by the J_2 obtained for the ε ring model, Nicholson et al. (1978) obtained a model for the α ring. The parameters for the α and ε rings are given in Table 5. Further observations of the ring occultations over a period of several years should allow similar models to be established for the other rings.

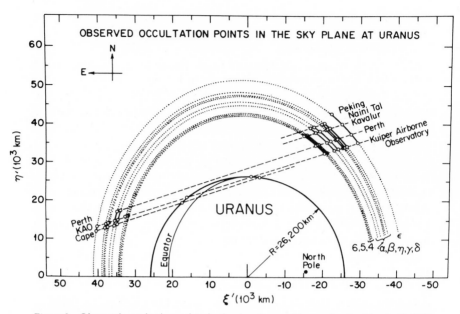

Figure 9   Observed occultation points in the sky plane at Uranus for the March 10, 1977, occultation of SAO 158687. The observed points are indicated by open circles, pre-immersion points to the *right* and post-emersion points to the *left*. The dashed lines show the tracks of the observatories in the sky plane. The rings are indicated by dotted lines and the solid segments between the observed occultation points. The radius of Uranus corresponds to a number density level of 6×10^{13} cm^{-3} (after Elliot et al. 1978).

The first model for the Uranian rings was a three-body resonance model proposed by Dermott & Gold (1977). In this picture, the ring particles would slowly spiral toward Uranus due to the Poynting-Robertson effect until they would reach a radius corresponding to a three-body resonance with two of the (known) Uranian satellites. This resonance would then trap the particles in narrow rings. Aksnes (1977) and Goldreich & Nicholson (1977) criticized this model on the grounds that the resonance strengths would not be sufficient to account for the widths of the rings. Furthermore, they pointed out that Dermott & Gold had neglected resonances involving Miranda, which would be the strongest series due to the proximity of Miranda to the rings. However, Aksnes (1977) and Goldreich & Nicholson (1977) pointed out that the Miranda-Ariel resonances lying within the ring system closely matched the preliminary radii for rings 5, α, γ, and ε. The radii obtained by Elliot et al. (1978) confirm this agreement. It appears that the Miranda-Ariel resonances have some influence on the placement of these rings, although the mechanism is more subtle than suggested by Dermott & Gold. The

Table 6   Stellar diameters (obtained from occultations by bodies other than the moon)

Star	V_0	Sp.	Angular diameter (in milli-arc sec)	Occulting body	References
β Sco C	+4.83[a]	B2V	0.16	Io	Batholdi & Owen 1972
β Sco A$_1$	+2.22	B0V	0.422 ± 0.026	Jupiter (atmosphere)	Elliot et al. 1976b
β Sco A$_2$	+3.48	B2V	0.264 ± 0.019	Jupiter (atmosphere)	Elliot et al. 1976b
SAO 158687	+8.82	K2III	0.49 ± 0.05	rings of Uranus	Millis et al. 1977c
SAO 120774	+5.92	M2III	2.9 ± 1.5	532 Herculina	Bowell et al. 1978

[a] Hardie & Crawford (1961) give $V_0 = +4.29$ for β Sco C.

other rings do not appear connected with three-body resonances (Elliot et al. 1978). To explain the existence of some of the rings and the large eccentricity of the ε ring, additional satellites have been suggested by Marsden,[3] Ip (1978), and Steigmann (1978). A dynamical model in which each ring is stabilized by two small satellites, one interior and one exterior to each ring, has been constructed by Goldreich & Tremaine (1979). Dermott et al. (1979) have proposed that each ring is stabilized by a single small satellite within it. In their model, the satellite also would be the source of the ring material.

7   STELLAR DIAMETERS

Since the observed light curve for a stellar occultation is the convolution of that which would be observed for a point source and the brightness distribution across the disk of the occulted star (Elliot et al. 1975a), these data contain information about the diameter of the occulted star. The advantage gained through using occultations by distant bodies is that the angle subtended by the Fresnel scale is over an order of magnitude smaller than that for the moon (see Table 1), with a corresponding gain in angular resolution. Hence the angular resolution achievable is comparable to that of the intensity interferometer (Hanbury Brown et al. 1974), but the stars whose angular diameters are measurable can be at least seven magnitudes fainter than for the intensity interferometer.

Three different situations arise for which it has been possible to obtain stellar diameters for occultations by bodies other than the moon: (a) occultations by limbs of satellites and asteroids; (b) fitting a model to the narrow spikes (the method described by Elliot et al. 1976b; see Jokipii & Hubbard 1977 for an alternate interpretation); (c) fitting models to the occultation profiles of the Uranian rings. The stellar diameters obtained by these methods are given in Table 6.

[3] Remark at Division for Planetary Sciences Meeting, Boston, Massachusetts, October 1977.

From measurements of the separation of spike pairs, formed by the two components of the spectroscopic binary β Scorpii A, Elliot et al. (1975a) determined the angular separation of the two stars, which they used to estimate their masses.

8   PROSPECTS FOR FUTURE WORK

The results summarized in this review reveal the potential of occultations for increasing our knowledge about the dimensions and atmospheres of solar system bodies. Particularly interesting information obtainable with this technique includes: temperature profiles for the atmospheres of Saturn and Titan; the composition of Titan's atmosphere; upper limits on (or detections of) atmospheres for Pluto and Triton; precise diameters for Pluto, Titan, Triton, other satellites, and the larger asteroids; and precise orbital parameters for the rings of Uranus. The value of this information will depend on what will have already been learned through spacecraft exploration, and in turn, information obtained from occultations should prove useful in planning future missions.

Two outstanding problems remain in the study of upper atmospheres with occultations. First, the nonisothermal features in atmospheres that cause the "spikes" in the light curves need to be determined. The second and related problem is to determine the dynamical origin of the "waves" that appear in the temperature profiles obtained for Uranus, Neptune, and Jupiter.

Use of the occultation technique is limited by the frequency of occultations by sufficiently bright stars, the availability of adequate observing facilities within the occultation track, and the weather. Improved observational techniques would allow more occultation events to be classified as "useful" (O'Leary 1972). For example, a network of portable telescopes of the 0.3-m class could be used to obtain numerous asteroid and satellite diameters, since these could be set up to adequately cover narrow occultation tracks (Millis & Elliot 1979). The Space Telescope will permit observations in the far uv and the use of small focal plane apertures so that the amount of background light from the occulting body can be greatly reduced. Also, it may become possible to use the velocity of an orbiting telescope to "slow down" the shadow speeds of occultations, thereby improving their signal-to-noise ratios (Elliot 1978).

ACKNOWLEDGMENTS

Several colleagues, notably E. Dunham, R. G. French, R. L. Millis, and L. H. Wasserman, provided helpful discussions, and partial support was provided by NASA Grant NSG-2174 and NSF Grant AST-76-14832.

Literature Cited

Aksnes, K. 1977. *Nature* 269:783
Aksnes, K., Franklin, F. A. 1976. *Astron. J.* 81:464–81
Aksnes, K., Franklin, F. A. 1978. *Icarus* 34:194–207
Albers, S. C. 1979. *Sky Telesc.* 57:220–22
Anderson, J. D. 1976. In *Jupiter*, ed. T. Gehrels, pp. 113–21. Tucson: Univ. Ariz. Press. 1254 pp.
Barnard, E. E. 1890. *MNRAS* 50:107–10
Barnes, T. G. III, Evans, D. S., Moffett, T. J. 1978. *MNRAS* 183:285–304
Bartholdi, P., Owen, F. 1972. *Astron. J.* 77:60–65
Baum, W. A., Code, A. D. 1953. *Astron. J.* 58:108–12
Baum, W. A., Thomsen, B. A., Morgan, B. L. 1977. *Bull. Am. Astron. Soc.* 9:499
Bhattacharyya, J. C. 1972. *Nature Phys. Sci.* 238:55–56
Bhattacharyya, J. C., Bappu, M. K. V. 1977. *Nature* 270:503–6
Bhattacharyya, J. C., Kupposwamy, K. 1977. *Nature* 267:331–32
Binzel, R. P., Van Flandern, T. C. 1979. *Science* 203:903–5
Bixby, J. E., Van Flandern, T. C. 1969. *Astron. J.* 74:1220–22
Bowell, E., McMahon, J., Horne, K., A'Hearn, M. F., Dunham, D. W., Penhallow, W., Taylor, G. E., Wasserman, L. H., White, N. M. 1978. *Bull. Am. Astron. Soc.* 10:594
Brinkmann, R. T. 1971. *Nature* 230:515–16
Brinkmann, R. T., Millis, R. L. 1973. *Sky Telesc.* 45:93–95
Buarque, J. A. 1978. *Publ. Astron. Soc. Pac.* 90:117–18
Cameron, R. M. 1976. *Sky Telesc.* 52:327–31
Carlson, R. W., Bhattacharyya, J. C., Smith, B. A., Johnson, T. V., Hidayat, B., Smith, S. A., Taylor, G. E., O'Leary, B., Brinkmann, R. T. 1973. *Science* 182:53–55
Combes, M., Vapillon, L., Lecacheux, J. 1975. *Astron. Astrophys.* 45:399–403
Dermott, S. F., Gold, T. G. 1977. *Nature* 267:590–93
Dermott, S. F., Gold, T. G., Sinclair, A. T. 1979. *Astron. J.* 84: In press
de Vaucouleurs, G., Menzel, D. H. 1960. *Nature* 188:28–33
de Vaucouleurs, G., Nather, R. E., Young, P. J. 1976. *Astron. J.* 81:1147–52
Duncombe, R. L., Seidelmann, P. K. 1977. *J. Inst. Nav.* 24:160–65
Dunham, D. W. 1971. PhD thesis. Yale Univ.
Dunham, E., Elliot, J. L., Gierasch, P. J. 1979. *Astrophys. J.* Submitted
Dütsch, H. U. 1971. *Adv. Geophys.* 15:219–322

Elliot, J. L. 1977. *Astron. J.* 82:1036–38
Elliot, J. L. 1978. *Icarus* 35:156–64
Elliot, J. L., Dunham, E. W. 1979. *Nature* 279:307–8
Elliot, J. L., Veverka, J. 1976. *Icarus* 27:359–86
Elliot, J. L., Wasserman, L. H., Veverka, J., Sagan, C., Liller, W. 1974. *Ap. J.* 190:719–30
Elliot, J. L., Rages, K., Veverka, J. 1975a. *Ap. J. Lett.* 197:123–26
Elliot, J. L., Veverka, J., Goguen, J. 1975b. *Icarus* 26:387–407
Elliot, J. L., Wasserman, L. H., Veverka, J., Sagan, C., Liller, W. 1975c. *Astron. J.* 80:323–32
Elliot, J. L., Dunham, E., Church, C. 1976a. *Sky Telesc.* 52:23–25
Elliot, J. L., Rages, K., Veverka, J. 1976b. *Ap. J.* 207:994–1001
Elliot, J. L., Dunham, E., Millis, R. L. 1977a. *Sky Telesc.* 53:412
Elliot, J. L., Dunham, E., Mink, D. 1977b. *IAU Circ. No. 3048*
Elliot, J. L., Dunham, E., Mink, D. 1977c. *Nature* 267:328
Elliot, J. L., Dunham, E., Mink, D. 1977d. *Bull. Am. Astron. Soc.* 9:498
Elliot, J. L., French, R. G., Dunham, E., Gierasch, P. J., Veverka, J., Church, C., Sagan, C. 1977e. *Ap. J.* 217:661–79
Elliot, J. L., French, R. G., Dunham, E. W., Gierasch, P. J., Veverka, J., Church, C., Sagan, C. 1977f. *Science* 195:485–86
Elliot, J. L., Veverka, J., Millis, R. L. 1977g. *Nature* 265:609–11
Elliot, J. L., Dunham, E. W., Wasserman, L. H., Millis, R. L., Churms, J. 1978. *Astron. J.* 83:980–92
Elliot, J. L., Dunham, E., Churms, J. 1979. In preparation
Eshleman, V. R., Haugstad, B. S. 1978. *Icarus* 34:396–405
Fabry, Ch. 1929. *J. Obs.* 12:1–10
Fairall, A. P. 1972. *Nature* 236:342
Fjeldbo, G., Kliore, A., Seidel, B., Sweetnam, D., Woiceshyn, P. 1976. In *Jupiter*, ed. T. Gehrels, pp. 238–46. Tucson: Univ. Ariz. Press. 1254 pp.
Freeman, K. C., Lyngå, G. 1970. *Ap. J.* 160:767–80
Freeman, K. C., Stokes, N. R. 1972. *Icarus* 17:198–201
French, R. G., Elliot, J. L. 1979. *Ap. J.* 229:828–45
French, R. G., Gierasch, P. J. 1974. *J. Atmos. Sci.* 31:1707–12
French, R. G., Gierasch, P. J. 1976. *Astron. J.* 81:445–51
French, R. G., Elliot, J. L., Gierasch, P. J. 1977. *Icarus* 33:186–202

474 ELLIOT

French, R. G., Goguen, J. D., Duthie, J. G. 1978. *Icarus* 34: 182–87

Goldreich, P., Nicholson, P. 1977. *Nature* 269: 783–85

Goldreich, P., Tremaine, S. 1979. *Nature* 277: 97—99

Goldsmith, D. W. 1963. *Icarus* 2: 341

Groth, E. J., Klopfenstein, J. B., Wickes, W. C., Caldwell, J. J. 1978. *Astron. J.* 83: 442–46

Hanbury Brown, R., Davis, J., Allen, L. R. 1974. *MNRAS* 167: 121–36

Hardie, R. H., Crawford, D. L. 1961. *Ap. J.* 133: 843–59

Harris, A. W. 1978. *Bull. Am. Astron. Soc.* 10: 583

Hubbard, W. B. 1977. *Nature* 268: 34–35

Hubbard, W. B. 1978. In *The Mars Reference Atmosphere*, p. 133. Pasadena, Calif.: JPL (first draft)

Hubbard, W. B. 1979. *Ap. J.* 229: 821–27

Hubbard, W. B., Jokipii, J. R. 1977. *Icarus* 30: 531–36

Hubbard, W. B., Van Flandern, T. C. 1972. *Astron. J.* 77: 65–74

Hubbard, W. B., Nather, R. E., Evans, D. S., Tull, R. G., Wells, D. C., van Citters, G. W., Warner, B., Vanden Bout, P. 1972. *Astron. J.* 77: 41–59

Hubbard, W. B., Jokipii, J. R., Wilking, B. A. 1978. *Icarus* 34: 374–95

Hunten, D. M., McElroy, M. B. 1968. *J. Geophys. Res.* 73: 4446–48

Hunten, D. M., Veverka, J. 1976. In *Jupiter*, ed. T. Gehrels, pp. 247–83. Tucson: Univ. Ariz. Press. 1254 pp.

Ip, W.-H. 1978. *Nature* 272: 802–3

Jokipii, J. R., Hubbard, W. B. 1977. *Icarus* 30: 537–50

Klemola, A. R., Marsden, B. G. 1977. *Astron. J.* 82: 849

Klemola, A. R., Liller, W., Marsden, B. G., Elliot, J. L. 1978. *Astron. J.* 83: 205

Kliore, A. J., Woiceshyn, P. M. 1976. In *Jupiter*, ed. T. Gehrels, pp. 216–37. Tucson: Univ. Ariz. Press. 1254 pp.

Kovalevsky, J., Link, F. 1969. *Astron. Astrophys.* 2: 398–412

Larson, S. M. 1972. *Contrib. Bosscha Obs. No. 45*

Lecacheux, J., Combes, M., Vapillon, L. 1973. *Astron. Astrophys.* 22: 289–92

Liller, W., Papaliolios, C., French, R. G., Elliot, J. L., Church, C. 1978. *Icarus* 35: 395–99

Martynov, D. Ya. 1961. *Sov. Astron. J.* 4: 798–804

Matthews, K., Neugebauer, G., Nicholson, P. D. 1978. *Bull. Am. Astron. Soc.* 10: 581

McElroy, M. B. 1968. *J. Geophys. Res.* 73: 1513–21

Millis, R. L., Elliot, J. L. 1979. In *Asteroids*, ed. T. Gehrels. Tucson: Univ. Ariz. Press. In press

Millis, R. L., Wasserman, L. H. 1978. *Astron. J.* 83: 993–98

Millis, R. L., Birch, P., Trout, D. 1977a. *IAU Circ. No. 3051*

Millis, R. L., Wasserman, L. H., Birch, P. 1977b. *Nature* 267: 330–31

Millis, R. L., Wasserman, L. H., Elliot, J. L., Dunham, E. 1977c. *Bull. Am. Astron. Soc.* 9: 498

Morrison, D., Cruikshank, D. P., Burns, J. A. 1977. In *Planetary Satellites*, ed. J. A. Burns, p. 11. Tucson: Univ. Ariz. Press. 598 pp.

Mulholland, J. D. 1972. *Publ. Astron. Soc. Pac.* 84: 357–64

Nather, R. E., Evans, D. S. 1970. *Astron. J.* 75: 575–82

Nicholson, P. D., Persson, S. E., Matthews, K., Goldreich, P., Neugebauer, G. 1978. *Astron. J.* 83: 1240–48

Nier, A. O., McElroy, M. B. 1977. *J. Geophys. Res.* 82: 4341

O'Leary, B. 1972. *Science* 175: 1108–11

O'Leary, B., Van Flandern, T. C. 1972. *Icarus* 17: 209–15

O'Leary, B., Marsden, B. G., Dragon, R., Hauser, E., McGrath, M., Backus, P., Robkoff, H. 1976. *Icarus* 28: 133–46

Osawa, K., Ichimura, K., Shimizu, M. 1968. *Tokyo Astron. Bull.* 183: 2183–87 (2nd ser.)

Owen, G. T., Biemann, K. 1976. *Science* 193: 801

Pannekoek, A. 1904. *Astron. Nachr.* 164: 5–10

Rages, K., Veverka, J., Wasserman, L., Freeman, K. C. 1974. *Icarus* 23: 59–65

Sander, W. 1973. *Stern* 49: 242–46

Seiff, A., Kirk, D. B. 1977. *Science* 194: 1300–3

Shelus, P. J., Benedict, G. F. 1978. *Proc. IAU Symp. No. 81*. In press

Sinton, W. M. 1977. *Science* 198: 503–4

Smart, W. M. 1962. *Textbook on Spherical Astronomy*, pp. 368–403. Cambridge: Cambridge Univ. Press. 430 pp. 5th ed.

Smith, B. A. 1977. *Nature* 268: 32

Smith, D. W., Greene, T. F., Shorthill, R. W. 1977. *Icarus* 30: 697–729

Steigmann, G. A. 1978. *Nature* 274: 454–55

Taylor, G. E. 1962. *Observatory* 82: 17–20

Taylor, G. E. 1963. *R. Obs. Bull.* 72: E355–66

Taylor, G. E. 1970. *MNRAS* 147: 27–33

Taylor, G. E. 1972. *Icarus* 17: 202–8

Taylor, G. E. 1974. *Nautical Almanac Office Tech. Note No. 34*. Hailsham, Sussex, England: Royal Greenwich Obs.

Taylor, G. E. 1976. *Nature* 264: 160–61

Taylor, G. E., Dunham, D. W. 1978. *Icarus* 34: 89–92

Texas-Arizona Occultation Group. 1977. *Ap. J.* 214:934–45

Thorpe, T. E. 1977. *J. Geophys. Res.* 82: 4151–59

Vapillon, L., Combes, M., Lecacheux, J. 1973. *Astron. Astrophys.* 29:135–49

Veverka, J., Wasserman, L. 1974. *Icarus* 21: 196–98

Veverka, J., Wasserman, L. H., Elliot, J. L., Sagan, C., Liller, W. 1974a. *Astron. J.* 79: 73–84

Veverka, J., Wasserman, L. H. Sagan, C. 1974b. *Ap. J.* 189:569–75

Vilas, F., Millis, R. L., Wasserman, L. H. 1977. *Pap. L08-2.* Presented at DPS Meet., Honolulu, Jan. 1977

Wallace, L. 1975. *Ap. J.* 197:257–61

Wasserman, L. H., Veverka, J. 1973a. *Icarus* 18:599–604

Wasserman, L. H., Veverka, J. 1973b. *Icarus* 20:322–45

Wasserman, L. H., Millis, R. L., Williamson, R. M. 1977. *Astron. J.* 82:506–10

Wasserman, L. H., Millis, R. L., Franz, O. G., Bowell, E., White, N. M., Giclas, H. L., Martin, L. J., Elliot, J. L., Dunham, E., Mink, D., Baron, R., Honeycutt, R. K., Henden, A. A., Kephart, J. E., A'Hearn, M. F., Reitsema, H., Radick, R., Taylor, G. E. 1979. *Astron. J.* 84:259–68

Young, A. T. 1976. *Icarus* 27:335–58

Zurek, R. W. 1976. *J. Atmos. Sci.* 33:321–37

Ann. Rev. Astron. Astrophys. 1979. 17:477–511

INFRARED EMISSION OF ⋇2158
EXTRAGALACTIC SOURCES

G. H. Rieke[1] and M. J. Lebofsky

Steward Observatory, University of Arizona, Tucson, Arizona 85721

1 INTRODUCTION

Because of rapid developments in instrumental techniques (see, for example, Soifer & Pipher 1978), infrared observations have burst upon the scene of extragalactic astronomy with an array of interesting and sometimes startling results. From determining the nature of the nuclei of nearby galaxies (including our own—see, for example, Oort 1977) to measuring cosmological parameters, infrared measurements are proving of fundamental importance. This review describes the observational material in this field with comments on its general implications. We hope that it will lead to a closer integration of infrared measurements into theoretical treatments of the behavior of extragalactic sources at all wavelengths.

We have tried to include all relevant material published before September 1978. Through the generosity of many of our colleagues in making material available before publication, many important results are discussed that will not appear until well after this date. We first discuss galactic nuclei whose infrared output is dominated by direct radiation by stars, then those that contain heavily obscured luminosity sources whose presence is manifested through strong thermally reradiated fluxes, and finally sources that emit nonthermally in the infrared.

2 INFRARED EMISSION FROM ELLIPTICAL GALAXIES

A General Properties

Most of the study of ellipticals has been carried out at near-infrared wavelengths (e.g. Frogel et al. 1978 and references therein) and has been

[1] Alfred P. Sloan Foundation Fellow.

477

0066-4146/79/0915-0477$01.00

concerned with the stellar content of these galaxies. The light of ellipticals is dominated by late-type stars (Johnson 1966, Penston 1973, Frogel et al. 1975b) whose energy distributions peak in the near-infrared, and luminosity and temperature sensitive spectral features at these wavelengths (see Frogel et al. 1978, Aaronson et al. 1978a) have aided the determination of what spectral types contribute to the emission. Information about the stellar population is important for testing theories on the evolution of ellipticals (e.g. Tinsley & Gunn 1976, Larson 1975a, Frogel et al. 1978, Aaronson et al. 1978a) and for using ellipticals in cosmological tests (Tinsley 1973).

Elliptical galaxies apparently contain little dust and gas. With the early observations of intense infrared radiation centered on dust and gas-rich spirals and Seyfert galaxies, elliptical galaxies have not been extensively studied at mid-infrared wavelengths. However, the limited observations already made show that some ellipticals have strong infrared excesses (Rieke & Low 1972a). Further study of these ellipticals is needed to learn whether the infrared excess is nonthermal or related to dust; the latter possibility needs careful checking because of the bearing of such a discovery on the evolution of ellipticals.

B   Near-Infrared Studies of Late-Type Stars

Systematic surveys of elliptical galaxies have been carried out only from 1 μm to 3.5 μm using both broad- (designated J, H, K, L) and narrow-band filters. These wavelengths sample the stellar composition and can distinguish between dwarf- and giant-rich populations. The first survey at these wavelengths of extragalactic objects (Johnson 1966) revealed that, at least for his small sample, the near-infrared colors of spirals and ellipticals are essentially the same, and the longer the wavelength, the later the spectral type observed. His crude stellar synthesis model demonstrated that stars with spectral types as late as M10 were required to match the $K-L$ colors. Surveys of broad-band colors that expanded the sample Johnson studied include Frogel et al. (1978), Persson et al. (1979), Frogel et al. (1976, 1975a), Grasdalen (1975), and Pacholczyk & Tarenghi (1975). Low accuracy and, in Grasdalen's work, a discrepancy in the apparent beam size, compromise the usefulness of the last two surveys mentioned. Surveys measuring narrow-band indices include Aaronson et al. (1978a) and Frogel et al. (1975b, 1978). Other studies of the stellar population include Strom et al. (1976) and Glass (1976) and measurements of NGC 2768 and NGC 3115 by Strom et al. (1978), of M32 by Penston (1973), and of Maffei 1 by Spinrad et al. (1971).

The main thrust of these observations has been to study the composition of the stellar population with several goals in view. The most obvious goal was to use near-infrared data, which is sensitive to the presence of late-type stars, to test stellar synthesis models and to study the metal content of ellipticals. The above surveys reach conclusions similar to the early results of Johnson (1966). The composite nature of the spectral energy distribution is confirmed and the H_2O index (at 2.0 μm) supports the requirement from the $K-L$ color for inclusion of very late M stars in the stellar mix (Aaronson et al. 1978a). The surveys have also confirmed the small dispersion in near-infrared colors for ellipticals: $V-K$ ranges from 2.9 to 3.5 mag as M_V ranges from -18 to -24, with bluer colors tending to be associated with fainter galaxies. Through the CO index (at 2.3 μm), a sensitive indicator of the relative number of dwarfs and giants (Frogel et al. 1978 and references therein), these surveys also show that the population is dominated by late-type giants unlike the predictions from early elliptical population models such as those of Spinrad & Taylor (1971) which indicated a population rich in red dwarfs and with correspondingly high M/L ratios. The low values for M/L estimated by Frogel et al. must be taken as lower limits since unobservable mass in the form of very low-luminosity stars or other objects could be present. Penston (1973) reached a similar conclusion using data for M 32.

The detailed comparison of more recent models with near-infrared data finds no model completely satisfactory. Frogel et al. (1978) compare their colors with those from models by Tinsley & Gunn (1976) and by O'Connell (1976), neither of which were in complete agreement with the observations but which did point out that only models with rich giant branches and relatively flat main-sequence luminosity functions could be consistent with the observations. Aaronson et al. (1978b) constructed metal-rich models with initial mass functions similar to the Salpeter function with slopes similar to that found in the solar neighborhood (see Scalo 1978). These models reproduce the infrared colors accurately but cannot fit the observed $U-V$ color, probably because of difficulties in modeling and blanketing in the U filter. The net results of the model comparison are that ellipticals are giant-rich with perhaps smaller M/L ratios than suspected in the past and that the initial mass function appears to be adequately described as a Salpeter-type function with a power-law slope of ~ 2.

These surveys also attempt to measure color gradients in ellipticals, which, as discussed later in this section, may be related to metallicity. The broad-band colors, studied both by multiaperture measurements (Frogel et al. 1976, 1978, Persson et al. 1979) and by scans across a galaxy (Strom et al. 1976, 1978), show that ellipticals become redder toward their nuclei. The $V-K$ gradient is small and a few galaxies in the Frogel et al. (1978) sample actually seem to have $V-K$ colors bluer toward their nuclei. The largest and statistically most significant gradient seen in the Frogel et al. data is in the $J-K$ color, which also appears redder toward the nucleus. Frogel et al. also attempted to measure CO gradients, but they found none. Aaronson et al. (1978a) found no gradient in the H_2O index. Both groups also found no significant variation in indices between galaxies. However, since the broad-band gradients are quite small, and since the CO gradients estimated from a comparison of CO and broad-band color spreads in globular clusters (Aaronson et al. 1978b) are very small, the multiaperture technique is not sufficiently accurate to show a gradient in the CO index.

The color gradients and differences in color between galaxies may reflect metallicity differences between and within ellipticals. Current theories about the formation of ellipticals (e.g. Larson 1975a,b) predict a metallicity gradient with the nucleus having the highest metal content. The $V-K$ color should become redder with increasing metal content because of increased line blanketing suppressing V, a shift of the horizontal branch population to the red, a cooler giant branch, and a redder turn-off from the main sequence (Aaronson et al. 1978b, Strom et al. 1978). Several different methods of establishing the relationship between metallicity and $V-K$ have been attempted. Strom et al. (1976, 1978) argue that color differences between galaxies appear to result from luminosity-dependent metallicity differences. Strom et al. (1976) use a metallicity-$(V-K)$ calibration based on unpublished measurements of K giants. They also showed that $U-B$, known to be a metallicity indicator (e.g. Faber 1973), is well correlated with $V-K$ for the sample of galaxies they studied (NGC 3115, NGC 3377, and NGC 4762), and that the $V-K$ gradient is in the sense predicted by collapse/enrichment models (Larson 1975a,b). Strom et al. (1978) demonstrated that the color versus metallicity relation within a galaxy is the same as the relation for a group of galaxies, strengthening the case for using $V-K$ as a metallicity indicator. Aaronson et al. (1978b) attempted to calibrate metallicity against $V-K$ in globular clusters. They found that the near-infrared colors, including $V-K$, and the CO index are correlated with metallicity in the globular clusters they studied. They then showed that the same colors for ellipticals correlated with $V-K$ and therefore also with metallicity. As mentioned earlier, the CO index is virtually the same for all ellipticals and shows no relationship with $V-K$, possibly because of the saturation of the 2.3-μm CO feature at the metallicity levels encountered in ellipticals (Frogel et al. 1975b). Frogel et al. (1978) tried to relate near-infrared colors to absolute luminosities to test for color-luminosity dependences similar to those seen at shorter wavelengths, which would also presumably reflect metallicity differences. Their plots of M_V versus color have much more scatter than would be predicted from the observational error. S. E. Persson et al. (1979, unpublished) studied the M_V–color relations further and noted that the relations flattened out for the brightest galaxies.

Although using $V-K$ as a metallicity indicator seems tempting, the above discussion demonstrates that this color must be used with caution, particularly when some of the difficulties with the methods of calibrating $V-K$ are taken into account. Besides its sensitivity to metal content, the $V-K$ color for an elliptical also measures other factors such as the stellar composition. The studies by Strom et al. (1976, 1978) do not have enough observational constraints to disentangle metallicity effects from other differences. For example, Struck-Marcell & Tinsley (1978) show that $V-K$ may be used as an age discriminant. The galaxies used by Strom et al. (1976) to correlate $V-K$ and $U-B$ cover only a very limited range in absolute magnitude. The work by Frogel et al. (1978), Persson et al. (1979), and Aaronson et al. (1978b) studies a much larger sample of galaxies and measures more colors, but this sample shows only a general trend with metallicity. The flattening of M_V–color relations noted by these authors is similar to the flattening of M_V–color relations at optical wavelengths, which also suggests that broad-band colors may not provide straight-forward metallicity indicators. Aaronson et al. (1978b) noted that the globular clusters used to calibrate $V-K$/metallicity are low-luminosity, metal-poor systems, and a linear relation derived from their properties may become saturated at the luminosity and metallicity levels encountered in ellipticals. To contribute to understanding the evolution of ellipticals, infrared measurements will have to demonstrate how metallicity, popu-lation, and perhaps also environment (e.g. field or cluster galaxy) affect the colors of ellipticals.

One interesting galaxy measured by Persson et al. (1979), NGC 5102, has near-infrared colors, including $V-K = +2.59$ mag, that are much bluer than any other elliptical measured. Freeman (1975) also noted that its optical colors and spectrum are more similar to those of a globular cluster than to those of an elliptical. NGC 5102 has an absolute magnitude of $M_V \sim -19.2$ and is lenticular in shape and appears to be a galaxy. Further study of this galaxy, which appears to be quite metal-poor compared to other ellipticals, might provide insight into the role of metal content in galactic evolution.

One of the most significant uses of infrared observations of elliptical galaxies lies in determining the evolutionary correction to the cosmo-logical deceleration parameter, q_0. The most direct determination of q_0 comes from searching for nonlinearities in the magnitude-redshift relation at high z (see Sandage 1975). This method relies on observation of the light from elliptical galaxies at ages substantially earlier than the current epoch when galaxies would appear more luminous than at present (e.g. Spinrad 1977 and references therein), and hence requires a correction. As demonstrated in Tinsley (1973), measurements of the CO index in ellipticals at the current time can be used to determine

$$\Delta q_0 = q_{0_{\text{apparent}}} - q_{0_{\text{actual}}}.$$

Although this use of the CO index was realized early, spurious CO gradients (Frogel et al. 1975b) made it appear that calculation of Δq_0 would have to wait until measurements of CO could be made across an entire galaxy. However, since the CO index appears quite uniform within and between galaxies (Frogel et al. 1978), its use in calculating Δq_0 seems warranted. Tinsley (1973) gives a relation between Δq_0 and G, the ratio of giant-produced to dwarf-produced light, which can be expressed as follows:

$$\Delta q_0 = \frac{2}{2-\alpha} \frac{1}{Ht} G \frac{t_\odot l_\odot}{l_g t_g} \left(\frac{t_\odot}{t}\right)^{\gamma-1}$$

where H = Hubble constant in inverse time units, t is the age of the dominant old stars, $t_\odot$ is the main sequence lifetime of the sun, γ appears in the relation between luminosity and main-sequence lifetime, l_g is a mean luminosity for giant stars of $\sim 1\ M_\odot$, and t_g is their mean lifetime. The term $2/2-\alpha$ represents the aperture correction discussed in Gunn & Oke (1975) and $\alpha \sim 0.7$. From the evolutionary tracks of Sweigart & Gross (1978) and the work of Tinsley & Gunn (1976) l_g is at most $\sim 100\ L_\odot$ while t_g is at most $\sim 8 \times 10^8$ yr and more probably $\sim 2 \times 10^8$ yr. Since $t < t_0$, the age of the universe, and $Ht_0 < 1$, setting $Ht = 1$ will result in a lower limit to Δq_0. The value for γ shall be taken as 1.3, the maximum plausible (Tinsley 1973), and $t_\odot/t$ shall be set equal to 0.5 as about the minimum possible. From the data of Frogel et al. (1978), the maximum allowable dwarf contribution to the 2-μm radiation is 25% or a $G \sim 3$. Larger values of G up to ~ 10 are possible depending on how the CO index is converted to a giant-dwarf ratio. The above values lead to $\Delta q_0 \gtrsim 0.5$ with $\Delta q_0 \sim 1.5$ being most plausible. It is difficult to assign a formal error to Δq_0 computed in this fashion, but the values used in the computation were all chosen to give the smallest possible Δq_0. Since q_0 as determined from the magnitude-redshift curve is of order 1 (Sandage 1973), this evolutionary correction leads to $q_{0_{\text{actual}}} \sim 0$, in agreement with less direct but evolutionary-independent determinations of q_0 (Sandage 1975). Gunn & Oke (1975) derive a value of $q_{0_{\text{apparent}}}$ near 0; the Δq_0 derived above would then lead to negative values of q_0 with consequences as discussed in Gunn & Tinsley (1975). Although this calculation of Δq_0 is not yet very precise, it does demonstrate that the evolutionary correction is indeed large as suggested by Tinsley (1973) and that when a more accurate conversion of CO index to giant-dwarf ratio can be performed, the evolutionary correction can be known accurately. A further difficulty lies in possible dynamical corrections to q_0 (e.g. Tinsley 1977), which are difficult to estimate and may have the opposite sign from the Δq_0 calculated here. This process can also be inverted and used to study the evolution of ellipticals directly—observations of galaxies at z's higher than $z \sim 0.02$ covered by existing surveys will be particularly valuable.

C Mid-Infrared Studies of Ellipticals

Although no extensive survey of elliptical galaxies has been made at 10 μm, the sample included in the survey of bright galaxies by Rieke & Lebofsky (1978) indicates that infrared excesses in ellipticals are not rare. A total of 10 elliptical and lenticular galaxies were measured in this survey; a total of 4 galaxies (NGC 205, NGC 221, NGC 4486, and NGC 4526) were detected, although the fluxes from NGC 205 and NGC 221 are undoubtedly of stellar origin, leaving 2 galaxies or 20% of the sample with infrared excesses. The measurement of Maffei 1 by Rieke & Low (1972a) is another case where only the stellar component is seen at 10 μm. The 10-μm luminosities of NGC 4486 and NGC 4526 are near the low end of the range for spirals but are substantially larger than the extrapolated stellar flux.

The most luminous elliptical at 10 μm is NGC 1052 with a 10-μm luminosity several times that of NGC 253 (Rieke & Low 1972a). Unpublished near-infrared observations also by the authors yield an $H-K$ color of 0.33 mag and $K-L$ color of 0.85, both redder than the normal colors for ellipticals (see Frogel et al. 1978, Johnson 1966). Determination of the infrared emission mechanism in NGC 1052 and its relationship to the optical emission lines and weak radio source in the galaxy will await further observations such as a search for the silicate feature at 10 μm.

3 INFRARED EMISSION FROM SPIRAL AND IRREGULAR GALAXIES

Infrared investigators originally found intense emission from galaxies such as M82 and NGC 253 that show evidence of large amounts of dust on optical photographs. However, in an unbiased survey of bright galaxies, Rieke & Lebofsky (1978) found that 40% of the nuclei of bright spirals are strong sources of infrared radiation and that this radiation dominates their energy output and leads to very low M/L ratios. Dust plays a significant role in producing the radiation from spiral nuclei as evidenced by silicate absorption (Lebofsky & Rieke 1979). A variety of other galaxies such as irregular galaxies and more distant spirals with emission-line spectra have also been detected at 10 μm and indicate that the questions raised by M82 and spiral nuclei may represent common difficulties in understanding galactic nuclei.

Near-infrared measurements have shown that the direct output of stars is observed in the 1-μ to 2-μm region (Johnson 1966, Aaronson 1977); high-resolution maps have shown that the infrared nucleus may be displaced from the optical nucleus due to heavy extinction (Lebofsky & Rieke 1979). Since the near-infrared properties of most spiral and irregular galaxy nuclei are similar to those of ellipticals, we discuss near-infrared observations before addressing the infrared excesses of these sources.

A   Near-Infrared Observations of Late-Type Stars

The output of most spirals at near-infrared wavelengths is dominated by late-type stars. Johnson (1966) observed a sample of spirals from J(1.25 μm) to L(3.6 μm) and discovered that a composite stellar spectrum including late M stars was required to match the colors. Baldwin et al. (1973) observed the CO bands near 2.3 μm, which are strong in late-type giants but not present in the spectra of M dwarfs (Frogel et al. 1978 and references therein). Their data demonstrated that giants predominate and that mass-to-luminosity ratios estimated from stellar synthesis models may be near 10 rather than the much higher values predicted from dwarf-rich models. These first studies were expanded by Glass (1973b), Grasdalen (1975) (note that the beam size in this work must have been smaller than stated), Pacholczyk & Tarenghi (1975), Glass (1976), and Aaronson (1977). Smaller studies include Spinrad et al. (1973), Glass (1973a), Penston (1973), and Aaronson (1978a). Aaronson (1977) confirmed Johnson's suggestion of late M stars in the stellar mix by observing H_2O absorption at 2 μm.

The work by Glass (1973b, 1976) contains L(3.6 μm) measurements which indicate the possible presence of dust in some spirals. He noted that the J(1.25 μ), H(1.6 μ), and K(2.2 μ) colors matched those expected from a composite stellar population, but the K–L color indicated an excess at L relative to stars, and that the K–L color was not useful for studying the stellar population. All of these K–L excess galaxies observed at 10 μm have been detected (Becklin et al. 1971, Rieke & Low 1972a, Kleinmann & Wright 1974), which lends weight to the suggestion that the L-excess is caused by thermal dust emission.

The study by Aaronson (1977) includes measurements of flux versus aperture for both broad-band colors and CO and H_2O indices. He found no radial gradients in the near-infrared colors except for his group of "2-μm excess" galaxies whose spiral members are highly inclined or heavily obscured such as NGC 253. The $V–K$ color does show a radial gradient with a redder color toward the nucleus which Aaronson suggests may reflect a metallicity gradient, although this does not seem as obvious as for the similar gradients in ellipticals (Frogel et al. 1978). This interpretation of the color gradient is particularly difficult for late-type spirals that exhibit a large range of gradient strengths as well as for a few galaxies that are bluer toward their nuclei. Aaronson also found no correlation between absolute magnitude and $V–K$, or infrared colors.

The spread of $V–K$ color between galaxies has been considered by Grasdalen (1975), Pacholczyk & Tarenghi (1975), and Aaronson (1978b). The first two papers do not separate the spirals into morphological classes but note that for spirals taken as a whole the dispersion in $V–K$ is substantially larger than for ellipticals. Aaronson considered morphological classes separately, where he found a dispersion within a class consistent with the dispersion for optical colors. His results also show that all spirals have a larger $V–K$ dispersion than ellipticals.

The sample of spirals measured by Aaronson (1977) is large enough to check for relations between colors and morphological class. The near-infrared colors are virtually the same for all spirals. The mean $V–K$ color is the same for spirals through type Sbc as for ellipticals and later types through Scd have increasingly bluer colors. The Magellanic irregulars have $V–K$ colors that continue the spiral sequence while the I0-type galaxies appear to be dusty early-type galaxies. These color differences are revealed in plots of $U–V$ versus $V–K$ where spirals are seen to form a sequence separate from the sequence formed by ellipticals. The difference between ellipticals and spirals can be seen roughly in the graph in Pacholczyk & Tarenghi (1975) but is seen more clearly in Aaronson (1978b). The elliptical sequence is probably a metallicity sequence while the spiral sequence is probably a population sequence (Aaronson 1977, 1978b). Supporting evidence for the population sequence hypothesis comes from the CO index, which is essentially constant through type Sbc and then decreases for later types indicating a decrease in the amount of radiation from giants. This population change may reflect different star formation histories (Aaronson 1977, Struck-Marcell & Tinsley 1978), but more data are required to check this possibility.

B   Examples

M31   M31 has been studied at wavelengths from 1 μm to 10 μm and at a variety of spatial resolutions, and the net result of this study demonstrates that M31 behaves quite predictably. The main contribution has been the discovery that the stellar composition observed in the near-infrared is giant-dominated (Baldwin et al. 1973), and that the dwarf-rich models of Spinrad & Taylor (1971) are not appropriate. The nucleus of M31 appears to be quite dust-free; the scans of Sandage et al. (1969) at B and K have identical profiles with no evidence for extinction. Iijima et al. (1976) and Matsumoto et al. (1977) also obtained near-infrared scans that demonstrate the lack of extinction in the central region of M31. The latter authors also found some evidence for a bar-like structure in the central bulge.

The most powerful evidence for the lack of dust and other sources of infrared excess comes from data taken at mid- and far-infrared wavelengths. Rieke & Lebofsky (1978) present 5-μm and 10-μm measurements that lie on the stellar continuum extrapolated from 2 μm. If the nucleus of M31 contained a 10-μm source such as that in the center of our galaxy, it would have been easily detected at a level of at least ~ 0.3 Jy based on the measurement in Rieke et al. (1978) of the Galactic center. At 100 μm, where cold dust in molecular clouds should be observable, Telesco (1977) found an upper limit of 30 Jy while ~ 200 Jy would be expected if the nucleus of M31 were analogous to the Galactic center. In short, the nucleus of M31 is free from dust and has no apparent nonthermal source in the infrared. It is radically different from the nucleus of our galaxy.

M82   The spectral flux distribution of M82 illustrated in Figure 1 is from observations by Gillett et al. (1975a), Willner et al. (1977), Hildebrand et al. (1977), Telesco (1977), and unpublished data by W. F. Forrest et al. and by R. I. Thompson et al. The luminosity of M82 emerges almost entirely in the infrared and is $3 \times 10^{10} L_\odot$. The most striking aspect of the spectrum of M82 is the many features also found in galactic thermal sources. These include unidentified broad features at 3.3, 6.2, 7.6, and 11.3 μm, strong "silicate" absorption at 10 μm, the Brackett α and γ lines, and fine structure lines of [Ar II] at 6.98 μm, of [Ne II] at 12.8 μm, and of [S III] at 19 μm.

Gillett et al. (1975a) fit the 8–13-μm spectrum of the nucleus of M82 with a renormalized spectrum of BD 30°3639 (which includes the unidentified features at 8.7 and 11.3 μm) and a "silicate" absorption curve with optical depth $\tau = 1.5$. Assuming a "normal" reddening law, the corresponding visible extinction, A_V, is at least 20 magnitudes (Rieke 1974, Gillett et al. 1975b), and could be substantially more if there were an underlying emission feature. At this same position the relative strengths of the Brackett α and γ lines also indicate $A_V \gtrsim 15$ (Willner et al. 1977, Simon et al. 1979, R. I. Thompson et al., unpublished). Again, the

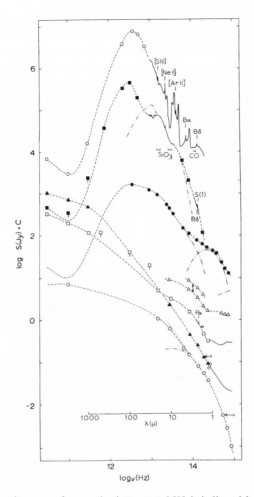

Figure 1 Infrared spectra of extragalactic sources. M82 is indicated by open circles and has $C = 3.77$; NGC 1068 by filled squares and $C = 3.07$; NGC 4151 by filled circles and $C = 2.5$; Mrk 509 by open triangles and $C = 1.77$; 3C273 by open squares and $C = 1$; 3C279 by filled triangles and $C = 2$; and $0235 + 164$ by open hexagons and $C = 0$. For 3C279, the infrared spectrum is normalized to join onto the spectrophotometric spectrum from Neugebauer et al. (1978); for $0235 + 164$, the spectrum is at maximum light (Rieke et al. 1976). Other references are in the text. Redshifts for the last three sources are indicated by horizontal arrows drawn from the rest wavelength to the observed wavelength. Spectrophotometry is drawn as a full line and dashed lines connect photometric points. The dash-dot line for NGC 1068 shows the results of model calculations by Jones et al. (1977). For NGC 4151, dash-dot lines illustrate separation of the spectrum into ultraviolet, stellar, and infrared components. Possible variability of MrK 509 is shown by two spectra connected by a double-headed arrow. For $0235 + 164$, the spectral shape early in the outburst is shown with the dash-dot line and that late in the outburst by the dotted line.

extinction may be significantly higher given the large optical depth already implied for Brackett γ by the average extinction computed here. Although the reddening law in an object like M82 may deviate from the "normal" behavior found in the solar neighborhood, there is little question that the nucleus of the galaxy is extremely strongly obscured. The strong obscuration and close spectral resemblance to galactic thermal sources leave no doubt that the infrared emission of M82 is of thermal origin. Because of the obscuration, observations in the optical and even in the near infrared must be interpreted with caution.

The strength of Brackett α implies a free-free flux of ~ 0.8 Jy at 3.5 mm roughly equal to the total radio flux observed at this wavelength (Willner et al. 1977). However, the radio flux also lies on a power-law extrapolation of the nonthermal spectrum, so that the Bα line may be generated in part in compact H II regions with a high enough emission measure to be optically thick at 3.5 mm. This suggestion may be supported by the large optical depth near 100 μm, where a source size of $10'' \times 40''$ and plausible range of temperatures require $\tau > 0.1$ (Telesco 1977). The average near-infrared obscuration would be even stronger than observed unless there are large fluctuations in density such as would be produced if a significant part of the far-infrared flux arises in compact H II regions. The ratio of 10-μm flux to free-free flux in M82 is ≥ 40, substantially larger than is found for large Galactic H II regions. This discrepancy would also be ameliorated if M82 contained many compact H II regions.

Peimbert & Spinrad (1970) deduced from the ratio n(He II)$/n$(He I) ≈ 1 that the temperature of the ionizing source in M82 is $\leq 30,000$ K. Further support for this conclusion comes from the ratio n(Ar III)$/n$(Ar II) < 0.7 (Willner et al. 1977). The Bα strength requires 6×10^{53} Lyman continuum photons per second, assuming that dust does not compete effectively with gas in absorbing ionizing photons. This estimate of the required Lyman continuum should therefore be considered a lower limit. Being mindful of the upper limit on the temperature, a model for the ionizing source is then 2×10^6 or more B0V stars (Willner et al. 1977).

The relative strengths of the [Ne II], [Ar II], and Bα lines (none of which suffer a large amount of extinction) are consistent with "normal"— i.e. solar—abundance ratios in the nucleus of M82 (Willner et al. 1977, Simon et al. 1979).

M82 has been mapped at 10 μm by Kleinmann & Low (1970) and at 10 μm and 2 μm by G. H. Rieke and F. J. Low (unpublished). At the latter wavelength, the flux comes from a compact (diameter $\sim 3''$) nucleus, and from a bright nuclear disk that is symmetric about the nucleus and lies along the galactic plane with dimensions of $\sim 20'' \times 5''$. At 10 μm the source lies along the plane with dimensions of $\sim 40'' \times 8''$. The structure does not resemble that at 2 μm—in particular, the nucleus is not prominent at 10 μm—but there is a very close correspondence between the map at 10 μm and the nonthermal radio structure (Hargrave 1974, Kronberg & Wilkinson 1975), except that the very compact radio source 41.9 + 58 has no counterpart at 10 μm. Telesco and Harper (Telesco 1977) scanned M82 at 58 μm and showed that the size of the source at least along the plane of the galaxy is similar at this wavelength to the size at 10 μm. The extent and distribution of Bα (Simon et al. 1979) are also similar to the 10-μm source but not to that at 2 μm. The distribution of the [Ne II] emission (Beck et al. 1978) is roughly similar to the 10-μm source except that the [Ne II] emission is less extended to the west; since Bα is detected from this area, the absence of [Ne II] indicates that the ionizing sources are of lower temperature than elsewhere in the region.

An upper limit to the mass of the $40'' \times 8''$ infrared-emitting region can be estimated from the rotation curve under the approximation that the mass is distributed spherically symmetrically. The optically determined rotation curve (e.g. Burbidge et al. 1964) indicates a velocity differential of 180 km s^{-1} across this region. In view of the obscuration, this estimate may have systematic errors. The CO line is $\sim$40 km s^{-1} wider than the Hα line (Rickard et al. 1977); we therefore adopt a velocity differential of 220 km s^{-1}. This estimate agrees well with an extrapolation of the [Ne II] rotation curve measured by Beck et al. (1978), who also found evidence for noncircular motions possibly associated with mass ejection from the nucleus. The mass of the infrared-emitting region is then $\leq 8 \times 10^{8} M_{\odot}$. Nearly $3 \times 10^{8} M_{\odot}$ is in neutral (Rickard et al. 1977) and ionized (Willner et al. 1977) gas. Thus, M/L for the stellar population is <0.02.

It would be attractive to attribute the very low M/L for M82 to rapid star formation with an initial mass function similar to that found in the solar neighborhood. Such models have been considered by Struck-Marcell & Tinsley (1978). However, a number of constraints must be met. First, the light of the galaxy is strongly red-giant dominated at 2 μm (Willner et al. 1977, R. I. Thompson et al., unpublished), contrary to the predictions of models for a population of the youth implied by the M/L. Second, it will be difficult to meet the excitation conditions for the ionized gas with such a model. Third, the different spatial distribution of the red giants compared with the ionized gas and heated dust, together with the very large flux near 2 μm, implies that a significant part of the mass of this region is in stars from a previous bust of star formation that has now evolved to a relatively lower rate of energy generation. The requirements on the current dominant burst of star formation may therefore be made even more demanding. Considering these constraints, it seems likely that stars are forming in M82 with an initial mass function weighted substantially toward massive stars (if stars account for the luminosity of M82).

NGC 253   The extreme behavior deduced from the infrared observations of M82, as contrasted with M31, may not seem surprising in view of the bizarre properties and morphology of M82. However, from the infrared point of view M82 has a virtual twin in the Sc spiral galaxy NGC 253. As will be discussed in the following section, many other apparently normal spiral galaxies also have properties similar to, although less extreme than, those of M82 and NGC 253.

The spectrum and structure of the infrared source in NGC 253 can be determined from the work of Becklin et al. (1973b), Glass (1973b), Rieke et al. (1973), Rieke & Low (1975c), Gillett et al. (1975a), Hildebrand et al. (1977), and Russell et al. (1977). Rather than reproduce the spectrum here, we remark that it is virtually identical to that of M82, even including the strengths of the absorption and emission features, except that the spectrum of NGC 253 rises slightly more steeply into the far infrared. The luminosity is 2.8×10^{10} $L_\odot$ (Telesco 1977), within errors equal to that of M82, and the ratio of M/L is comparable with that for M82 (Rieke & Low 1975c, Telesco 1977) or, if a mass estimate by Ulrich (1978) is adopted, may be even lower. The 10-μm source coincides with a compact nonthermal radio source (Rieke & Low 1975c). The 2-μm spectrum is dominated by red giants (R. I. Thompson et al., unpublished). Thus, the discussion of M82 can be transferred virtually without modification to NGC 253.

C   Mid- and Far-Infrared Observations of Bright Spiral Galaxies

The two galaxies discussed above as examples of infrared-bright galaxies are not isolated cases. Rieke & Lebofsky (1978) observed most of the galaxies with $M_{pg} < 11.0$ and detected 16 of the 39 spirals in the sample at 10 μm. All of these galaxies have infrared excesses above the output of stars since the stellar output can be detected only for very nearby galaxies such as M31. These observations used small ($\sim 6''$) beams that typically correspond to ~ 300 pc so the measurements apply to the nuclei only. Other measurements of some of these galaxies are presented in Rieke & Low (1972a), Kleinmann & Wright (1974), and Dyck et al. (1978); references to earlier work can be found in these papers.

As discussed in Rieke & Lebofsky (1978), an average M/L ratio for the spirals computed using the mass for the entire sample but only the luminosity from the sixteen detected galaxies is ~ 0.2. This M/L ratio is close to the minimum that can be maintained by thermonuclear reactions for the lifetime of the universe (Rieke & Low 1975c). Therefore, to maintain an average M/L ratio this low may require a large-scale exchange of relatively unprocessed material from areas surrounding the nucleus with heavily processed nuclear material. The extremely low M/L ratios in a large fraction of individual spiral nuclei may be explained in the same manner as for M82 and NGC 253.

In addition to M82 and NGC 253, a number of other bright galaxies have silicate absorption features and 3.3-μm emission (Lebofsky & Rieke 1979), and thermal emission by dust is most likely to be the dominant mechanism in producing the infrared radiation. These measurements indicate much larger column densities of dust than previously suspected. The sources responsible for the high luminosities were probably not detected visibly because of the very heavy obscuration. Detailed mapping at 2.2 μm of NGC 6946, a nearly face-on spiral, indicates that the nucleus is so obscured that it does not appear as the brightest spot at optical wavelengths (Lebofsky & Rieke 1979). Similar effects have been found for other galaxies (G. Neugebauer 1978, private communication). If many spirals have such heavy extinction, then the lack of correlation between infrared and optical properties such as emission-line strength may result since different regions are being observed at the two wavelengths. The relatively small levels of obscuration deduced from ultraviolet and visible colors may only reflect inhomogeneities in the dust clouds or that the sources are optically thick at these wavelengths so that one sees only to optical depth unity (corresponding to $A_\lambda \sim 1$). Whenever an extinction $A_\lambda \gtrsim 1$ is encountered in an extragalactic source, one should be aware that these effects may be present.

Other characteristics of these galaxies that are similar to those of M82 and NGC 253 are the typical sizes, 100–600 pc, of the 10-μm emitting regions (Rieke 1976a), the fact that these galaxies follow the radio-infrared relation discussed in Section IV, and an absence of free-free radio emission at the level typical for large H II regions of comparable brightness in the infrared (Rieke 1976a).

The total luminosity has been measured only for a few of these galaxies, but all are of the same order, $\sim 10^{10} L_\odot$, as M82 and NGC 253. E. E. Becklin et al. (1977, unpublished) have observed IC 342 and Telesco (1977) has observed NGC 2903, NGC 5194, NGC 5236, and NGC 6946 through 100 μm and have shown them all to be similar to M82 and NGC 253 in their far-infrared properties. NGC 5194 is of special interest since it has not been detected at 10 μm (Rieke & Lebofsky 1978) and has an unusual ratio of $L_{IR}/L_{Lyman\alpha}$ (Telesco 1977). The presence of interstellar CO emission may be correlated with strong far-infrared fluxes; additional measurements are needed to confirm this possibility.

In addition to the nearby, normal-appearing spiral galaxies discussed above, strong infrared excesses have been found in a broad variety of irregular galaxies and in many slightly more distant spiral galaxies with bright emission-line spectra from their nuclei. Surveys of these types of extragalactic infrared source can be found in Rieke & Low (1972a), Neugebauer et al. (1976), and Allen (1976); additional surveys are being readied for publication by J. A. Frogel and by G. H. Rieke, M. Tarenghi, and M. J. Lebofsky.

A number of these sources, e.g. NGC 1614 and NGC 7714, have a strong resemblance to type 2 Seyfert galaxies with regard to morphology, emission-line spectrum, and infrared luminosity. With broader lines, they would be typical of this class of Seyfert galaxy. An additional resemblance between galaxies with high infrared luminosity and Seyfert galaxies is suggested by spectroscopic studies of NGC 253 (Ulrich 1978) and M82 (Beck et al. 1978), which show large velocities in the nuclear region roughly perpendicular to the line of sight. The narrow lines in these sources may in part result from the high spatial resolution made possible since they are close to the local group and from the fact that they are viewed edge-on. However, NGC 1614 and NGC 7714 are viewed more nearly face-on and still have narrow lines (Demoulin et al. 1968, Ulrich 1972). The infrared observations strongly support suggestions of a close resemblance and possible evolutionary relation between type 2 Seyfert galaxies and some strong emission-line galaxies.

Compact irregular galaxies in interacting systems, e.g. NGC 5195 and Mrk 171, tend to be strong infrared sources (G. H. Rieke, M. Tarenghi, and M. J. Lebofsky, unpublished). The companion galaxy is frequently also found to have an infrared excess. It has been suggested that tidal interactions between galaxies may trigger bursts of star formation (see, for example, Larson & Tinsley 1978); the infrared measurements appear to support this hypothesis.

The giant radio elliptical galaxies Cen A and Cyg A also contain strong nuclear infrared sources (Becklin et al. 1971, Rieke & Low 1972a). The former galaxy has been studied in some detail. Grasdalen & Joyce (1976a) observed the galaxy spectroscopically from 2.9–4.1 μm and from 8–13 μm. There is an absorption feature at 10 μm, from which they deduce a visual extinction to the nucleus of $\sim$22 magnitudes. However, unlike M82 and NGC 253, there are no emission features in either spectral range. The spectrum of Cen A rises to a peak near 100 μm (D. A. Harper, unpublished) and the general behavior is similar to the spectra of other galaxies with strong thermal infrared emission. There is no evidence for variability in the infrared (Grasdalen & Joyce 1976a). The source contains a compact (diameter $< 3''$) central component (Grasdalen & Joyce 1976a) but emission is also detectable over distances of $\pm 1'$ along the obscuring lane across the galaxy (Telesco 1978).

The dwarf emission-line galaxies II Zw40 and He2-10 also have infrared excesses (Rieke & Low 1972a, Allen et al. 1976). Their infrared properties are generally consistent with the suggestion that they are dominated by H II regions, although II Zw40 may be rather faint at 100 μm (Harvey et al. 1978).

4 SEYFERT GALAXIES

Strong infrared emission is virtually a universal characteristic of Seyfert galaxies (Rieke 1978). For type 2 (Weedman 1977) Seyfert galaxy nuclei, the infrared flux accounts for most of the total luminosity. For type 1 nuclei, the sample of known Seyfert galaxies may be significantly biased against those with strong infrared fluxes (see below); nonetheless, the infrared luminosity clearly is a significant part of the total. Even for NGC 4151, which is much brighter than average in the X-ray region [if all type 1 Seyfert galaxies were as bright, the total X-ray flux would exceed the observed diffuse background (Gursky & Schwartz 1977)], the infrared luminosity is as large or larger than that in any other spectral region, excepting gamma ray (Di Cocco et al. 1977).

Despite the important role of the infrared region in the emission of Seyfert galaxies, there has been disagreement even on the mechanism producing the infrared fluxes. Recently some consensus has emerged that type 2 nuclei emit predominantly through thermal reradiation by dust grains heated by the central ultraviolet source (see, for example, Stein & Weedman 1976, Neugebauer et al. 1976, Rieke 1978). However, for type 1 nuclei it has been argued that one detects the nuclear nonthermal source directly in the infrared (e.g. Neugebauer et al. 1976, Stein & Weedman 1976, Weedman 1977), or that one sees a combination of the nonthermal source and thermal reradiation by dust (W. A. Stein 1978, private communication), or that one sees thermal reradiation by dust with the role of nonthermal emission so far not determined (Rieke 1978).

A *Variability*

A decisive argument in favor of nonthermal mechanisms for the infrared emission could be made if sufficiently rapid variations in flux level occur. Dust grains should not survive at temperatures above ∼ 1000 K, so that a minimum angular diameter for a thermally reradiating source is that of a 1000 K blackbody capable of emitting the observed flux. Without a contrived geometry, the fastest timescale for variability in such a source is of the order of the radius divided by the speed of light. At 2 μm, these minimum time scales range from a few weeks to a few years, depending on the luminosity of the source.

A number of observers have investigated the near-infrared variability of Seyfert galaxies (e.g. Pacholczyk 1971, Penston et al. 1974, O'Dell et al. 1978a, Rieke & Lebofsky 1979). Nearly every study has found some evidence for changes in one or more sources. The behavior of two of the most intensively studied galaxies is illustrated in Figure 2. Among all galaxies studied to date 3C 120 is unique in that a change by more than a factor of two has occurred, many times larger than the possible measurement errors. In all other cases, doubling the estimated measurement errors would seriously reduce the confidence in the apparent variations. Therefore, in these cases some skepticism may still be justified regarding the reality of the changes. After 3C 120, the strongest case for variations can be made for NGC 4151, where small ($\sim 30\%$) changes have been reported by a number of observers (Pacholczyk 1971, Penston et al. 1974, O'Dell et al. 1978a). Variability is also suspected for NGC 1275 (Rieke & Lebofsky 1979), NGC 3783 (Glass 1979a), and Markarian 509 (see below).

The minimum time scales for thermal sources are ~ 1 year and ~ 1

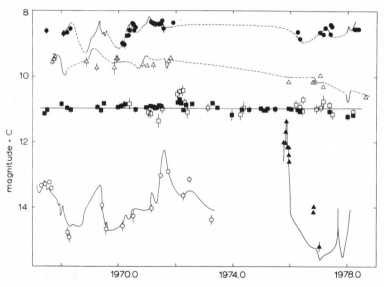

Figure 2   Infrared variations of extragalactic sources. NGC 4151 at 2.2 μm is indicated by filled circles and has $C = 0$; 3C120 at 2.2 μm by open triangles and $C = -0.4$; 3C273 at 2.2 μm by filled squares and $C = 1.3$; 3C273 at 10 μm by open squares and $C = 5.8$; 3C345 at 2.2 μm by open circles and $C = 1.6$; and 0235+164 by filled triangles and $C = 1.9$. References are in the text. For 3C273, the solid line is horizontal for comparison with possible variations; for all the other sources the solid lines show the behavior in the blue (renormalized for comparison with the infrared). Particularly for sources like 3C345 and 0235+164, spurious deviation between the blue and infrared light curves may appear during rapid variations not sampled simultaneously at both wavelengths.

month respectively for 3C 120 and NGC 4151. There are no observations as yet of near-infrared variations on scales shorter than these. In the long term, the infrared light curves seem to follow the behavior in the blue; however, there is some evidence that the infrared flux does not follow the most rapid variations in the blue and ultraviolet (see, for instance, Penston et al. 1974). This behavior is all consistent with the infrared flux originating in dust clouds that absorb the output of the central source and reemit at a level that reflects the output of the central source, averaged by time delays over the cloud complex. The behavior is also consistent with a nonthermal infrared source.

This ambiguity might be removed by additional observations of high accuracy and with thorough coverage in time. Near-infrared photometry has become sufficiently sensitive during the past five years that both of these goals could be achieved easily.

Because cooler dust grains would dominate the emission and the minimum size of a thermal source would be correspondingly larger, monitoring at longer wavelengths can in principle distinguish more easily between thermal and nonthermal sources. Observations at 10 μm have, in fact, produced some evidence for variations (e.g. Rieke & Low 1972b, Stein et al. 1974). However, there are contradictions in the measurements by different groups. In addition, the apparent changes are relatively small, such that doubling the estimated errors would significantly reduce the confidence at which the changes are detected. A review of the Arizona measurements (Rieke & Low 1972b) shows that part, but not all, of the apparent changes resulted from unsuspected inconsistencies in the system of standard stars, which were exacerbated by the data reduction techniques. In summary, the available data suggest variations in Seyfert galaxies at 10 μm, but confirmation is needed. Recent measurements of 3C 120 at 10 μm (M. J. Lebofsky and G. H. Rieke, unpublished) indicate a flux 0.86 ± 0.15 times that observed in 1971–1972 (Rieke & Low 1972a), consistent with the expected behavior of a thermal source.

B *Examples*

NGC 1068   The spectral flux distribution from the type 2 Seyfert galaxy NGC 1068 is illustrated in Figure 1, drawn from Hildebrand et al. (1977), Telesco et al. (1976), Lebofsky et al. (1978), Kleinmann et al. (1976), and W. F. Forrest et al. (1978, unpublished). Particularly in the 8–34-μm region, it has been necessary to adjudicate discordances to arrive at this figure: between 8 and 13 μm, the spectrophotometry by Kleinmann et al. (1976) and photometry by Rieke & Low (1975a) agree well and delineate a spectrum that differs from the results of Jameson et al. (1974); the spectrum from 16–25 μm is from Lebofsky et al. (1978) and W. F. Forrest et al. (1978, unpublished) and also differs from Jameson et al. (1974); at 34 μm, the ground-based measurements by Rieke & Low (1975a) agree well with airborne measurements by Telesco et al. (1976) and Telesco (1977) and with a small extrapolation of those by W. F. Forrest et al. (1978, unpublished).

The infrared luminosity of NGC 1068 is 3×10^{11} $L_\odot$ [assuming a distance of 18 Mpc (Sandage & Tammann 1975)], nearly two orders of magnitude larger than the visual (0.3–1 μm) luminosity. This energy is produced in two source components. At least half of the luminosity comes from a warm, compact nuclear source, for which Becklin et al. (1973a) measured a diameter at 10 μm of $\sim 1''$ (90 pc). Taking the cold, far-infrared source to be thermal, it must have a diameter of at least 5″ (Hildebrand et al. 1977). There is marginal evidence for an increase of flux with aperture at 60 μm between 18 and 50″ (Telesco 1977).

A number of attempts have been made to model theoretically the infrared properties of NGC 1068, the most successful being by Jones et al. (1977). They carried out detailed calculations of the radiative transfer in a spherically symmetric cloud containing a mixture of silicate and graphite dust grains and surrounding an ultraviolet source of luminosity $\sim 2 \times 10^{11}$ $L_\odot$. The apparent diameter of the source at 10 μm was constrained to agree with the measured value. As shown in Figure 1, Jones et al. (1977) were able to fit reasonably well the spectrum between 4 and 40 μm. Note that the silicate emission feature is shifted in their calculation to ~ 22 μm because they assumed that the peak emissivity of the grains occurred at this wavelength, whereas measured spectra of interstellar and circumstellar dust indicate that the peak emissivity is near 18 μm. The spectrum longward of 40 μm required a second, much larger source component, which they suggest is associated with dust in extensive molecular clouds surrounding the nucleus out to a distance of ~ 3 kpc. The spectrum shortward of 3 μm had to arise in a third source component, possibly associated with the nuclear core.

It is unlikely that the necessary simplifying assumptions for model calculations, e.g. spherical symmetry, correspond to the actual source in NGC 1068, but the close correspondence of the model of Jones et al. with the middle infrared spectrum appears to establish certain important assumptions underlying their calculations, i.e. that the middle infrared flux is generated predominantly thermally in a dense cloud containing dust with properties roughly similar to those of interstellar dust in the solar neighborhood.

One of the most interesting implications of the middle infrared measurements and the model by Jones et al. is that the amount of dust corresponds to a visual extinction of ~ 10 magnitudes, which scattering effects will reduce to ~ 4 magnitudes (Jones & Stein 1975). Thus, the nuclear core would be virtually invisible except for scattered light and inhomogeneities in the extinction. Beyond 3 μm, the output of the core is dominated by the thermal emission of the surrounding cloud, so that the core may be detectable directly only in the near infrared, radio, and X-ray. Thus, the near infrared is of great importance to study this source.

A spectrum of NGC 1068 from 1.9 to 2.5 μm has been obtained by Thompson et al. (1978). It shows He, H_2, Brackett γ, and unidentified weak lines detected previously in NGC 7027 and MWC 349. All of the lines have a width comparable with those of the lines in the visible (~ 1000 km s^{-1}). Brackett γ is somewhat weaker than expected from the predictions of recombination theory and the strength of $H\alpha$, even assuming only the level of extinction needed to account for the relative line strengths in the visible, and is substantially weaker if the extinction deduced from the middle infrared is correct. Either the near-infrared extinction differs significantly from our expectations, or the line ratios are not predicted accurately by recombination. The detection of molecular hydrogen in its vibrational-rotational states, which are believed to be excited by shock heating (e.g. Hollenbach & Schull 1977), is at least qualitatively consistent with the existence of large-scale mass motions within the nucleus. Kleinmann et al. (1976) detected the 12.8 μm Ne II line at a level expected from the strength of the Balmer lines, assuming "normal" abundances and an effective extinction $A_V \sim 3.4$.

Little is known about the continuum spectrum of the central, near-infrared source. The spectrum of Thompson et al. shows it to be smooth and featureless. Models like those of Jones et al. cannot give the continuum shape uniquely because of uncertainties in the geometry and in the effective extinction level and extinction law. However, the continuum is strongly polarized (Knacke & Capps 1974, Dyck & Jones 1976, Lebofsky et al. 1978). Lebofsky et al. have tried to deduce the continuum shape from the wavelength dependence of the near-infrared polarization. A large part of the polarization appears to arise through scattering in the dust shell surrounding the nucleus; although the model is not unique, there is also evidence for an underlying source with a roughly power-law spectrum and wavelength-independent polarization of $\sim 2.5\%$. Additional polarimetric measurements could test and refine this picture of the nuclear source, particularly if the measurements at longer wavelengths can be improved to show whether the polarization at 10 μm (Knacke & Capps 1974) arises from extinction by aligned silicate grains or demonstrates that a polarized central source contributes significantly to the flux at this wavelength.

Hildebrand et al. (1977) considered the far-infrared source in NGC 1068.

From the steepness of the spectrum between 1 mm and 100 μm, and from other arguments, they believe that the flux is emitted thermally. They show that the amount of dust required is consistent with the amount of interstellar matter deduced from the CO observations by Rickard et al. (1977), between 10^9 and $10^{10} M_\odot$. The source must be larger than 5", and presumably consists of large molecular clouds surrounding the nucleus (see also Elias et al. 1978).

NGC 4151   The infrared luminosity of the type 1 Seyfert galaxy NGC 4151 is $\sim 1 \times 10^{10} L_\odot$, significantly larger than the radio, optical, and ultra-violet luminosity and comparable with the output in the X-ray region. The spectral flux distribution from NGC 4151 is illustrated in Figure 1, drawn from Rieke & Low (1975b) and unpublished measurements by M. J. Lebofsky and G. H. Rieke and by D. A. Harper. Simultaneous or nearly simultaneous observations covering the regions 0.3–4 μm and 1–15 μm have been combined to eliminate possible errors in the spectrum because of variability.

Two possibilities have been suggested for the infrared spectrum of galaxies like NGC 4151. In the first case, it is assumed that single power law extends from the ultraviolet into the infrared and that the infrared flux is generated predominantly by the same nonthermal source respon-sible for the ultraviolet excess (Neugebauer et al. 1976, Stein & Weedman 1976). In the second case, the ultraviolet and infrared sources are distinct and the spectrum of the infrared sources falls steeply near 1 μm (Penston et al. 1974, Rieke 1978).

Both models can fit the spectrum in Figure 1. The first one requires that stars contribute about 40% of the light at V while the second one would increase this percentage to $\sim 75\%$. The small percentage of stars in the first class of model is difficult to reconcile with the dependence of flux on aperture (Zasov & Lyutyi 1973, Penston et al. 1974), which which shows that $\sim 50\%$ of the flux in Figure 1 (taken with an 18" aperture) lies more than 2".5 from the nucleus. In addition, the under-lying assumption in this class of model is violated by the color changes with variations in flux level (Lyutyi 1973, Penston et al. 1974), which would require that the slope of the ultraviolet continuum change (note that the slope of the power-law continuum is similar in the blue-to-red region to that of the stellar flux). Therefore, models of the second type are favored, in which the ultraviolet and infrared have different slopes and may well be generated by different mechanisms. Such a model is illustrated in Figure 1.

The infrared continuum of NGC 4151 is only weakly polarized (Kemp et al. 1977), with a wavelength dependence similar to the interstellar polarization law found for stars in our own galaxy. The wavelength dependence does not appear to be consistent with a model where the flux from a polarized, nonthermal, power-law source is diluted by unpolarized stellar flux, since the polarization peaks near 0.5 μm, near the wavelength where the stars contribute the largest percentage of the flux. A plausible explanation for the polarization behavior would be the presence of aligned dust grains in the nucleus.

The spectrum of NGC 4151 does not show any 10-μm absorption or emission feature (Lebofsky & Rieke 1979, Figure 1). There is therefore no direct evidence that the infrared flux is generated thermally by heated dust. However, this interpretation is consistent with the spectrum of this source component, which appears to steepen shortward of 2 μm and to peak near 60 μm, and with the possibility of dust within the nucleus, as suggested by the polarization.

MRK 231 AND MRK 509   The spectrum of Markarian 231 (Rieke 1976b, Joyce et al. 1975, Allen 1976) illustrates a type 1 Seyfert galaxy with strong absorption at 10 μm and presumably with infrared emission predominantly by thermal mechanisms. The presence of dust in the nucleus of this galaxy is also indicated by the polarimetric behavior (Kemp et al. 1977) and the strongly reddened emission-line system (Boksenberg et al. 1977). The luminosity of Markarian 231 is $4 \times 10^{12} \, L_\odot$, larger than that of any other known Seyfert galaxy and comparable with the more luminous quasistellar objects (Rieke & Low 1975b).

Markarian 509 is included in Figure 1 to show the spectrum of a type 1 Seyfert galaxy for which the infrared output does not dominate the visible and ultraviolet luminosity (the spectrum is from Allen 1976, Rieke 1978, and Glass 1979a). In preparing this figure, a total range of flux level of a factor of 1.6 was found in the published measurements, indicating variability as shown by the two spectra plotted. Examples like this one may well be the most likely candidates to exhibit clearly nonthermal infrared sources in Seyfert galaxy nuclei, since their spectra are distinctly different from those of most galaxies with thermal emission. Polarimetry, 10-μm spectroscopy, and studies of variability could presumably test for this possibility. Nonetheless, note that Figure 1 implies that there is a spectral inflection near 1 μm and that the infrared source is distinct from the visible-ultraviolet one.

C General Properties

Extensive surveys of the infrared properties of Seyfert galaxies and closely related sources have been carried out by Stein & Weedman (1976), Rieke (1978), McAlary et al. (1978) and G. L. Grasdalen (1978, private communication). Smaller surveys can be found in Penston et al. (1974), Allen (1976), Neugebauer et al. (1976), and Glass (1979a). Individual galaxies in addition to those already described have been studied by Kleinmann & Wright (1974) [NGC 3783 and IC 4329A], Andrews et al. (1974) [PKS 0521-36], Puschell (1978) [Arakelian 120], Glass (1978a) [NGC 5506], Kemp et al. (1977) [NGC 1275], Joyce & Simon (1976) [Mrk 3, 348], O'Dell et al. (1978c) [3C 227 and 3C 382], Lebofsky & Rieke (1979) [NGC 1275, 3227, 4051, 7469, IZw I], references to earlier work can be found in the papers already cited.

From this body of observations, most type 2 galaxies behave very similarly to NGC 1068 (exceptions are Mrk 176 and NGC 1275), with steep spectra that presumably arise from thermal reradiation by dust. Type 1 Seyfert galaxies, on the other hand, exhibit a very broad range of behavior, as already illustrated by the examples above. Their spectra tend to be less steep than those of type 2 galaxies. Both from the presence of 10-μm absorption features in the spectra of individual galaxies (NGC 3227, 4051, and 7469 in addition to Mrk 231; Lebofsky & Rieke 1979), from an anticorrelation of the strengths of the infrared and ultraviolet excesses, and from a weak correlation of the strength of the infrared excess and the reddening indicated by the Balmer decrements, it seems likely that thermal radiation by dust is responsible for a large part of the infrared radiation from the type 1 galaxies with steep infrared spectra (Rieke 1978). There is little evidence with a direct bearing on the radiation mechanism of the galaxies with relatively flat infrared spectra. As in the case of Mrk 509, the measurements of most of these galaxies are not consistent with single power laws extending from the ultraviolet through the infrared. In addition the dust accompanying the large amounts of excited gas in these sources could be expected to produce a thermal source with luminosity at least within an order of magnitude of that observed in the infrared (Rieke 1978).

There appears to be a correlation of infrared and radio fluxes from Seyfert and other types of galaxy (Rieke 1978), as originally suggested by van der Kruit (1971). The significance of this empirical relation is not yet known. Correlations of infrared with X-ray properties have been suggested by some authors (e.g. Glass 1979a). Possible relations will be difficult to evaluate definitively without more X-ray measurements of Seyferts.

In any general discussion of the infrared sample of Seyfert galaxies, the possibility of selection effects must be considered (Rieke 1978). In contrast to the radio, ultraviolet, and X-ray regions, where most, if not all, of the sky has been surveyed uniformly to a sensitivity adequate to detect a reasonably large number of the brightest Seyfert galaxies, there is as yet no infrared survey adequate for this purpose. Instead, infrared studies have been possible only of galaxies discovered in other spectral regions, primarily the ultraviolet. Since the strengths of the infrared and ultraviolet excesses are anticorrelated (Rieke 1978), the known sample must be biased against galaxies with relatively strong infrared sources.

5   QUASI-STELLAR OBJECTS

Unlike the situation for most Seyfert galaxies, there is compelling evidence for some types of QSO that a nonthermal continuum extends through the visible and well into the infrared. The continua of these sources typically rise into the infrared as $v^{-1.2}$. There is a tendency for the most variable QSOs to have the redder spectra. Infrared observations are therefore needed to define the total luminosity of these sources which in many of the most enigmatic objects lies predominantly in that spectral region. In addition, by greatly expanding the wavelength coverage of studies of the continuum, infrared observations have discovered properties of QSOs that were not apparent previously. The infrared region is also proving valuable for spectroscopic studies, particularly since lines that lie in the visible in the rest frame can be observed in the infrared for highly redshifted objects.

A   *Examples*

3C273   The behavior of 3C273 with time is shown in Figure 2, taken from Rieke & Low (1972b), Neugebauer et al. (1979), and our unpublished measurements. The early Arizona data in this case are not affected by the calibrational difficulties mentioned with regard to the observations of Seyfert galaxies. Note that a mild outburst appears to have occurred at 10 μm in early 1972. Support for the reality of this change can be found in the 2.2-μm light curve, from Neugebauer et al. (1979) and also shown in Figure 2. The probable variability of 3C273 is not compatible with any reasonable thermal model for its entire infrared flux; at least a significant component of the infrared spectrum is probably generated nonthermally.

The spectrum of 3C273 is illustrated in Figure 1, taken from Neugebauer et al. (1979), Hildebrand et al. (1977), Rieke & Low (1972b), Grasdalen (1976), and unpublished measurements. Neugebauer et al. (1979) suggest that the broad bulge near 3 μm may arise from heated dust. Although there is no independent evidence in support of this hypothesis, it seems reasonable in view of the large amounts of excited gas in the source. Thus, 3C273 may be an example of an object emitting in the infrared by a combination of thermal and nonthermal mechanisms. If a power-law nonthermal spectrum is estimated from the points at 1.25 and 10 μm, a lower limit to the thermal contribution to the flux at 2 μm is $\sim 35\%$. The decrease between 10 and 2 μm in the amplitude of the probable variability may reflect dilution of the nonthermal flux at the latter wavelength.

The inflection in the spectrum near 1 μm implies that the dominant visible-ultraviolet source is different from the one(s) responsible for most of the infrared flux. Thus, there is evidence for two or three distinct continuum sources being responsible for the ultraviolet-to-infrared spectrum of 3C273, as appears to be the case for Seyfert galaxies as well. In principle, additional information about the continuum could be obtained from polarimetry, but the polarization of 3C273 is small in both the visible and the infrared (Kemp et al. 1977).

Grasdalen (1976) and Puetter et al. (1978) obtained spectra of 3C273 that show Paschen α in emission at very nearly the strength predicted from recombination theory and the strength of the Hβ line (which arises from the same upper level). They conclude that the reddening of the lines is small and that the steep Balmer decrement is an intrinsic property of the source. Puetter et al. (1978) have obtained a similar result for 0026 + 129. These measurements provide one of the most stringent constraints on attempts to fit the relative hydrogen line strengths in QSOs (see Krolik & McKee 1978 and references therein).

3C279 AND 3C345   The near-infrared variability of 3C345 is shown in Figure 2 (data from Neugebauer et al. 1979). Not only are large-amplitude changes apparent, but the infrared and visible regions appear to change in synchronism.

The spectrum of the similar QSO 3C279 is shown in Figure 1, from the data of Elias et al. (1978), Rieke et al. (1977), and Neugebauer et al. (1979). There is some difficulty in matching the infrared and visible spectra, because of the rapid variability of the source. However, both spectra have the same slope and it is almost certain that they join in a single power law of form $v^{-1.5}$. Within the measurement errors, this shape is invariant from 0.3 to 10 μm during changes of flux level by a factor of ~ 2.5.

The behavior of these extremely variable QSOs (and BL Lac sources— see below) poses extremely difficult theoretical problems, which are summarized by Stein et al. (1976). The most salient of these problems is exacerbated substantially by the high infrared luminosities of these sources. The general source characteristics (e.g. continuum, polarization) can be explained in terms of incoherent synchrotron radiation. However, the high surface brightnesses deduced from the variations, particularly in the infrared, would lead in conventional synchrotron-radiating models to an unacceptably high rate of Compton collisions, both with respect to the energetics of the sources (see, for example, Hoyle et al. 1966) and to the upper limits to the X-ray fluxes (see, for example, Margon et al. 1976).

AO 0235 + 164   The BL Lac-type sources appear to be very closely related to violently variable QSOs like 3C279 and 3C345. The most extensively studied object of this kind in the infrared is AO 0235 + 164, the variations of which are plotted in Figure 2 (data from Rieke et al. 1976, 1977, and O'Dell et al. 1978a). The spectral behavior during the variations is illustrated in Figure 1, from the same data sources already cited.

To first order, the infrared spectrum of AO 0235 + 164 has remained the same while the luminosity has varied from $\sim 3 \times 10^{13}$ $L_\odot$ to $\sim 3 \times 10^{14}$ $L_\odot$. However, closer inspection reveals significant variation of the spectrum, particularly during the outburst in 1975 (see Figure 1). At the beginning, the spectrum had a strong downward curvature toward longer wavelengths; toward the end of the outburst, the spectrum had become much more nearly a simple power law. Roughly similar behavior has been seen in 1308 + 326, another exceptionally luminous BL Lac-type source (O'Dell et al. 1978a, Puschell et al. 1979). Even at its maximum, AO 0235 + 164 showed infrared variations by $\sim 30\%$ in two days. It therefore poses in the most extreme form the theoretical problems mentioned with regard to 3C279 and 3C345.

B   General

Infrared surveys of emission-line QSOs have been carried out by Oke et al. (1970) and Neugebauer et al. (1979). Additional photometry of sources not discussed above can be found in Glass (1979a) [2251 − 178, 2154 − 18, 0537 − 441], Rieke & Low (1975b) [3C48], and O'Dell et al. (1978b) [0736 + 017]. Data on BL Lac-type sources have been spread over a large number of publications, with no comprehensive survey. Table 1 summarizes the information available on these objects. The paper by O'Dell et al. (1978a) includes a republication of results from O'Dell et al. (1977a,b, 1978b) and Baldwin et al. (1977). The earlier references are therefore excluded from the table.

In the survey by Neugebauer et al. (1979), QSOs were selected for infrared observations purely on the basis of visual magnitude brighter than 17. As will be discussed later, this sample may be significantly biased against infrared-bright sources. The spectra of the QSOs rise into the infrared roughly as $v^{-1.2}$. The visual-to-infrared luminosities range from

Table 1   Infrared observations of BL Lac sources

Source	References[a]
0109 + 22	16
3C66A	13
0235 + 164	12, 14, 15
0735 + 178	8, 14, 15
OI 090.4	14
OJ 287	2, 4–6, 8, 14, 15, 17
0912 + 29	8
1101 + 38 = Mrk 421	9–11, 14, 15
ON 325	5, 8, 13
ON 231	5, 8, 16
1308 + 326	15, 19
1400 + 162	15
1418 + 54	16
AP Lib	5, 7, 18
1652 + 398 = Mrk 501	9, 10, 11, 15
BL Lac	1, 3–5, 8, 15, 17

[a] References for Table 1:

1. Oke et al. 1969	11. Joyce & Simon 1976
2. Dyck et al. 1971	12. Rieke et al. 1976
3. Stein et al. 1971	13. O'Dell et al. 1977a
4. Epstein et al. 1972	14. Rieke et al. 1977
5. Rieke 1972	15. O'Dell et al. 1978a
6. Strittmatter et al. 1972	16. O'Dell et al. 1978b
7. Andrews et al. 1974	17. Rudnick et al. 1978
8. Rieke & Kinman 1974	18. Glass 1979
9. Ulrich et al. 1975	19. Puschell et al. 1979
10. Allen 1976	

$\sim 10^{11}$ to $\sim 3 \times 10^{13}$ $L_{\odot}$ (for an exceptionally luminous example, see Wright & Kleinmann 1978). The power in the 1–10-μm region dominates the contribution from 0.3–1 μm by a factor of two or three. There is evidence for at least two general types of source. For those analogous to 3C273, a power law or other simple spectral shape is unable to fit the data; separate source components dominating in the visible and infrared are indicated. The curvature of the spectrum near 3 μm seen in 3C273 is not as clearly present in any other source (quite possibly simply because of the lower accuracy achieved on other sources), but there are indications of a similar curvature in a number of cases. In many respects, the behavior of this class of object is analogous to that of type 1 Seyfert galaxies, except that the QSOs on the average have higher luminosities and may have infrared spectra with a more dominant power-law behavior. For the second type of source such as 3C279 and 3C345, a single power law is a reasonable approximation to the continuum, the spectrum frequently rises steeply (power-law index $\alpha \sim -1.5$) into the infrared, and rapid and large-amplitude variability is seen. These objects are similar to BL Lac-type sources. However, to our knowledge there is as yet no example of a low-luminosity, strong emission-line object with these properties, in analogy with the BL Lac-type sources in the nuclei of Mrk 421 and Mrk 501 (Ulrich et al. 1975). The two-part division of QSOs suggested by the behavior of the continua is also suggested by other lines of evidence, such as polarimetry (see, for example, Stockman & Angel 1978).

The characteristics described above do not apply in all cases. For example, the exceptionally red (in the visible) QSO 3C68.1 has a more normal slope of $\alpha \sim -1.8$ in the near infrared. For other apparent exceptions, such as PHL 957 and Mrk 132, the spectrum does not rise into the near infrared. However, these sources are at high redshift ($z = 2.69$ and 1.76 respectively), so that the infrared observations refer to the visible-to-ultraviolet in the rest frame, where many other QSOs have similar spectra.

Balmer lines shifted into the infrared have been measured in the high-redshift QSOs 0237−23 (Hyland et al. 1978) and 1225+317 (Soifer et al. 1979, Puetter et al. 1979). In agreement with the conclusions of Baldwin (1977) for a composite QSO spectrum and of Davidsen et al. (1977) and Boksenberg et al. (1978) from direct observation of Lα in the spectrum of 3C273, it was found that the ratio of intensities Lα/Hα is much lower (1–3) than predicted by recombination theory (8–11). The anomalous hydrogen line intensity ratios in QSOs remain an important and unsolved theoretical problem. For a discussion, see Krolik & McKee (1978), and references therein. Soifer et al. (1978) also showed that the Hα/Hβ ratio for 1225+31 is in the range found by Baldwin (1977) for low redshift QSOs. Thus, the infrared measurements argue strongly against the existence of significant spectroscopic differences between low- and high-redshift QSOs.

As already noted, BL Lac-type objects have many similarities to the highly variable QSOs. However, a significant difference in spectral characteristics (in addition to the absence of emission lines) is suggested by combined infrared and visible observations. The emission-line sources all have spectra that can be fitted reasonably well by a single power law; in the case of 3C446, for which $z = 1.40$, this behavior persists at least to wavelengths as short as 0.14 μm in the rest frame (Neugebauer et al. 1979). Although the infrared spectra of BL Lac-type sources are similar in slope to those of the emission-line sources, the former class frequently exhibits a steepening of the spectrum toward shorter wavelengths, i.e. in the visible and ultraviolet. This behavior is seen in all three of the BL Lac-type sources known to have large redshifts—0235+164 (already discussed), 0735+178 (Rieke & Kinman 1974, Rieke et al. 1977, and O'Dell et al. 1978a), and 1308+326 (Puschell et al. 1979). It should be mentioned that, in the latter two cases, extreme fluctuations in the ultraviolet spectrum (Rieke & Kinman 1974, O'Dell et al. 1978a) may significantly reduce or even eliminate the spectral steepening on a few occasions, but in an average sense it appears to be a permanent characteristic of the sources. A steepening is also seen for Mrk 421 (Ulrich et al. 1975, Boksenberg et al. 1978), ON231, ON325 (= B21215+30), and OJ 287 (Rieke & Kinman 1974), and BL Lac (Oke et al. 1969, O'Dell et al. 1977a).

Two explanations might be considered for the continuum behavior of the BL Lac-type sources: that it is a result of reddening by the material responsible for the absorption lines in their spectra, or that it is an intrinsic property of the sources. Emission-line QSOs exhibit little correlation between the presence of absorption lines and the colors in the visible and ultraviolet; the first possibility therefore seems unlikely. The measurements suggest that the lack of emission lines in some BL Lac-type sources results at least in part from a deficiency of ultraviolet flux. The relatively strong ultraviolet flux from B21101+38 (= Mrk 421) (Boksenberg et al. 1978) is a countervailing indication that the lack of lines results from an absence of material near the ionizing source. Further measurements of the infrared-to-ultraviolet spectra of these sources are needed to resolve this dilemma.

Knacke et al. (1976) measured BL Lac in the visible and near infrared to show that the degree and direction of polarization were very nearly the same from 0.4 through 3.6 μm. This behavior confirms independently of the variability and spectral behavior that a single, nonthermal source component dominates the output of the source over this entire spectral range. Rieke et al. (1977) measured the polarization of the BL Lac-type sources 0735+178 and OI 090.4 and showed that the position angle rotated continuously (although the amount of polarization remained the same) between 2.2 μm and the visible. The measurements by Knacke et al. (1976) suggested a small position angle rotation for BL Lac also, which unpublished measurements by M. J. Lebofsky et al. confirm. From the visual polarimetry by Tapia et al. (1977), it appears that the degree of rotation may be variable in these sources. The observed rotations do not vitiate the basic conclusion by Knacke et al. (1976), but they provide a new parameter to be matched against models of the emission mechanism and source geometry. Additional infrared polarimetry of BL Lac-type sources has been carried out by Rudnick et al. (1978) and Puschell et al. (1979).

Finally, we point out that the known sample of QSOs is probably biased against those with steep infrared spectra. If QSO spectra are represented approximately by $Cv^{-\alpha}$ from 0.3 to 30 μm, the volume of space sampled for QSOs of a given luminosity by searches to a given visual magnitude is a strong function of α (since the visual band lies near one end of the assumed spectral range). For constant luminosity, Euclidean space, and α of 1, 1.5, and 2.5 the volumes would be in the ratio 1:0.2:0.003. Known QSOs have a range of α roughly as $\alpha = 1.2 \pm 0.5$; it seems very likely that the upper limit to α is set by selection effects rather than by an intrinsic property of the sources. A possible confirmation of this hypothesis is our discovery of an extremely red infrared source at the position of the high frequency radio source 0026 + 34 (to be published).

6 OTHER INFRARED STUDIES

Observations at infrared wavelengths have been used to study sources other than nuclei in galaxies and to determine distances and Hubble's constant. To estimate H_0, Aaronson et al. (1979) combined measurements of galaxies at H (1.6 μm) with the relation between absolute magnitude and H I velocity profile found by Tully & Fisher (1977). Because the near-infrared measurements are not subject to uncertainties in the extinction to as large a degree as optical data and because the infrared detects the light of the old red stellar population rather than the young blue component, the infrared relation will allow the distances to galaxies to be calculated and values of the Hubble constant to be estimated. A preliminary estimate of H_0 from this technique is 60 ± 4 km/s/Mpc (Aaronson et al. 1979).

Other extragalactic studies in the infrared include programs such as that of Glass (1979b) where near-infrared measurements of late-type stars in the Magellanic Clouds lead to improved bolometric corrections for these stars, and programs such as that of Gatley et al. (1978) who showed that the far-infrared properties of extragalactic H II regions are similar to galactic H II regions. Another study of extragalactic H II regions (Grasdalen & Joyce 1976b) demonstrates that phases of H II region evolution not seen in our galaxy can be observed in other galaxies. Such studies will not be discussed in this review.

7 CONCLUSION

Of the thousands of galactic nuclei and QSOs studied in the radio and optical, only a small fraction have been measured in the infrared: about 200 near 2 μm, slightly more than 100 at 10 μm, and barely 10 near

100 μm. Nonetheless, many important conclusions have already emerged from the infrared work, including

1. The near-infrared fluxes from the nuclei of elliptical and spiral galaxies are dominated by late-type giants.

2. Very strong mid- and far-infrared fluxes are observed from a broad variety of extragalactic sources, including normal spirals, ellipticals, interacting galaxies, Seyfert galaxies, dwarf emission-line galaxies, giant elliptical radio galaxies, emission-line QSOs, and BL Lac-type sources. The infrared emission dominates the luminosity of many of these types of source.

3. There is a relation between the nonthermal radio and the infrared sources in galactic nuclei, seen both as a proportionality of flux and a similarity in structure between the two spectral regions. The nature of this relation (which is not apparent for Galactic nonthermal radio sources—see Price & Walker 1976) is not understood.

4. The nuclei of many spiral and irregular galaxies are much more luminous than previously suspected. The energy outputs are difficult to reconcile with current ideas regarding star formation in these regions.

5. Most type 2 and at least some type 1 Seyfert galaxies emit thermally in the infrared. The infrared fluxes of the most variable QSOs are produced by the same nonthermal sources responsible for their visible radiation. For most type 1 Seyfert galaxies and QSOs, however, it is not yet possible to determine unambiguously the dominant emission mechanism.

6. Heavy obscuration in the nuclei of many spiral, irregular, and Seyfert galaxies makes it hard to interpret observations in the near infrared, visible, and ultraviolet.

7. The theoretical difficulties in understanding the most variable QSOs are exacerbated by their strong fluxes and other characteristics in the infrared.

8. The spectroscopic characteristics of low- and high-redshift QSOs are similar. Both types exhibit large and not fully explained departures from the line intensity ratios predicted by recombination theory.

9. The known samples of Seyfert galaxies and QSOs are probably biased against infrared-bright sources.

10. Infrared observations can be used to determine fundamental cosmological parameters.

In most cases where infrared observations are available, they have brought some fundamental property of the source to our attention. Indeed, few extragalactic sources can be correctly understood on even an elementary basis without knowledge of their properties in the infrared.

ACKNOWLEDGMENTS

We thank many colleagues who sent preprints in order that we could include their work in advance of publication. Special thanks are due to M. Aaronson, I. Glass, D. A. Harper, J. Houck, T. Kinman, G. Neugebauer, C. Telesco, and M. Werner, who permitted us to make use of their work before it was prepared for publication. We thank H. Butcher, I. Glass, S. Strom, B. Tinsley, R. Weymann, and R. Williams for their comments on early versions of the manuscript. E. Meeden, C. Thompson, and P. Van Buren helped to prepare the article. This work was supported by the National Science Foundation.

Literature Cited

Aaronson, M. 1977. *Infrared Observations of Galaxies.* PhD thesis. Harvard Univ., Cambridge

Aaronson, M. 1978a. *Publ. Astron. Soc. Pac.* 90:28

Aaronson, M. 1978b. *Ap. J. Lett.* 221:L103

Aaronson, M., Cohen, J. G., Mould, J., Malkan, M. 1978b. *Ap. J.* 223:824

Aaronson, M., Frogel, J. A., Persson, S. E. 1978a. *Ap. J.* 220:442

Aaronson, M., Huchra, J., Mould, J. 1979. *Ap. J.* 229:1

Allen, D. A. 1976. *Ap. J.* 207:367

Allen, D. A., Wright, A. E., Goss, W. M. 1976. *MNRAS* 177:91

Andrews, P. J., Glass, I. S., Hawarden, T. G. 1974. *MNRAS* 168:7p

Baldwin, J. A. 1977. *MNRAS* 178:67p

Baldwin, J. A., Wampler, E. J., Burbidge, E. M., O'Dell, S. L., Smith, H. E., Hazard, C., Nordsieck, K. H., Pooley, G., Stein, W. A. 1977. *Ap. J.* 215:408

Baldwin, J. R., Danziger, I. J., Frogel, J. A., Persson, S. E. 1973. *Astrophys. Lett.* 14:1

Beck, S. C., Lacy, J. H., Baas, F., Townes, C. H. 1978. *Ap. J.* 226:545

Becklin, E. E., Fomalont, E. B., Neugebauer, G. 1973b. *Ap. J. Lett.* 181:L27

Becklin, E. E., Frogel, J. A., Kleinmann, D. E., Neugebauer, G., Ney, E. P., Strecker, D. W. 1971. *Ap. J. Lett.* 170:L15

Becklin, E. E., Matthews, K., Neugebauer, G., Wynn-Williams, C. G. 1973a. *Ap. J. Lett.* 186:L69

Boksenberg, A., Carswell, R. F., Allen, D. A., Fosbury, R. A. E., Penston, M. V., Sargent, W. L. W. 1977. *MNRAS* 178:451

Boksenberg, A., Snijders, M. A., Wilson, R., Benvenuti, P., Clavell, J., Macchetto, F., Penston, M., Boggess, A., Gull, T. R., Gondhalekar, P. 1978. *Nature* 275:404

Burbidge, E. M., Burbidge, G. R., Rubin, V. C. 1964. *Ap. J.* 140:942

Davidsen, A. F., Hartig, G. F., Fastie, W. G. 1977. *Nature* 269:203

Demoulin, M.-H., Burbidge, E. M., Burbidge, G. R. 1968. *Ap. J.* 153:31

DiCocco, G., Boella, G., Perotti, F., Stiglitz, R., Villa, G., Baker, R. E., Butler, R. C., Dean, A. J., Martin, S. J., Ramsden, D. 1977. *Nature* 270:319

Dyck, H. M., Becklin, E. E., Capps, R. W. 1978. *Bull. Am. Astron. Soc.* 10:422

Dyck, H. M., Jones, T. J. 1976. *Bull. Am. Astron. Soc.* 8:568

Dyck, H. M., Kinman, T. D., Lockwood, G. W., Landolt, A. U. 1971. *Nature Phys. Sci.* 234:71

Elias, J. H., Ennis, D. J., Gezari, D. Y., Hauser, M. G., Houck, J. R., Lo, K. Y., Matthews, K., Nadeau, D., Neugebauer, G., Werner, M. W., Westbrook, W. E. 1978. *Ap. J.* 220:25

Epstein, E. E., Fogarty, W. G., Hackney, K. R., Hackney, R. L., Leacock, R. J., Pomphrey, R. B., Scott, R. L., Smith, A. G., Hawkins, R. W., Roeder, R. C. 1972. *Ap. J. Lett.* 178:L51

Faber, S. M. 1973. *Ap. J.* 179:731

Freeman, K. C. 1975. *Stars and Stellar Systems, IX: Galaxies and the Universe,* ed. A. Sandage, M. Sandage, J. Kristian, p. 409. Chicago: Univ. Chicago Press

Frogel, J. A., Persson, S. E., Aaronson, M., Becklin, E. E., Matthews, K. 1976. *R. Greenwich Obs. Bull. No. 182:*111

Frogel, J. A., Persson, S. E., Aaronson, M., Becklin, E. E., Matthews, K., Neugebauer, G. 1975a. *Ap. J. Lett.* 200:L123

Frogel, J. A., Persson, S. E., Aaronson, M., Becklin, E. E., Matthews, K., Neugebauer, G. 1975b. *Ap. J. Lett.* 195:L15

Frogel, J. A., Persson, S. E., Aaronson, M., Matthews, K. 1978. *Ap. J.* 220:75

Gatley, I., Harvey, P. M., Thronson, H. A. 1978. *Ap. J. Lett.* 222:L133

Gillett, F. C., Forrest, W. J., Merrill, K. M.,

Capps, R. W., Soifer, B. T. 1975b. *Ap. J.* 200:609
Gillett, F. C., Kleinmann, D. E., Wright, E. L., Capps, R. W. 1975a. *Ap. J. Lett.* 198:L65
Glass, I. S. 1973a. *MNRAS* 162:35p
Glass, I. S. 1973b. *MNRAS* 164:155
Glass, I. S. 1976. *MNRAS* 175:191
Glass, I. S. 1978. *MNRAS* 183:85p
Glass, I. S. 1979a. *MNRAS* 186:29p
Glass, I. S. 1979b. *MNRAS* 186:317
Grasdalen, G. L. 1975. *Ap. J.* 195:605
Grasdalen, G. L. 1976. *Ap. J. Lett.* 208:L11
Grasdalen, G. L., Joyce, R. R. 1976a. *Ap. J.* 208:317
Grasdalen, G. L., Joyce, R. R. 1976b. *Astron. Astrophys.* 50:297
Gunn, J. E., Oke, J. B. 1975. *Ap. J.* 195:225
Gunn, J. E., Tinsley, B. M. 1975. *Nature* 257:454
Gursky, H., Schwartz, D. A. 1977. *Ann. Rev. Astron. Astrophys.* 15:541
Hargrave, P. J. 1974. *MNRAS* 168:491
Harvey, P. M., Gatley, I., Thronson, H. A. 1978. *Bull. Am. Astron. Soc.* 9:629
Hildebrand, R. H., Whitcomb, S. E., Winston, R., Stiening, R. F., Harper, D. A., Moseley, S. H. 1977. *Ap. J.* 216:698
Hollenbach, D. J., Shull, J. M. 1977. *Ap. J.* 216:419
Hoyle, F., Burbidge, G. R., Sargent, W. L. W. 1966. *Nature* 209:751
Hyland, A. R., Becklin, E. E., Neugebauer, G. 1978. *Ap. J. Lett.* 220:L73
Iijima, T., Ito, K., Matsumoto, T., Uyama, K. 1976. *Publ. Astron. Soc. Jpn.* 28:27
Jameson, R. F., Longmore, A. J., McLinn, J. A., Woolf, N. J. 1974. *Ap. J. Lett.* 187: L109
Johnson, H. L. 1966. *Ap. J.* 143:187
Jones, T. W., Leung, C. M., Gould, R. J., Stein, W. A. 1977. *Ap. J.* 212:52
Jones, T. W., Stein, W. A. 1975. *Ap. J.* 197:297
Joyce, R. R., Knacke, R. F., Simon, M., Young, E. 1975. *Publ. Astron. Soc. Pac.* 87:683
Joyce, R. R., Simon, M. 1976. *Publ. Astron. Soc. Pac.* 88:870
Kemp, J. C., Rieke, G. H., Lebofsky, M. J., Coyne, G. V. 1977. *Ap. J. Lett.* 215:L107
Kleinmann, D. E., Gillett, F. C., Wright, E. L. 1976. *Ap. J.* 208:42
Kleinmann, D. E., Low, F. J. 1970. *Ap. J. Lett.* 161:L203
Kleinmann, D. E., Wright, E. L. 1974. *Ap. J. Lett.* 191:L19
Knacke, R. F., Capps, R. W. 1974. *Ap. J. Lett.* 192:L19
Knacke, R. F., Capps, R. W., Johns, M. 1976. *Ap. J. Lett.* 210:L69
Krolik, J. H., McKee, C. F. 1978. *Ap. J. Suppl.* 37:459

Kronberg, P. P., Wilkinson, P. N. 1975. *Ap. J.* 200:430
Larson, R. B. 1975a. *MNRAS* 166:585
Larson, R. B. 1975b. *MNRAS* 173:671
Larson, R. B., Tinsley, B. M. 1978. *Ap. J.* 219:46
Lebofsky, M. J., Rieke, G. H. 1979. *Ap. J.* 229:111
Lebofsky, M. J., Rieke, G. H., Kemp, J. C. 1978. *Ap. J.* 222:95
Lyutyi, V. M. 1973. *Sov. Astron. AJ* 16:763
Margon, B., Bowyer, S., Jones, T. W., Davidsen, A., Mason, K. O., Sanford, P. W. 1976. *Ap. J.* 207:359
Matsumoto, T., Murakami, H., Hamajima, K. 1977. *Publ. Astron. Soc. Jpn.* 29:583
McAlary, C. W., McLaren, R. A., Crabtree, D. R. 1978. *Bull. Am. Astron. Soc.* 10:389
Neugebauer, G., Oke, J. B., Becklin, E. E., Matthews, K. 1979. *Ap. J.* 230:79
Neugebauer, G., Becklin, E. E., Oke, J. B., Searle, L. 1976. *Ap. J.* 205:29
O'Connell, R. W. 1976. *Ap. J.* 206:370
O'Dell, S. L., Puschell, J. J., Stein, W. A. 1977a. *Ap. J.* 213:351
O'Dell, S. L., Puschell, J. J., Stein, W. A., Owen, F., Porcas, R. W., Mufson, S., Moffett, T. J., Ulrich, M.-H. 1978b. *Ap. J.* 224:22
O'Dell, S. L., Puschell, J. J., Stein, W. A., Warner, J. W. 1977b. *Ap. J. Lett.* 214:L105
O'Dell, S. L., Puschell, J. J., Stein, W. A., Warner, J. W. 1978a. *Ap. J. Suppl.* 38:267
O'Dell, S. L., Puschell, J. J., Stein, W. A., Warner, J. W., Ulrich, M.-H. 1978c. *Ap. J.* 219:818
Oke, J. B., Neugebauer, G., Becklin, E. E. 1969. *Ap. J. Lett.* 156:L41
Oke, J. B., Neugebauer, G., Becklin, E. E. 1970. *Ap. J.* 159:341
Oort, J. H. 1977. *Ann. Rev. Astron. Astrophys.* 15:295
Pacholczyk, A. G. 1971. *Ap. J.* 163:449
Pacholczyk, A. G., Tarenghi, M. 1975. *Mem. Soc. Astron. Italiana* 46:199
Peimbert, M., Spinrad, H. 1970. *Ap. J.* 160:429
Penston, M. V. 1973. *MNRAS* 162:359
Penston, M. V., Penston, M. J., Selmes, R. A., Becklin, E. E., Neugebauer, G. 1974. *MNRAS* 169:357
Persson, S. E., Frogel, J. A., Aaronson, M. 1979. *Ap. J.* Submitted
Price, S. D., Walker, R. G. 1976. *AFGL TR-76-0208*
Puetter, R. C., Smith, H. E., Willner, S. P. 1979. *Ap. J. Lett.* 227:L5
Puschell, J. J. 1978. *Publ. Astron. Soc. Pac.* 90:652
Puschell, J. J., Stein, W. A., Jones, T. W., Warner, J. W., Owen, F., Rudnick, L.,

Aller, H., Hodge, P. 1979. *Ap. J. Lett.* 227:L11

Rickard, L. J., Palmer, P., Morris, M., Turner, B. E., Zuckerman, B. 1977. *Ap. J.* 213:673

Rieke, G. H. 1972. *Ap. J. Lett.* 176:L61

Rieke, G. H. 1974. *Ap. J. Lett.* 193:L81

Rieke, G. H. 1976a. *Ap. J. Lett.* 206:L15

Rieke, G. H. 1976b. *Ap. J. Lett.* 210:L5

Rieke, G. H. 1978. *Ap. J.* 226:550

Rieke, G. H., Grasdalen, G. L., Kinman, T. D., Hintzen, P., Wills, B. J., Wills, D. 1976. *Nature* 260:754

Rieke, G. H., Harper, D. A., Low, F. J., Armstrong, K. R. 1973. *Ap. J. Lett.* 183:L67

Rieke, G. H., Kinman, T. D. 1974. *Ap. J. Lett.* 192:L115

Rieke, G. H., Lebofsky, M. J. 1978. *Ap. J. Lett.* 220:L38

Rieke, G. H., Lebofsky, M. J. 1979. *Ap. J.* 227:710

Rieke, G. H., Lebofsky, M. J., Kemp, J. C., Coyne, G. V., Tapia, S. 1977. *Ap. J. Lett.* 218:L37

Rieke, G. H., Low, F. J. 1972a. *Ap. J. Lett.* 176:L95

Rieke, G. H., Low, F. J. 1972b. *Ap. J. Lett.* 177:L115

Rieke, G. H., Low, F. J. 1975a. *Ap. J. Lett.* 199:L13

Rieke, G. H., Low, F. J. 1975b. *Ap. J. Lett.* 200:L67

Rieke, G. H., Low, F. J. 1975c. *Ap. J.* 197:17

Rieke, G. H., Telesco, C. M., Harper, D. A. 1978. *Ap. J.* 220:556

Rudnick, L., Owen, F., Jones, T. W., Puschell, J. J., Stein, W. A. 1978. *Ap. J. Lett.* 225:L5

Russell, R. W., Soifer, B. T., Merrill, K. M. 1977. *Ap. J.* 213:66

Sandage, A. R. 1973. *Ap. J.* 183:711

Sandage, A. R. 1975. *Ap. J.* 202:563

Sandage, A. R., Becklin, E. E., Neugebauer, G. 1969. *Ap. J.* 157:55

Sandage, A. R., Tammann, G. A. 1975. *Ap. J.* 196:313

Scalo, J. M. 1978. *Protostars and Planets*, ed. T. Gehrels, p. 265. Tucson: Univ. Ariz. Press

Simon, M., Simon, T., Joyce, R. R. 1979. *Ap. J.* 227:64

Soifer, B. T., Oke, J. B., Matthews, K., Neugebauer, G. 1979. *Ap. J. Lett.* 227:L1

Soifer, B. T., Pipher, J. L. 1978. *Ann. Rev. Astron. Astrophys.* 16:335

Spinrad, H. 1977. *The Evolution of Galaxies and Stellar Populations*, eds. B. M. Tinsley, R. B. Larson, pp. 301–32. New Haven: Yale Univ. Obs.

Spinrad, H., Bahcall, J., Becklin, E. E., Gunn, J. E., Kristian, J., Neugebauer, G., Sargent, W. L. W., Smith, H. 1973. *Ap. J.* 180:351

Spinrad, H., Sargent, W. L. W., Oke, J. B., Neugebauer, G., Landau, R., King, I. R., Gunn, J. E., Garmire, G., Dieter, N. H. 1971. *Ap. J. Lett.* 163:L25

Spinrad, H., Taylor, B. J. 1971. *Ap. J. Suppl.* 22:445

Stein, W. A., Gillett, F. C., Knacke, R. F. 1971. *Nature* 231:254

Stein, W. A., Gillett, F. C., Merrill, K. M. 1974. *Ap. J.* 187:213

Stein, W. A., O'Dell, S. L., Strittmatter, P. A. 1976. *Ann. Rev. Astron. Astrophys.* 14:173

Stein, W. A., Weedman, D. W. 1976. *Ap. J.* 205:44

Stockman, H. S., Angel, J. R. P. 1978. *Ap. J. Lett.* 220:L67

Strittmatter, P. A., Serkowski, K., Carswell, R., Stein, W. A., Merrill, K. M., Burbidge, E. M. 1972. *Ap. J. Lett.* 175:L7

Strom, K. M., Strom, S. E., Wells, D. C., Romanishin, W. 1978. *Ap. J.* 220:62

Strom, S. E., Strom, K. M., Goad, J. W., Vrba, F. J., Rice, W. 1976. *Ap. J.* 204:684

Struck-Marcell, C., Tinsley, B. M. 1978. *Ap. J.* 221:562

Sweigart, A. V., Gross, P. G. 1978. *Ap. J. Suppl.* 36:405

Tapia, S., Craine, E. R., Gearhart, M. R., Pacht, E., Kraus, J. 1977. *Ap. J. Lett.* 215:L71

Telesco, C. M. 1977. *Galaxies as Far-Infrared Sources.* PhD thesis. Univ. Chicago

Telesco, C. M. 1978. Preprint

Telesco, C. M., Harper, D. A., Lowenstein, R. F. 1976. *Ap. J. Lett.* 203:L53

Thompson, R. I., Lebofsky, M. J., Rieke, G. H. 1978. *Ap. J. Lett.* 222:L49

Tinsley, B. M. 1973. *Ap. J. Lett.* 184:L41

Tinsley, B. M. 1977. *IAU Colloq. No. 37, Decalages vers le Rouge et Expansion*, p. 223

Tinsley, B. M., Gunn, J. E. 1976. *Ap. J.* 203:52

Tully, R. B., Fisher, J. R. 1977. *Astron. Astrophys.* 54:661

Ulrich, M.-H. 1972. *Ap. J.* 178:113

Ulrich, M.-H. 1978. *Ap. J.* 219:424

Ulrich, M.-H., Kinman, T. D., Lynds, C. R., Rieke, G. H., Ekers, R. D. 1975. *Ap. J.* 198:261

van der Kruit, P. C. 1971. *Astron. Astrophys.* 15:110

Weedman, D. W. 1977. *Ann. Rev. Astron. Astrophys.* 15:69

Willner, S. P., Soifer, B. T., Russell, R. W. 1977. *Ap. J. Lett.* 217:L121

Wright, E. L., Kleinmann, D. E. 1978. *Nature* 275:298

Zasov, A. V., Lyutyi, V. M. 1973. *Sov. Astron. AJ* 17:169

Ann. Rev. Astron. Astrophys. 1979. 17:513–49

MODEL ATMOSPHERES ✱2159
FOR INTERMEDIATE- AND
LATE-TYPE STARS

Duane F. Carbon
Kitt Peak National Observatory,[1] P.O. Box 26732, Tucson, Arizona 85726

1.0 INTRODUCTION

The intermediate- and late-type stars are central to a number of exciting research topics in contemporary astronomy. For example, the surface compositions of post–main-sequence stars in the red-giant domain afford a unique and crucial testing ground for theoretical predictions concerning nucleosynthesis and large-scale mixing in stellar interiors. Our understanding of the chemical history of the galactic halo and disk follows in large part from the compositions of intermediate- and late-type stars. Since giant branch stars contribute a major component of the radiative flux from galactic nuclei, they play a key role in investigations of the stellar content of external galaxies. Closer to home, the solar atmosphere provides crucial information on important dynamical processes difficult or impossible to study in stars. In all of the examples just cited progress depends to a large extent on our ability to correctly interpret stellar spectra. This, in turn, requires that we be able to compute accurate models for the atmospheric layers that emit the observed line and continuum radiation. The following review examines the current status of our ability to construct model atmospheres for intermediate- and late-type stars.

"Line blanketing" refers to the presence in the stellar spectrum of atomic and/or molecular absorption lines. The occurrence of pronounced line blanketing in the visual and infrared portions of the spectrum is a distinguishing characteristic of intermediate- and late-type stars. For many years satisfactory model atmospheres for these stars could not be com-

[1] Operated by the Association of Universities for Research in Astronomy, Inc., under contract with the National Science Foundation.

513

0066-4146/79/0915-0513$01.00

puted primarily because no suitable method existed for treating atomic and molecular line opacities. This problem has now been largely solved through efforts over the last dozen years. Model atmospheres with a reasonably accurate treatment of lines can now be computed for most intermediate- and late-type stars. Because the blanketing is well in hand, attention can now be focused on other problems. Chief among these are departures from local thermodynamic equilibrium, convection, and the failure of the standard assumption of plane-parallel geometry. These three problems plus a detailed discussion of atomic and molecular line blanketing are the major topics of this review.

For the present purposes intermediate-type stars are considered to be those with effective temperatures between 8000 and 4000 K; late-type stars are those cooler than 4000 K. The boundary at 4000 K roughly marks the transition from atomic to predominantly molecular line blanketing. The boundary at 8000 K roughly coincides with the transition from hydrogen to H^- as the dominant continuous opacity and with the onset of atmospheric convection. Because of space limitations only photospheric modeling is discussed; the related topic of stellar chromospheres has a rich literature and is best treated in a separate review. Model atmospheres for degenerate stars are not covered. Finally, a number of technical subjects central to the practice of model building are not treated here (e.g. temperature correction procedures, solution of equations of state, solution of transfer equation, atomic, and molecular parameters). The omission is deliberate as these topics are not of great importance to investigators primarily interested in using rather than building atmospheres. For a treatment of these technical points the reader is directed to the following general references on model building: Mihalas (1967), Kurucz (1970), Gustafsson (1971, 1973), Mihalas (1978), and Kurucz (1979). Valuable discussions of atomic and molecular parameters can be found in Tsuji (1964), Grevesse & Sauval (1973), Tsuji (1973), Kurucz & Peytremann (1975), Lambert (1978), and Bell & Gustafsson (1978).

A brief remark concerning the general philosophy adopted in preparing this review is appropriate. The review is not oriented toward the small group of investigators actively engaged in stellar atmosphere research and in computing model atmospheres. Rather the intent is to approach the subject as much as possible from the standpoint of the considerable number of astronomers who use model atmospheres to interpret stellar spectra. The fundamental information provided by a model atmosphere calculation is the atmospheric structure. This structure generally is represented by the run of temperature (T) plus other basic thermodynamic quantities (e.g., gas pressure, P_g, and electron concentration, n_e) as a function of depth in the atmosphere. (The independent depth variable is generally the continuum optical depth,[2] τ_c, at some appropriate frequency although gas pressure or geometric height are occasionally employed.) The model structure is necessary for calculating the emergent spectrum that is to be compared with observations. Different problems in stellar spectroscopy require different levels of model atmosphere sophistication. While some problems are perfectly soluble using a simple, two-layer approximation, others require the most careful application of the best available modeling techniques. Each researcher must determine how dependent the particular analysis to be undertaken is on the adopted model structure. In order to make this determination one must have some feeling for the factors that influence model structure. The aim in this review is to inform the model atmosphere user by emphasizing two themes: 1. How are uncertainties or approximations in the input physics likely to influence the model structure? and 2. How sensitive is the computed model structure to changes in the input parameters?

2.0 LINE BLANKETING

It has long been recognized that line blanketing alters not only the emergent spectrum of a stellar atmosphere but also the structure of the atmosphere itself (Chandrasekhar 1935, Münch 1946). The computational difficulty of treating line blanketing has been the principal obstacle to constructing model atmospheres for intermediate- and late-type stars. Although some technical difficulties still remain, this obstacle has been essentially overcome. The solution of the blanketing problem represents a major advance in our ability to analyze the spectra of intermediate- and late-type stars and opens up exciting new research areas. The following sections provide an introduction to the effects of line blanketing on atmospheric structure and a brief review of the commonly used techniques for constructing blanketed atmospheres.

2.1 The Physics of Line Blanketing

The specific effects of line blanketing on atmospheric structure depend in a complex way on the strength of the line opacities, their depth dependence, and their distribution in frequency. Although the quantitative effects of

---

[2] Frequent reference is made in the text to the behavior of physical parameters on various optical depth scales. These optical depth scales are denoted as τ_x where x specifies the particular optical depth scale used. If x is a number ($\neq 0$) it refers to the continuum optical depth at a wavelength of x μm. The quantity τ_R is the Rosseland mean optical depth; τ_c refers to a characteristic (but unspecified) continuum optical depth—any continuum optical depth scale within the general vicinity of the model's peak emergent flux would usually suffice. τ_0 is equivalent to τ_c.

line blanketing on atmospheric structure can be determined only by detailed numerical computations, the following brief description will serve to illustrate the relevant physics. The reader is encouraged to carefully study the papers by Athay & Skumanich (1969), Athay (1970), Mihalas & Luebke (1971), and Mihalas et al. (1976) for particularly good, detailed discussions of the physics of blanketing.

When line opacities are added to a previously line-free atmosphere the atmospheric temperature structure can be altered dramatically. The deeper, continuum-forming layers ($\tau_c \gtrsim 0.1$) generally will be heated relative to the line-free case, a phenomenon frequently referred to as "backwarming." The physics of this heating is straightforward. In intermediate- and late-type stars the line opacity responsible for the backwarming generally is present right in the continuum-forming layers. Because of their high opacity the line frequencies cannot contribute significantly to the outward flow of radiation. The temperature of the continuum-forming layers must increase relative to the unblanketed case in order to preserve radiative equilibrium. Heating will also occur even if the line opacity is found only in the layers *superficial* to the continuum-forming layers. In this circumstance the net outward flux is reduced at line frequencies by the increased inward-directed flux from the superficial layers. Once again the continuum-forming layers must heat (relative to the unblanketed case) in order to preserve radiative equilibrium.

The behavior of the surface layers ($\tau_c \gtrsim 0.03$) in the presence of line blanketing can be quite complex. In the most commonly encountered circumstances the line-blanketed atmosphere is *substantially* cooler in its surface layers than an unblanketed atmosphere *with the same temperature near* $\tau_c = 1$. This "surface cooling" can be produced by either of two processes. The first of these acts only when the opacity at the line frequencies is appreciably larger than the continuous opacity everywhere in the atmosphere except at the surface; this cooling mechanism is not strongly dependent on the frequency distribution of the line opacity. The second requires more restricted circumstances and is discussed later. In the first process cooling occurs because line frequencies that are optically thick throughout the deeper layers of the atmosphere become optically thin near the surface owing to the decreasing particle densities. As the line frequencies become optically thin they begin contributing to the outward flow of radiation. The atmosphere regains radiative equilibrium by reducing the temperature, hence the emissivity, of the surface layers.

This cooling mechanism can be seen more clearly by considering the radiative equilibrium constraint (e.g. Mihalas 1978):

$$\int_0^\infty \kappa_\nu^T(\tau_0)[J_\nu(\tau_0) - B_\nu(\tau_0)] \, d\nu = 0, \tag{1}$$

where κ_v^{T} is the total (line plus continuum) absorption at frequency v and at depth τ_0 on a standard optical depth scale, J_v is the mean intensity, and B_v is the Planck function; scattering opacities do not contribute to the integral. Equation (1) may be rewritten separating line-blanketed frequencies from continuum frequencies:

$$\int_{\Delta v_c} \kappa_v^c(\tau_0)[J_v(\tau_0) - B(\tau_0)]\, dv + \int_{\Delta v_L} [\kappa_v^c(\tau_0) + \kappa_v^L(\tau_0)]$$
$$[J_v(\tau_0) - B_v(\tau_0)]\, dv = 0; \qquad (2)$$

κ_v^c is the continuous opacity and κ_v^L is the line opacity. The first integral is taken over all the unblanketed frequencies ($\kappa_v^L \equiv 0$), the second integral is taken over all frequencies for which $\kappa_v^L > 0$. Deep in the atmosphere where the lines are opaque, $J_v \approx B_v$ at the line frequencies. The temperature profile is then controlled by the first term in Equation (2). Approaching nearer the surface, the continuum frequencies become optically thin first. When this occurs $J_v(\tau_0)$ in the continuum becomes nearly constant and radiative equilibrium can be maintained by a shallow temperature gradient. (This explains the almost isothermal surface temperature profile of unblanketed models.) At sufficiently small optical depths the line frequencies also become transparent. Since the temperature profile is generally rather shallow at this point, $J_v < B_v$ at all the line frequencies. This, in turn, makes the second term in (2) negative. Radiative equilibrium is restored by reducing the surface temperature.

The above form of surface cooling is the most commonly encountered. The process begins as soon as the line frequencies become optically thin. If there is a considerable range of line strengths then the cooling will occur over a similar range in optical depths. This is the usual case and produces a temperature profile that keeps dropping until the most opaque lines become optically thin. If, on the other hand, a significant number of lines all become optically thin at approximately the same depth in the atmosphere the cooling can be quite abrupt. This frequently occurs with strong molecular transitions like the CO fundamental bands and is responsible for the abrupt temperature drops often seen in the shallow layers of cool star models (e.g. Querci & Querci 1974, 1975).

It is important to recognize that the lines which produce the back-warming are not necessarily the same lines which control the surface temperature structure. An excellent example is provided by the CO opacities in oxygen-rich K giants. In the atmospheres of these stars the backwarming is produced primarily by the myriad of atomic and molecular lines in the blue, visual, and near-infrared portions of the spectrum. The CO fundamental and first-overtone vibrational-rotational bands (near 5 μm and 2.5 μm) occupy such a small fraction of frequency space that they do not appreciably affect the net flux (and consequently the backwarming) even though these bands are extremely opaque (Gustafsson et al. 1975). They do, however, produce a marked cooling of the surface layers. This occurs because the CO lines are sufficiently numerous and strong that they dominate the emissivity of the surface layers. Their great strength in K giants more than compensates for their restricted frequency range. CO cooling is discussed quantitatively in the next section.

The behavior of the surface layers in the presence of line blanketing is not as straightforward as described above when the dominant line opacities concentrate near the surface. Under such circumstances the blanketing can produce either heating or cooling depending on the frequency distribution of the line opacity. A very interesting study by Dumont & Heidmann (1973, 1976) is relevant here. Dumont & Heidmann examine the behavior of the surface temperature structure when a major shift in opacity occurs in the shallow layers. Although their analysis deals with continuous opacity sources, the same principles apply if bound-bound opacities are considered. Consider Equation (2) again only now assume that the line opacity is negligible below some continuum optical depth, τ_0^*; above this depth the line opacity appears, i.e.

$$\kappa_v^T = \kappa_v^c \qquad \tau_0 > \tau_0^*$$

$$\kappa_v^T = \kappa_v^c + \kappa_v^L \qquad \tau_0 < \tau_0^*$$

At small optical depths in grey or nearly grey atmospheres there is a critical frequency, v_c, such that $J_v(\tau_0) > B_v(\tau_0)$ at all frequencies $v > v_c$ and $J_v(\tau_0) < B_v(\tau_0)$ for all $v < v_c$. This frequency lies somewhat shortward of the frequency of the maximum of the Planck function evaluated for the T_{eff} of the atmosphere (Dumont & Heidmann 1973). (This behavior results from the very strong temperature dependence of the Planck function in the Wein limit.) Unblanketed nongrey atmospheres of intermediate- and late-type stars are sufficiently grey that they also have a critical frequency, v_c. This is the case for the example we have been considering at depths below τ_0^*.

Assume that the atmosphere is in radiative equilibrium below τ_0^*. We now wish to examine how the temperature structure above τ_0^* will differ from the temperature structure that would be obtained by extrapolating the $T(\tau_0)$ from $\tau_0 > \tau_0^*$. The differences will be due to the presence of the line opacities. If the line opacity that appears for $\tau_0 < \tau_0^*$ is predominantly in the frequency range where $J_v < B_v$ (a "long-wavelength" opacity) then the second term in Equation (2) becomes negative for $\tau < \tau_0^*$ and the atmosphere must cool to restore radiative equilibrium. This is the second surface cooling process mentioned earlier. If, on the other hand, the line opacity is predominantly in the region where $J_v > B_v$, then the second term in (2) becomes positive for $\tau_0 < \tau_0^*$. The surface layers must then increase in temperature in order to preserve radiative equilibrium. If only the single line opacity source were present, the surface temperature would rise toward shallower depths much like the temperature inversions found in the continuum studies by Dumont & Heidmann, by Linsky (1966), and by Gingerich et al. (1966). In practice, however, the line opacity that occurs only at small optical depths is one of many line-blanketing opacities present in the atmosphere. Since these other line transitions are generally quite optically thick everywhere *except* at the surface, they produce strong surface cooling. The heating effect noted above serves to moderate this cooling rather than actually producing a temperature inversion.

Marked surface heating has been noted in model atmospheres when TiO lines are added to the opacity set (Mould 1975, Lengyel-Frey 1977, Tsuji 1978, Krupp et al. 1978). Lengyel-Frey and Tsuji also reported heating from the vibrational-rotational lines of H_2O. The TiO and H_2O opacities are relatively concentrated toward the shallowest layers because both species are preferentially formed in a low temperature environment. Since TiO and H_2O have strong bands in the spectral region $v > v_c$, both species produce surface heating when their lines are included in the opacity set. It should be noted that the far-infrared H_2O bands cool rather than heat the surface layers (Tsuji 1978). This is in keeping with the above for opacities in the region $v < v_c$.

2.2 Some Quantitative Examples of Line-Blanketing Effects

In this section I present some examples that demonstrate quantitatively the characteristic effects of line blanketing on model atmosphere structure. Elemental and isotopic abundances, microturbulence, and atmospheric velocity gradients all influence the line blanketing and can alter the temperatures in model atmospheres by hundreds of degrees. Strictly speaking each combination of effective temperature, gravity, composition, and velocity field leads to a unique model structure. Ideally, every analysis of stellar spectra should use model atmospheres constructed with parameters identical to those of the stars under investigation. Since model atmosphere grids are available for only a limited set of atmospheric parameter combinations it is not possible to satisfy this requirement except approximately. An investigator must consider whether or not the model structure adopted differs significantly from that of the star to be analyzed. The following discussion will give some rough guide to how the composition and atmospheric velocity field can influence the model structures. Every user of model atmospheres will have to make a decision on whether or not a particular effect is relevant and important based on the nature of the particular application. *This decision should not be made casually.*

The quantitative effects of line blanketing to be described here are based on the best available model atmosphere computations. Nevertheless, it is important to recognize that advances in modeling or improvements in atomic and molecular constants may alter the magnitudes of some effects. Also, the following discussion deals only with changes in model structure. Most abundance variations will also produce marked changes in the emergent stellar spectrum. While these spectrum changes are obviously invaluable as diagnostics, they are simply too varied and complex to be adequately treated here. The reader should examine the relevant discussions in Querci & Querci (1976), Tsuji (1976b, 1978), Manduca et al. (1977), Gustafsson & Bell (1979), Buser & Kurucz (1978), Relyea & Kurucz (1978), and in references cited therein.

2.2.1 INFLUENCE OF [A/H] ON ATMOSPHERIC STRUCTURE   The computations of Kurucz (1979) and Gustafsson et al. (1975) demonstrate that in atmospheres dominated by atomic line blanketing, overall metallicity variations can produce significant structural changes. Increasing the metal abundances increases both backwarming and surface cooling. This point is nicely illustrated by Gustafsson et al. (1975) who find that the following relation holds for their giants and supergiants, in the metallicity[3] range $-1.0 \leqq [A/H] \leqq 0.0$:

$$\Delta T(\tau_R) \approx f(\tau_R) \cdot \Delta [A/H],$$

where ΔT is the temperature change produced at Rosseland optical depth τ_R by changing the metallicity by $\Delta [A/H]$. The scale factor $f(\tau_R)$ is comparatively insensitive to $\log g$ and $[A/H]$. While the shape of $f(\tau_R)$ varies somewhat with T_{eff}, it is always negative near the surface ($\tau_R \approx 10^{-3}$) and positive in the continuum-forming layers ($\tau_R \approx 1$). In the range $6000 \leqq T_{eff} \leqq 5000$ K, the surface layers cool relative to the continuum-forming layers by roughly 200 K when $[A/H]$ changes from -1.0 to 0.0. Although the effects are somewhat smaller in the range $5000 > T_{eff} \geqq 4000$ K, they

---

[3] The following conventions, standard to stellar atmosphere work, are used: All elements heavier than He are designated metals. [X/H] represents the usual logarithmic element (X) to hydrogen ratio taken relative to the sun. [A/H] refers collectively to all metals. Unless otherwise stated all metals are assumed to vary in abundance together; in Sections 2.2.2 and 2.2.3 C, N, and O will be distinguished from the other metals. Model atmospheres are identified by their effective temperature (T_{eff}), surface gravitational acceleration (g) and [A/H] expressed in the form ($T_{eff}, \log g, [A/H]$); [A/H] may be omitted when the abundances are solar, i.e. [A/H] = 0.0.

are not negligible. For example, at $T_{eff} = 4500$ K, the temperatures in the shallow line-forming layers decrease by ≈ 250 K relative to the continuum-forming layers when [A/H] increases from -2.0 to 0.0. The analysis by Peterson (1976) demonstrates that temperature gradient changes as large as these can have significant impact on the predicted line strengths of many atomic and molecular species (also see Gray 1976). This point should be borne in mind whenever intermediate-type stars of nonsolar composition are being analysed.

It also should be noted that the gas pressure at a given T_{eff} and log g can be quite sensitive to [A/H] (see Section 4.1.1). The pressure changes can be substantial (e.g. Bell et al. 1976, Kurucz 1979) and can affect dissociation and ionization equilibria, line broadening, and even the structure of the convectively unstable layers of the atmosphere.

2.2.2 THE EFFECTS OF CNO VARIATIONS ON INTERMEDIATE-TYPE ATMOSPHERES   The discussion of the preceding section assumed that all the metals varied together. It has long been recognized that this is not the case for late-type stars where *inter alia* variations in the elemental abundances are thought to give rise to the M, S, and C spectral sequences (e.g. Vardya 1970). Until fairly recently, however, it was not generally appreciated that substantial *inter alia* abundance variations also occurred in the K-giant domain. Strong evidence is now accumulating in support of such variations particularly in the case of the light elements C, N, and O (e.g. Sneden 1974, Lambert & Ries 1977, Kraft 1979). Since the CNO abundances can be decoupled from those of the heavier metals it is advisable to consider the effects of the CNO group on blanketing separately. The following discussion refers exclusively to the effects of the red and infrared molecular transitions. Even though molecular bands lying in the shorter wavelength regions may affect atmospheric structure, particularly at high metallicities, their significance has not yet been discussed in the literature. It also should be noted that no investigations of blanketed model atmospheres for intermediate-type carbon stars (e.g. CH stars, R stars) have yet been reported. Since these stars have rather different blanketing than K giants, they certainly merit future attention.

Backwarming   The lines of the CN red system are sufficiently weak in intermediate-type stars that they do not produce an appreciable amount of backwarming when [A/H] ≤ 0.0. At (4000, 2.25, 0.0) the CN lines heat the continuum-forming layers by only ≈ 60 K according to the calculations of Gustafsson et al. (1975). The heating effect of CN will be more important for the metal-rich ([A/H] > 0) models that may represent the "super-metal-rich" (SMR) stars found in the field and in the nuclei of spirals and massive ellipticals (e.g. Deming 1979, Faber 1977, Pritchet 1977). Very little model atmosphere work has been done in this important domain of high metallicity.

Surface cooling   Surface cooling by CO was first recognized by Alexander & Johnson (1972) and discussed in detail by Johnson (1973). The most complete study of CO cooling in K giants has been carried out by Gustafsson et al. (1975). It should be noted that Johnson (1973) obtained more than twice the temperature drop found by Gustafsson et al. (1975). The difference apparently lies in the different treatment of atomic and molecular opacities. The Gustafsson et al. calculations should be the more accurate (see Sections 2.3.1, 2.3.2) and the following numbers are drawn from their results. CO cooling depresses the surface temperatures ≈ 200 K for giants in the range $4500 \geqq T_{\text{eff}} \geqq 4000$ K. The extent of cooling decreases with increasing T_{eff}, to ≈ 140 K at $T_{\text{eff}} = 5000$ K and to ≈ 90 K at 5500 K. Both Johnson and Gustafsson et al. found that CO becomes substantially less effective as a cooling agent with increasing gravity. This presumably occurs because the ratio of CO line opacity to continuous (H^-) opacity varies roughly as $(P_{\text{gas}})^{-1}$. Thus as P_{gas} increases with increasing gravity the CO cooling moves deeper into atmospheric layers where the cooling by atomic lines becomes important. This process reduces the relative contribution of CO to the surface emissivity. The gravity and temperature dependences of CO cooling are important because they weaken the homologous nature (e.g. Carbon & Gingerich 1969, Gustafsson et al. 1975) of the $T(\tau_0)$ relation.

As part of their study Gustafsson et al. examined the metallicity dependence of the CO cooling. A particularly interesting result was their finding that: 1. the CO cooling is both large (≈ 200 K) and comparatively insensitive to metallicity for K-giant models with $T_{\text{eff}} \gtrsim 4500$ K and $[A/H] \gtrsim -2$, and 2. the CO cooling in this temperature range rapidly diminishes for $[A/H] < -2.0$. The change in behavior at $[A/H] = -2$ occurs because the CO concentration finally drops so low that even the strongest CO lines do not contribute significantly to the emissivity of the surface layers. The failure of CO to cool at very low C abundances provides the interesting circumstance that the C-over-deficient ($[C/H] < [A/H]$) giants in metal-poor clusters like M15 and M92 (e.g. Norris & Zinn 1977, Carbon et al. 1977, Kraft 1979) will have substantially higher surface temperatures than cluster stars with the same T_{eff} and log g but normal $[C/H](=[A/H])$. These temperature structure differences must be considered when analyzing the line spectra of these stars. Similar effects can also be encountered in metal-poor field stars with low $[A/H]$ and processed atmospheric compositions (e.g. HD122563; Sneden 1974).

2.2.3   EFFECTS OF CNO VARIATIONS ON LATE-TYPE ATMOSPHERES   Because the necessary molecular parameters are not yet available, current opacity codes lack many of the species that contribute significantly to the line blanketing below $T_{eff} \approx 2800$ K (e.g. HCN, C_2H_2, C_2H, VO). This deficiency leads to considerable uncertainty in computing the effects of composition on atmosphere structure at low T_{eff}. Consequently the discussion in this section deals with abundance effects in the hotter classes of late-type stars. The atmospheric structure of these stars depends primarily on only a few, comparatively well-understood, molecular species: C_2, CN, CO, TiO, and H_2O. The relative abundances of these are sensitively dependent on the abundances of C, N, and O. Since a large range of CNO abundances are found in late-type stars, it is especially important to be fully aware of the changes in model structure that can accompany composition changes.

$C < O$ versus $C > O$ Substantial changes in cool star model structures are produced by changing the sign of the ratio (C–O)/H, where C, O, and H represent the respective elemental abundances. It is well known, of course, that increasing C/O through unity moves an atmosphere into the carbon-star domain. If (C–O)/H changes sign by increasing C with O and H fixed at solar values, the dominant blanketing opacities change from TiO, CO, and H_2O to CN, C_2, and CO. Since CN and C_2 are much more effective than TiO and H_2O at producing backwarming, there is an extensive heating of the deeper layers of the atmosphere. This can be appreciated by comparing the carbon-star models of Querci & Querci (1975) with the solar abundance models of Tsuji (1978). At (3800, 1.0), for example, changing from solar abundances (C/O = 0.6, N/O = 0.14) to a carbon- and nitrogen-enriched composition (C/O = 1.3, N/O = 0.7, O/H $\approx$ solar) representative of a plume-mixed carbon star (Scalo & Ulrich 1973) produces a heating of ≈ 500 K in the continuum-forming layers; the temperatures at $\tau_{0.8} = 10^{-3}$ are essentially unchanged. The same composition change at (3400, 1.0) produces ≈ 750 K heating in the continuum-forming layers again without affecting the surface temperatures.

Effects of changing C, N, O when $C > O$ Within the carbon-star domain itself substantial structural changes can be affected by varying the CNO abundances, hence the CN and C_2 concentrations. Querci & Querci (1975) illustrated this point by comparing three models at (3000, 1) computed with different carbon-star mixtures. Two of these are plume-mixed compositions with different carbon enhancements (O/H $\approx$ solar, N/O = 0.14, C/O = 1.3 and 5.0). The model with C/O = 5.0 is substantially more blanketed, and more heated, than the model with $C/O = 1.3$. The temperature differences are ≈ 900 K, ≈ 800 K, and ≈ 300 K at $\tau_{0.8} = 10^{-3}$, 10^{-2}, and 1.0, respectively. A third model with reduced carbon and oxygen and enhanced nitrogen ($O/H \approx$ one-tenth solar, $N/O = 120$, $C/O = 3.3$) was ≈ 150 K, ≈ 500 K and ≈ 500 K cooler at $\tau_{0.8} = 10^{-3}$, 10^{-2}, and 1.0, respectively, than the plume-mixed model with $C/O = 1.3$. Temperature effects of this magnitude cannot be ignored when analyzing the atomic and molecular spectra of carbon stars.

Effects of changing C, N, O when C < O   The relative C and O abundances strongly affect the atmospheric structure when $C < O$. Particularly interesting is the effect on atmospheric structure when the difference (O–C) approaches zero at constant oxygen abundance. This behavior would be expected for a red giant undergoing surface carbon enrichment as a result of plume-mixing (Scalo & Ulrich 1973). Tsuji (1978) computed the atmospheric structures of giants in the range 3400 K $\leq T_{eff} \leq$ 2600 K for two compositions, solar and solar except for C, which is increased so that $O/C = 1.05$. The models with enhanced carbon are cooler throughout the line- and continuum-forming layers. At (3000, -1), for example, the carbon-enriched model is ≈ 120 K cooler than the solar abundance model at $\tau_{0.8} = 10^{-3}$; at $\tau_{0.8} = 1$ the difference is ≈ 150 K. Reducing the (O–C) difference reduces the amount of free oxygen in the atmosphere. When O/C reaches ≈ 1.05 so little oxygen is left that the TiO and H_2O concentrations in the atmospheres drop sharply. Since these species dominate the backwarming, the carbon-enriched atmospheres become cooler in the continuum-forming layers. In Section 2.1 it was noted that TiO and, except for the reddest bands, H_2O acted to heat the shallow atmospheric layers in the late-type stars. This heating effect is nearly absent in the carbon-enriched models; the temperature of their surface layers is largely controlled by CO. It is interesting that stars with O/C ≈ 1 have minimal line blanketing. Stars either more carbon- or oxygen-rich exhibit (except for very low CNO abundances) strong molecular blanketing. This strong composition dependence of line blanketing and, therefore, atmospheric structure must be considered when analyzing stars of the M-MS-S-SC spectral sequence.

Effects of changing the $^{12}C/^{13}C$ ratio   When a significant amount of the blanketing arises from CN and C_2, as in the carbon stars, the atmospheric structure becomes somewhat dependent upon the $^{12}C/^{13}C$ ratio. Determinations of the $^{12}C/^{13}C$ ratio in stellar atmospheres yield values ranging from greater than solar ($^{12}C/^{13}C \simeq 90$) down to the CNO bicycle equilibrium value ($^{12}C/^{13}C \simeq 3.4$) (e.g. Lambert & Ries 1977). The lines of ^{13}C-containing molecules contribute significantly to the blanketing at low ^{12}C/^{13}C. Carbon (1974) constructed two carbon-star models at (3500, 0) blanketed only by CN and atomic lines. One model has ^{12}C/^{13}C = 90, the other ^{12}C/^{13}C = 10. The model with ^{12}C/^{13}C = 10 was ≈ 250 K hotter than the low ^{13}C model throughout the continuum-forming layers ($\tau_{1.0} = 0.1 - 1$). The temperature differences decrease to ≈ 150 K at $\tau_{1.0} = 10^{-4}$. The differences are less for smaller isotopic variations. Querci & Querci (1975) find negligible changes ($\lesssim 100$ K) in carbon-star model structure in going from ^{12}C/^{13}C = 11 to 5. Isotope effects on CN blanketing have not been investigated for stars with C < O although such effects might be significant in some CN-strong K giants. The isotopic sensitivity of CO cooling has not been studied.

2.2.4 BLANKETING EFFECTS OF STELLAR VELOCITY FIELDS   The adopted value of the microturbulence parameter has only moderate effect on atmospheric structure. For giant models (log g = 1.5) with 5500 $\leqq T_{eff} \leqq$ 3750 K, Gustafsson et al. (1975) found that increasing the total (thermal + microturbulence) Doppler broadening velocity from 2 to 5 km s^{-1} produced ≈ 75 K heating at $\tau_R = 1$ and ≈ 25 K cooling at $\tau_R = 10^{-3}$. Peytremann (1974) found somewhat smaller changes when he increased the microturbulence parameter in a (7000, 4, 0) model from 2 to 5 km s^{-1}. Tsuji (1976a, 1978) obtained very similar results ($\gtrsim 100$ K heating in continuum-forming layers) for turbulence changes in M giant and supergiant models. Carbon (1974) found ≈ 150 K heating in the continuum-forming layers when the turbulence parameter was increased from 3 to 6 km s^{-1} in a carbon-star model. The turbulence parameter should have greatest influence on atmospheres whose blanketing lines reside primarily on the flat portion of the curve of growth. This would explain the increased importance of turbulent broadening in carbon stars and in K giants of higher metallicity (Gustafsson et al. 1975).

Mihalas (1969) and Mihalas et al. (1976) have demonstrated that a steep velocity gradient in the line- and continuum-forming layers of an atmosphere can produce large structural changes by purely radiative interactions. The velocity gradient could be the result of either a shock front or a differential atmospheric expansion (or contraction). It is only necessary that the gradient shift spectral lines by more than roughly a Doppler width. When sufficient velocity shifts occur, frequencies that are comparatively transparent at one depth become opaque at other depths due to the Doppler shift of the line opacities. Such velocity shifts in the line-forming layers produce both increased backwarming and substantial heating of the shallow layers. These studies were carried out using a very schematic picket representation of the line opacity; the problem has not been examined in realistic cases representing late-type atmospheres. Nevertheless, since shock phenomena, large-scale material motions, and mass loss are observed to be present in many late-type stars (e.g. Tsuji 1971a, Merrill & Ridgway 1979, Hall et al. 1979), velocity gradient effects should prove to have considerable importance for model building.

2.3 Techniques for Computing Blanketed Models

In order to impose the radiative equilibrium constraint on a model atmosphere it is necessary to determine the frequency integrated radiative flux and its derivative (or equivalent quantities) at each depth in the atmosphere (e.g. Mihalas 1978). These integrals require evaluation of the monochromatic flux, $F_\nu(\tau_0)$, and mean intensity, $J_\nu(\tau_0)$, at each depth, τ_0. The integrals are evaluated by dividing the spectrum into N subintervals of width $\Delta\nu_i$ centered on the frequencies ν_i. At each ν_i, the radiative transfer equations are solved to determine $F_\nu(\tau_0)$ and $J_\nu(\tau_0)$. The values of N, ν_i, and $\Delta\nu_i$ must be carefully selected to ensure accurate evaluation of the integrals (e.g. Mihalas 1967). The entire procedure is comparatively straightforward for unblanketed atmospheres since one needs only a frequency spacing fine enough to follow the rather gentle contours of the spectrum between bound-free edges. The required frequency spacing is generally of the order of 10^3–10^5 Doppler widths (e.g. Kurucz 1969). For a line-blanketed atmosphere, however, the problem is considerably more difficult. Since line opacities vary significantly over frequency intervals of order of a Doppler width, the intervals $\Delta\nu_i$ must be quite small to perform the frequency quadrature accurately. Such a tight frequency mesh leads to very large values of N ($\approx 10^5$–10^6). The problem is compounded by the requirement that the monochromatic opacity be determined at each depth in the model by summing over all the spectral lines that contribute significant opacity at ν_i. These circumstances quickly make the direct approach prohibitively time-consuming on all but the fastest machines currently available. This simple economic fact has dominated the construction of model atmospheres for intermediate- and late-type stars.

Model builders have sought to deal with this problem by adopting one of several approximate representations for the line opacities in a model atmosphere. These representations have a common feature: they all permit the value of N to be reduced to manageable size ($\gtrsim 10^3$) generally by allowing $\Delta\nu_i$ to increase significantly ($\Delta\nu_i \approx 10^3$–10^5 Doppler widths). Clearly, these changes make the treatment of line-blanketed models substantially more tractable. To be useful for model building a representation of the line opacities must satisfy one important criterion. It must lead to a model structure that is a reasonable approximation to that of the model one would have obtained by computing the radiation field monochromatically at all frequencies. Judged by this criterion some of the line opacity representations are significantly more successful than others. In the next sections I review from a user's standpoint the important characteristics of the various line representations. The reader should take special notice of the weaknesses of each opacity representation.

2.3.1 THE STRAIGHT AND HARMONIC MEAN OPACITIES   The straight mean opacity represents the line opacity in Δv_i by a single number for a given set of thermodynamic and composition parameters. It is the most easily calculated representation for line opacities.

The deficiencies of the straight mean all stem from the fact that it smears the opacity of each line uniformly over the interval Δv_i. Since the smearing fills in the opacity minima that are responsible for carrying the bulk of the flux in the spectral interval Δv_i, the straight mean leads to an underestimate of the flux in each layer of the model (Carbon 1974). This produces converged model atmospheres that are systematically too hot throughout (Carbon 1974, Querci et al. 1974, Johnson & Krupp 1976, Johnson et al. 1977). The errors can amount to hundreds of degrees in the line- and continuum-forming layers of heavily blanketed atmospheres. Since the emergent fluxes are also underestimated, spectral features computed with the straight mean approximation are generally much too strong (Carbon 1974, Querci et al. 1974, Tsuji 1976a). Other deficiencies of the straight mean are discussed by Carbon (1974).

The opinion has occasionally appeared in the literature that the over-blanketing produced by the straight mean could be eliminated if only the mean were taken over a sufficiently small interval (e.g. van Paradijs & Vardya 1975). It is clear that this can be correct in the general case only when Δv_i approaches a Doppler width in size. However, if the blanketing lines are of roughly equal strength and so closely spaced as to form a quasi-continuum and/or are predominantly quite weak (i.e. on the linear portion of the curve of growth), then the straight mean could prove reasonably accurate even for Δv_i much larger than a Doppler width. The band systems of TiO apparently satisfy these criteria to a sufficient extent since Krupp et al. (1978) find comparatively little difference between atmospheres computed with the straight mean for TiO and ones computed using a more elaborate approximation.

Straight mean opacity codes have been published for the following line-blanketing species: atomic lines (Mutschlecner & Keller 1970), CN (Johnson et al. 1972, Bailey 1977), CO (Kunde 1968), TiO (Collins & Faÿ 1974), and H_2O (Auman 1967). See Tsuji (1966a,b, 1969) for a derivation of straight mean opacities for CO, TiO, CaH, MgH, SiH, OH, and H_2O.

Like the straight mean, the harmonic mean represents the opacity in $\Delta \nu_i$ by a single number. Although the harmonic mean systematically underestimates the effective opacity in a spectral interval it does tend to yield appreciably more accurate fluxes than does the straight mean, particularly in the deeper layers (Carbon 1974). The only harmonic mean opacity code to appear in the literature is one for H_2O developed by Auman (1967). Model builders have generally abandoned the mean opacity representations in favor of the two representations discussed next.

2.3.2 MULTIPLE PICKETS BASED ON OPACITY DISTRIBUTION FUNCTIONS

The concept of the opacity distribution function (ODF) has been discussed many times (e.g. Strom & Kurucz 1966, Querci et al. 1971, Carbon 1973, Querci & Querci 1975, Mihalas 1978). A particularly well-illustrated description may be found in Kurucz et al. (1974) and in Kurucz (1979). Basically, the ODF is an array of numbers that specifies the fraction of spectral interval $\Delta \nu_i$ occupied by an opacity greater or equal to some particular value. In model atmosphere calculations the ODF is represented by a histogram with N_i steps of fractional widths $\Delta \omega_l$ ($l = 1, \ldots, N_i$). This histogram is referred to as a "multiple picket" in the literature. The values for $\Delta \nu_i$, N_i, and $\Delta \omega_l$ are chosen to provide the best compromise between accuracy and the expenditure of computer time. The transfer equation for the interval $\Delta \nu_i$ is solved N_i times, once for each picket step. The multiple picket based on ODFs is not only substantially more accurate than the straight and harmonic means but it is, unlike the means, sensitive to line broadening mechanisms and variations in isotopic composition (e.g. Carbon 1973, 1974, Querci & Querci 1975). Although ODFs require a substantial initial investment of computer time to prepare, they can be used repeatedly in model calculations at comparatively little expense. This aspect makes them particularly attractive when an extensive grid of models with fixed composition is to be computed.

2.3.3 FAILURE OF THE ODF APPROXIMATION

All information regarding the frequency distribution of opacity, κ_ν^l, within the interval $\Delta \nu_i$ is lost in the process of computing the ODF. This means that the ODF-based picket is a strictly correct representation of the monochromatic opacity only if each step of the ODF histogram corresponds to the opacity of specific frequency subintervals of the monochromatic opacity within $\Delta \nu_i$ *regardless* of temperature and pressure (i.e. independent of depth in the atmosphere). This requirement will not be satisfied whenever circumstances act to change either the relative strengths or the positions of lines within $\Delta \nu_i$ as a function of depth in the atmosphere. Unfortunately, such circumstances are not unusual (e.g. Carbon 1974, Mihalas 1978). In a typical atmosphere, for example, the temperature and pressure change markedly with depth producing relative line strength variations whenever lines of different species or having different excitation potentials occur together in the same interval. To be useful, the ODF-based picket must be able to handle at least this commonly encountered situation. When circumstances are such that the ODF approximation is invalid, the ODF picket will underestimate the amount of backwarming produced by the line blanketing. Qualitatively, the resulting converged models will have a temperature gradient that is too shallow in the continuum-forming layers. As will be seen shortly, the ODF picket appears to adequately represent the atomic and molecular opacities in stars with $C < O$. When $C > O$, the ODF approximation may fail; the seriousness of this failure can be judged only when the necessary model calculations have been made.

In intermediate-type atmospheres Gustafsson et al. (1975) carried out a comparison between models computed with ODF-based pickets (for 100 Å intervals) and models in which the monochromatic opacity was evaluated at high resolution (0.1 Å mesh, over 40,000 frequency points per model). The models used in the comparison were giants ranging from 4000 to 6000 K in temperature; only atomic and molecular line blanketing shortward of 7200 Å was considered. In all cases the temperature differences between models calculated with the two methods were negligible (<20 K). Using the same opacity set, Johnson & Krupp (1976) found that the temperature-pressure structures of two (4000, 2.25, 0.0) atmospheres, one calculated with ODFs and one calculated by opacity sampling (Section 2.3.4), differed by less than 30 K. These results strongly suggest that the relative line strength variations produced by atmospheric temperature and pressure gradients may not have great practical significance, at least for atmospheres dominated by atomic line blanketing. The ODF approximation appears to be adequate for atmospheres of intermediate spectral type. [While these studies did not include CO opacity, Carbon's (1974) results suggest that CO surface cooling will not be sensitive to the ODF approximation.]

In late-type atmospheres the situation is more uncertain. For oxygen-rich stars the blanketing is dominated by TiO, H_2O, and CO, the principal bands of which do not extensively overlap (e.g. Tsuji 1971b). In this circumstance the ODF approximation should be as accurate as it is for the hotter stars since only excitation potential differences will be acting to compromise the results. However, until test calculations are carried out this is only a conjecture. For carbon stars, the blanketing is dominated by CN, C_2, and CO. The lines of CN and C_2 are interwoven throughout the spectrum and both species are important for determining the atmospheric structure (e.g. Querci & Querci 1974, Sneden et al. 1976). Furthermore, the CN/C_2 ratio can vary by substantially more than an order of magnitude through the line- and continuum-forming layers of a carbon star (Carbon 1974, Querci & Querci 1975). Because of these conditions it is possible that the ODF-based picket may be an inadequate opacity representation for carbon stars. Unfortunately, this problem has not yet been properly examined by anyone computing carbon-star atmospheres. A systematic comparison of cool star atmospheres computed with ODF pickets and with some direct technique like opacity sampling would be quite valuable. The comparison should embrace a range of elemental compositions, T_{eff}, and gravities. Until such a study is carried out, carbon-star models based on ODF opacities must be regarded as tentative since they may be appreciably underblanketed. This point should be considered when ODF-based model structures are employed to analyze the atomic and molecular spectra of carbon stars.

2.3.4   OPACITY SAMPLING   The method of evaluating frequency integrals by statistical sampling was introduced by Peytremann (1974). This approach has been adopted, with some modification, by the Indiana University model atmospheres group in a number of investigations (Sneden et al. 1976, Johnson & Krupp 1976, Johnson et al. 1977, Krupp et al. 1978). The sampling technique is quite straightforward: the depth-dependent monochromatic radiation field is explicitly computed for a set of effectively random frequencies; the number of such frequencies is made sufficiently large that a frequency quadrature over the selected points yields accurate values for the frequency integrals at each depth in the atmosphere. The number of frequency points required is not particularly large—600 to 1000 points appear to be sufficient depending upon the nature of the blanketing (see Peytremann 1974, Gustafsson et al. 1975, Sneden et al. 1976, Johnson & Krupp 1976).

Although it is expensive computationally, the sampling technique enjoys a number of advantages when compared with ODFs (Sneden et al. 1976). The most important of these, from the user's standpoint, is its freedom from any assumption concerning the behavior of the line opacity. Provided enough sampling points are selected, the technique will yield the correct atmospheric structure including those cases where the ODF approximation fails. The sampling technique can deal more directly with atmospheric velocity gradients than the ODF picket (e.g. Mihalas 1969, Carbon 1974). This may be an important capability in modeling late-type atmospheres with large-scale mass flows.

The sampling approach must be used with care if an accurate

temperature structure is to be obtained for shallow layers. A frequency mesh that is too coarse will not properly sample the comparatively few lines that often control the energy balance in the shallowest atmospheric layers (Sneden et al. 1976, Johnson & Krupp 1976). In the case of CO blanketing the model will converge to a temperature structure that is too warm for radiative equilibrium (Heasley et al. 1978). This problem will require some care particularly if atmospheres generated by sampling are to be used to investigate such shallow layer phenomena as chromospheric heating (e.g. Ulmschneider et al. 1977, Cram & Ulmschneider 1978, Linsky & Ayres 1978, Kelch et al. 1978). A hybrid technique in which the strongest lines are represented by an ODF may prove to be the most practical in such circumstances.

3.0 DEPARTURES FROM LOCAL THERMODYNAMIC EQUILIBRIUM (LTE)

The focus in this section is on non-LTE effects that can influence atmospheric structure either through the continua or the principal line-blanketing species. The reader is urged to study the excellent general review of non-LTE in stellar atmospheres by Mihalas & Athay (1973); the current discussion emphasizes results not stressed by Mihalas and Athay, as well as more recent investigations. It is important to recognize that this is a relatively undeveloped research area especially when compared to the efforts expended in studying non-LTE in early-type stars. As will be seen, considerable work remains to be done particularly for the late-type stars where departures from LTE may be the rule rather than the exception.

3.1 *Departures from LTE in the Continua*

3.1.1 H⁻ FORMATION   The bound-free and free-free transitions of H^- are the dominant continuous absorptive opacities in intermediate- and late-type stars. Lambert & Pagel (1968) and Praderie (1971) carefully considered the reaction network that determines the concentrations of the critical species H^- and H_2. They find that the reactions $3H \rightleftarrows H_2 + H$ (three-body collision) and $H^- + H \rightleftarrows H_2 + e^-$ (associative detachment) dominate in nearly all intermediate-type atmospheres; for the hotter intermediate-type stars in layers where the H^+ concentration is appreciable, the charge-exchange reaction $H^+ + H^- \rightleftarrows H(1s) + H$ (*ns*, p, or d) becomes significant. Strom (1967) computed a grid of unblanketed dwarf and giant atmospheres in the range $6000 \leqq T_{eff} \leqq 4000$ K in which the statistical equilibrium equations were solved for H^- and the first three levels of H. Associative detachment ($H^- + H \leftrightarrows H_2 + e^-$) was assumed to be the principal mechanism for forming H^-; H_2 was assumed to have its LTE concentration, an assumption verified by Lambert & Pagel (1968). The calculations were made for various values of the associative detachment rate, $R_{H^-,H}$. ($R_{H^-,H} n_{H^-} n_H$ is the reaction rate per unit volume.) Strom found that the non-LTE effects were negligible when $R_{H^-,H} \gtrsim 10^{-11}$ cm^3 s^{-1}. For lower reaction rates H^- was found to be overpopulated relative to LTE (departure coefficient $b > 1$). This overpopulation was predicted qualitatively by Kalkofen (1968) for F, G, and K stars. The departures from LTE altered the model temperature-pressure structures and decreased the fluxes in the bound-free continua ($\lambda < 1.65$ μm). The departures from LTE increased with decreasing gravity. Strom argued that these non-LTE effects were detectable in the observed $(R - K, R - I)$ diagram for field dwarfs and giants.

Subsequent to the study by Strom, laboratory and theoretical investigations indicated an associative detachment rate of $10^{-8.9}$ cm^3 s^{-1} (Schmeltekopf et al. 1967, Dalgarno & Browne 1967). For this high rate Strom's calculations predict insignificant departures from LTE in G and K stars. This result is confirmed by Gustafsson & Bell (1979) and Vernazza et al. (1976). Lambert & Pagel suggest that Strom's observational evidence for non-LTE is inconclusive since his dwarf sample was too small to be statistically significant. The question of departures from LTE in H^- formation in late-type stars has not yet been directly examined. This is unfortunate since the low temperatures and densities of cool giants and supergiants will favor non-LTE. An analysis such as Strom's is required since H^- remains a major continuous opacity source in all but the coolest stars.

3.1.2 METAL IONIZATION EQUILIBRIA The preceding studies of H^- statistical equilibrium all assumed an LTE value for the electron concentration. Recent investigations have suggested significant departures from LTE in the ionization equilibria of the principal electron donors in intermediate- and late-type stars. Si, Mg, and Fe are the dominant metallic sources of electrons in the solar-type photospheres (e.g. Gingerich et al. 1971). Over-ionizations relative to LTE are found for these species in the shallow layers of the solar photosphere (Athay & Lites 1972, Athay & Canfield 1969, Vernazza et al. 1976). These departures are not important for photospheric modeling because they occur at small depth ($\tau_{0.5} \gtrsim 10^{-2}$) and the species involved are already nearly completely ionized in LTE. Lites & Cowley (1974) report some over-ionization of Fe at small depths in giant atmospheres of G and K spectral type. Auman & Woodrow (1975) carried out an investigation of departures from LTE in the ionization equilibria of principal electron donors (K, Na, Al, Ca, and Mg) in late-type giants. They computed a grid of atmospheres assuming radiative detailed balance in the lines; red and infrared molecular line opacities were treated by straight means, atomic line blanketing was not considered. At $(4000, 2, 0)$ and $(3500, 1.5, 0)$ Ca and Na proved to be substantially over-ionized (five to ten times) relative to LTE. At lower T_{eff} large over-ionizations also occurred for K, Al, and Mg; at $(2000, -1, 0)$, for example, K, Na, and Ca are $\approx 10^4$ times over-ionized at $\tau_c \lesssim 10^{-2}$. These departures are overestimates, however, because the computed photoionization rates are larger than they would be in the presence of heavy blanketing. This point is carefully noted by Auman & Woodrow. Interestingly, even the large departures predicted by Auman & Woodrow's calculations *do not* lead to significant changes in model structure. In the hotter atmospheres the dominant electron contributors are already almost entirely ionized in LTE and in the cooler atmospheres Rayleigh scattering rather than H^- is the dominant opacity in the layers where the departures are the largest. The non-LTE effects in the ionization equilibria should be carefully considered when analyzing neutral atomic lines in late-type stars.

Ramsey (1977) attempted to detect departures from LTE in the ionization equilibrium of Ca by comparing the predicted LTE CaII/CaI ratios with those deduced from the spectra of late-type stars. Although he interpreted his results as indicating large departures from LTE (over-ionizations), this may not be the case. Ramsey's theoretical predictions are based on the effective temperature scale of Dyck et al. (1974). More recent studies (Tsuji 1976b, 1978, Ridgway et al. 1979) suggest that the Dyck et al. temperature scale is systematically too cool by ≈ 200 K for the early M stars. Adopting the higher T_{eff} for Ramsey's stars leads to substantially closer agreement between the LTE predictions and the observed CaII/CaI ratios. When this change is made Ramsey's results appear to be consistent with Auman & Woodrow's results after taking into account the effects of line blanketing on the latter.

3.1.3 H AND METAL BOUND-FREE CONTINUA Relatively little work has been done on departures from LTE in the H and metal bound-free continua in intermediate type stars. The known departures are usually regarded as negligible from a model-building standpoint since the bound-free opacities involved are not important for determining the photospheric structures. The reader should consult Vernazza et al. (1973, 1976) for a detailed description of the non-LTE calculations in the solar case. Kalkofen (1968) and Strom (1969) discuss non-LTE in H in the hotter intermediate-type stars.

3.2 Departures from LTE in the Atomic Lines

Because atomic lines are a dominant source of opacity in intermediate-type stars, departures from LTE in line formation could influence atmospheric structure by changing the blanketing. Unfortunately, the importance of non-LTE effects remains uncertain even though a number of investigations have addressed the question of departures from LTE in the primary atomic blanketing species, Fe I. Athay & Lites (1972) used a 15-level model atom to study non-LTE in Fe I in the solar atmosphere. As noted earlier, they found significant departures from LTE in the ionization equilibrium for $\tau_c \gtrsim 10^{-2}$. Departures from LTE in the excitation of Fe I levels were not significant deeper than the temperature minimum. Lites & Cowley (1974) extended this work to G and K giants using a simplified Fe I atom. Once more departures from LTE in the ionization equilibrium were found to dominate, departures from LTE in excitation were important only at small depths. The departures made the equivalent widths of most lines decrease although by less than a factor of two even in the worst case. Smith (1974a,b) attempted to check the predictions of Athay, Lites, and Cowley for the sun and α Boo. Smith found rough agreement although the observations did indicate somewhat smaller ($\approx$one-half) departures than predicted.

Very little work has been done to examine the effects on model structure of departures from LTE in the line spectrum. Athay & Skumanich (1969) and Athay (1970) in two important papers discuss the atmospheric heating and cooling produced by schematic model atoms in non-LTE. They find no striking difference between the backwarming when lines are formed in LTE and when they are controlled by radiative processes (non-LTE). However, the effectiveness of surface cooling is appreciatively reduced by non-LTE effects. Athay's (1970) study is the first attempt to compute a non-LTE line-blanketed solar model. He used a simple atomic model for the principal blanketing species in order to examine the effect of non-LTE on atmospheric structure. Unfortunately, since he did not tabulate departure coefficients for his model atoms, it is not possible to compare his results with those from the more detailed study by Athay & Lites (1972). Athay's non-LTE solar model generally agrees with LTE models (e.g. Gustafsson et al. 1975, Vernazza et al. 1976) for $\tau_c \gtrsim 10^{-4}$. The model has a higher boundary temperature than LTE models because the non-LTE effects temper the surface cooling. Nissen & Gustafsson (1978) give a preliminary report of a study in which they used the Fe I atom of Lites & Cowley (1974) to examine non-LTE effects in dwarfs with $6800 \leq T_{eff} \leq 5780$ K and $-0.5 \leq [M/H] \leq +0.5$. Like earlier investigators they found substantial over-ionization of Fe I (departures from LTE in the excitation occurred only in the shallowest layers). The departures were most pronounced in the models of lowest metallicity since these had the strongest UV radiation fields [e.g. in the $(6800, 4.5, -0.5)$ model Fe I was over-ionized relative to LTE by five times at $\tau_R \gtrsim 10^{-2}$]. While Nissen & Gustafsson indicate that these large departures do not lead to significant changes in model structure for the atmospheres, no details are given. A more complete description of their study is in preparation. The preceding investigations suggest that departures from LTE in the atomic spectrum do not appreciably affect the photospheric structure of intermediate-type dwarfs except in the shallowest layers. More work remains to be done on non-LTE in the giants and supergiants, however, particularly at low metallicities. The results of Lites & Cowley suggest that current LTE models may be too hot in their photospheres (i.e. have too much backwarming) since the departures from LTE generally reduce Fe I equivalent widths.

3.3   Departures from LTE in the Molecular Equilibria

The most thorough study to date of departures from LTE in molecular equilibria under stellar atmosphere conditions was carried out by Tsuji (1964). Tsuji found that bimolecular exchange reactions, $A + BC \rightleftarrows AB + C$, were very effective in rapidly establishing chemical equilibrium even under conditions characteristic of cool supergiants. Three body reactions, $A + B + C \rightleftarrows AB + C$, proved to be comparatively ineffective except at high densities. Photochemical reactions (see Dalgarno 1975) were also found to be ineffective. Similar conclusions have been reached by Thompson (1973) and by Hinkle & Lambert (1975). All these investigations indicate that chemical equilibria can be treated in LTE under most conditions found in intermediate- and late-type stars. A possible exception may arise if the temperature of low density atmospheric gas is appreciably perturbed by some time-dependent process. Stellar pulsation, shocks, and convection can all induce substantial time-dependent changes in the temperatures of atmospheric gas parcels (e.g. Tsuji 1971a, Schwarzschild 1975, Slutz 1976, Nordlund 1976, Hinkle 1978, Hinkle & Barnes 1979). If extensive molecular dissociation occurs (i.e. all abundant molecules, including H_2, are dissociated) then the recombination must proceed, at least initially, by relatively slow three-body and photochemical reactions. In this circumstance the time scale for regaining chemical equilibrium can be quite long; in the case of cool, low density atmospheres many days or weeks may be required as noted by Tsuji (1964). The question of departures from chemical equilibrium in late-type atmospheres deserves closer study particularly since new observational techniques have now opened the important infrared region to high resolution spectroscopic examination (e.g. Merrill & Ridgway 1979, Hall et al. 1979).

3.4   Departures from LTE in the Molecular Line Spectrum

Several studies have compared the rates of various radiative and collisional processes in an attempt to delimit the stellar atmosphere conditions under which LTE molecular line formation might be expected. Auman (1969a) found that collisional processes are likely to control both the vibrational and the rotational level populations of the H_2O ground electronic state in cool dwarfs. Radiative rates begin to exceed the collisional rates for vibrational levels in cool giants and supergiants ($T_{eff} \lesssim 3500$ K). The likelihood of departures from LTE increases with decreasing T_{eff} and log g; at (2000, -1) radiative rates exceed the collisional for vibrational levels over most of the atmosphere—even the rotational levels become radiatively controlled at small depths. Similar results were obtained for CO vibrational and rotational ground state levels by Thompson (1973). Hinkle & Lambert (1975) carried out a particularly useful survey of the relative collisional and radiative transition rates (rotational, vibrational, and electronic) for a variety of molecules encountered in intermediate- and late-type atmospheres. Their results for vibrational and rotational ground state levels agree with those of Tsuji and Thompson. More interesting is their demonstration that although collisional channels may be faster for vibrational and rotational transitions within an electronic state (as discussed above), transitions between electronic states are usually more likely to be radiative rather than collisional. This finding leads Hinkle & Lambert to argue that the Planck function may not be the correct source function for most electronic transitions in intermediate- and late-type atmospheres, even in the solar case. They suggest that scattering is the mechanism by which most lines of electronic transitions are formed.

While the preceding investigations are helpful in deciding whether or not LTE is a valid assumption, they do not give any quantitative information on the consequences of departures from LTE. Unfortunately, little work has been done on this problem, primarily because of the difficulty of solving the coupled radiative transfer and statistical equilibrium equations for a molecule. Mount & Linsky (1975) and Mount et al. (1975) solved the non-LTE radiative transfer problem for the (0, 0) band of the CN $B^2\Sigma - X^2\Sigma$ ("blue") system in the case of the sun and α Boo. LTE was assumed in calculating the relative populations of the vibrational-rotational levels within each electronic state and for the molecular concentrations. In the solar case the line source function is greater than the Planck function over the line-forming region leading to lines weaker than for LTE. In α Boo the source function drops below the Planck function in the line-forming layers and the CN lines are strengthened relative to LTE. Although non-LTE effects do not produce especially dramatic changes in the emergent spectrum in these particular examples, the detailed calculations do confirm Hinkle & Lambert's contention regarding the unsuitability of a simple absorption treatment of molecular electronic transitions.

Carbon et al. (1976) solved the non-LTE transfer problem for the first four vibrational levels of the CO ground state assuming an LTE distribution for rotational levels and for electronic states. They found significant departures from LTE (departure coefficients < 1) in the populations of the vibrational states with $v > 0$. Because these departures reduce the line source function, the CO $\Delta v = 1$ lines in the emergent spectra of cool giants and supergiants are strengthened relative to LTE. The extent of the departures is dependent on the collisional relaxation rates, which, unfortunately, are not presently well established.

The preceding investigations have important implications for constructing model atmospheres. With the exception of the vibrational-rotational bands of H_2O, molecular electronic transitions are the primary source of backwarming in all but the coolest late-type stars; electronic transitions are important sources of backwarming in the cooler intermediate-type stars as well. Under a wide range of conditions, the backwarming produced by scattering lines is quantitatively similar to that produced by lines formed in pure absorption (e.g. Athay & Skumanich 1969, Mihalas & Luebke 1971). *If* the molecular lines are essentially formed in scattering *and if* the ground state populations do not deviate significantly from LTE, then current LTE models may not be seriously in error at least with respect to the backwarming. Unfortunately not enough work has been carried out on molecular non-LTE to indicate if these are valid assumptions generally. Since departures from LTE decouple the gas from the radiation field, non-LTE is likely to reduce the surface cooling and heating produced by molecular lines. More work must be done before the consequences of non-LTE in the molecular spectrum are adequately understood. Until this work is carried out, LTE models for late-type stars must be used with a certain degree of caution.

4.0 CONVECTION IN STELLAR ATMOSPHERES

The atmospheres of all intermediate- and late-type stars become convectively unstable at some depth, most often within or just beneath the continuum-forming layers. No really acceptable procedure has yet been devised to model these unstable layers. The physics of convective energy transport is extremely complex and only partially understood. As a result almost all efforts to model the convective process in stellar atmospheres and interiors have been based on some variation of the mixing length formalism. The present discussion is restricted to treatments of mixing length convection used extensively in modeling stellar atmospheres. Those who wish to investigate the mixing length formalism itself more closely are strongly urged to read the thorough review by Gough (1977). Recently, several new treatments for convective energy transport in stellar atmospheres not based on the mixing length approach have appeared in the literature. These treatments are derived from much improved physical descriptions of convective transport in stellar atmospheres and open the way to exciting new developments. These non–mixing-length treatments are described at the end of this section.

4.1 *Local Mixing Length Theories*

The *local* mixing length formalism is by far the most commonly adopted representation for convective transport in modeling atmospheres. The local mixing length approach is popular primarily for two reasons. First, the relevant equations are easily programmed and present no particular computational difficulties; the computer time spent in evaluating the convective flux compares favorably with that spent evaluating the radiation field and calculating temperature corrections. Second, since a widely accepted theoretical description for convective energy transport does not yet exist, researchers engaged in model atmosphere work generally have opted for the simplest available representation. The local mixing length formalism is viewed as a convenient tool for approximating the effects of convection when computing model temperature structure. It has been considered adequate for model atmosphere work largely because it produces, with a suitable choice of mixing length, a temperature structure that agrees with the observed *solar* mean temperature structure. It is clear, nevertheless, that local mixing length theory is not suitable for serious modeling of granulation structures and their associated velocity fields in stellar atmospheres (including the solar case).

In the local mixing length approach (e.g. Cox & Giuli 1968, Mihalas 1978) the convective flux $F_c(z)$ at a geometrical depth z is proportional to $[\rho(z) \cdot v(z) \cdot \alpha \cdot (\nabla - \nabla')]$ where $\rho(z)$ is the mass density of convecting elements, $v(z)$ is their velocity, $\alpha (\equiv l/H)$ the ratio of mixing length, l, to pressure scale height, $H[H = H(z)]$, $\nabla(z)$ is the temperature gradient of the ambient medium at z and $\nabla'(z)$ is the temperature gradient experienced by convective elements in passing through z; $v(z)$ is proportional to $\alpha(\nabla - \nabla')^{1/2}$. It should be stressed that $F_c(z)$ is determined exclusively in terms of the conditions at z. This follows from the assumption, central to the local mixing length approach, that physical parameters are constant over the interval $z \pm l/2$. Different investigators have chosen to adopt somewhat different approaches for specifying the mixing length, the convective velocity, and the radiative losses from convecting elements (e.g. Henyey et al. 1965, Gough 1977). These variations do not lead to substantive changes in the predictions of the local mixing length theory. Numerous calculations have been carried out with various versions of this theory. The next subsections describe, in general terms, the principal results of these computations.

4.1.1 STRUCTURAL CHANGES PRODUCED BY LOCAL MIXING LENGTH CONVECTION   If energy is transported by both radiation and convection, flux constancy can be maintained with a shallower temperature gradient than if radiation alone carries the energy. If the convection is efficient the final ambient gradient, $\nabla(z)$, will be substantially shallower than the radiative gradient, $\nabla_R(z)$. The flattening of the ambient temperature gradient by convection will produce a flux deficiency in the adjacent superficial radiative layers since the latter now experience a smaller incident radiative flux from below. Flux constancy is preserved by increasing the temperature somewhat in these interacting layers. This effect was apparently first noted by Mihalas (1965) in his study of convection in unblanketed models with $T_{eff} > 7200$ K. For a $(7750, 4, 0)$ model Mihalas found that the temperature at $\tau_{0.5} = 1.0$ increased by ≈ 200 K and ≈ 500 K in going from $\alpha = 0$ to $\alpha = 1.0$ and 2.0, respectively. Deeper in the atmosphere the convective models are markedly cooler than the radiative. For example, at $\tau_{0.5} = 3.2$, Mihala's convective models are ≈ 200 K and ≈ 550 K cooler than the radiative model when $\alpha = 1.0$ and 2.0, respectively. Heating of the superficial radiative layers in response to convection also was reported for intermediate spectral type dwarf models by Carbon & Gingerich (1969), Gustafsson & Nissen (1972), and Bell (1973). Auman (1969a,b) reported substantial heating induced by convection in late-type dwarf models. The cool, low metallicity giants computed by Gustafsson et al. (1975) also show evidence of heating in response to shallow convection.

Temperature and gravity dependence   Along the main sequence the hydrogen convection zone occurs nearest the surface in the hotter, intermediate type atmospheres ($T_{eff} \approx 7500$ K, e.g. Carbon & Gingerich 1969). Although the region of partial hydrogen ionization recedes deeper into the atmosphere as one goes down the main sequence, convection never becomes unimportant in the deeper ($\tau_{0.5} \gtrsim 2.5$) layers. This is due largely to compensating increases in atmospheric density, $\rho(z)$ (which determines the total heat capacity of convecting elements), as one goes to lower effective temperatures (e.g. Carbon & Gingerich 1969, Auman 1969a,b). For a fixed T_{eff} and mixing length/scale height ratio the effectiveness of convective energy transport decreases with decreasing gravity. This occurs because $\rho(z)$ decreases with decreasing gravity. This effect is readily apparent in the model grids by Auman (1969a,b), Carbon & Gingerich (1969), Gustafsson et al. (1975), and Kurucz (1979). Convective fluxes for cool giants and supergiants of roughly solar metallicity computed using the local mixing length theory are generally so small that they have no significant effect on the temperatures of the line- and continuum-forming layers. This fact has led most workers modeling late-type giants and supergiants to simply ignore convection altogether. As noted in the next section, this may be a valid approximation only in the context of the local mixing length theory.

Metallicity dependence The efficiency of convective energy transport varies appreciably with metallicity. This phenomenon has been examined in some detail by Auman (1969a), Böhm-Vitense (1971), Gustafsson & Nissen (1972), and Gustafsson et al. (1975). In the atmospheres of intermediate- and late-type stars, H^- is the dominant continuous opacity source near the flux peak. The mass absorption coefficient of H^- varies as the electron pressure, which in turn varies as the metal abundance when H is not appreciably ionized. At a given (T_{eff}, log g), decreasing the metal abundance, hence the continuous opacity, results in increased atmospheric densities when the equation of hydrostatic equilibrium is integrated. These higher densities increase the energy-transporting effectiveness of convective elements. The density enhancement can be strong enough to offset the effects of decreased gravity noted earlier. Thus, for example, Gustafsson et al. (1975) find significant changes in the atmospheric structure of the continuum-forming layers due to convection in their lowest metallicity cool giants.

4.1.2 EFFECTS OF MIXING LENGTH CONVECTION ON SPECTRUM The most straightforward observable manifestation of convection in the integrated disk spectrum is its effect on the emergent continuum. Since local mixing length convection usually alters just the deeper atmospheric layers, only the fluxes through the most transparent spectral intervals are likely to be sensitive to convection. If the spectral interval is sufficiently transparent one may see into the layers where convective transport dominates and the model is substantially cooler than in the radiative case; less transparent intervals will have their continua formed either in layers heated (relative to a radiative model) by convection or in layers completely unaffected by convection.

Effects on the spectrum of hotter stars In the hotter intermediate-type stars the spectral region most transparent in the continuum while at the same time remaining comparatively free from line blanketing lies redward of the Balmer discontinuity, i.e. roughly in the region of the B-filter passband (Strömgren v,b filter regions). Relyea & Kurucz (1978) found that the Strömgren indices predicted by Kurucz's (1979) blanketed (8000, 4, 0) and (7500, 4, 0) models changed substantially between the purely radiative ($\alpha = 0$) and strongly convective ($\alpha = 2$) cases. In going from $\alpha = 0$ to $\alpha = 2.0$, ($b–y$) increased (reddened) by ≈ 0.077 magnitudes, m_1 decreased by ≈ 0.067 magnitudes, and c_1 decreased by 0.14 magnitudes for these dwarf models. These changes correspond to substantial changes in model parameters; the purely radiative model would appear ≈ 700 K hotter and ≈ 1.0 dex higher in $\log g$ than the corresponding convective model with $\alpha = 2.0$. The $(b-y)$ colors of intermediate type stars with $T_{eff} < 7500$ K are less affected by local mixing length convection primarily because the hydrogen convection zone recedes to greater depth with decreasing T_{eff}. Gustafsson & Nissen (1972) found that the $\Delta(b-y)$ produced by going from a purely radiative to a convective case ($\alpha = 1.5$) decreased by more than 50% over the interval $7600 \leq T_{eff} \leq 6700$ K. Olson (1974) found similar results for the $(b-y)$ colors of Carbon & Gingerich's (1969) grid of blanketed models.

The hot intermediate-type stars span much of the transition from purely radiative atmospheres to atmospheres markedly influenced by convection. It was noted above that $(b-y)$ increases by ≈ 0.08 magnitudes when convective transport is included in models of these hotter stars. More significantly, perhaps, the calibration of $(b-y)$ by Relyea & Kurucz (1978) shows a sharp increase in $\Delta(b-y)/\Delta T_{eff}$ for the T_{eff} range where convection first appears in going down the main sequence. Böhm-Vitense (1970, 1977, 1978) and Böhm-Vitense & Canterna (1974) suggested that the gap in the distribution of stars along the main sequence observed near $B-V \approx 0.25$ ($b-y \approx 0.18$) might represent observational evidence for a transition from radiative to convective atmospheres. This is an extremely interesting possibility worth considerably more attention since it may provide a particularly accessible testing ground for convection theory including an opportunity for studying the interaction between convection and stellar rotation (Böhm-Vitense 1978).

Effects on the solar continuum   Solar abundance stars cooler than $T_{eff} \approx$ 6500 K are too heavily blanketed for the region just redward of the Balmer discontinuity to be significantly affected by convection. The opacity minimum in H^- at 1.65 μm becomes the critical region for detecting convection. Recent empirical studies of the solar temperature distribution have relied on this spectral region to establish the temperatures for $\tau_{0.5} \gtrsim 2.0$ (e.g. Gingerich et al. 1971, Vernazza et al. 1976, Avrett 1977, Allen 1978). The effects of convection are readily apparent in the empirical models. For example, the model M of Vernazza et al. (1976) is cooler than a purely radiative model (scaled from the Carbon & Gingerich grid) by ≈ 140 K, ≈ 360 K, and ≈ 870 K at $\tau_{0.5} = 2.0$, 3.0, and 5.0, respectively. These differences are substantially larger than the uncertainties in the empirical models. The empirical temperature structures are readily reproduced by theoretical models using the local mixing theory (for examples see Vernazza et al. 1976).

Effects on the colors of cooler stars   Surprisingly little has been done with regard to examining the effects of convection on the 1.65-μm fluxes in cool stars. While Bell et al. (1976) noted that the 1.65-μm flux peak could be used as a diagnostic for convection, they did not themselves explore this possibility. Mould (1976) and Mould & Hyland (1976) point out that the *J–H* broadband colors of late-type dwarfs are sensitive to convection. The *J–H* color for dwarfs increases with decreasing T_{eff} until $T_{eff} \approx 4000$ K after which *J–H* becomes smaller (bluer) with decreasing T_{eff} (Glass 1975, Mould & Hyland 1976). This behavior is not observed in giants and supergiants. The explanation (Mould 1976) lies in the sensitivity of the *H* band ($\lambda_{eff} \approx 1.6$ μm) to the flux arising from the deepest photospheric layers. At $T_{eff} \approx 4000$ K the association of H_2 brings the region of strong convection sufficiently close to the surface that the temperatures in the photosphere are depressed. This produces less flux in the *H* filter region and, consequently, a decrease in *J–H*. The effect is not observed in giants and supergiants simply because convection is substantially less efficient at the lower densities (Section 4.1.1). Mould's calculations show that lower metallicity models not only have smaller *J–H* indices when convection is important but also change the sign of $\Delta(J–H)/\Delta T_{eff}$ at a higher temperature. This behavior can be attributed to the increased efficiency of convection at low metallicities (Section 4.1.1). Mould & Hyland (1976) note that Mould's models accurately describe the variation of *J–H* for the metal-poor dwarfs in their sample.

4.2 *Nonlocal Mixing Length Theories*

In the local mixing length formalisms all the quantities used in evaluating the convective flux and velocity, $F_c(z)$ and $v_c(z)$, are evaluated at the depth z. An underlying assumption of this approach is that physical parameters do not change markedly over a mixing length. This assumption is not a particularly good one for stellar atmospheres simply because conditions do change rapidly with depth; the transition from radiative to convective transport can occur within just one or two pressure scale heights. In addition, the local mixing length formalisms have no way of accounting for convective overshoot into stable layers. Nonlocal mixing length formalisms attempt to overcome these deficiencies. Ulrich (1976) emphasizes that the nonlocal mixing length theories discussed next are all similar in that each determines the convective flux at depth z by performing some averaging of the atmospheric properties over a region about z. The details of the averaging process distinguish one theory from another. The distance over which the averaging is done, called the "diffusion length" by Ulrich, enters these nonlocal theories as another free parameter in addition to the mixing length.

Parsons (1967, 1969) made the first model atmosphere computations with a nonlocal formalism in his study of yellow supergiants. He determined $v_c(z)$ by integrating the depth-dependent buoyancy forces over the whole path of the convective bubble. This procedure relaxed the local mixing length approximation that physical parameters are constant over the bubble trajectory. Ulrich's (1970a, b) nonlocal mixing length theory not only provided for nonlocal determination of convective velocities and fluxes but also introduced a convective thermal model for the convecting elements themselves. Ulrich's $F_c(z)$ and $v_c(z)$ are computed by taking a weighted average over all convecting elements that pass through z and originate within some specified range of depths about z. This averaging process allows for the fact that the convecting elements passing through z vary both in place of origin and stage of development. Spiegel (1963) presented a mixing length formalism analogous to the transfer equation for radiation. In this approach the convecting cells are treated much like photons diffusing through a stellar atmosphere. An approximate convective source function is derived by requiring that the flux deep within the convecting region be given by the local mixing length theory. Travis & Matshushima (1973) have developed a convenient differential equation representation of Spiegel's formalism.

These nonlocal mixing length formalisms have not been widely adopted for model atmosphere work. The principal objection seems to be that the nonlocal formalisms, while more physically correct, have not led to model structures significantly different from those obtained with the computationally simpler local theory. A careful critical comparison of the local and the nonlocal mixing length theories has been carried out by Nordlund (1974). Nordlund computed solar model atmospheres based on the above nonlocal formalisms using the same composition and opacities for each so the results could be accurately intercompared. The local mixing length theory and Parson's theory were found to agree with one another and with empirical solar temperature structures. The theories of Ulrich and of Travis and Matshushima were found to yield excessively shallow gradients when used with standard choices of the mixing length and diffusion length. Nordlund suggests a modification of Ulrich's derivation that leads to essentially the same convective fluxes and model structure as Parson's nonlocal theory. The results obtained with Travis & Matshushima's theory are in better accord with solar observations when a sufficiently small ($\gtrsim 0.35$) mixing length/scale height ratio is chosen (Travis & Matshushima 1973).

4.3 Non–Mixing-Length Convection

4.3.1 TWO-STREAM CONVECTION THEORIES

In the local and nonlocal mixing length theories discussed above, the object is to deduce the (single) ambient temperature structure of a medium in which heat transport by convecting bubbles is significant. An alternative approach is to assume that the convection can be represented by multiple streams of rising and falling material, each stream with a steady state structure compatible with energy, mass, and momentum conservation in the atmosphere. Such a model is suggested by the temperature fluctuations and motions associated with solar granulation (e.g. Beckers & Canfield 1976, Böhm 1977). The development of multistream convective models represents one of the most important theoretical advances in the treatment of convection in stellar atmospheres in recent years. Ulrich (1970c) and Nordlund (1976) have developed two-stream convection theories. In Ulrich's approach conservation of energy constraints are used to establish the temperature differences of rising (hot) and descending (cool) streams relative to a (given) unperturbed mean atmosphere. The velocity distribution in the streams is treated as a free parameter. Nordlund presents a fully deductive theory that, by imposing conservation of mass, energy, and vertical momentum, permits determination of both temperature and velocity structure.

Solar models   The temperature structures of Nordlund's two-stream solar model may be briefly described as follows [$\tau_{0.5}$ refers to the optical depths in the individual streams (Allen 1978)]. Both streams have nearly the same temperatures ($\Delta T \lesssim 50$ K) at $\tau_{0.5} \lesssim 0.2$, deeper than this they begin to diverge. At $\tau_{0.5} = 1$ they differ by ≈ 250 K. The descending stream has a substantially shallower temperature gradient than the rising stream over the range $0.7 \lesssim \tau_{0.5} \lesssim 3.0$. At $\tau_{0.5} = 3$, the rising stream has a temperature of ≈ 8900 K and the descending stream only ≈ 7100 K. Between $\tau_{0.5} = 3$ and $\tau_{0.5} = 10$ the gradient in the rising stream becomes shallower while that in the descending stream becomes steeper. Deeper than $\tau_{0.5} \approx 10$ the streams begin to converge. (This two-stream temperature structure of Nordlund is qualitatively very like the one predicted by Ulrich's two-stream theory.) Nordlund's theoretical model is quantitatively similar to an empirical two-stream solar model developed by Allen (1978) based on very accurate limb-darkening and central intensity measurements. Since Allen's empirical model is derived completely independently of Nordlund's calculations, it constitutes an important confirmation of the theoretical results. This excellent agreement between theory and observation for the sun is an especially exciting development because Nordlund's theory is readily applicable to nonsolar cases as well.

Nonsolar models   Nordlund (1976) computed two-stream convective models for cool giants and dwarfs as well as for main sequence stars hotter than the sun. Nissen & Gustafsson (1978) extended Nordlund's calculations for the latter. A most important aspect of these calculations and the solar calculations described above is that they predict substantial (hundreds of degrees) photospheric temperature fluctuations near $\tau_c \approx 1$ from penetrating convection. This is in sharp contrast to the local mixing length results particularly in the case of cool giants and supergiants. Two-stream convective models coupled with diagnostic techniques for analyzing the effects of convection on (flux) line profiles (Dravins 1974, 1975, 1976, Nordlund 1978, Bray & Loughhead 1978) should prove to be powerful tools for increasing our understanding of stellar convection.

5   EXTENDED ATMOSPHERES

Researchers have been aware for quite some time that the atmospheres of highly luminous intermediate- and late-type stars are often so distended that the standard plane-parallel geometry is inappropriate. It is only recently, however, that the theoretical and computational techniques have been developed to model such atmospheres (see Mihalas 1978 for an excellent discussion); earlier investigators had to limit themselves to very rough estimates of the effects of atmospheric distension (e.g. Auman 1969a, Böhm-Vitense 1972, Gustafsson et al. 1975, Tsuji 1976a). Schmid-Burgk & Scholz (1975) constructed an extensive grid of *grey* spherically symmetric atmospheres in an effort to define the range of atmospheric parameters for which the planar approximation fails. They found that stars less massive than $\approx 7\ M_\odot$ generally become significantly distended only near their highest luminosities on the red-giant branch while stars with masses $\gtrsim 1.3\ M_\odot$ can be appreciably distended at intermediate spectral types as well.

While only a little work has been done to date on nongrey extended atmospheres for intermediate- and late-type stars, some results based on radiative and hydrostatic equilibrium, LTE and spherical symmetry are available. Schmid-Burgk & Scholz (1977) constructed an unblanketed yellow supergiant model with $M/M_\odot = 0.19$, $T_{\rm eff} = 6550$, and $\log g = 0.7$ using a computer code (Hundt et al. 1975) based on Schmid-Burgk's (1975) integral equation technique. The resulting atmosphere is moderately extended with a thickness $\approx 20\%$ of the stellar radius. This atmosphere has lower temperatures and higher gas pressures in its superficial layers than a comparable plane-parallel atmosphere (at $\tau_R = 10^{-3}$, $\Delta T \approx -400$ K, and $\Delta \log P_g \approx 0.6$); the differences decrease with increasing depth and the spherically symmetric and plane-parallel atmospheres merge for $\tau_R \gtrsim 0.10$. These structural differences produce detectable effects in the computed emergent spectrum (see Schmid-Burgk & Scholz 1977). Watanabe & Kodaira (1978) carried out a very interesting study of extended nongrey atmospheres for late-type stars using a code based on Schmid-Burgk's formulation. They assumed a solar composition and included the effects of molecular line blanketing using straight mean opacities. Two sequences of models were constructed. In the first, T_{eff} and gravity were held fixed $(3200, -0.5)$ while the stellar mass was reduced from $20\,M_\odot$ to $1\,M_\odot$. In the second sequence of models, T_{eff} and luminosity (hence radius) were held constant $(T_{eff} = 3800,\ R/R_\odot = 600,$ $\log L/L_\odot = 4.8)$ while the stellar mass was reduced from $15\,M_\odot$ to $1.5\,M_\odot$ ($\log g$ from 0.0 to -1.0). The atmospheric extent increases with decreasing mass in both sequences. Significant differences in atmospheric structure $(T\text{-}P\text{-}\tau_c)$ were found to occur between spherically symmetric and plane-parallel models when the atmospheres exceeded $\approx 5\%$ of the stellar radius. At $(3200, -0.5)$ increasing the extent of the atmosphere from 5% to 23% lowers the temperature and gas pressure in the surface layers (e.g. at $\tau_R = 10^{-2}$, $\Delta T \approx -200$ K, and $\Delta \log P_g \approx -0.5$) but does not affect the atmosphere structure appreciably deeper than $\tau_R \approx 0.10$. In the sequence with constant T_{eff} and luminosity, decreasing $\log g$ from 0.0 to -1.0 increases the atmospheric extent from 3% to 54% of the stellar radius. This behavior produces a lowering of the temperature and gas pressure near the surface (at $\tau_R = 10^{-2}$, $\Delta T \approx -200$ K, and $\Delta \log P_g \approx -0.75$). Watanabe & Kodaira computed low resolution spectra for their models. In both sequences there are significant differences between the spectra of plane-parallel and spherically symmetric models (see Figures 6 and 11 in Watanabe & Kodaira 1978). Particularly interesting is the finding that the bands of chemical species formed preferentially near the surface (TiO and H_2O) are enhanced by atmospheric extension while the bands of species like CO are weakened. Since this result has important implications for abundance analyses of cool, highly luminous stars, more work in this area is clearly needed.

ACKNOWLEDGMENTS

I wish to thank Drs. R. A. Bell, A. Bernat, B. Gustafsson, D. N. B. Hall, K. H. Hinkle, H. R. Johnson, R. L. Kurucz, J. L. Linsky, Å. Nordlund, F. Querci, and T. Tsuji for kindly providing preprints. I am pleased to acknowledge valuable discussions with Drs. A. Bernat, D. N. B. Hall, H. R. Johnson, R. W. Milkey, and L. W. Ramsey. I am particularly grateful to Drs. S. T. Ridgway, R. L. Kurucz, R. C. Peterson, I. Furenlid, and W. Romanishin for their many thoughtful comments on the original manuscript. I wish to thank Ms. A. Haynes for careful preparation of the typescript. Finally, I especially wish to thank the Editors, Dr. G. Burbidge and Ms. L. Brower, for their encouragement and forbearance.

Literature Cited

Alexander, D. R., Johnson, H. R. 1972. *Ap. J.* 176:629–43
Allen, R. G. 1978. *An Infrared Investigation of the Temperature Structure of the Solar Atmosphere.* PhD thesis. Univ. Arizona, Tucson. 204 pp.
Athay, R. G. 1970. *Ap. J.* 161:713–35
Athay, R. G., Canfield, R. C. 1969. *Ap. J.* 156:695–706
Athay, R. G., Lites, B. W. 1972. *Ap. J.* 176:809–31
Athay, R. G., Skumanich, A. 1969. *Ap. J.* 155:273–94
Auman, J. 1967. *Ap. J. Suppl.* 14:171–206
Auman, J. R. 1969a. *Ap. J.* 157:799–826
Auman, J. R. 1969b. In *Low Luminosity Stars*, ed. S. S. Kumar, pp. 483–92. New York: Gordon & Breach. 542 pp.
Auman, J. R., Woodrow, J. E. J. 1975. *Ap. J.* 197:163–73
Avrett, E. H. 1977. In *The Solar Output and Its Variation*, ed. O. R. White, pp. 327–48. Boulder: Colorado Associated University Press. 526 pp.
Bailey, W. L. 1977. *Ap. J.* 211:596–99
Beckers, J. M., Canfield, R. C. 1976. In *Physique des Mouvements dans les Atmosphères Stellaires, Colloques Internationaux du Centre National de la Recherche Scientifique, No. 250*, ed. R. Cayrel, M. Steinberg, pp. 207–58. Paris: editions du Centre National de la Recherche Scientifique. 478 pp.
Bell, R. A. 1973. *MNRAS* 164:197–210
Bell, R. A., Eriksson, K., Gustafsson, B., Nordlund, Å. 1976. *Astron. Astrophys. Suppl.* 23:37–95
Bell, R. A., Gustafsson, B. 1978. *Astron. Astrophys. Suppl.* 34:229–40
Bell, R. A., Gustafsson, B., Nordh, H. L., Olofsson, S. G. 1976. *Astron. Astrophys.* 46:391–96
Böhm, K. H. 1977. In *Lecture Notes in Physics, Problems of Stellar Convection*, ed. E. A. Spiegel, J. P. Zahn, 71:102–18. New York: Springer. 363 pp.
Böhm-Vitense, E. 1970. *Astron. Astrophys.* 8:283–98
Böhm-Vitense, E. 1971. *Astron. Astrophys.* 14:390–95
Böhm-Vitense, E. 1972. *Astron. Astrophys.* 17:335–53
Böhm-Vitense, E. 1977. See Böhm 1977, pp. 63–85
Böhm-Vitense, E. 1978. *Ap. J.* 223:509–25
Böhm-Vitense, E., Canterna, R. 1974. *Ap. J.* 914:629–35
Bray, R. J., Loughhead, R. E. 1978. *Publ. Astron. Soc. Pac.* 90:609–14
Buser, R., Kurucz, R. L. 1978. *Astron.*

Astrophys. 70:555–63
Carbon, D. F. 1973. *Ap. J.* 183:903–21
Carbon, D. F. 1974. *Ap. J.* 187:135–45
Carbon, D., Butler, D., Kraft, R. P., Nocar, J. L. 1977. In *CNO Isotopes in Astrophysics, Astrophysics and Space Science Library, Vol. 67*, ed. J. Audouze, pp. 33–38. Dordrecht, Holland: Reidel. 195 pp.
Carbon, D. F., Gingerich, O. 1969. In *Theory and Observation of Normal Stellar Atmospheres; Proceedings of the Third Harvard-Smithsonian Conference on Stellar Atmospheres*, ed. O. Gingerich, pp. 377–400, 401–72. Cambridge, Mass.: MIT Press. 472 pp.
Carbon, D. F., Milkey, R. W., Heasley, J. N. 1976. *Ap. J.* 207:253–62
Chandrasekhar, S. 1935. *MNRAS* 96:21–42
Collins, J. G., Fay, T. D., Jr. 1974. *J. Quant. Spectrosc. Radiat. Transfer* 14:1259–72
Cox, J. P., Giuli, R. T. 1968. *Principles of Stellar Structure.* New York: Gordon & Breach. 1327 pp.
Cram, L. E., Ulmschneider, P. 1978. *Astron. Astrophys.* 62:239–44
Dalgarno, A. 1975. In *Atomic and Molecular Processes in Astrophysics, Swiss Society of Astronomy and Astrophysics Fifth Advanced Course*, ed. M. C. E. Huber, H. Nussbaumer, pp. 1–98. Sauverny, Switzerland: Geneva Observatory. 308 pp.
Dalgarno, A., Browne, J. C. 1967. *Ap. J.* 149:231–32
Deming, D. 1979. *Ap. J.* In press
Dravins, D. 1974. *Astron. Astrophys.* 36:143–45
Dravins, D. 1975. *Astron. Astrophys.* 43:45–50
Dravins, D. 1976. See Beckers & Canfield 1976, pp. 456–66
Dumont, S., Heidmann, N. 1973. *Astron. Astrophys.* 27:273–79
Dumont, S., Heidmann, N. 1976. *Astron. Astrophys.* 49:271–75
Dyck, H. M., Lockwood, G. W., Capps, R. W. 1974. *Ap. J.* 189:89–100
Faber, S. M. 1977. In *The Evolution of Galaxies and Stellar Populations*, ed. B. M. Tinsley, R. B. Larson, pp. 157–91. New Haven: Yale Univ. Obs. 449 pp.
Gingerich, O., Latham, D. W., Linsky, J., Kumar, S. S. 1966. In *Colloquium on Late-Type Stars*, ed. M. Hack. pp. 291–312. Trieste: Obs. Astron. Trieste. 465 pp.
Gingerich, O., Noyes, R. W., Kalkofen, W., Cuny, Y. 1971. *Solar Phys.* 18:347–65
Glass, I. S. 1975. *MNRAS* 171:19p–23p
Gough, D. 1977. See Böhm 1977, pp. 15–56
Gray, D. F. 1976. *The Observation and*

Analysis of Stellar Photospheres, pp. 343–44. New York: Wiley. 471 pp.

Grevesse, N., Sauval, A. J. 1973. *Astron. Astrophys.* 27:29–43

Gustafsson, B. 1971. *Astron. Astrophys.* 10:187–92

Gustafsson, B. 1973. *Uppsala Astr. Obs. Ann.* 5:No. 6, pp. 1–31 and appendices

Gustafsson, B., Bell, R. A. 1979. *Astron. Astrophys.* 74:313–52

Gustafsson, B., Bell, R. A., Eriksson, K., Nordlund, Å. 1975. *Astron. Astrophys.* 42:407–32

Gustafsson, B., Nissen, P. E. 1972. *Astron. Astrophys.* 19:261–82

Hall, D. N. B., Hinkle, K. H., Ridgway, S. T. 1979. In *Proc. I.A.U. Colloq. No. 46, Changing Trends in Variable Star Research.* In press

Heasley, J. N., Ridgway, S. T., Carbon, D. F., Milkey, R. W., Hall, D. N. B. 1978. *Ap. J.* 219:970–78

Henyey, L., Vardya, M. S., Bodenheimer, P. 1965. *Ap. J.* 142:841–54

Hinkle, K. H. 1978. *Ap. J.* 220:210–28

Hinkle, K. H., Barnes, T. G. 1979. *Ap. J.* 227:923–34

Hinkle, K. H., Lambert, D. L. 1975. *MNRAS* 170:447–74

Hundt, E., Kodaira, K., Schmid-Burgk, J., Scholz, M. 1975. *Astron. Astrophys.* 41:37–40

Johnson, H. R., Marenin, I. R., Price, S. D. 1972. *J. Quant. Spectrosc. Radiat. Transfer* 12:189–205

Johnson, H. R. 1973. *Ap. J.* 180:81–89

Johnson, H. R., Collins, J. G., Krupp, B., Bell, R. A. 1977. *Ap. J.* 212:760–67

Johnson, H. R., Krupp, B. M. 1976. *Ap. J.* 206:201–7

Kalkofen, W. 1968. *Ap. J.* 151:317–32

Kelch, W. L., Linsky, J. L., Basri, G. S., Chiu, H.-Y., Chang, S.-H., Maran, S. P., Furenlid, I. 1978. *Ap. J.* 220:962–79

Kraft, R. P. 1979. *Ann. Rev. Astron. Astrophys.* 17:309–43

Krupp, B. M., Collins, J. G., Johnson, H. R. 1978. *Ap. J.* 219:963–69

Kunde, V. G. 1968. *Ap. J.* 153:435–50

Kurucz, R. L. 1969. See Carbon & Gingerich 1969, pp. 375–76, 401–72

Kurucz, R. L. 1970. *Smithsonian Astrophysical Observatory Special Report No. 309.* Cambridge, Mass.: Smithsonian Institution Astrophysical Observatory

Kurucz, R. L. 1979. *Ap. J. Suppl.* 40:1–340

Kurucz, R. L., Peytremann, E. 1975. *Smithsonian Astrophysical Observatory Special Report No. 362,* Cambridge, Mass.: Smithsonian Institution Astrophysical Observatory

Kurucz, R. L., Peytremann, E., Avrett, E. H. 1974. *Blanketed Model Atmospheres for Early-Type Stars.* Washington DC.: Smithsonian Institution. 186 pp.

Lambert, D. L. 1978. *MNRAS* 182:249–72

Lambert, D. L., Pagel, B. E. J. 1968. *MNRAS* 141:299–315

Lambert, D. L., Ries, L. M. 1977. *Ap. J.* 217:508–20

Lengyel-Frey, D. A. 1977. *Titanium Oxide in Cool Star Atmospheres.* PhD thesis. Univ. Maryland, College Park. 159 pp.

Linsky, J. 1966. *Astron. J.* 71:863

Linsky, J. L., Ayres, T. R. 1978. *Ap. J.* 220:619–28

Lites, B. W., Cowley, C. R. 1974. *Astron. Astrophys.* 31:361–69

Manduca, A., Bell, R. A., Gustafsson, B. 1977. *Astron. Astrophys.* 61:809–13

Merrill, K. M., Ridgway, S. T. 1979. *Ann. Rev. Astron. Astrophys.* 17:9–41

Mihalas, D. 1965. *Ap. J.* 141:564–81

Mihalas, D. 1967. In *Methods in Computational Physics, Vol. 7: Astrophysics*, ed. B. Alder, S. Fernbach, M. Rotenberg, pp. 1–52. New York: Academic. 262 pp.

Mihalas, D. 1969. *Ap. J.* 157:1363–67

Mihalas, D. 1978. *Stellar Atmospheres.* San Francisco: Freeman. 632 pp. 2nd ed.

Mihalas, D., Athay, R. G. 1973. *Ann. Rev. Astron. Astrophys.* 11:187–218

Mihalas, D., Kunasz, P. B., Hummer, D. G. 1976. *Ap. J.* 203:647–59

Mihalas, D., Luebke, W. R. 1971. *MNRAS* 153:229–39

Mould, J. R. 1975. *Astron. Astrophys.* 38:283–88

Mould, J. R. 1976. *Astron. Astrophys.* 48:443–59

Mould, J. R., Hyland, A. R. 1976. *Ap. J.* 208:399–413

Mount, G. H., Ayres, T. R., Linsky, J. L. 1975. *Ap. J.* 200:383–91

Mount, G. H., Linsky, J. L. 1975. *Sol. Phys.* 41:17–33

Münch, G. 1946. *Ap. J.* 104:87–109

Mutschlecner, J. P., Keller, C. F. 1970. *Sol. Phys.* 14:294–309

Nissen, P. E., Gustafsson, B. 1978. In *Astronomical Papers Dedicated to Bengt Strömgren*, ed. A. Reiz, T. Anderson, pp. 43–92. Copenhagen: Copenhagen Univ. Obs. 428 pp.

Nordlund, Å. 1974. *Astron. Astrophys.* 32:407–22

Nordlund, Å. 1976. *Astron. Astrophys.* 50:23–39

Nordlund, Å. 1978. See Nissen & Gustafsson 1978, pp. 95–114

Norris, J., Zinn, R. 1977. *Ap. J.* 215:74–88

Olson, E. C. 1974. *Publ. Astron. Soc. Pac.* 86:80–89

Parsons, S. B. 1967. *Ap. J.* 150:263–71

Parsons, S. B. 1969. *Ap. J.* 18: Suppl. 159, pp. 127–65

Peterson, R. 1976. *Ap. J. Suppl.* 30:61–83
Peytremann, E. 1974. *Astron. Astrophys.* 33:203–14
Praderie, F. 1971. *Astrophys. Lett.* 9:27–31
Pritchet, C. 1977. *Ap. J. Suppl.* 35:397–418
Querci, F., Querci, M. 1974. *Highlights of Astronomy, Vol. 3*, ed. G. Contopoulos, pp. 341–56. Dordrecht: Reidel. 574 pp.
Querci, F., Querci, M. 1975. *Astron. Astrophys.* 39:113–25
Querci, F., Querci, M. Kunde, V. G. 1971. *Astron. Astrophys.* 15:256–74
Querci, F., Querci, M., Tsuji, T. 1974. *Astron. Astrophys.* 31:265–82
Querci, M., Querci, F. 1976. *Astron. Astrophys.* 49:443–62
Ramsey, L. W. 1977. *Ap. J.* 215:827–35
Relyea, L. J., Kurucz, R. L. 1978. *Ap. J. Suppl.* 37:45–69
Ridgway, S. T., Joyce, R. R., White, N., Wing, R. F. 1979. *Ap. J.* In press
Scalo, J. M., Ulrich, R. K. 1973. *Ap. J.* 183:151–75
Schmeltekopf, A. L., Fehsenfeld, F. C., Ferguson, E. E. 1967. *Ap. J.* 148:L155–56
Schmid-Burgk, J. 1975. *Astron. Astrophys.* 40:249–55
Schmid-Burgk, J., Scholz, M. 1975. *Astron. Astrophys.* 41:41–45
Schmid-Burgk, J., Scholz, M. 1977. *MNRAS* 179:563–67
Schwarzschild, M. 1975. *Ap. J.* 195:137–44
Slutz, S. 1976. *Ap. J.* 210:750–56
Smith, M. A. 1974a. *Ap. J.* 190:481–86
Smith, M. A. 1974b. *Ap. J.* 192:623–27
Sneden, C. 1974. *Ap. J.* 189:493–507
Sneden, C., Johnson, H. R., Krupp, B. M. 1976. *Ap. J.* 204:281–89
Spiegel, E. A. 1963. *Ap. J.* 138:216–25
Strom, S. E. 1967. *Ap. J.* 150:637–45
Strom, S. E. 1969. *Ap. J.* 156:177–82
Strom, S. E., Kurucz, R. L. 1966. *J. Quant.*

Spectrosc. Radiat. Transfer 6:591–607
Thompson, R. I. 1973. *Ap. J.* 181:1039–54
Travis, L. D., Matsushima, S. 1973. *Ap. J.* 180:975–85
Tsuji, T. 1964. *Ann. Tokyo Astron. Obs., Ser.* 2 9:1–110
Tsuji, T. 1966a. *Publ. Astron. Soc. Jpn.* 18:127–73
Tsuji, T. 1966b. See Gingerich et al. 1966, pp. 260–90
Tsuji, T. 1969. See Auman 1969b, pp. 457–81
Tsuji, T. 1971a. *Publ. Astron. Soc. Jpn.* 23:275–312
Tsuji, T. 1971b. *Publ. Astron. Soc. Jpn.* 23:553–65
Tsuji, T. 1973. *Astron. Astrophys.* 23:411–31
Tsuji, T. 1976a. *Publ. Astron. Soc. Jpn.* 28:543–65
Tsuji, T. 1976b. *Publ. Astron. Soc. Jpn.* 28:567–86
Tsuji, T. 1978. *Astron. Astrophys.* 62:29–50
Ulmschneider, P., Schmitz, F., Renzini, A., Cacciari, C., Kalkofen, W., Kurucz, R. 1977. *Astron. Astrophys.* 61:515–21
Ulrich, R. K. 1970a. *Astrophys. Space Sci.* 7:71–86
Ulrich, R. K. 1970b. *Astrophys. Space Sci.* 7:183–200
Ulrich, R. K. 1970c. *Astrophys. Space Sci.* 9:80–96
Ulrich, R. K. 1976. *Ap. J.* 207:564–73
van Paradijs, J., Vardya, M. S. 1975. *Astrophys. Space Sci.* 33:L9–L12
Vardya, M. S. 1970. *Ann. Rev. Astron. Astrophys.* 8:87–114
Vernazza, J. E., Avrett, E. H., Loeser, R. 1973. *Ap. J.* 184:605–31
Vernazza, J. E., Avrett, E. H., Loeser, R. 1976. *Ap. J. Suppl.* 30:1–60
Watanabe, T., Kodaira, K. 1978. *Publ. Astron. Soc. Jpn.* 30:21–38

AUTHOR INDEX

SUBJECT INDEX

A

Ammonia
  emission regions of
    H II regions and, 376
Andromeda galaxy
  globular clusters in, 256–59
Asteroids
  occultation of, 458,466
Astronomical photography
  advances, 43–69
  current hypersensitization,
    52–65
    ammonia, 53
    autoradiography, 64, 65
    chemical pre-exposure
      treatment, 52, 53
    combined nitrogen-
      hydrogen treatment,
      61, 62
    environment during
      exposure, 65
    evacuation, 54, 55
    hydrogen treatment, 59–61
    latensification, 63
    moisture and oxygen
      during exposure, 65
    nitrogen baking, 56–59
    nitrogen treatment, 55
    oxygen interference, 54
    pre-flash, 62, 63
    push development and
      intensification, 63–65
    silver nitrate bathing, 53
    temperature during
      exposure, 65
    vacuum-treatment, 54
    water bathing, 53
  evaluating photographic
    response, 47–52
    application classes, 51
    characteristic curve, 47–49
    density, 48
    detective quantum
      efficiency, 51
    dynamic range, 50
    exposure, 48, 49
    signal-to-noise, 49, 50
    speed, 49
    test light sources, 48
    uniformity of response, 51,
      52
  introduction, 43–47
    chemical model of latent
      image formation, 44,
      45

development, 45
evolution of photographic
  sensitivity, 46, 47
"fog" grains in photo-
  graphic process,
  45
formation of latent image,
  44, 45
gold sensitization, 45
Gurney-Mott theory
  explaining "recipro-
  city law", 45, 46
latent image and
  reciprocity failure, 44
low-intensity reciprocity
  failure, 46
photographic emulsion and
  information storage,
  43, 44
reciprocity failure, 45, 46
reduction sensitivity
  centers, 45
reviews of emulsion
  function, 44
silver halides in
  photography, 44, 45
sulfur "ripened" in image
  formation, 45
prospects, 66–69
  applications, 66, 68
  automatic direct star
    images, 68
  computer analysis of plate
    content, 68
  emulsion technology, 66
  hypersensitization
    treatments summary
    of Kodak plates, 67
  microphotomatic stellar
    photometry on
    electrograms, 68
  photographic measures, 68
  photographic projects, 68

B

Betelgeuse
  highly magnified photograph
    of, 131
Big Bang
  cluster near pregalactic
    values, 325
BL Lac-type objects
  similarity to quasistellar
    objects, 505

C

Capella
  highly magnified photograph
    of, 131
Carbon
  isotopes of
    star content of, 28
Carbon monoxide
  analysis of bands of, 26
  prediction of emission cores
    and, 18
  presence in cool IR stars, 11,
    14
  surface cooling in stars, 522
Cloud complexes
  observation of stars by, 15
Cluster domain
  chemical composition same,
    309
Clusters, globular
  Milky Way mass and, 139
  radial velocities of, 140
  see also Globular clusters
Comets
  "grazing comets", 5
  silicate features in dust of, 104
Commission 29 (Stellar Spectra) of the International
  Astronomical Union, 4, 5
Compact H II regions and OB
  star formation, 345–80
  infrared observation, 364–69
  amount of dust in H II
    region, 368
  BN object discovery
    identified, 365
  chemical elements
    identified, 365
  cool circumstellar dust
    shells absorbing
    near-infrared, 367
  dense dust clouds around
    H II regions, 368, 369
  dust role, 366
  far-infrared observations,
    367–69
  models for infrared
    emission, 365
  near-infrared map, 366
  near-infrared observations,
    364–67
  objects of radio compact H
    II regions, 367
  total far-infrared flux
    densities, 368

CUMULATIVE INDEXES

CONTRIBUTING AUTHORS, VOLUMES 13–17

CHAPTER TITLES, VOLUMES 13–17

Please list on the order blank on the reverse side the volumes you wish to order
and whether you wish a standing order (the latest volume shipped to you automatically
upon publication each year). Volumes not yet published will be shipped in month and
year indicated. Out of print volumes subject to special order.

PRICE CHANGE NOTICE

All volumes of the Annual Reviews published on or after July 1, 1980 will increase
in price by $3.00 per copy (USA) and $3.50 per copy (elsewhere). This price change
is effective regardless of when the order is placed. Volumes published prior to
July 1, 1980 will not be affected by the price increase.

NEW SERIES.... Volume 1 to be published May 1980

Annual Review of PUBLIC HEALTH $17.00 (USA), $17.50 (elsewhere) per copy

SPECIAL PUBLICATIONS

ANNUAL REVIEW REPRINTS: CELL MEMBRANES, 1975-1977 (published 1978)
A collection of articles reprinted from recent Annual Review series.
    ISBN 0-8243-2501-X   Soft cover: $12.00 (USA), $12.50 (elsewhere) per copy
----
THE EXCITEMENT AND FASCINATION OF SCIENCE (published 1965)
A collection of autobiographical and philosophical articles by leading scientists.
    ISBN 0-8243-1601-0   Clothbound: $6.50 (USA), $7.00 (elsewhere) per copy
----
THE EXCITEMENT AND FASCINATION OF SCIENCE, VOLUME 2:
Reflections by Eminent Scientists (published 1978)
    ISBN 0-8243-2601-6   Hard cover: $12.00 (USA), $12.50 (elsewhere) per copy
    ISBN 0-8243-2602-4   Soft cover: $10.00 (USA), $10.50 (elsewhere) per copy
----
THE HISTORY OF ENTOMOLOGY (published 1973)
A special supplement to the Annual Review of Entomology series.
    ISBN 0-8243-2101-7   Clothbound: $10.00 (USA), $10.50 (elsewhere) per copy

ANNUAL REVIEW SERIES

Annual Review of ANTHROPOLOGY   ISSN 0084-6570
    Vols. 1-7 (1972-78) now available
    Vol. 8 available Oct. 1979 $17.00 (USA), $17.50 (elsewhere) per copy
----
Annual Review of ASTRONOMY AND ASTROPHYSICS   ISSN 0066-4146
    Vols. 1-16 (1963-78) now available
    Vol. 17 available Sept. 1979 $17.00 (USA), $17.50 (elsewhere) per copy
----
Annual Review of BIOCHEMISTRY   ISSN 0066-4154
    Vols. 28-48 (1959-79) now available $18.00 (USA), $18.50 (elsewhere) per copy
----
Annual Review of BIOPHYSICS AND BIOENGINEERING   ISSN 0084-6589
    Vols. 1-8 (1972-79) now available $17.00 (USA), $17.50 (elsewhere) per copy
----
Annual Review of EARTH AND PLANETARY SCIENCES   ISSN 0084-6597
    Vols. 1-7 (1973-79) now available $17.00 (USA), $17.50 (elsewhere) per copy
----
Annual Review of ECOLOGY AND SYSTEMATICS   ISSN 0066-4162
    Vols. 1-9 (1970-78) now available
    Vol. 10 available Nov. 1979 $17.00 (USA), $17.50 (elsewhere) per copy
----
Annual Review of ENERGY   ISSN 0362-1626
    Vols. 1-3 (1976-78) now available
    Vol. 4 available Oct. 1979 $17.00 (USA), $17.50 (elsewhere) per copy
----
Annual Review of ENTOMOLOGY   ISSN 0066-4170
    Vols. 7-24 (1962-79) now available $17.00 (USA), $17.50 (elsewhere) per copy
----
Annual Review of FLUID MECHANICS   ISSN 0066-4189
    Vols. 1-11 (1969-79) now available $17.00 (USA), $17.50 (elsewhere) per copy
----
Annual Review of GENETICS   ISSN 0066-4197
    Vols. 1-12 (1967-78) now available
    Vol. 13 available Dec. 1979 $17.00 (USA), $17.50 (elsewhere) per copy
----
Annual Review of MATERIALS SCIENCE   ISSN 0084-6600
    Vols. 1-8 (1971-78) now available
    Vol. 9 available Aug. 1979 $17.00 (USA), $17.50 (elsewhere) per copy
----

(continued on reverse side)

Annual Review of MEDICINE: Selected Topics in the Clinical Sciences
    Vols. 1-3, 5-15, 17-30 (1950-52,   ISSN 0066-4219
    1954-64, 1966-79) now available $17.00 (USA), $17.50 (elsewhere) per copy

Annual Review of MICROBIOLOGY   ISSN 0066-4227
    Vols. 15-32 (1961-78) now available
    Vol. 33 available Oct. 1979 $17.00 (USA), $17.50 (elsewhere) per copy

Annual Review of NEUROSCIENCE   ISSN 0147-006X
    Vols. 1-2 (1978-79) now available $17.00 (USA), $17.50 (elsewhere) per copy

Annual Review of NUCLEAR AND PARTICLE SCIENCE   ISSN 0066-4243
    Vols. 10-28 (1960-78) now available
    Vol. 29 available Dec. 1979 $19.50 (USA), $20.00 (elsewhere) per copy

Annual Review of PHARMACOLOGY AND TOXICOLOGY   ISSN 0362-1642
    Vols. 1-3, 5-19 (1961-63, 1965-79)
    now available $17.00 (USA), $17.50 (elsewhere) per copy

Annual Review of PHYSICAL CHEMISTRY   ISSN 0066-426X
    Vols. 10-21, 23-29 (1959-70,
    1972-78) now available $17.00 (USA), $17.50 (elsewhere) per copy
    Vol. 30 available Nov. 1979

Annual Review of PHYSIOLOGY   ISSN 0066-4278
    Vols. 18-41 (1956-79) now available $17.00 (USA), $17.50 (elsewhere) per copy

Annual Review of PHYTOPATHOLOGY   ISSN 0066-4286
    Vols. 1-16 (1963-78) now available
    Vol. 17 available Sept. 1979 $17.00 (USA), $17.50 (elsewhere) per copy

Annual Review of PLANT PHYSIOLOGY   ISSN 0066-4294
    Vols. 10-30 (1959-79) now available $17.00 (USA), $17.50 (elsewhere) per copy

Annual Review of PSYCHOLOGY   ISSN 0066-4308
    Vols. 4, 5, 8, 10-30 (1953, 1954,
    1957, 1959-79) now available $17.00 (USA), $17.50 (elsewhere) per copy

Annual Review of SOCIOLOGY   ISSN 0360-0572
    Vols. 1-4 (1975-78) now available
    Vol. 5 available Aug. 1979 $17.00 (USA), $17.50 (elsewhere) per copy

To ANNUAL REVIEWS INC., 4139 El Camino Way, Palo Alto, CA 94306 USA (415-493-4400)

Please enter my order for the following publications:
(If a standing order, indicate which volume you wish order to begin with)

_____, Vol(s). _____ Standing order ____

_____, Vol(s). _____ Standing order ____

_____, Vol(s). _____ Standing order ____

_____, Vol(s). _____ Standing order ____

Amount of remittance enclosed $_____   California residents please add sales tax.
Please bill me _____   Prices subject to change without notice.

SHIP TO (Include institutional purchase order if billing address is different)

Name _____

Address _____

_____

_____ Zip code _____

Signed _____ Date _____

____ Send free copy of annual Prospectus for current year

____ Send free brochure listing contents of recent back volumes for Annual Review(s)

    of _____